T0406669

Leipziger Altorientalistische Studien

Herausgegeben von
Michael P. Streck

Band 15

2023
Harrassowitz Verlag · Wiesbaden

Maria Teresa Renzi-Sepe

The Perception of the Pleiades in Mesopotamian Culture

2023
Harrassowitz Verlag · Wiesbaden

Publication of this book was supported by a grant of Gerda Henkel Stiftung. .

GERDA HENKEL **STIFTUNG**

Bibliografische Information der Deutschen Nationalbibliothek
Die Deutsche Nationalbibliothek verzeichnet diese Publikation in der Deutschen Nationalbibliografie; detaillierte bibliografische Daten sind im Internet über http://dnb.dnb.de abrufbar.

Bibliographic information published by the Deutsche Nationalbibliothek
The Deutsche Nationalbibliothek lists this publication in the Deutsche Nationalbibliografie; detailed bibliographic data are available in the Internet at http://dnb.dnb.de.

For further information about our publishing program consult our website http://www.harrassowitz-verlag.de

Printed on permanent/durable paper.
Printing and binding: Memminger MedienCentrum AG
Printed in Germany

ISSN 2193-4436
ISBN 978-3-447-12053-1
e ISSN 2751-7608
e ISBN 978-3-447-39423-9

To Gina and Maria

Table of Contents

Figures and Tables

Acknowledgements

The present dissertation was conducted under the supervision of Prof. Dr. Michael P. Streck (Leipzig) and Dr. Jeanette C. Fincke (Leiden) between July 2018 and July 2022. I am indebted to the Gerda Henkel Foundation, which awarded me a PhD Scholarship in October 2017. That also allowed me to photograph and copy cuneiform sources at the British Museum in February 2019 and February 2020, and at the Vorderasiatisches Museum in November 2018. I thank Dr. Jonathan Taylor, Dr. Irving Finkel and the staff of the Arched Room of the British Museum for their support during my research on the tablets, and all the staff of the Vorderasiatisches Museum for allowing me to see the original tablets from the collection of Aššur. The tablets are published in this manuscript with the kind permission of the Trustees of the British Museum.

Over the years, I have received help from many people, to whom I would like to express my gratitude. My utmost appreciation goes to my supervisors, Prof. Dr. Michael P. Streck and Dr. Jeanette C. Fincke, who constantly guided and encouraged me. I am indebted to Prof. Dr. Michael P. Streck, who welcomed me to the Altorientalisches Institut of Leipzig and gave me an academic *Heimat*. He trusted me and offered me the opportunity to grow and improve in the scholarly environment by giving me the right advice at the right time. Without his guidance, this dissertation could not have been completed. I owe a particular thanks to Dr. Jeanette C. Fincke for our meetings, her generosity of lessons, corrections, suggestions, and for allowing me to explore her database of photos of tablets and fragments housed in the British Museum, to complete and improve the philological editions.

My research has benefited from stimulating conversations and feedback from other scholars. Special thanks in this respect are due to Prof. Dr. Hermann Hunger for reading my manuscript and giving me precious comments and suggestions. Likewise, I thank Prof. Dr. Mathieu Ossendrijver and Dr. Marvin Schreiber, who offered to read and give me insightful feedback on chapters 4 and 6. I thank Prof. Dr. Takayoshi Oshima for his insights into Akkadian literature and for reading chapter 2. I thank Dr. Henry Stadhouders for providing secondary literature on the series *Šumma Sîn ina tāmartišu* Dr. Marco Stockhusen for providing secondary literature on the Pleiades in the Ancient Near East, and Prof. Dr. Marten Stol for his notes on the Pleiades. I am furthermore very thankful to Prof. Dr. Johannes Hackl for the advice, the coffees, and constant support. Special thanks are also due to Alma Agostini and Libby Stevenson who took care of the English proofreading of this manuscript.

I want to express heartfelt gratitude to all the colleagues I was lucky to meet in Leipzig. In this regard, I especially thank Beatrice Dalla Volta, Giorgio Papitto, Dr. Anna Perdibon, Dr. Antonia Pohl, and Tommaso Scarpelli, with whom I shared suggestions and experiences over these years.

Most of all, I am thankful to Ludovica Cecilia for her friendship and the daily academic and emotional support which goes back to the beginning of our Assyriological journey in 2011.

Finally, my thanks go to Gina, Maria, Carla, and Lorenzo for patiently asking, listening, giving, and understanding in a way only those who truly love can do.

Abbreviations

The abbreviations – including excavation, registration, and collection numbers of cuneiform sources – follow the "Reallexikon der Assyriologie und Vorderasiatischen Archäologie" (RlA), searchable online at the following website: https://rla.badw.de/reallexikon/abkuerzungslisten/literatur-und-koerperschaften.html accessed 25.01.2023.

Other abbreviations are listed as follows:

Alb B	Astrolabe B, see Horowitz (2014).
App. A	Appendix A.
App. B	Appendix B.
CCP	Cuneiform Commentaries Project, https://ccp.yale.edu/ accessed 25.01.2023.
Commentary EAE 53	Reconstructed commentary of *Enūma Anu Enlil* tablet 53 (see 5.1.5.1.).
EAE 52	Reconstructed tablet 52 of *Enūma Anu Enlil* (see 5.1.4.).
GKAB	The Geography of Knowledge, a research project within the Corpus of Ancient Mesopotamian Scholarship (CAMS), http://oracc.museum.upenn.edu/cams/gkab/ accessed 25.01.2023.
SAD	Streck, M. P. (ed.) 2018–forthcoming. *Supplement to the Akkadian Dictionaries* (2 volumes). Wiesbaden: Harrassowitz.
SIT	The serialised commentary *Šumma Sîn ina tāmartišu*, lit. "If the Moon at its appearance" (see 5.1.2.).
SIT 6	Reconstructed tablet 6 of *Šumma Sîn ina tāmartišu* (see 5.1.6.).

Conventions

Texts in italics are in Akkadian (e.g. *um-ma-tu*$_4$), whereas texts in capitals are logograms with determinatives in superscript (e.g. ^{d}UDU.IDIM). Texts in simple lowercase are unilingual Sumerian quotes (e.g. ur-saĝ). In Akkadian texts, texts in lowercase also indicate uncertain readings of cuneiform signs. When lines of text are separated by rulings on the clay tablet, that is accordingly indicated in the transliteration. When referring to a series of tablets (e.g. *Enūma eliš*, MUL.APIN, etc.), capitalised Roman numerals (e.g. I, II, III, etc.) are used to indicate a particular tablet of the series. The entries in compositions like MUL.APIN, or in lexical lists, are indicated with the sign DIŠ (i.e. one vertical wedge), which is translated using the symbol ¶ (e.g. see Hunger-Steele 2019: 3).

With the exception of the philological editions of the present author, all the quotations in the Sumerian and Akkadian language follow the main or most recent edition, or the main sources used by the author of the edition. In quotations of more than one line reconstructed through various sources or with many parallels, the numbering of the lines of the quotation follows the main or most recent edition. In other Sumerian and Akkadian texts, the beginning and the end of a line on tablets correspond to the beginning and the end of a line in this manuscript. In short quotations or entries from series of omens, the sign / indicates, whenever necessary, the end and the beginning of each line in the cuneiform text. The bibliographical references and other information (e.g. sigla of sources, additional editions whenever necessary) are provided in brackets below the quotations. The translations are by the present author, unless differently stated; they are deliberately as literal as possible to reflect our understanding of Akkadian and Sumerian grammar. All sources mentioned in this study can be found on the CDLI,[1] unless differently stated.

The author has also provided a literal translation for the Akkadian and Sumerian names of stars, groups of stars (i.e. asterisms), and constellations (i.e. groups of stars that form a specific pattern, like animals or objects).[2] For their modern identifications, see Gössmann (1950), Reiner-Pingree (1981: 10–16), Hunger-Pingree (1999: 271–277), and Kurtik (2007). Modern identifications are given in this study only for the Pleiades (MUL.MUL, *zappu*) and the planets (e.g. *dilbat*, "Venus", *ṣalbatānu*, "Mars). More specifically, since planets have many names in cuneiform texts (see A.4.1.), modern identifications are given in brackets whenever necessary (e.g. if ^{d}ŠUL.PA.È, "Šulpaea", is mentioned in an explanation of an omen, that is a name for Jupiter).

All the dates mentioned in this study are according to Mesopotamian middle chronology, which places the fall of Babylonia around 1595 BC (Liverani 2014: 9–16). The dates used in this study conform with the periodisation used in CDLI.[3]

...-...-..., ...	Registration number of tablets excavated at Nineveh and taken to the British Museum (e.g. 79-7-8, 271). For the list of registration numbers, see http://www.finckecuneiform.com/nineveh/ accessed 25.01.2023.
... + ...	A fragment that joins to another fragment (e.g. K 3567 + K 8588).
...+	A fragment that joins to one or more other fragments without giving all the registration or museum numbers involved (e.g. K 3242+).
... (+) ...	(+) indicates that fragments belong to the same tablet but do not join physically (e.g. K 5981 (+) K 11867).
– (var.)	Translation of variants within the cuneiform texts (e.g. *Glossenkeil*).
: ⋮	*Glossenkeil*, triple *Glossenkeil*.
$x^{!}$; $ma^{!}$(GAL)	Erroneous use of a cuneiform sign; the wrong sign is given in brackets in capital letters after the correct sign.
$x^{?}$	Uncertain reading of a cuneiform sign.
(-*a*)	Cuneiform sign (e.g. phonetic complements, determinatives) attested only in a few sources within the reconstructed entries.
(and)	Words added in the English translation to enhance the understanding.
[x]	Lost cuneiform sign.
[...]	Lacuna of uncertain number of cuneiform signs.
{x}	Erased cuneiform sign.
x°	Cuneiform sign written over an incomplete erasure.
x*	Collated cuneiform sign where the reading is changed against the primary publication.

1 http://cdli.ucla.edu/ accessed 25.01.2023.

2 For a discussion on the issue on celestial bodies' identifications, see Hunger and Steele (2019: 6–7).

3 https://cdli.ox.ac.uk/wiki/doku.php?id=adopted_periodisation_in_cdli accessed 25.01.2023.

⸢x⸣	Damaged cuneiform sign.
x[...], [...]x	Traces of the beginning or the end of a not identifiable sign.
/	Indicates the end and the beginning of a line in the cuneiform text.
//	Indicates a duplicate.
<x>	Cuneiform sign omitted by the ancient scribe.
<<x>>	Redundant cuneiform sign.
ad si	Uncertain readings of cuneiform signs are written in lowercase.
^{d}*a-n*[*im* ÚŠ^{meš}]	Reconstructed cuneiform signs are written in square brackets.
i, ii, iii, etc.	Columns of the cuneiform text.
MA-RAB	Uncertain readings of Sumerograms are written in capitals.
MAR:TU	The colon indicates that two cuneiform signs were written by the ancient scribe in the opposite order.
(var.: …)	Variation in the texts among parallel sources.
– (var.)	Transposition of a *Glossenkeil* within the translations.

The names and the numbers of the Babylonian months (Months I to XII) mentioned in this study conform with Hunger-Steele (2019: 8; see previously Hunger 1980). This calendar is based on the beginning of the Babylonian year (Month I) around the vernal equinox (i.e. March-April according to the Gregorian calendar):

I	Nisannu
II	Ajaru
III	Simanu
IV	Du'uzu
V	Abu
VI	Ululu
VII	Tešritu
VIII	Araḫsamnu
IX	Kislimu
X	Ṭebetu
XI	Šabaṭu
XII	Adaru

Chapter 1

Introduction: Research Questions, Methodology, Structure, and Terminology

This study explores how the Pleiades[4] were regarded within the Mesopotamian cultural framework. The idea to focus on a specific group of stars, or asterism, arose from the following core question: How were the stars perceived in Mesopotamia?

In ancient Mesopotamia, stars were not only seen as celestial bodies,[5] but also as divine entities *per se*, or associated with gods who bore the same names. Moreover, celestial bodies were responsible for producing natural phenomena and for being mediators between the gods and humankind. This notion permeates the entire Mesopotamian culture, and it has sometimes been labelled as astralisation of ancient Near Eastern religions.

Astralisation literally means the "conversion" of an anthropomorphic being into an astral one, by describing it as a star, asterism, constellation, or planet. However, the term generically refers to the increasing overlap between the divine and the celestial domains, towards an assumed universality or cosmology (Pongratz-Leisten 2011). Among Assyriologists, astralisation has become a synonym for fallacy, because it is tied to the so-called "Pan-Babylonism" theory.[6] The Pan-Babylonists at the beginning of the twentieth century assumed the existence of an "astral religion". They based their assumption on chronologically and scientifically inaccurate associations between mythology, astronomy, and astrology. Recently, the study of how celestial bodies were perceived in Mesopotamia was reintegrated into the Assyriological debate. Despite being wrong in their analysis and conclusions, Pan-Babylonists foresaw a pattern in Mesopotamian mythology; that is, all the main gods of the Mesopotamian panthea have astral features. But, how does astralisation work in Mesopotamia? Can we really talk about a conversion of the gods into stars? These questions are still in the process of being answered.

1.1. Research Question

The present study aims to introduce a new case-study for the topic of celestial bodies in Mesopotamia. In this respect, a case-study approach was chosen to capture the complexities of this topic, as portrayed by the perception[7] of the Pleiades in Mesopotamia, which has

---

4 According to the Cambridge English Dictionary, "Pleiades" is a *plurale tantum*, so it is used with both plural and singular verbs, the latter especially in astronomy, where "Pleiades" stays for the star cluster M45. In this study, the author chose to use "Pleiades" as a plural noun, also in the English translations of the Sumerian and Akkadian language.

5 In this study, the term "celestial" is used for everything related to the sky.

6 E.g. see Jensen (1900; 1928) and Weidner (1915). A recent study on Pan-Babylonism has been published by Weichenhan (2016).

7 In this study, the perception of the Pleiades signifies what Mesopotamian culture believed – and thus expressed through language (Lakoff-Johnson 2003: 3–6) – the Pleiades to be and how they functioned.

never undergone a thorough analysis. Thus, this study aims to answer the following research questions: How were the Pleiades perceived in Mesopotamia? What is their relationship with the concept of "god"?

The Pleiades were chosen as a case-study because they are among the oldest examples of a divine asterism known as a group of seven stars, not only in Mesopotamia, but also in a huge array of ancient cultures around the world.[8] In Mesopotamia they are called MUL.MUL (lit. "stars") or *zappu*, "Bristle", and they are depicted as seven stars (e.g. Figure 4). They played a primary astronomical role at least from the end of the second millennium BC onwards, because they were a reckoning device for the Mesopotamian calendar. They are the protagonist of several Mesopotamian celestial omens, which are not merely a list of observations, but expressions of the relatedness of heaven and earth. Moreover, the Pleiades are identified with seven gods (^{d}IMIN.BI, lit. "divine Seven"), several different forms of whom are known. The most famous are the Sebettu, characters from the poem of "Erra and Išum", seven warlike entities accompanying a god or the king on warpaths.[9] Besides, the presence of a huge variety of divine heptads scattered throughout Mesopotamian mythology contributes to the creation of a complex scenario: there are attestations of seven heroes, seven sages, seven demons, and seven gods of fates, and they all have astral features.

The need for a new definition of astralisation in Mesopotamia provokes the following questions: Why did the Pleiades come to be identified with seven gods? Are the Mesopotamian heptads all related to each other and to the Pleiades? In order to answer these questions, the primary focus of this study is on attempting to trace a possible tradition, or more traditions, in the way the Pleiades were perceived.

1.2. A Short History of the Study of the Pleiades in Mesopotamia

There is no comprehensive study or analysis of the role of the Pleiades, the seven stars, or seven gods in ancient Mesopotamia, yet there are useful short studies and collections of references. For instance, in their astronomical role, the Pleiades in Mesopotamia have been identified and studied by Schaumberger (1935: 336–344), who defined the so-called

8 The common characteristic in the perception of the Pleiades around the world is being a cluster of stars, usually a group of seven (Urton 1987–2005: 2865). For instance, they are seven in Mesopotamia, in Greece (i.e. seven daughters of Atlas) (see 2.1.), in the Aboriginal and Oceanic culture (i.e. seven sisters) (Orchiston 1996: 320; Kelley-Milone 2005: 344), in China (i.e. seven sisters), in North America (i.e. seven women), in India (i.e. seven wives) (Young 1987: 8734–8735), Japan (i.e. seven stars) (Renshaw 2012), and Mesoamerican cultures (i.e. seven stars) (Aveni 1996).

9 It is interesting to note that, where myths have been developed around the Pleiades, they are usually associated with women (see fn. 8). Whereas in Mesopotamia, the Pleiades are associated with the Sebettu, a name from the feminine of the numeral seven (*sebe*, fem. *sebēt*), but indicating seven male entities (see fn. 87). The only possible reference to a group of seven divine feminine entities is an element of the Pre-Sargonic (ca. 2700–2350 BC) onomasticon, PN-^{d}*si-bí*, which would be the masculine singular of the numeral seven (Wiggermann 2011a: 460). There are also "seven (and) seven daughters" (DUMU.MUNUS IMIN IMIN) of Anu in a few incantations dating to the Old Babylonian period (ca. 2000–1500 BC) (Farber 1990: 306–308, 2.2, 2.3, 2.4, 2.6).

Plejaden-Schaltregel,[10] and Weidner (1967), who identified the famous *Gestirndarstellung* text depicting the Pleiades (see Figure 4). A summary about the Pleiades in Mesopotamia was published by Hunger (2005), and Verderame (2016) who first discussed the association of the Pleiades with two different divine heptads. Stockhusen (2019: 44–45, 53–55, 204–205) in his dissertation provides references, comments, and insights on the Pleiades within the study of celestial bodies in the Ancient Near East.

The Sebettu and their association with the Pleiades and other divine heptads were first mentioned by Jean (1924), who hypothesised the existence of three simultaneous traditions: the Pleiades, the seven sons of the god Enmešarra – who later became the Sebettu – and the seven sages. A more specific study on the Sebettu within the poem of "Erra and Išum" was published by Graziani (1979), who assumed that the origin of the Sebettu could be west Semitic because the Pleiades were depicted as seven dots on the Syrian and Cappadocian iconography dating to the second millennium BC. Nevertheless, van Buren (1939–1941), who collected the seals in which the Pleiades are drawn as seven stars or dots, had already noted how the Pleiades are more unambiguously depicted on seals only from the first millennium BC onwards. Indeed, the presence of the seven dots in the earlier glyptic is not necessarily related to the Pleiades. [11] More recently, the Sebettu were studied by Wiggermann (2011a), Verderame (2017), and Konstantopoulos (2015) in her PhD dissertation. Konstantopoulos focused on the demonic values of the gods, and her study provides valuable and helpful insight into the difficulty of grasping the identity of demons and divine heptads scattered through the Mesopotamian literary tradition.

Regarding the association of the number seven with the Pleiades, a useful collection of essays was compiled and edited by Reinhold and Golinets (2008), who highlighted textual and iconographic references about the importance of the number seven in the Ancient Near East.

1.3. Methodological Approaches

In order to answer the research questions, several quotes from individual sources are commented on and compared with each other within this study. The analysis of the textual sources starts from what is preserved in the Mesopotamian lexical lists about the stars, the Pleiades, and the heptads; then it continues to their practical, divinatory, religious, and magical role, as reflected by Mesopotamian textual culture. The corpus of sources at the basis of the investigation comes from the vast panorama of Mesopotamian cultural tradition, dating from the second until the end of the first millennia BC (i.e. the end of the Late Babylonian period, ca. 30 BC). The texts are mainly in Akkadian (Assyrian and Babylonian language), though several Sumerian literary texts dating to the Old Babylonian period (ca. 2000–1500 BC) are discussed as well. The great majority of sources comes

10 The *Plejaden-Schaltregel* is an "astronomical" rule used to establish the Mesopotamian calendar (see 4.2.2. § 1a).

11 More specifically, he suggested that the seven sparse dots of the early glyptic are the representation of seven casting lots related to the seven cities of Sumer (see 2.3.3.).

from the library of Ashurbanipal (ca. 668–627 BC) in Nineveh;[12] a more limited number of discussed sources comes from Nippur dating to the Old Babylonian period (ca. 2000–1500 BC), from Aššur dating to the Middle and Neo-Assyrian period (ca. 1400–612 BC), from Sultantepe dating to Neo-Assyrian period (ca. 911–612 BC), and from Uruk dating to the Hellenistic period (ca. 323–63 BC).

Since in this study there is such a varied collection of textual evidence from ancient Mesopotamia – not only in time but also in genre – the chapters are structured according to the types of sources and contents; within the chapters, the sources are organised chronologically. Arranging the structure of the chapters on the basis of content first and then on chronology of sources was dictated by the methodology adopted while conducting the research: that is a historical and philological approach to the sources, complemented with two more interdisciplinary perspectives that are explained below according to their field of application, the micro-level (i.e. individual textual sources), and the macro-level (i.e. relationship between various textual genres).

– On a micro-level, the methodological approach points towards tracing a tradition in the perception of the Pleiades in Mesopotamia. The textual sources are chronologically organised, presented with transliterations and translations of Sumerian and Akkadian exemplars into English, and commented on philologically whenever needed. Next, they are compared according to the concept of intertextuality.[13] Intertextuality means that any text of any kind can be read and thus interpreted through other texts which shaped and encoded its meaning. Traces of intertextuality have been already found in Mesopotamian literature.[14] For instance, Hallo (2010: 607–622) discussed intertextual relationships in the framework of Sumerian proverbs. He noticed that Sumerian epics have a proverbial character, and the proverbs recur, or are even quoted, in later sources: "Rather we have here the apparently deliberate harking back from one genre to another or from one context to a thoroughly different one, with at least the implication that the source of the allusion is familiar to the 'author', perhaps even to the audience." And then: "The study of intertextuality in cuneiform literature cannot begin and end with Sumerian proverbs" (Hallo 2010: 611, 622). Annus (2016) investigated intertextuality in the myth of Adapa, the king of Eridu, dating to the second and first millennium BC: "The comparisons that my intertextual research will develop are considered as interrelated visual patterns, having a complex pictorial and metaphoric imagery" (Annus 2016: 5). More recently, Wisnom (2020: 1–23) has built her study by drawing on the concept of intertextuality, searching for similarities in three different myths: *Anzû*, *Enūma eliš*, and the poem of "Erra and Išum". She has

12 What is known as Ashurbanipal's library is a label for a great amount of Assyrian and Babylonian cuneiform texts excavated in Kuyunjik, and which are dated between ca. 800 and 612 BC. See online at the website *Ashurbanipal Library Project* (http://oracc.museum.upenn.edu/asbp/ accessed 25.01.2023), and Fincke (2003–2004) for the Babylonian texts.

13 For an overview of intertextuality, see Allen (2000); for its meaning in semiotics, see Kristeva (1986: 34–61) and Eco (1979); for its meaning in linguistics, see Plett (1991).

14 For a list of references to Assyriologists who mentioned or shortly discussed intertextuality, see Wisnom (2020: 10–11).

focused on structural allusions, i.e. intertextuality in the plot and characters of the above-mentioned myths.

In this study, the idea of adopting an intertextual approach is to go beyond the identification of simple parallelisms between the texts, and to understand their "meaning", i.e. trying to answer the research question going beyond the formal aspects of the texts. Intertextuality has been traced in quotations and repetition of names, epithets, or *topoi*, i.e. features of the Pleiades that recurrently appear in the divinatory and astral compositions and, less frequently in myths, prayers, inscriptions, and rituals, which likely triggered a specific idea, tradition, or reference in the ancient reader's mind. Once the features of the Pleiades had been established (i.e. parameters), they have been sought in the sources, where patterns arose (i.e. intertextuality). The major advantage of intertextuality is that, whenever the textual sources are compared, they disclose a persistent network of quotes and analogies, which can be then chronologically organised, and which allow new interpretations. Unfortunately, as Annus (2016: 3) put it, "in studies of intertextuality, one has to reckon that some links might be missing, which at times can make the approach more speculative." Wisnom (2020: 9–10) also underlined the absence of explicitness in cuneiform literature, or a "guide" to understand allusions and analogies between texts; in this respect, the situation of Mesopotamian culture is, for instance, different from that of the Classical one, in which ancient authors often deliberately explained allusions.

– On a macro-level, the methodological approach points towards understanding the relationship between the celestial realm (i.e. Pleiades) and the divine realm (i.e. seven gods). To achieve this, this study benefits from the definition of metaphor in cognitive science. According to this definition, the metaphor is not merely a literary tool, but the means through which human beings conceive and describe reality (i.e. metaphorical thought). The metaphors exist between different domains of knowledge, being at the basis of human reasoning and, consequently, shaping language and culture (i.e. metaphorical language) (Ortony 1993; Lakoff-Johnson 2003). When this way of reasoning is applied to Mesopotamian literacy, one suddenly realises that cuneiform knowledge is quite unique, because it was built by a scholarly effort to match two cultural units – or different languages (i.e. Sumerian and Akkadian) – in one writing system.

The definition of metaphor can be applied in two directions. First, a "metaphorical language" arises especially when comparing lexical lists and ancient commentaries to other textual sources (Maul 1999; Frahm 2011: 70–76). The gist of the Mesopotamian literary system is what Rochberg (2016: 92) defined as the "orthographic-semantic method", which is a tendency of Mesopotamian scribal culture to shape cuneiform knowledge around metaphors and analogies built through orthography and semantics. In this respect, the concept of analogy as a strategic tool of knowledge (also scientific knowledge) is relevant too (Gentner-Holyoak-Kokinov 2001). Second, the concept of "metaphorical thought", first addressed by Rochberg (1996) in the framework of celestial divination, is useful in investigating the many analogies between the physical and metaphysical domains in the Mesopotamian textual sources (i.e. naturalistic, divine, astral, chthonic domain, etc.). The idea of a metaphorical language and a metaphorical thought is particularly useful in studying

Mesopotamian literacy: this approach was chosen to allow a deeper content analysis of the sources, to understand why an asterism like the Pleiades (i.e. astral domain) was identified with seven gods (i.e. divine domain). However, methodologies like this one are relatively new in Assyriological studies and therefore less refined.[15] Annus (2014; 2016: 111–122) discussed the need to find a more "rigid" validity for the comparative methods used in historical humanities. He refers to the types of comparisons used in cognitive science, according to which the patterns and the metaphors are inherent to human reasoning. As Annus argued, by looking beyond the borders of historical humanities, combined approaches may be a fruitful way to re-think and improve the categories we use in historical humanities, even if they only produce hypotheses at first.[16]

1.4. Chapter Structure and Remarks on Terminology

This book is composed of seven chapters (including this introduction as chapter 1 and a conclusion as chapter 7), two appendices (App. A and B), plates, and an index of logograms and Akkadian words. Each chapter is based on the above-mentioned methodological preliminaries (see 1.3.). The structure of each chapter is explained below (see 1.4.1.–1.4.7.), with remarks on the aim of individual chapters and specific choices in terminology.

1.4.1. Chapter 2

Chapter 2 introduces the key-issues related to the perception of the Pleiades in Mesopotamian culture. The first part of the chapter discusses and compares the characteristics of the celestial bodies and the gods, on both a philological and ontological level. The theories of Pan-Babylonism are also introduced against the new perspectives on the relationship between the celestial and divine realms. The second part of the chapter establishes the features of the Pleiades as both a celestial and a divine heptad in the lexical lists of Mesopotamia, as well as their iconography. In the third part of the chapter, there is an excursus regarding the importance and the meaning of the number seven in Mesopotamia. The number seven, which has puzzled scholars since the earliest epochs, was considered a sort of mystical number in the whole of the Ancient Near East.[17] It is the fixed

15 A thorough study with similar approaches is conducted by the ERC project "REPAC. Repetition, Parallelism and Creativity: An Inquiry into the Construction of Meaning in Ancient Mesopotamian Literature and Erudition", directed by Nicla De Zorzi in Vienna.

16 "Scholars are very often short of historical evidence, working in the realm of hypotheses. How certain situations may have evolved historically is very often beyond our epistemic reach. However, this fact should not serve as the excuse for not exploring the problems of importance." (Annus 2014: 369).

17 See the famous and interesting article by the psychologist Miller (1955) who noticed that the short-term memory span for an adult comprises at maximum seven items and – only by coincidence – the same number is the limit of the one-dimensional absolute judgement in adults. Miller also humorously commented upon the fact that the "magical" sense of the number seven had haunted him for decades, as well as inspiring people and scholars since the dawn of human history.

number for the Pleiades (i.e. seven stars), for the corresponding gods (i.e. seven gods), and it has several empirical and attributive usages.

Chapter 2 focuses primarily on lexical lists, the most powerful tool for defining linguistics and conceptual categories. On a philological level, lexical lists underpin the coexistence of two different languages (i.e. cultural units) with two different encoding systems,[18] Sumerian and Akkadian. On an ontological level, they shed light on the potential meanings of the words, through which metaphors and analogies are built in the textual sources. As a result, the Mesopotamian taxonomy for what is divine and what is a celestial body may appear ambiguous to a modern scholar attempting to build working modern categories. In this respect, the chapter follows the arguments of Rochberg (2009; 2010: 317–338; 2011) who presented three perspectives based on which the relationship between gods and stars is potentially built. These three perspectives find their application in every textual genre of Mesopotamia, and accordingly in this study.

1.4.2. Chapter 3

Mesopotamian literature is littered with information about the relationship between the Pleiades and divine heptads. The first part of chapter 3 discusses the myths in which heptads (i.e. seven stars, heroes, gods, demons, weapons, etc.) with an astral component feature, in order to make clear whether they were associated with the Pleiades. Whilst, the second part of the chapter discusses the role of the Pleiades in Akkadian prayers, rituals, and royal inscriptions, that are rich in analogies and intertextuality. As a whole, the chapter deals with religious texts, such as myths, prayers, or rituals, labelled with the broader "literary texts".

Conventional terms are used in this study when presenting literary evidence for the Pleiades. Yet, the textual sources of Mesopotamia went – and still go – through a problem of a taxonomy of genres.[19] Under these circumstances, a few choices in chapter 3 need to be addressed. First, the word "myth" is used in its broader meaning, because "myth is not a literary genre; it is a generic category of the created world reflected in literature but not confined to it" (George 2007: 45).[20] Consequently, narrative poems like "Gilgameš and Ḫuwawa" or "Erra and Išum", but also the "Hymn of Ḫendursaĝa" and the UDUG.ḪUL incantations belong to the category of myths, even though their ancient rubrics (i.e. purposes or performances) are all different. Second, as a matter of consistency with the history of studies, and because the issue cannot be solved in the present context, all the compositions mentioned in the chapter are quoted with their commonly known titles.

---

18 See Peirce (1991), according to whom the signs of a language are "arbitrary signs" which refer to their object only by means of a code or, as Pierce puts it, of a "law". It is essential to note that only the code gives the signs their status and significance.

19 George (2007) discussed how the approach of the Assyriologists was mainly philological and "positivistic", instead of "critic" in modern terms: the modern taxonomy of literary genres of Mesopotamia is still borrowed from the Classical one, hence it is not always adequate when defining Mesopotamian textual genres.

20 George (2007: 37–48) also discussed the issue of defining the meaning of "myth", as applied to the Mesopotamian sources.

Regarding prayers and incantations, there is a similar issue as the one discussed above. For instance, the term *Gebetsbeschwörungen*, "incantation-prayers", was used by Kunstmann (1932) and Mayer (1976) to define the corpus of *šuʾilla* prayers, i.e. "hand-lifting prayers", dating to the first millennium BC. The term is still in use today to define a considerable variety of compositions with the incipit ÉN, *šiptu*, "incantation", and whose content is similar to a generic "prayer" addressed to the gods; at the same time, their rubric displays instructions for a ritual with magic purposes. For instance, the same *šuʾilla* prayers are included in different tablets with different rubrics.[21] The plain *šuʾilla* rubrics (ka-inim-ma šu-íl-lá, "word of the hand-lifting") meant "not simply a gesture of salutation or greeting performed before beginning to pray, but instead a preliminary prayerful act intended to demonstrate one's submission and piety in order to call for divine presence/intervention or request permission to approach the deity" (Oshima 2013: 112). This means that a *šuʾilla* is simply a prayer, and there was probably no differentiation between prayers and incantations in Mesopotamia, but rather prayers whose wording could have been used for magico-religious purposes (Oshima 2011: 7–8). However, in this study, to keep a reference to the history of the studies, a differentiation is made between the prayers, called *šuʾilla*, and other prayers with more specific magical[22] purposes, usually called "incantations", i.e. spells with targeted ritual enactments (anti-witchcraft, love, or purification).

1.4.3. Chapter 4

The role of the Pleiades within compositions about celestial topics (i.e. astral compositions), whose tradition dates to the late second or the first half of the first millennium BC, is discussed in chapter 4. The relevance of these compositions resides in the fact that they collect heterogeneous information and a descriptive astronomy, i.e. they give schemes and rules for the periodicity of celestial phenomena. The first part of the chapter analyses the MUL.APIN and the Three-stars-each tradition, together with other related sources. The second part of the chapter focuses on the use of the Pleiades as a reckoning device for the Mesopotamian calendar. It also discusses rules for the intercalation involving the Pleiades, i.e. the insertion of an extra month in the calendar, as preserved in several sources.

Although commonly used in past decades, the modern terms "astronomy"[23] and "astrology"[24] are not totally adequate to designate these sources, because they presume concepts which were not developed at that time. Several Assyriologists nowadays agree in

21 E.g. see Lambert (1974–1977: 198).

22 Magic is commonly intended as the power to control various forces through performative acts, often with a negative connotation of "irrational". Regarding the use and the meaning of the term "magic" applied to Mesopotamian texts, Schwemer (2011: 420) and Rochberg (2016: 160) argued that it is still the only term suitable to describe several exorcist texts, despite its disadvantageous implications.

23 "Astronomy" is an exact science which deals with space and celestial objects.

24 "Astrology" is not an exact science, and it deals with the assumption that the relative position of celestial bodies influences life on earth. "Astrology", in its modern meaning, presumes the concept of the zodiac and zodiacal belt of 12 signs.

using the broader term "astral science"[25], despite that term also being often debated. That is because of the ongoing argument regarding the inclusion of Mesopotamian knowledge within the history of science, and whether texts such as MUL.APIN, or the Astrolabe B could be defined as "scientific".[26] Ultimately, it is safe to say that the goal of these astral compositions is the "knowledge of the sky" (Hunger 2011: 62), a knowledge which cannot be easily labelled as modern "science", yet it implies ideal schemes and models (e.g. list of dates for the first visibility of stars, intercalation schemes, schemes for the length of day and night, etc.), based on analogical reasoning (see 1.3.), which cannot be downgraded to something apart from science. As Rochberg (2016: 102, 274–284) put it, the astral compositions, together with celestial divination (see 5.), attest to attempts to "put nature to the question", which is the essence of science, also the modern one.

1.4.4. Chapter 5

Although celestial divination is sometimes considered part of the astral science of Mesopotamia (see fn. 25), in this study it is discussed separately in chapter 5. Before explaining the structure of the chapter, a few remarks on divination in Mesopotamia are necessary. Divination was an essential practice, because with its working principles, and its relevance as a decisional tool for the country, it is embedded in both the royal court and the cultural system.[27] Indeed, the celestial divination was practised by the diviners, who were part of an erudite lore, shared by both the redactors of the written celestial omens and the interpreters of ominous signs. At the same time, celestial divination is tied to mythology and religion.

Our knowledge of Mesopotamian celestial divination comes from the huge number of written celestial omens dating to the first millennium BC. In this context, a written "omen" is a conditional statement built by a protasis, which includes a phenomenon, and an apodosis, which includes its prediction[28] ("If X" = protasis ↔ "then Y" = apodosis = phenomenon ↔ prediction). Whereas the "ominous signs" are the phenomena observed in the sky.

Chapter 5 aims to discuss the role of the Pleiades within the Mesopotamian divinatory corpus. The first part of the chapter shows the role of the Pleiades in the omens from the celestial divinatory series *Enūma Anu Enlil*, "When Anu (and) Enlil". Regardless of already edited omens, which are collected and analysed in this chapter, there is also the critical edition of two reconstructed tablets from *Enūma Anu Enlil*, the assumed tablets 52 and one commentary to tablet 53, and tablet 6 of the serialised commentary *Šumma Sîn ina tāmartišu*, lit. "If the Moon at its appearance". These tablets – in which the Pleiades are the

25 "The term 'astral sciences' is a catch-all to refer to scholarly activity that falls under the modern categories of astronomy, astrology and celestial divination, cosmology, and certain aspects of meteorology" (Hunger-Steele 2019: 1 fn. 1).
26 The debate about the Mesopotamian knowledge within the history of science is discussed in A.1.–A.1.2.
27 For a guide to divination and the divinatory sources of Mesopotamia, see Maul (2005) and Koch (2015).
28 In this study, the predictions of omens are meant as predictions *from* celestial phenomena, i.e. prognostications, and not predictions *of* phenomena, like in astronomy (Rochberg 2016: 232).

protagonists – are presented in three score transliterations with translation, commentaries to the text, and further comments. The second part of the chapter deals with omens from reports of Assyrian and Babylonian scholars: despite these reports being very well edited and studied,[29] they show the reasoning of Mesopotamian scholars who dealt with the omens.

The reconstruction of the *Enūma Anu Enlil* series started in the first half of the twentieth century,[30] but it still faces issues in the numbering systems, due to the poor preservation of the sources.[31] The tablets 52 and 53 of *Enūma Anu Enlil* are only "assumed", because these are the numbers of tablets displayed only by a few sources. Such sources do not represent the overall organisation of the series *Enūma Anu Enlil* as a whole, but only the organisation of the series in the time and place the sources were written. The individual omens from tablet 52 and the commentary of tablet 53 of *Enūma Anu Enlil* are reconstructed from a huge variety of sources, including ancient commentaries, which are the first attestations of hermeneutics in history.[32] A complete study on commentaries and hermeneutical techniques was published by Frahm (2011), whose work allowed the reconstruction of several commentaries. Among them, there is *Šumma Sîn ina tāmartišu*, the serialised commentary on *Enūma Anu Enlil*. While looking for fragments belonging to omens about the Pleiades, the author noticed that the content of the reverse of one fragment, LB 1321, was previously misunderstood as *Šumma Sîn ina tāmartišu* tablet 2, while it belongs to the assumed *Šumma Sîn ina tāmartišu* tablet 6 (Renzi-Sepe 2021). Part of tablet 6 of *Šumma Sîn ina tāmartišu* is very similar to the commentary to tablet 53 of *Enūma Anu Enlil*, and the great majority of entries in tablet 6 has the Pleiades as the main subject. A thorough study on *Šumma Sîn ina tāmartišu* was conducted by Wainer in his unpublished PhD dissertation;[33] still, tablet 6 of this series is reconstructed and edited in chapter 5 by the present author.

1.4.5. Chapter 6

Chapter 6 aims to discuss the role of the Pleiades within the astronomical and astrological texts dating to the Neo- and mostly Late Babylonian period (ca. 626–30 BC). These texts share their main topic about celestial matter with the sources in chapters 4 and 5. Yet, they include new concepts, like the zodiac, and presume significant changes in the predictability of planetary phenomena, at least from the sixth century BC onwards. However, these texts will not be treated in this study as the expression of a "scientific revolution" compared to the "non-scientific" past of divination and astral compositions.[34] Chapter 6 relies on

29 See the comprehensive edition by Hunger (1992) in SAA 8.

30 The groundwork in reconstructing the numbering system of *Enūma Anu Enlil* was made by Weidner (1941–1944; 1954–1956; 1968–1969).

31 See Fincke (2001; 2013), Gehlken (2005: 235–268) and Rochberg (2018).

32 If one intends hermeneutics narrowly "as pertaining to the interpretation of written texts" (Frahm 2011: 3).

33 "The Series 'If the Moon at Its Appearance' and Mesopotamian Scholarship of the First Millennium BCE". PhD dissertation, Brown University (2016).

34 E.g. see the Kuhnian's methodology used by Brown (2000: 3, 9, 126, 153–161), and critiques by

Rochberg's arguments on this topic: as she widely discussed, divination, astrology, and astronomy share many predictive modes and theories, not necessarily based on the physics of nature, but rather on analogical reasoning to understand the meaning of things (what she refers to as "putting nature to question") (Rochberg 2016: 102, 274–284). According to this point of view, influenced by historical epistemologists and cognitive studies, there will be no distinction between "non-scientific" divination and "scientific" astrological and astronomical texts. Instead, efforts will be put on drawing parallels between them, through a comparative and intertextual approach, to understand continuities and changes in the perception of the Pleiades during the Late Babylonian period.

The first part of chapter 6 deals with an overview of the astronomical texts of ancient Mesopotamia, whereas the second part deals with astrological texts. In chapter 6, the term "astronomical" is used only to describe the content of texts which preserve computed astronomical data (e.g. Astronomical Diaries). While the term "astrological" is used to describe the texts related to the concept of the zodiac (e.g. astral medicine, micro-zodiac texts, etc.), a concept fully adopted only from the end of the fifth or beginning of the fourth century BC onwards (Britton 2010). Such distinction has only been made by modern scholars *a posteriori* for a matter of practicality. Indeed, "Babylonian texts know no difference between astronomy and astrology, neither in terminology nor in concept. There was only knowledge of the heavens and that was used in several ways" (Hunger 2020: 272).

1.4.6. Appendix A

The celestial omens about the Pleiades are collected in chapter 5, whereas what was conceived as ominous or not, and the working principles of the celestial divination, is a matter discussed in appendix A. More specifically, appendix A aims to introduce the celestial omens within new perspectives about divination. The inclusion of appendix A within this study arose from the necessity to explain the sense of puzzlement that modern scholars often have while reading Mesopotamian omens. The first part of appendix A deals with the ongoing debate regarding the inclusion of Mesopotamian divination, with its techniques of predictions based on deductive logic, in the history of science, whereas the second and the third part of appendix A discuss the working principles and the rationale of omens through several examples – sometimes also not specifically related to the Pleiades.

A structuralist approach[35] was adopted to categorise the working principles of the celestial omens in question. Other scholars in the framework of divinatory practices have referred to structuralism,[36] but the works of Brown (2000: 105–207) and Rochberg (2010a) are particularly discussed and taken into account in this book.

Rochberg (2016: 247–249).

35 See de Saussure (1916) and all the Assyriologists who already adopted this approach (see A 1.1.).

36 See Böck (2010) De Zorzi (2011) and Winitzer (2017).

1.4.7. Appendix B and the Plates

Appendix B provides editions of selected divinatory sources whose text has not been included in the reconstructed texts of *Enūma Anu Enlil* tablet 52, the commentary of tablet 53, and *Šumma Sîn ina tāmartišu* tablet 6 (see 5.1.4.–5.1.6.), and sources whose content is relevant to the Pleiades. The plates include digital hand copies of all the sources given in the Appendix B.

The author personally collated and photographed 41 sources at the British Museum of London, and at the Vorderasiatische Museum of Berlin, in February 2019 and February 2020. Between 2018 and 2020, Jeanette C. Fincke allowed the author to explore her database of photographs of tablets and fragments housed in the British Museum to complete and improve the editions. She also provided the author with additional photographs of the following fragments: BM 44005, K 2118, K 2138, K 2170, K 2254, K 2301, K 3923+, K 6484, K 12425, Rm 477, Sm 247, and Sm 259. In total, 38 sources were used for the reconstructed tablets of *Enūma Anu Enlil* and *Šumma Sîn ina tāmartišu*. The sources K 3123, K 2254, and K 6686 + K 9234 (see App. B § 24–28) do not belong to the reconstructed tablets but to other tablets of *Šumma Sîn ina tāmartišu*, and they have been included in the appendix B due to their references to intercalation (see 4.2.2.).

Out of all 41 sources, 15 sources have been recopied, while 26 have been copied for the first time. All copies are included and numbered in the plates (see pp. 385–423), and they are in scale 1,5: 1, except for a few examples explicitly stated to be in scale 1: 1. The necessity of a re-edition of several sources was due to one major reason. The edition of *Enūma Anu Enlil*'s fragments started at the end of the nineteenth century; most of the texts and the hand copies were published in the publications of Craig (1899) and Virolleaud (1908–1912), but they are now considered out of date. At that time, the hand copies of clay tablets did not provide any information regarding the dimension, the provenance, the ductus, and the shape of the cuneiform signs. That is, the copies were idealised and often combined using several fragments which only ideally corresponded to the original texts.

Chapter 2

The Pleiades in Their Context: Perception of Celestial Bodies, Taxonomy, Relationships with Gods, and the Number Seven in Mesopotamian Culture

This chapter introduces the Pleiades as they were conceived in the Mesopotamian cultural framework, a framework that dictates the rules of relationships between gods and celestial bodies. It also presents the necessary groundwork for interpreting the textual sources pertaining to the Pleiades mentioned in the following chapters. The first part of the chapter, after a preliminary paragraph about the nature of the Pleiades as a cluster of stars (see 2.1.) has an overview of the study of celestial bodies in Mesopotamia. First, the issue of the binomial god-star, or the Mesopotamian taxonomy of gods and celestial bodies, is discussed through many examples (see 2.2.–2.2.2.). Second, both gods and celestial bodies are considered in the light of old and new perspectives on their relationship (see 2.2.3.–2.2.3.1.). Third, the binomial god-star is framed in a broader methodological setting that draws on the concept of metaphor in cognitive science and semiotics (see 2.2.3.2.–2.2.3.3.). In the second part of the chapter, the Pleiades are presented as an inherent element of the complex Mesopotamian cultural scenario. The focus is on the lexical repertoire (see 2.3.–2.3.2.), and the iconography (see 2.3.3.) of the Pleiades, which disclose the basic information on the nature of the Pleiades in Mesopotamia as a celestial body associated with a divine heptad. A last section deals with the association between the Pleiades and the number seven, with its widespread metaphorical meaning in Ancient Near East (see 2.4.–2.4.2.).

2.1. Preliminaries: the Pleiades with Naked Eye

Before discussing how the Pleiades were perceived in Mesopotamia, it is necessary to have in mind what the Pleiades look like and what their characteristics are. The informative words of the astronomer Stephen James O'Meara (1998) given below will introduce the appearance of the Pleiades with the naked eye:

> "With the naked eye, the Pleiades looks like a tiny dipper, a forest of starlight bathed in moonlit mist, or a distant gathering of veiled brides. (...) Just how many stars in the Pleiades are truly visible with the naked eye is the subject of some debate. Traditionally, the number has been seven. (...) Although largely symbolic, the age-old association of the Pleiades with the number seven remains fixed to this day – to the point that some observers swear they cannot see more than seven members, even though the Pleiades contains 10 stars brighter than 6th magnitude" (O'Meara 1998: 145).

As shown in Figure 1, the Pleiades, astronomically referred as M45,[37] are an open cluster of very bright stars visible in the constellation Taurus. The name "Pleiades" comes from the Greek mythology (Πλειάδες, *Pleiades*), and it designates the "seven daughters of Atlas", the nymphs of rain.

Although situated in the northern hemisphere, the Pleiades are sufficiently visible from every part of the globe. They are visible during winter months in the northern hemisphere; in April they disappear from the western sky, and then they "suddenly" reappear in the eastern sky, approximately 40 days later. From an ancient observer's point of view in Babylonia – unaware of the planetary motion[38] – the Pleiades' risings and settings recall a travel across the entire sky from east to west over one year – a year that Mesopotamians marked by observing the motion of the Sun and Moon (see 4.2.1.).[39] If the Pleiades can be visible up to 11 synodic months[40] and a few more days, which is roughly the time of 12 sidereal months,[41] the Mesopotamian ancient observers likely expected to see the Pleiades with the Moon about 12 times per year,[42] as Schaumberger (1935: 337–340) was the first to notice.

37 Charles Messier (1784) numbered to the Pleiades this way in 1781, in the final version of his "Catalogue des Nébuleuses et d'Amas d'Étoiles" of 110 astronomical objects, published in the annual publication of astronomical ephemerides in France.

38 For an introduction to naked-eye astronomy, see Aaboe (2001: 1–23) and Walker (ed.) (1996).

39 Patterns of visibility of the Pleiades in a sample site of Babylon, 700 BC, were compiled by Konstantopoulos (2015: 266–268).

40 A synodic month, or lunation, is the time it takes for the Moon to make one revolution around the earth with respect to the Sun, that is about 29,5 days (Aaboe 2001: 12).

41 A sidereal month is the time it takes for the Moon to make one revolution around the earth with respect to the fixed stars, that is about 27,3 days (Aaboe 2001: 12).

42 Theoretically 13 times, but the Moon is not visible when it is new, i.e. once a year.

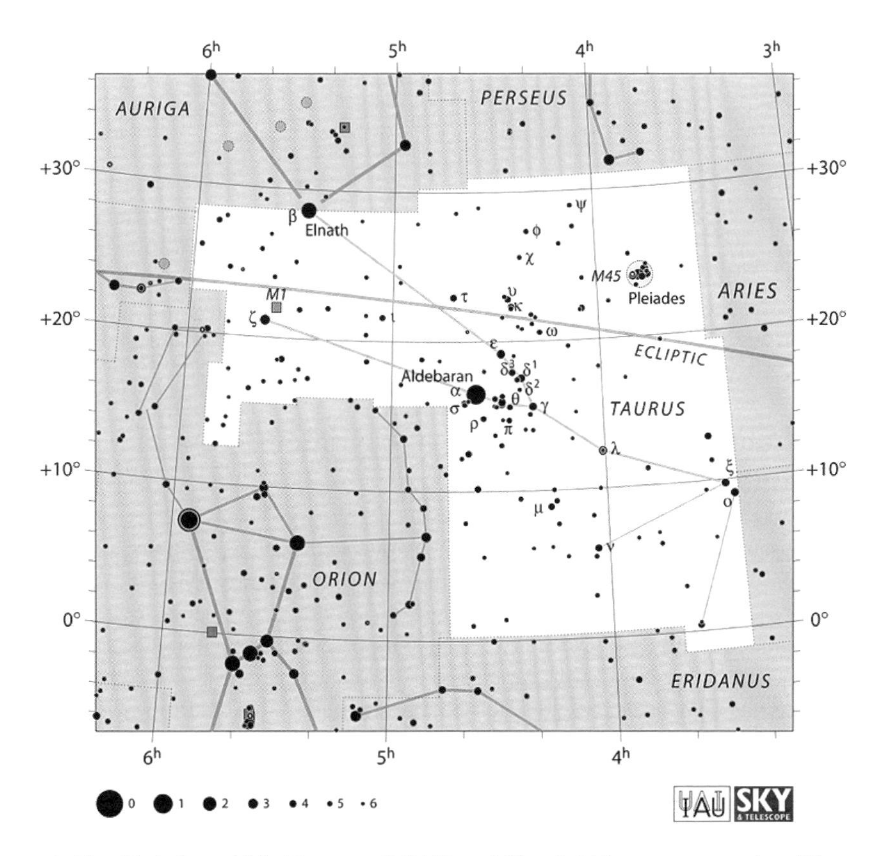

Figure 1. The Pleiades within Taurus - © IAU and Sky & Telescope magazine (Sinnott and Fienberg 2015, https://www.iau.org/public/images/detail/tau/ accessed 25.01.2023)

Due to their shape and visibility, the Pleiades were known in ancient times and several cultures as a group of "many" – probably because of their characteristic nebulosity (O'Meara 1998: 145–147) – or "seven" stars, even though unaided naked-eye observers claim to see only six stars, while sharp and expert eyes see from ten up to 17 stars (O'Meara 1998: 145; Kelley-Milone 2005: 143).[43] The Pleiades were one of the first asterisms used for an astronomical conception in history, especially related to the seasonal phenomena (in April-May and autumn, see Kelley-Milone 2005: 22–24). The best-known example for the Pleiades as seasonal markers comes from Hesiod's "Work and Days" (eighth century BC). In the poem, the moment in which the Pleiades are most visible determined the starting point of the agricultural activities (i.e. harvesting and ploughing):

43 For the number seven as the traditional number for the Pleiades, see the debate regarding the "lost Pleiad" in Kelley-Milone (2005: 141–143).

When the Pleiades, born of Atlas, rise before the Sun,
begin the reaping; the ploughing when they set.
For forty days and forty nights they are hidden.
(Works and Days 383–385; Translation by West 1988: 48)

This was a rule of thumb in Greece, but also somehow existing in Mesopotamia in a different form. As noted by Brown (2018: 34), there could be a close resonance to this passage in the "Farmer's Instructions",[44] a Sumerian collection of advice about farming and cultivation dating to the second millennium BC. Before delving into this any deeper (see 4.1.2.), it is vital to first contextualise the Pleiades within the broader framework of the study of stars in Mesopotamia.

2.2. The Context: The Perception of Celestial Bodies in Mesopotamia

The study of stars and planets in Mesopotamia started with the beginning of Assyriological studies, which was established at the end of the nineteenth century. Modern scholars were already aware – thanks to the Classical and Biblical sources – of the great influence of Mesopotamian astronomy and astrology on contemporary and later neighbouring cultures. Indeed, what has been known by the Classical literature as the "Chaldean sources" is nothing other than the legacy of the Babylonian astronomy and astrology.[45] Despite the many profiting studies conducted on the astronomical data, and on divinatory, astrological, and religious texts, the perception of celestial bodies in Mesopotamian cultures is still an open question. Before presenting any issue related to this topic, it is necessary to try defining what a celestial body according to the Mesopotamian perception is. In this respect, to "define" shall not mean to "label" in modern terms, but to shed light on the ancient taxonomy, and consequently on a possible ontology for stars and gods (see 2.2.1.–2.2.3.3.).

2.2.1. The Taxonomy of Gods and Stars

In Mesopotamia, gods are associated with stars, a statement that we first learn from the lexical repertoire. The lexical lists dating to the Old Babylonian period (ca. 2000–1500 BC) produced the first Mesopotamian catalogues of stars,[46] in which an individual celestial body was given many different names, and also names of specific gods. A good example for

44 u$_4$ mul an-na šu im-ma-ab-du$_7$-a-ta 10-àm á gud a-ša zi-zi-i-da-še igi-zu nam-ba-e-gíd-i, "When the constellations in the sky are right, do not be reluctant to take the oxen force to the field many times" (Civil 1994: 30–31 ll. 38–39; for the translation, see online at the website ETCLS 5.6.3.).

45 For instance, Ptolemy in the *Almagest* made use of astronomical data from the Astronomical Diaries of Mesopotamia (Hunger-Pingree 1999: 156–159). In *Tetrabyblos*, Ptolemy frequently mentioned the use of "Chaldean sources", and his theories on astrology share many similarities with Mesopotamian sources (Geller 2014: 69–71).

46 E.g. the thematic bilingual list Ura = *ḫubullu*, rooted in the second millennium BC. For a bibliography and short overview of star lists, catalogues and where to find them, see Horowitz (2015).

these associations with several names is the Great Star List, a star catalogue dating to the first millennium BC.[47] The names of the planet Mars according to the Great Star List are given below:

237. ^{mul4}MAN-*ma* ^{mul}*a-ḫu-ú* ^{mul4}*na-ka-ru*
238. ^{mul4}*sar*$_6$*-ru* ^{mul4}ḪUL ^{mul4}KA$_5$.A ^{mul4}ELAM.MA^{ki}
239. ^{mul4}*ṣal-bat-a-nu*
240. IMIN *zik-ru-šu*

237. The Other One, the Strange One, the Hostile One,
238. the Liar, the Evil One, the Fox, the star of Elam,
239. Mars.
240. Its seven names.
(Koch-Westenholz 1995: 198, 200 ll. 237–240)

In this passage, the planet Mars has seven names; additionally, Mars is usually associated with the god of war and death, ^{d}U.GUR, "Nergal" (von Weiher 1971: 76–83). The names of Mars are just one example for a long list of names that can be associated with specific celestial bodies, in a conspicuous number of textual sources. These textual sources also belong to different traditions; hence, they might have had different associations because of the different local panthea. The example of Mars epitomises the necessity to "untangle" textual evidence, from an ontological perspective too.

In cuneiform, the oldest sign form for "god", DINGIR (or AN = 𒀭), *ilu* in Akkadian, is a pictogram of a star formed by four wedges. The sign DINGIR means everything that is divine or has divine power. Whereas the sign for "star", MUL, *kakkabu* in Akkadian, is composed by three DINGIR (= 𒀯) signs, and it means any kind of celestial body, be it an individual star, a group of stars (i.e. asterism), a constellation, a meteor, or a planet (Borger 2004: 93 n. 247). The bilingual lexical lists (e.g. Proto Aa from Nippur) dating to the Old Babylonian period (ca. 2000–1500 BC) that have Akkadian equations to Sumerian logograms, give several synonyms for the sign MUL:

(139)	1.	DIŠ	mu-ul	3xAN	*ka-ak-ka-bu-um*	¶ star
	2.	DIŠ		3xAN	*ši-ṭi-ir-tum*	¶ inscription
	3.	DIŠ		3xAN	*na-pa-ḫu-um*	¶ to glow
	4.	DIŠ		3xAN	*na-ba-ṭù-um*	¶ to shine brightly

(MSL 14: 94 ll. 139:1–4)

According to the lines quoted above, the sign MUL is related to stars (*kakkabu*), brightness (*napāḫum, nabāṭum*),[48] but also to something written (*šiṭirtum*). These words point to a

47 For references to the Great Star List, see 4.1.3. and fn. 228.
48 For the equation of MUL with *napāḫum* (CAD N1 263–270) and *nabāṭum* (CAD N1 22–24), see the fragment BM 48659 dating to the Neo-Babylonian period (ca. 626–484 BC), a commentary on the 14th *pirsu*, "section", of the lexical list series Aa, whose edition is provided online (CCP 6.1.14,

well-established relationship between the stars and writing, also known from a famous poetic utterance: the starry sky is called in Akkadian *šiṭir šamê* or *šiṭirti šamāmī*, “writing of the sky”.[49]

In cuneiform sources, both the signs DINGIR and MUL are used as determinatives (i.e. semantic classifiers),[50] and they can alternate freely before the name of a celestial body. Nevertheless, the anthropomorphic gods of the panthea can only have the determinative DINGIR. Turning back to the example of Mars, the name of Nergal can be written ^{d}U.GUR but never ^{mul}U.GUR; the planet Mars, however, can be written with both determinatives, ^{d}*ṣalbatānu* and ^{mul}*ṣalbatānu*.[51] The name of Nergal (^{d}U.GUR) can designate the planet Mars,[52] but the name of Mars (^{d}*ṣalbatānu* or ^{mul}*ṣalbatānu*) cannot designate the god Nergal.

Although the determinatives were not fixed for celestial bodies, there was a differentiation between stars and groups of stars, planets, and constellations. Indeed, different words were in use to write their names. For instance, “planet” is written $^{d/mul}$UDU.IDIM, *bibbu*, lit. “wild sheep”. The logograms $^{d/mul}$UDU.IDIM can also precede the name of a specific planet, e.g. Saturn, Mercury, and sometimes Mars.[53] Yet, there seems to be no consistency in the way planets were addressed. For example, these are the names and the determinatives of the planets given in the Great Star List:

241. ^{d}30 *u* ^{d}UTU ^{d}ŠUL.PA.È ^{mul}*dil-bat*
242. ^{mul}UDU.IDIM ^{mul}SAG.UŠ ^{d}UDU.IDIM.GU$_{4}$.U[D]
243. ^{mul4}*ṣal-bat-a-nu*
244. IMIN ^{mul}UDU.IDIM^{meš}

241. The Moon and the Sun, Jupiter, Venus,
242. (generic) planet, Saturn, Mercury,
243. Mars.
244. (These are) the seven planets.
(Koch-Westenholz 1995: 200 ll. 241–244)

---

https://ccp.yale.edu/P461243 accessed 05.03.2021). The sign NAB/P (i.e. two AN signs on top of each other) is the sign for the Elamite “god” (BM 48659 l. 3’). As Rochberg (2010: 322) noted, the reading *napāḫum* and *nabāṭum* for MUL (i.e. AN followed by NAB/P) may derive from the Elamite “god” (NAB/P) as connected to the meaning of “bright”.

49 For the *šiṭir šamê* or *šiṭirti šamāmī*, “writing of the sky”, see CAD Š3 144 b, and further discussions in 2.2.2. and 2.2.3.3.

50 The cuneiform determinatives classify words and not objects, mainly because, as Veldhuis (2014: 48–49) stated, the lexical lists and the determinatives find their roots in the earliest administrative needs. As a result, all the objects which were not useful to that purpose – like the stars – tend to be excluded from the lexical repertoire.

51 E.g. in celestial omens: ^{d}*ṣalbatānu* (EAE 52: 12) and ^{mul}*ṣalbatānu* (Commentary EAE 53: 3) are both used for Mars.

52 E.g. in Rm 192 rev. 1 (App. B § 21).

53 E.g. in celestial omens (SIT 6: 2) ^{d}UDU.IDIM.SAG.UŠ is attested for “Saturn”, and ^{d}UDU.IDIM.GU$_{4}$.UD for “Mercury”. Sometimes Mars is called just ^{d}UDU.IDIM (Brown 2000: 57).

In these lines, Jupiter is called by the name of its associated god (^{d}Šulpaea); Saturn, Venus, and Mars have the determinative MUL (or MUL_4), while Mercury has the designation "planet" (^{d}UDU.IDIM, *bibbu*).[54] The Sun and the Moon, because of their higher status within individual local panthea, can solely have the determinative DINGIR (Koch 2015: 155).[55]

The constellations usually have the determinative MUL, though the Akkadian word *lumāšu*, LÚ.MAŠ in Sumerian, can designate a constellation (CAD L 245). The word *lumāšu* could be understood as the "pattern" (Lambert 2013: 98 V 2, 477), or the image that the gods "draw" in the night sky, as can be seen in the concluding paragraph of the celestial omen series *Enūma Anu Enlil* tablet 22:[56]

> [...] *ú-za-i-zu ḫar-*⌜*ra-ni*⌝ / MUL^{meš} *tam-ši-li-*[*šu-nu e-ṣ*]*i-ru lu-ma-a-*[*ši*]
>
> [...] They (i.e. Anu, Enlil, and Ea) distributed the paths / they [dre]w the constellatio[ns], the likeness of the stars.[57]
> (AfO 17: 89 ll. 4–5// Rochberg-Halton 1988: 271, 17'–18')

The word *lumāšu* is used in poetic, explanatory and zodiological contexts[58] to indicate either a particular constellation which rise near the solstices or equinoxes, or a zodiacal sign (Koch 2015: 155, 202–203).

From all the above-mentioned quotations, the binomial god-star arises as a unique characteristic of Mesopotamia. The stars, planets, and constellations in Mesopotamia bore different names and two different determinatives, "god" and "star", which were used alternatively. At the same time, the proper name of an anthropomorphic god was used to designate a celestial body as the god's representation in the night sky, but never the opposite. The fact that the taxonomy is varied, and the determinatives are interchangeable constituted an ontological problem to modern scholars who often tried to fathom cuneiform sources through a Western, Classical or Biblical approach. It is nowadays understood that trying to channel the concept of "god" and "celestial body" into what modern thinkers – shaped by the Scientific Revolution – conceive as "supernatural" and "natural" is a mistake.[59] The definition of what nature and religion signify in Mesopotamia is still an ongoing discussion which goes beyond the purpose of this study, yet it must be stressed

54 See also Hunger (2005a: 589). The designation appears alone at the beginning of l. 242, as a reference to the following Saturn, Mercury, and Mars, the only planets for which it is used (Brown 2000: 57).

55 Nevertheless, see Saturn designated as MUL ^{d}UTU, "the star of the Sun", in MUL.APIN II i 64.

56 The series *Enūma Anu Enlil* is discussed at length in chapter 5.

57 For the meaning of the "paths", see 4.1. For the construction and the translation of the sentence, see Lambert (2013: 22).

58 E.g. see Koch-Westenholz (1995: 198 ll. 226–230); Pingree-Walker (1988: 315 obv. 6, 8); Lambert (2013: 58 l. 2); see also 2.2.3.1. § 2.

59 For a discussion on what nature and religion signify in Mesopotamia, see Rochberg (2016: 17–37; 131–163), according to whom words such as "natural" and "supernatural" do not apply to cuneiform knowledge because they presume a Western-oriented cultural framework. Instead, a "god" in Mesopotamia is better described by his or her ability to determine "the designs of heaven and earth" (*uṣurāt šamê u erṣeti*) (Rochberg 2016: 164–190).

that the more traditional definition of the concept of "god",[60] or "celestial body", has been proven to be inadequate to describe the relationship between the gods and the world according to Mesopotamians, because cuneiform knowledge "is situated in history before either nature *or* God became terms of discourse" (Rochberg 2016: 281). More than on the categories of "natural" or "supernatural", cuneiform knowledge was always based on a unique framework, where a multiplicity of conceptions and ideas coexisted, such as the binomial god-star and the taxonomy being interchangeable without necessarily constituting an ontological problem.

2.2.2. The First Attestations of Celestial Knowledge

The earliest sources which properly attest to celestial observations in Mesopotamia – the celestial omens – date to the Old Babylonian period (ca. 2000–1500 BC).[61] Nevertheless, such *terminus post quem* does not necessarily mean that the observation of the skies started only in the Old Babylonian period. For instance, Sumerian literary texts dating to the Old Babylonian period point to an extensive literary activity that took place in earlier times and of which we have only a few traces. In these texts, the establishment of the Sumerian panthea likely went alongside the observation of celestial bodies. Above all, the Moon and the Sun were the main gods of the panthea: Nanna/Suen and his son Utu, called respectively Sin and Šamaš in Akkadian. The astral aspect of Nanna/Suen seems already well-established in the Pre-Sargonic period (ca. 2700–2350 BC), due to the cult of Nanna in Ur and the legacy transmitted by the hymns of the priestess Enḫeduanna (Hall 1985: 395–401).[62] There was also Nisaba, the goddess of grain, writing, and stars, who is often described as consulting a "tablet of the stars of the sky" (dub mul-an) in early sources.[63] The tablet of Nisaba might be considered the oldest metaphor for the firmament (Sjöberg-Bergmann-Gragg 1969: 138b), or the oldest reference to the interpretation of messages from the stars (Rochberg 2004: 64; Fincke 2016: 111–112). Finally, the association between the goddess Inanna and the planet Venus is probably one of the oldest. As early as in the Late Archaic Uruk period (ca. 3500–3000 BC), the goddess of love, Inanna, was called ^{d}inanna-ud, "Inanna of the day",[64] and ^{d}inanna-sig, "Inanna of the evening",[65] in administrative texts. This subdivision recalls the two phases of Venus in the sky: when she is to the east (i.e. as a morning star) or to the west (i.e. as an evening star) of the Sun (Szarzyńska 1993; Kurtik 2016).

---

60 "God concepts are concepts about supernatural agents. Supernatural agents are often (although not invariably) beings with minds like ours but no bodies." (McNamara 2009: 193).

61 For a chronological overview of all the Mesopotamian sources concerning celestial observation, see Koch-Westenholz (1995: 32–53), Hunger-Pingree (1999: 7–31) and Fincke (2016: 107–119).

62 See the Sumerian temple hymn (n. 8) to the Ekisnuga: nun-zu nun ka-as-bar men-an-dagal-la / lugal-an-kam ^{d}áš-im-babbar-e, "Your prince, the prince who makes decisions, the crown of the wide heaven, / the king of heaven, Ašimbabbar (i.e. Nanna)" (Sjöberg-Bergman-Gragg 1969: 23 ll. 115–116).

63 See the Gudea Cylinder A v 21–25 (Edzard 1997: 72).

64 VAT 15307 obv. iii 1 (ATU 5 pl. 5, W 6288).

65 VAT 15246 rev. i 1 (ATU 5 pl. 2, W 5233, b).

Under these circumstances, the origin of celestial knowledge – likely established in oral tradition – presumably goes back earlier than the first textual sources at our disposal. Even though the first explicit attestations about celestial observation date to the Old Babylonian period (ca. 2000–1500 BC), the binomial god-star existed already in the third millennium BC.

2.2.3. Approaches and Perspectives on Astralisation

After establishing the issue regarding the taxonomy and an ideal *terminus post quem* for Mesopotamian celestial knowledge (see 2.2.1.–2.2.2.), there is another open question about the relationship between the gods and the celestial bodies: are the celestial bodies gods? We shall assume that the sign DINGIR, *ilu*, "god", when used as a determinative, defines everything that inherently has a divine essence or divine agency. The determinative DINGIR is used for anthropomorphic divine entities (e.g. gods depicted as humans, such as ^{d}Enki, ^{d}Enlil, ^{d}Nergal, ^{d}Adad, etc.),[66] but also for celestial bodies (e.g. ^{d}30, the Moon, and ^{d}20, the Sun), and even for other inanimate objects that have divine power (e.g. rivers, mountains, and plants).[67] Yet, only the main anthropomorphic gods of the Mesopotamian panthea are associated with celestial bodies (see 2.2.1.).

These premises underline the issue in defining something which is never explicitly mentioned, or explained, in Mesopotamian textual sources. Due to this ambiguity, the existence of an assumed "astral religion" composed of "astral gods" has often been debated. More specifically, the concept of *Astralmythen* was developed at the end of the nineteenth and beginning of the twentieth century by a group of Orientalists influenced by comparisons in mythography and history of religions. Their approach, called "Pan-Babylonistic", was pursued by Assyriologists such as Peter Jensen and Ernst Weidner, who interpreted Mesopotamian mythology based on planetary and cosmic phenomena, also extending their interpretation to the whole Ancient Near East.[68] Already then, contemporary scholars pinpointed their approach as methodologically and scientifically inaccurate.[69] For instance, the Pan-Babylonists often referred to the zodiac in order to explain Mesopotamian mythology, but the concept of zodiac is attested only in texts dating to the end of the fifth or beginning of the fourth century BC onwards (Sachs in Neugebauer 1969: 140; Britton 2010), later than the time when the main Mesopotamian myths were composed. After Pan-Babylonism, the concept of an astral religion in Assyriology became a taboo, a synonym for misconception, and later scholars carefully avoided similar approaches.[70] On these conditions, modern scholars began to look with suspicion at the

66 For anthropomorphism and non-anthropomorphism in iconography, see Ornan (2009).

67 For the deification of mountains, rivers, and trees, and the concept of non-anthropomorphism in divine agency, see Perdibon (2019). The concept of "divine" itself is not strictly related to anthropomorphism in Mesopotamia (Porter 2009).

68 E.g. see Stucken (1896), Jensen (1910; 1928: 305–309) and Weidner (1915).

69 E.g. See Kugler (1909).

70 See Brown (2010) for a discussion on how Pan-Babylonism and the schools of thought of the last century shaped the genre of astral and mathematical cuneiform texts.

broader concept of the association of an anthropomorphic deity with a celestial body. The term "astralisation" also (see 3.3.1.), if not rejected at all, was associated only with a religious trend dating to the first millennium BC, due to the rising importance of the systematic celestial observations in Mesopotamia (Pongratz-Leisten 2011: 153–187). Nevertheless, even if presuming an astral religion and mythology in Mesopotamia was methodologically and scientifically incorrect, the Pan-Babylonists foresaw a pattern in Mesopotamian mythology; only from the 1980s onwards, did Assyriologists like Erica Reiner[71] and Francesca Rochberg (see 2.2.3.1.) redeem the study of gods and stars in Mesopotamia and assert the necessity of reconsidering and redefining their meaning.

2.2.3.1. Perspectives on Celestial Bodies

Being a pioneer in this respect, Francesca Rochberg started an interesting new way of looking at the relationship between the celestial bodies and the gods in Mesopotamia, profiting from the studies on cognitivism (Ortony 1993). In her articles "'The Stars Their Likenesses': Perspectives on the Relation between Celestial Bodies and Gods in Ancient Mesopotamia" (Rochberg 2009) and "The Heavens and the Gods in Ancient Mesopotamia: the View from a Polytheistic Cosmology" (Rochberg 2011; see also Rochberg 2010: 317–338), she distinguishes three perspectives through which this relationship can be defined.

§ 1. Gods as Celestial Bodies

The first perspective (Rochberg 2009: 48–64) is drawn from mythology. Anthropomorphic gods, identified by their proper names, often have astral features in myths. The Sumerian mythological narrative of Lugalbanda is a case in point (see 3.1.1.3.): in the myth "Lugalbanda in the Wilderness", Lugalbanda, the hero of Uruk, is escorted by the major gods of the pantheon, Nanna/Suen, i.e. the Moon god, Utu, i.e. the Sun god, and Inanna as Venus. These gods are said to be starlike and to shine across the sky, as shown in the following lines from the myth referring to Nanna/Suen:

> mul-amar kug en-nu-un-šè àm-ši-ri / mul-ud-zal-le-da-ke$_4$ an-e im-sar-e[72]
>
> The Holy Calf took up his guard. / Like the morning star he glows in the sky.
> (Wilcke 1969: 75 ll. 197–198; Vanstiphout 2003: 114 ll. 202–203).

In this myth, the Moon god, the Sun god, and Inanna/Venus all have a ME.LÁM, *melammū*, "(awe-inspiring) radiance" (CAD M2 9–12), an attribute which likely derives by abstraction from the concept of the brightness of stars in the night sky (Rochberg 2009: 49–50; Thavapalan 2020: 34–36). This is not unusual: in both ancient and modern religions, light symbols (e.g. auras, halos, beings of light, etc.) usually describe divine entities (Mohr 2006: 1104). This first perspective is close to the concept of catasterism in Classical

71 E.g. see the famous book "Astral Magic in Babylonia" (Reiner 1995).
72 As Wilcke (1969: 76 fn. 313) assessed sar stands here for *napāḫu*, "to glow", or *ṣarāru*, "to flare".

studies, the transformation of anthropomorphic deities into constellations. It can also be associated with the term "astralisation", if we consider it narrowly as the "conversion", or description of an anthropomorphic being with astral attributes, and not only as a generic overlapping between astral and anthropomorphic attributes.

§ 2. Celestial Bodies as the Representations of Gods

In the second perspective, planets and stars addressed by their celestial names are regarded as images of the anthropomorphic gods, they represent them (Rochberg 2009: 64–75). For instance, the stars represent the gods in the cosmogony of the Akkadian poem *Enūma eliš* (V 1–2):

> *ú-ba-áš-šim man-za-za ana* DINGIR^{meš} GAL^{meš} / MUL^{meš} *tam- šil-šu-nu lu-ma-ši uš-zi-iz*
>
> He (i.e. Marduk) created the positions for the great gods, he set up the constellations, the likeness of the stars.
> (Lambert 2013: 98 ll. 1–2)

In *Enūma eliš*, the stars and the constellations represent the gods, like in celestial divination – further discussed at length in the chapter 5 – where the stars produce ominous signs to be interpreted as messages of the gods (see 5.1., 5.3.). According to this second perspective, the stars are the celestial images of the gods, their "likenesses", and celestial phenomena are the actions, or the will, of the gods.

§ 3. Personified Celestial Bodies

The third perspective (Rochberg 2009: 75–83), which Rochberg discussed only in her article from 2009, emerges mainly from celestial omens (see 5.), astrological texts (see 6.2.), and prayers dating to the first millennium BC (see 3.2.1.–3.2.4.). According to this perspective, which can be partly included in the second one (§ 2), celestial phenomena are described through anthropomorphic terms and typical attributes of the gods with whom the celestial bodies are associated. One can talk about the adoption of a metaphorical language, which led to the personification of celestial bodies. Rochberg (1996; 2004: 167–181) discussed and proved the use of a metaphorical language in lunar omens: the word for "(lunar) eclipse", AN.GE$_6$, *attalû* in Akkadian,[73] is often replaced by expressions like "the god disappeared in distress" (*adāru*) or "mourns" (i.e. expressed by *lumun libbi*, "grief"), metaphors for the Moon perceived as an anthropomorphic entity.

This third perspective is opposed to the theory of "mythopoetic thought" of Frankfort (Frankfort-Frankfort-Wilson-Jacobson 1946), a theory which influenced more than one generation of Assyriologists. According to Frankfort, the ancient thought was speculative and apart from a rational thought:

---

73 But also *namtallû(m)* and *nantallû(m)* in texts dating to the Old Babylonian texts (Khait 2014: 80–82).

> "The ancients expressed their 'emotional thought' (as we might call it) in terms of cause and effect; (...) They could reason logically; but they did not often care to do it." (Frankfort-Frankfort-Wilson-Jacobson 1946: 11).

However, the omens testify the opposite. The use of metaphorical language in celestial omens only shows that such a use of personification, when applied to a celestial phenomenon, would imply the use of analogical reasoning, which is a form of inductive or deductive reasoning. Indeed, it seems that Mesopotamians had a clear conceptual distinction between what is divine and human (i.e. "the god disappeared in distress"), and what is divine and astral (i.e. "eclipse"),[74] or non-human.[75] Therefore, it could be assumed that divine agency presumed both anthropomorphism and asterism in Mesopotamia, and the binomial god-star was built on that.

2.2.3.2. The Concepts of Metaphorical Language and Metaphorical Thought Applied to Mesopotamian Culture

The metaphor is not intended as mere ornamental speech in this study. Cognitive linguists have developed different and profitable perspectives on the metaphorical language, based on the assumption that metaphorical language comes from metaphorical thought, the human and analogical thought par excellence.[76] A theory of metaphor was proposed by Lakoff and Johnson (2003), according to whom the metaphor is a strategic tool of our minds to understand and communicate reality. According to their cognitive studies, one starts from a domain of experience (e.g. nature, stars, death, etc.) and arrives at another one in order to describe something. In other words, one domain of experience can be understood in terms of another domain.[77] When this line of reasoning is applied to the culture of Mesopotamia, one suddenly realises that Mesopotamians consistently used metaphor as a tool to understand and express reality. In addition to the example of the lunar eclipse (see 2.2.3.1. § 3), another good example is the *šiṭir šamê* or *šiṭirti šamāmī*, "writing of the sky", an idiom to describe the starry sky (Rochberg 2004: 64). The metaphor implies that the sky is a writing medium, and the stars are cuneiform signs: one domain of

74 "If the heavenly bodies were thought of as gods – not manifestations of gods, but identical to and synonymous with gods – we ought not regard the anthropomorphic descriptions of their movements and appearances as metaphorical" (Rochberg 2004: 169). Furthermore: "The Assyro-Babylonian sciences of celestial divination and astral magic are furthermore predicated on an anthropomorphic notion of deity. Each requires that the heavenly bodies, as gods or as the images of gods, communicate with human beings, hear their prayers and answer them" (Rochberg 2010: 333).

75 The concept of the non-human or "other-than-human" person, as introduced by Hallowell (1960), was developed in the case of deified mountains, rivers, and trees in Mesopotamia by Perdibon (2019: 19–30). As she suggested, the concept of "other-than-human" personhood applies to deified natural elements, which can be seen "as deities and as cosmic entities participating in the divine and relational cosmos of the ancient Mesopotamians" (Perdibon 2019: 4).

76 See Gentner-Holyoak-Kokinov (2001) for a collection of studies on analogy in cognitive science.

77 The more basic example is the utterance "love is a journey", where one domain of experience (i.e. love) is understood in terms of a more concrete domain (i.e. journey) (Tendhal-Gibbs 2008: 1826).

experience (i.e. the sky, the stars) is understood in terms of a more concrete domain of experience (i.e. the tablet, the cuneiform signs).

The theory of the metaphor is also related to the concept of metaphor as understood by semiotics. For instance, Eco (1971; 1998) considered the metaphor as the result of semiotic connections, not only on a semantic level, but also on a cultural level. If the meaning of a word (i.e. the object that the word denotes) is a cultural unit, then cultural units are organised in different ways, according to the cultural system that codifies them. Thus, a semantic field may be more or less faceted in various cultures. One good example: the Sumerian sign TÙR, *tarbaṣu* in Akkadian covers two completely different semantic fields, at least to our modern perception, that is the terrestrial as "cattle-pen", and the astronomical as "halo (of light)". These two semantic fields can be seen in the context of three examples given below, the first from the UDUG.ḪUL incantations (see 3.1.2.2.), the second from a Neo-Assyrian report (see 5.2.), and the third from the series *Enūma Anu Enlil*:

> *re-ʾu* [*ina*] ⸢*áš*⸣-*ri* KÙ ⸢*tar*⸣-*ba-ṣa u* ⸢*su*⸣-*pu-ra ú-kin-ma*
>
> The shepherd placed the cattle-pen and the sheepfold [in] a pure place.
> (Geller 2016: 478 l. 167)
>
> DIŠ 30 TÙR NÍGIN-*ma* ^{d}*né-bé-ru ina* ŠÀ-*šú* GUB-*iz* / ŠUB-*tì* MÁŠ.ANŠE
> *na-maš-še-e šá* EDIN
>
> If the Moon is surrounded by a halo and Jupiter stands inside it, / (there will be) an epidemic among the herd of the wild animals of the steppe.
> (SAA 8, 147 obv. 5–6)
>
> [DIŠ ^{mul}Š]U.GI TÙR NÍGIN *ina* MU BI *ina* KUR DÙ.A.BI RI.RI.GA NAM.LÚ.U$_{18}$.LU *ana*° ÁB.GU$_{4}$$^{hi.a}$ *u* USDUḪA ⸢NU⸣ [TE]
>
> [If the O]ld Man is surrounded by a halo, in that year in the entire country (there will be) a downfall of people, (but the evil) [will] not [come close] to the livestock and the flock.
> (K 3099 + K 18689 rev. 12', see EAE 52: 46 source E$_1$ and parallels)

According to Eco (1998: 98–100), the semantic fields shape cultural units within a certain conception of the world, and many semantic fields can be complementary or even coexist for a long time. This is not different from what happened within the Sumerian and Akkadian language: two different systems, linguistically unrelated, used the same signs and came their way to coexist and proliferate for centuries. Ultimately, the use of Sumerian logograms alongside Akkadian is the tangible proof of the coexistence of two faceted semantic systems at the same time.

2.2.3.3. Unifying Perspectives

In light of the previous remarks (see 2.2.3.2.), it is possible to briefly sum up Rochberg's three perspectives (see 2.2.3.1.; Rochberg 2009: 83–91). The celestial phenomena were understood as the messages, utterances, or the will of the gods. Those phenomena were also expressed in human terms, using metaphors and personifications (e.g. the Moon god mourning is a metaphor for a lunar eclipse). The divine agency in Mesopotamia is characterised by both transcendence and immanence, without constituting an ontological problem. The three perspectives acquire more coherence when compared among each other (Rochberg 2009: 75):

> "Although ostensibly aimed at physical descriptions of phenomena, the omen texts and related material contribute to the evidence for the idea that celestial bodies were regarded as divine. From this point of view, they are entirely consistent with the perspective of the mythological and hymnic texts that make clear that some gods had astral aspects and could be referred to as heavenly bodies."

According to this unification of perspectives, there is no such thing as "astral religion". There is a kind of inherent ambiguity for us, but in Mesopotamia it was not perceived as such on a conceptual level (Rochberg 2010: 337):

> "From the emergence of a theology concerning the heavenly bodies, reflected in Sumerian mythological works and Akkadian divinatory scholarship, to the expressions of religious and cosmological philosophy in the first century C.E., the relation between gods (or God) and the heavens was seen in a multiplicity of ways."

The association between gods and stars comes from the idea of divine agency being scattered, and the cosmos being multifaceted. Consequently, the taxonomy of gods and stars also seems to be interchangeable on a certain level (see 2.2.1.). The association between different aspects of cosmos is reflected in the Mesopotamian writing system at first, in which one sign can be read in many ways.[78] Frequently, in the scholarly tradition of Mesopotamian scribes – from the bilingual lists dating to the Old Babylonian period (ca. 2000–1500 BC) onwards – we find the concept of "revealing" the inner meanings of a word through synonyms and through the way a word is written (Maul 1999). The same concept is likely adopted for the cosmos: the reality is multifaceted, and every element "hides" more meanings, which can be "revealed" and thus associated with something else. First, this plurality is, on a philological level, well expressed by playing with logograms and syllabic writing: the logogram MUL can be understood as "star", but also as "inscription" (see 2.2.1.), or TÙR is either a "cattle-pen", or a "halo" (see 2.2.3.2.). Second, on an ontological level, the plurality is well expressed by the three perspectives, in which every celestial body can represent a god and *vice versa*, and by using metaphorical language (see 2.2.3.1. § 1–3).

The taxonomy and the ontology of gods and celestial bodies shed light on a cultural panorama – the Mesopotamian one – in which the analogy seems to have been the fundamental mean to organise knowledge. Analogy was – and still is today – one of the

78 This is the argument of Tallay Ornan in Porter (2009a: 195–210).

first cognitive tools employed in both literature and science.[79] The metaphor of the starry sky as the "writing of the sky" (*šiṭir šamê* or *šiṭirti šamāmī*), like the script on a tablet (see 2.2.3.1.), is the best example to describe analogy as a strategic tool for building the knowledge. This idiom is implicitly related to the sign MUL, which is associated with both "star" (*kakkabu*), and "script" (*šiṭirtu*) in bilingual lexical lists (see 2.2.1.). This way of reasoning through analogies is a method which Rochberg (2016: 92) defined as an "orthographic-semantic" method. The case of the "writing of the sky" attests, among the others, to techniques tied to etymology and etymography,[80] widely used in cuneiform hermeneutical texts, modalities in which the cuneiform knowledge was built by the Mesopotamian scribes. That is, to analyse the meaning of a word by associating it with other words or cuneiform signs that may be written alike but have a completely different meaning and *vice versa*.

2.3. The Pleiades in Context: Overview of Their Names, Religion, and Iconography

The key points and the methodological framework in defining the binomial god-star (see 2.2.–2.2.3.3.) should be now applied to the present case-study, the Pleiades in Mesopotamia. As many other celestial bodies, the Pleiades have their own names (see 2.3.1.), they are associated with divine entities, that is the divine heptad (see 2.3.2.), sometimes depicted as seven stars, as attested in iconography (see 2.3.3.).

2.3.1. MUL.MUL – *zappu*

The logogram for "Pleiades" is the reduplication of the sign for "star", MUL.MUL, literally "stars".[81] Other logographic writings are $MUL_4.MUL_4$, MÚL.MÚL, and ÁB.ÁB,[82] the latter being a standard form for denoting the plural of star, stars, or the Pleiades, in Seleucid Babylonian astrological and astronomical texts (Neugebauer-Sachs 1967: 183). In the lexical lists dating to the Old Babylonian period (ca. 2000–1500 BC), no reading of MUL.MUL as celestial bodies is attested. The Akkadian equivalents of the signs MUL.MUL show that these logograms are understood as forms of the verb *napāḫu* and *nabāṭu* (see 2.2.1.), as can be seen from the entries of the bilingual lexical list Proto Izi I:

79 See e.g. Lloyd (1966) for analogical reasoning in Greek philosophy, and Dunbar (2001) for analogy in modern science.

80 The term "etymography" was first employed in Assyriology by Frahm (2011: 70 f. 337 76), who adopted it from Assmann (2003). Examples of etymology and etymography in hermeneutics can be found in Frahm (2011: 70–76); more recently, Bennett (2021: 49–157) discussed the techniques of etymology and etymography in syncretistic texts.

81 The writing mul.mul is also attested in the acrographic list A from Ebla (Pettinato 1982: 126 vi 6'), and in a bilingual list with the reading *kà-ma-tù*, (Pettinato 1982: 72 iv 7'–8'). That is found in Hebrew (*kīmā*, "Pleiades") as well (Stol 1992: 268 fn. 69; Horowitz 2005: 173).

82 For all these writings, see Borger (2004: 302 n. 247, 378 n. 589, 396 n. 672, 402 n. 698). The logogram ÁB can also be read *lītu*, *littu*, "cow": see Horowitz (2014: 14–15) for references to stars as cows living in cattle-pens.

15. [MUL].MUL	*i-ta-an-pu-ḫu-um*	to constantly glow up
16. [MUL].MUL	*i-ta-an-bu-ṭù-um*	to constantly shine brightly
(MSL 13: 36 ll. 15–16)		

In the quoted entries, MUL.MUL is understood as the reduplication of the logogram MUL, which stands for the Ntn-stem form of the verbs *napāḫu* and *nabāṭu*, "to glow" and "to shine brightly".

The earliest attestation of MUL.MUL as the name of the Pleiades appears in the bilingual lexical list Ura = *ḫubullu* 22 dating to the Middle Assyrian period (ca. 1400–1100 BC):[83]

MUL.MUL	⸢*za*⸣-[*á*]*p*-⸢*pu*⸣	Bristle
(KAL 8, 97 rev. i 15)		

MUL.MUL is read *zappu*, "Bristle" in Akkadian. "Bristle" could be a reference to the "hair" of the Taurus, perceived as the image of a bull in the night sky. This association is secured by a clay tablet (VAT 7851) depicting the shape of the Pleiades in the sky. In this tablet, the drawing of a bull (i.e. Taurus) is placed next to seven stars, the Pleiades, named MUL.MUL (see 2.3.3. Figure 4). The name *zappu* may derive from the shape of Taurus, as an image of a bull in the sky. It is possible to assume that the Akkadian *zappu*, and thus the tradition according to which the Pleiades were perceived as the bristle of a bull, was at some point associated with the Sumerian MUL.MUL, which originally had another meaning. There may be only one reference, albeit implicit, to the Bristle in Sumerian language of the Old Babylonian period: in one of the "Inanna and Dumuzi" songs, literary compositions written in Sumerian, probably related to the celebration of a "sacred" marriage rite for the fertility of the king and the well-being of his offspring (Sefati 1998: 30–49). In one of these songs, Inanna is exalted as a beautiful maiden bringer of abundance; at the beginning of the text, she is referred to with the following words:

> [^{lú}ki-sikil kun-sìg m]ul-mul-la sig$_7$ sag$_9$-ga-à[m] / [^{d}inan]na kun-sìg mul-mul-la sig$_7$ sag$_9$-ga-àm
>
> [Maiden, shi]ning [bristle], perfect beauty, / [Inan]na, shining bristle, perfect beauty.
> (Sefati 1998: 236 ll. 1–2)

As noted by Sefati (1998: 90, 242), among the various interpretations that can be proposed for kun-sìg mul-mul-la, "shining bristle", or "starry bristle", there is that of an astral character of Inanna. In the song, she is described with animal metaphors, and called "ibex" (l. 3 taraḫ) and "stag" (l. 3 lu-lim); thus, her "bristle" could be a first, possible reference to the Pleiades as Inanna's bristle conceived as (the image of) an animal in the sky. Although it cannot be excluded entirely, this interpretation remains hypothetical since it would attest

83 The first edition of Ura = *ḫubullu* is given in MSL 11, while a new edition for the standard recension of Ura = *ḫubullu* 22 with a catalogue of stars has been published by Bloch and Horowitz (2015).

one of the few astral references in "Inanna and Dumuzi" songs, together with one epithet of Inanna as nin_9-mul-mul "shining sister" or "starry sister" (Sefati 1998: 134, 136, ll. 22, 26), and one attribute of the seven sages (*apkallū*), which will be further discussed in 3.1.1.1.

Regardless of their name, the fact that the Pleiades were perceived as seven in number is not unique, as it is attested in many other ancient cultures (see 1.2. fn. 8 and 9; 2.1). The number of the Pleiades is made clear not only by the iconography (see 2.3.3.), but also by celestial omens dating to the first millennium BC, which always use the same epithet to refer to their number:

MUL^{meš}-*šú-nu se-bet-ti-šú-nu*

Their stars, their seven (or: the seven of them).
(Gehlken 2012: 80 obv. 1; see 5.1.8.3.)

In Akkadian, MUL.MUL can also be read *kalītu*, "kidney", an unclear association sometimes found in literary texts dating to the first millennium BC (see 3.2.4.). The late first millennium BC standard recension of Ura = *ḫubullu* 22 has this reading, among others:[84]

268'. MUL.MUL	MIN (i.e. *kak-ka-bu*)	ditto (i.e. Stars)
269'. MUL.MUL	*ka-li-tu*$_4$	Kidney
270'. MUL.MUL	*za-ap-pu*	Bristle

(SpTU 3, 114a rev. ii 12'–14'; Bloch-Horowitz 2015: 104 ll. 268'–270')

A possible explanation to the equation of the Pleiades with *kalītu,* may reside in an analogy with the number seven: at the seventh day of its cycle, the Moon is, indeed, called *kalītu*, "kidney", because of its shape (Hätinen 2021: 102–103).[85] Hence, the word *kalītu* could have been associated with the Pleiades by means of their relationship with the seven.

At least from the Middle Babylonian period (ca. 1400–1100 BC) onwards, MUL.MUL can also be read *mulmullu*, "arrow" (CAD M2 190–191), one of the weapons of Marduk in *Enūma eliš* IV (Lambert 2013: 88 l. 36). That is an even more obscure association because the Pleiades are nowhere related to Marduk or to one of his weapons. Only as a speculative hypothesis, the word *mulmullu*, "arrow", might be referred to a celestial weapon that Marduk used in the cosmogony of *Enūma eliš*, and therefore called after the phonetic rendering of MUL.MUL, understood as "stars", "shining", or "starry".

To sum up, according to the lexical lists dating to the Old Babylonian period (ca. 2000–1500 BC) , MUL.MUL indicates a constant brilliance or a group of brilliant stars. Later, MUL.MUL is read as *zappu*, "Bristle", referring to the Pleiades within the sign Taurus. As previously discussed (see 2.2.3.3.), different Akkadian readings and synonyms

84 A lexical list dating to the Middle Babylonian period (ca. 1400–1100 BC) from Emar (Msk 74115 rev. iii 3'–5', see Arnaud 1985: 287; Rutz 2016: 40) preserves the same readings.

85 See e.g. the Neo-Assyrian scholarly text (*i-NAM-geš-ur-an-ki-a* type) K 170 + Rm 520 l. 2: AGA UD.7.KAM *ka*$_{15}$-*lit*(*u*), "the crown of the seventh day (is) the kidney" (Livingstone 1986: 30–31; Hätinen 2021: 137–139).

for Sumerograms are characteristics of cuneiform knowledge, which is often built on analogical association among signs, words, and meanings. The Pleiades are not apart from this reasoning: in the next chapters, it will be shown how the different readings of MUL.MUL, and the analogies created from them, play almost a hermeneutical role in literary (see 3.) and divinatory texts (see 5.–6.).

2.3.2. ^{d}IMIN.BI

Just like Venus is associated with Inanna, or Mars with Nergal, the Pleiades are associated with a divine heptad:

> DIŠ MUL.MUL ^{d}IMIN.BI DINGIR^{meš} GAL^{meš}
>
> ¶ The Pleiades, the divine Seven, the great gods.
> (MUL.APIN I i 44, see Hunger-Steele 2019: 39)

The divine heptad is written with the determinative for "god" (DINGIR) and the number "seven" (IMIN), written ^{d}IMIN.BI (lit. "these/their divine Seven").[86] The Akkadian reading, ^{d}*sebettu*, derives from the cardinal *sebe*.[87] As Konstantopoulos (2015: 15) remarked, the syllabic attestations of ^{d}IMIN.BI are scarce, and rendered with different final vowels as case endings.[88]

The God List *An* = *Anum* VI (Litke 1998: 211–213 ll. 149–184) dating to the first millennium BC attests not one, but several divine heptads (i.e. several ^{d}IMIN.BI), where ^{d}IMIN.BI seems to be an attribute for several types of entities:

- The UR.SAG ^{d}IMIN.BI, the "seven heroes" (*An* = *Anum* VI 150), lit. "the heroes, the divine Seven", a heptad well known from the Sumerian forerunners of the Standard Babylonian epic of Gilgameš. This early heptad has astral features and is further discussed in 3.1.1.1.
- The DINGIR^{meš} ^{d}IMIN.BI, the "seven gods" (*An* = *Anum* VI 151), lit. "the gods, the divine Seven", whose names are taken from the seven major gods of the Mesopotamian panthea who have an astral counterpart: Anu, Enlil, Ea, Sin, Šamaš, Ištar, Adad/Ninurta. This heptad is further discussed in 3.2.5.1.

86 The logogram BI stands for the Akkadian anaphoric third person plural masculine demonstrative pronoun *šunu*, "these", or for the third person plural masculine possessive pronominal suffix *-šunu*, "their" (Streck 2018: 33–38). In bilingual texts, imin-bi is usually rendered *sebettišunu*, "their seven" or "the seven of them", for a group of determined seven entities (i.e. *these* Seven) (Wiggermann 2011a: 460).

87 For references on the ambiguity of the phonetic spellings, see Wiggermann (2011a: 459–461) and Konstantopoulos (2015: 15). The correct form is *sebettu*, not *sibittu*, the feminine singular status rectus of the numeral seven, to designate a group of seven male divine entities (Streck 1995: 22 fn. 69).

88 E.g. DINGIR^{meš} *se-bet-ti* (STT 176 l. 7'; STT 230 rev. 12'); DINGIR^{meš} *se-bet-tú* (Jursa 2001: 79 iii 14'), ^{d}*se-bet-te* (or DINGIR$^{\langle meš\rangle}$ *se-bet-te*) (SAA 2, 5 rev. iv 5').

– The IMIN ^{d}IMIN.BI, the "seven seven" (*An* = *Anum* VI 152–184), lit. "the seven divine Seven", the divine heptads from Sumer (KI.EN.GI.NA.KE$_4$), Akkad (KUR URI^{ki}.KE$_4$), Gutium (KUR GU.TI^{ki}.KE$_4$) and Elam (KUR ELAM.MA^{ki}.KE$_4$). The latter heptad is tied to the Elamite goddess Narunde.[89]

In addition to these divine heptads, in texts dating to the first millennium BC there are also traces of "seven sons" of the old chthonian god Enmešarra,[90] and "seven brothers" of the Western Semitic goddess Išḫara.[91] Mesopotamian sources are rich in groups of seven, indicating that the number seven plays an important role that will be discussed in 2.4. Among all these groups, the most famous are the seven warlike gods known from the poem of "Erra and Išum" (see 3.1.2.1.–3.1.2.1.5.) dating to the first millennium BC, always written ^{d}IMIN.BI and usually translated "Sebettu". The warlike Sebettu are gods attested among the official Assyrian pantheon of Nineveh from the first millennium BC onwards, therefore they are sometimes called "great gods" (^{d}IMIN.BI DINGIR^{meš} GAL^{meš}, "the Sebettu, the great gods", see 3.2.7.). According to Mesopotamian mythology, the Sebettu have a truly chaotic nature because they are a sort of demonic warriors (see 3.1.3.). Although the provenance of the Sebettu cannot be proven, they always kept the position of "outsiders" in the Assyrian pantheon, even though they were sometimes called "great gods". Consequently, they must be distinguished from the seven major gods of the Mesopotamian panthea as a group (i.e. Anu, Enlil, Ea, Sin, Šamaš, Ištar, Adad, or Ninurta), who are also addressed as ^{d}IMIN.BI DINGIR^{meš} GAL^{meš}, "the divine Seven, the great gods". According to Konstantopoulos (2015), who studied the Sebettu specifically in their demonic role, their behaviour is quite atypical for higher gods, but not for demons, who are also considered divine beings (i.e. they have the determinative DINGIR, "god"). The Sebettu are, indeed, the expression of the "overall flexibility inherent in demonic beings" of ancient Mesopotamia (Konstantopoulos 2015: 312).[92] Another hint about the special nature of the heptads and the Sebettu is the fact that they are gods in both the west and the east – a statement made evident by the God List *An* = *Anum* (VI 152–184), and by their connection with Narunde and Išḫara. It cannot be excluded that this characteristic may be based on the analogy with the Pleiades, as the Pleiades are a very bright asterism, visible both in the western and the eastern sky. Roughly after the beginning of the Mesopotamian year the Pleiades appear in the eastern sky, and in the second part of the year they appear in the

---

89 ^{d}NA.RU.DI *a-ḫat-su-nu*, "Narunde (is) their (i.e. Sebettu's) sister" (Litke 1998: 213 ll. 184–184a). The connection to Narunde exists since the second millennium BC and is known from the so-called *tākultu* rituals, i.e. cerimonial or cultic meals (Frankena 1961: 201 x l. 30).

90 See Jean (1924), Wiggermann (1992: 115), Al-Rawi-George (2014: 74–75), Hätinen (2021: 355 fn. 2003), and SAA 3, 40 obv. 5 (^{d}IMIN.BI DUMU^{meš} ^{d}EN.ME.ŠÁR.RA, "The divine Seven, the sons of Enmešarra"). See also the myth "Enmešarra's Defeat" (Lambert 2013: 281–298). For a further discussion on Enmešarra and the Pleiades, see 5.1.4. further comments on the text, pp. 170–172.

91 See Wiggermann (1992: 115, 219 "Sebettu").

92 In Hittite culture, there is also a malignant "doubled" heptad called ^{d}IMIN.IMIN.BI, but its roots are perhaps in Anatolia. The Sebettu of Mesopotamia surely influenced the heptad, but it has dubious and little association with the Pleiades. A discussion about the divine heptad in Anatolia can be found in Polvani (2005) and Archi (2010).

western sky, where they disappear to be visible again to the east.[93] As suggested by Konstantopoulos, this feature "permeates the behaviour of the Sebettu: they are as far-ranging as their astral counterparts that were inherently connected to the horizons" (Konstantopoulos 2015: 288).

Konstantopoulos's thorough study made it obvious that a differentiation between, at least, the demonic and warlike Sebettu and the other divine heptads should be made. Their names were almost always written alike, so a terminological distinction is not always possible, and only the context might help to establish a differentiation. Whenever it is possible thanks to the context, in this study a distinction is made in the English translation: dIMIN.BI is translated "Sebettu", when it explicitly refers to the warlike heptad, or whenever the heptad is addressed through epithets or quotes from the poem of "Erra and Išum" (see 3.1.2.1.–3.1.2.1.5.). Alternatively, dIMIN.BI is literally translated "divine Seven", meaning a group of seven gods, a divine heptad whose identity will be defined whenever possible according to the context.

Ultimately, it must be stressed that the association of the Pleiades with a divine heptad is redundant in all the types of textual sources, mainly in literary texts (see 3.) and in divinatory texts (see 4.–6.), where the ancient scribes often played on the ambiguity of the binomial god-star. Yet, the varied panorama of heptads, all named after dIMIN.BI, may create an unclear picture: are all of them related to the Pleiades and *vice versa*? Thus, the association with a divine heptad will be analysed and discussed in the following chapters, in contextualisation with the chronology, the genre of the texts, and the new perspectives on celestial bodies.

2.3.3. The Iconography of the Pleiades

Even though, on average, up to ten of the brightest stars are visible with the naked eye (see 2.1.), the Pleiades are depicted as seven dots or stars in Mesopotamian glyptic. The seven dots are represented in the early glyptic: initially, they were depicted in a sparse order (Figure 2 example 1), sometimes with bulls or deers. Van Buren (1939–1941: 278) stressed how in early seals, like those dating to the Jemdet Nasr period (ca. 3100–2900 BC), "the symbol does not seem to have had a beneficent or apotropaic quality, but the meaning invariably attached to it was that of a sign to denote the mystical nature of the subject or scene depicted. That scene might be quite simple or mundane in appearance, but it was revealed as esoteric or magical by the presence of the seven dots." This statement is true because the meaning of the number seven has indeed been mystical since early times, as will be shown in 2.4. Nevertheless, van Buren associated the earliest seven dots to the seven casting lots used for giving oracles. The oracular practice of the casting lots is not very well attested in Mesopotamia, if not in a few instances dating to the first millennium BC (Koch 2015: 144). Therefore, van Buren's suggestion should remain speculative.

---

93 Since the Babylonian year began roughly around the vernal equinox (i.e. Nisannu, Month I), it should be assumed that the Pleiades started to be visible in the sky just after that time (see 2.1. and 4.1.1).

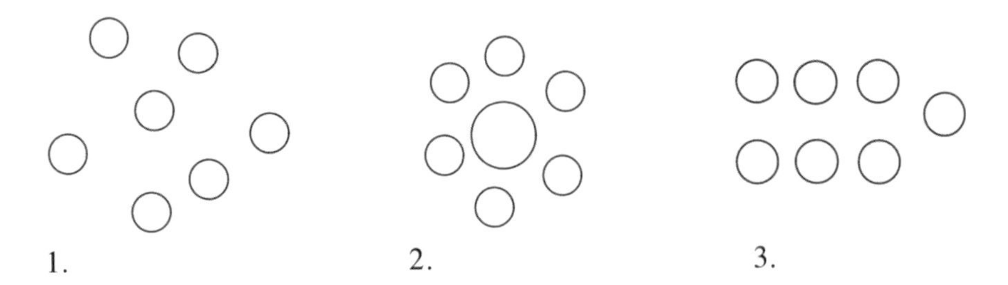

Figure 2. The three phases of depiction of the seven dots in the Mesopotamian glyptic, based on the catalogue of seals of van Buren (1939–1941). From left to right: the seven dots in a sparse order (1), the seven dots depicted as a rosette (2) and the seven dots depicted as the Pleiades (3).

In the Old Babylonian period (ca. 2000–1500 BC), the seven dots are depicted as a rosette, in a circle with one dot in the middle, as can be seen in Figure 2 example 2. In this shape, they always appear in a rather generic setting (van Buren 1939–1941: 280–282).

From a certain point onwards, the seven dots – arranged in two rows of three dots and one dot on their side – increasingly resemble the actual shape of the Pleiades (Figure 2 example 3). According to van Buren (1939–1941: 281), the first seals in which the seven dots appear in the shape of the Pleiades date already to the Old Babylonian period (ca. 2000–1500 BC) (von der Osten 1934: pl. XVI n. 200; Porada 1938: 9 pl. II n. 13), and others mainly come from Syria and Cappadocia and date to the second millennium BC (van Buren 1939–1941: 281 fn. 40, 44, 45). On account of that, Graziani (1979: 680–681) suggested that the shape of the Pleiades may have originated in the north-west Semitic area, where the seven dots are found in the glyptic of the second millennium BC. However, the main limitation of that suggestion is that it finds no parallels in textual evidence. Another early possible attestation of the seven dots as the Pleiades comes from Babylonia and dates to the second millennium BC during the Kassite period (ca. 1595–1155 BC). From that period, the most important sources available are the boundary stones, or Kudurru inscriptions, on which seven dots are sometimes depicted together with other astral symbols (mainly the crescent for the Moon and the eight-pointed star for Venus). Nevertheless, neither the Pleiades nor seven gods are mentioned in the oaths' sections of the inscriptions (Seidl 1989: 55–57, 101–103 n. 97, 98). Therefore, one should assume that the seven dots stand, in those instances, as a generic reference to the starry sky.

More secure attestations of the Pleiades as seven dots begin to appear on seals dating to the first half of the first millennium BC (Seidl 1989: 59–60, 63 n. 103, 110). According to the Catalogue of Seals of the Pleiades by Konstantopoulos (2015: 324–367), more than half (circa 64%) of the seals depicts the seven dots or stars together with the crescent in an astral setting. Out of those seals, only nine are older than the Neo-Assyrian period (ca. 911–612 BC) (Konstantopoulos 2015: n. 21, 22, 33, 44, 53, 79, 82, 125, 167). The seals in which the seven dots are depicted in the shape of the Pleiades, or even in the shape of stars (Moortgat 1940: taf. 71 n. 598), mainly come from the Neo-Assyrian period onwards (van

Buren 1939–1941: 284–289). In addition to the seals collected by Konstantopoulos, at least other three seal impressions of seven dots as the Pleiades on clay tablets dated to Neo-Assyrian period should be added.[94]

At any rate, there are a few cases where the seven dots are depicted in an astronomically related scenario, contributing to a more secure attestation for the Pleiades. For instance, a reference to the beginning of the ploughing under the presence of the Pleiades might be seen in a few Neo-Assyrian seals depicting a man with an ox and a plough, as can be seen in Figure 3 (e.g. von der Osten 1934: pl. XXVIII n. 415; Porada 1938: 16 pl. III n. 24).

Figure 3. Neo-Assyrian seal (Ward 1910: fig. 372). Over the man with the ox and the plough, the crescent, the eight-pointed star as Venus, and the Pleiades are depicted. The Pleiades and the Moon may represent a celestial conjunction expected around the beginning of the Mesopotamian year (i.e. Nisannu, Month I), or simply the first visibility of the Pleiades in Ajaru (i.e. Month II), the time for starting the ploughing (see 4.1.2., 4.2.–4.2.2.1.).

Moreover, another Neo-Assyrian seal impression with a cow and its calf, the seven dots, the crescent moon, and a star is reminiscent of a pastoral setting taking place under the starry sky (CTN 6, 106; see figure 103).

The Neo-Assyrian royal stelas sometimes show the seven dots or stars iconography for the Pleiades. The most famous of these attestations is perhaps the Stela of Sargon II from Kition on which the seven dots, depicted on the right top of the stone, correspond to the warlike Sebettu (^{d}IMIN.BI) mentioned in the first lines of the stele (VA 968 i 22, see Frame 2020: 405).

The most unambiguous drawing of the Pleiades as seven stars was identified by Weidner (1967 taf. 2) in the famous *Gestirndarstellung* (i.e. micro-zodiac) tablet (VAT 7851) from Uruk dating to the Hellenistic period (ca. 323–63 BC), as can be seen in Figure 4.

94 See CTN 6, 87 (figure 83), 102 (figure 97), and 106 (figure 103). In CTN 6, 94 (figure 86), there is a possible depiction of the Pleiades as seven doughnut-shaped spheres in a Neo-Assyrian stamp seal impression on a clay envelope.

Figure 4. Obverse of VAT 7851 (Weidner 1967: taf. 2). From left to right: the Pleiades as seven stars with the caption MUL.MUL, the "man in the Moon", and the Taurus as a bull (broken at the left bottom) with its bristles accentuated.

The seven stars, depicted on the left of the drawing, are labelled MUL.MUL. On the right of the drawing, a bull (i.e. Taurus) is depicted. Between the stars and the bull, there is the so-called "man in the Moon", which represents the *bīt niṣirti*, "secret house" of the Moon, i.e. where the Moon has specific importance, located near the Pleiades and Taurus (Beaulieu 1999; Horowitz-Andre-Kritsch 2018). The whole depiction on the *Gestirndarstellung* tablet likely symbolises an important celestial conjunction of the Moon and the Pleiades at the beginning of the Mesopotamian year (see 4.2.2.–4.2.2.1.), and the concept of *hypsoma*, "(planetary) exaltation", found in later Hellenistic astrology (see 6.2.1.).

On a final note, the earliest depictions of the seven dots in a sparse order and as a rosette seems to be more meaningful for the importance of the number seven than for the identification of the Pleiades. While sparse dots in the background can simply represent a starry sky, the seven dots represent an asterism more unambiguously only in sources dating to the first millennium BC. That is roughly the same period when the name *zappu*, "Bristle", for the Pleiades is found in the lexical lists (see 2.3.1.).

2.4. The Pleiades as Seven Stars: The Mystical Meaning of the Number Seven

The existence of the several divine heptads of Mesopotamia (see 2.3.2.), and the iconography for the Pleiades (see 2.3.3.) shed light on how the number seven plays a key role for understanding the perception of the Pleiades.[95] Indeed, in the Ancient Near East the number seven always had a strong symbolic meaning. The following overview (see 2.4.1.–2.4.2.) of the number seven in its empirical (e.g. mathematical, computational, etc.), and

95 For references about the number seven in Mesopotamia and Ancient Near East, see Reinhold-Golinets (2008).

attributive usage will show how its meaning shaped the perception of the Pleiades in Mesopotamia.

2.4.1. Empirical Usage

The computation in Mesopotamia is mainly based on a sexagesimal system (Friberg 1987–1990: 533–534),[96] which means it is based on the number 60. In the framework of a sexagesimal system, the number seven numerically represents the first number which is not a divisor in a sexagesimal system (Muroi 2014). The number seven in Sumerian is written IMIN (or UMUN$_x$, pronounced ú-mu-un), while in Akkadian is written *sebe*, *sebēt* in feminine.[97] The word IMIN is composed of the word for "five" (IÁ/Í, *ḫamiš*), and the word for "two", or "reduplication" (MIN, *šinā*) (Borger 2004: 430–431 n. 825; Halloran 2006: 126 "imin"). This means that the word for "seven" likely comes from the numeral system of five, that is from the fingers of one hand as the basic computation tool (i.e. 5+2 = 7).

On a practical level, the Mesopotamian ideal or standard calendar (see 4.2.1.) is based on a lunar cycle of 29 or 30 days, which is roughly divisible by seven days, i.e. one week.[98] Since the Mesopotamian calendar used lunar months, one lunar unit would consist of what we would call a "week". Yet, a division of time into seven-day weeks was never in use; on the contrary, as a literary feature, the hebdomadal structure was used to divide the Moon's phases into units,[99] as can be seen in the Akkadian poem *Enūma eliš*:

15. *i-na* SAG ITI-*ma na-pa-ḫi e-*[*l*]*i ma-a-ti*
16. *qar-ni na-ba-a-ta a-na ud-du-ú za-ka-ri u*$_4$*-mu*
17. *i-na* UD.7.KÁM *a-ga-a* [*maš*]*-la*

15. At the beginning of the month, o[v]er the land,
16. he (i.e. Marduk) appointed him (i.e. Sin, the Moon) to shine (with) horns in order to reveal the calling of the day,
17. and to (have), on the 7th day, [half] of the crown.
(Lambert 2013: 98 ll. 15–17)

---

96 The numbering system in Mesopotamia is not necessarily based on the number six only. For example, there are a few traces of a ternary system (i.e. 3 = a complete unit) (Friberg 1987–1990: 538–539), see Antagal C: [peš]-peš-g[i] (i.e. 3+3+1) = ⸢IMIN⸣ (MSL 17: 197 l. 58). A decimal system of numbers was also in place (Friberg 1987–1990: 537).

97 See Ea 2: ú-mu-un = *se-bet* (MSL 14: 253 l. 140).

98 This leads to a chicken and a egg issue, because the hebdomadal structure is challenging to postulate without an empirical phenomenon, and *vice versa* (Negretti 1973: 52 fn. 44).

99 The known units, or names of the Moon during specific days, were: "crescent" (U$_4$.SAKAR, *nannāru*) from the 1st to the 5th day, "kidney" (*kalītum*) from the 6th to the 10th day, and "glorious crown" (AGA, i.e. *agê, tašriḫti*) from the 11th to the 15th day; see also K 170 + Rm 520 l. 2: AGA UD.7.KAM *ka*$_{15}$-*lit*(*u*), "the crown of the seventh day is the kidney" (Livingstone 1986: 30–31; Lambert 2013: 186–187; Hätinen 2021: 137–139).

According to the cosmogony of *Enūma eliš* (V 15–17), the seventh is the day when the "crown" of the Moon (i.e. the illuminated section of the Moon) is half size.

2.4.2. Attributive Usage and Semantic Fields of the Number Seven

The Akkadian word *kiššatu*, "totality", sometimes equated to IMIN (K 2054 obv. ii 19', see CT 18, 29–30), even though *kiššatu* is usually logographically written with the number ten, U in Sumerian, and *ešer* in Akkadian.[100] The association between U, "ten", and *kiššatu*, "totality", is based on a decimal numeral system (i.e. when the basic computation tool is the ten fingers of two hands). The association between IMIN, "seven", and *kiššatu*, "totality", might be considered the product of a "speculative interpretation":[101] the sign IÁ/Í (5) could be reduplicated using the sign MIN (2), resulting in ten units (i.e. 5 × 2 = 10, instead of 5 + 2 = 7), which are the "totality" of fingers of two hands.

On a semantic and phonetic level there are even more associations with the number seven. For instance, the following entries are an extract from a two-column fragment[102] in which the associations with number seven appear:

19'. IMIN	*kiš-šá-tu*	Totality
20'. IMIN	IMIN-*et*	Seven
21'. IMIN	MUL	Star
22'. IMIN	*ú-ru-uk*	Uruk
23'. IMIN	*ki-ši*	Kiš
24'. IMIN	*ia-mut-ba-la*	Yamutbal

(CT 18, 29–30 obv. ii 19'–24')

The analogies are challenging to detect. Regardless of the above-mentioned association between "seven" and "totality", the equation IMIN = MUL may reflect a semantic field in which the word IMIN finds its application (i.e. the seven stars as the Pleiades, MUL.MUL). Even though purely speculative, the equation IMIN = Yamutbal may come from the etymology of IMIN (IÁ/Í + MIN), because the first syllable (IÁ) sounds like the first syllable of Yamutbal (written *ia-mut-ba-la*). At the same time, the equation IMIN = Kiš

100 According to the lexical list Ea, dating to the Old Babylonian period (ca. 2000–1500 BC), the Sumerian sign U (10) can be read both *ešeret*, "ten", and *kiššatum*, "totality", in Akkadian (MSL 9: 132 XLIX 399, 401).

101 Speculative interpretation has been recently studied by Bennett (2021: 49–158) and it means the etymological extrapolation of latent meanings of names. Following Bennett's study, ancient scholars could have based the equation IMIN = *kiššatu* on the etymology of IMIN by syllabifying the sign into IA/Í and MIN, and then following Sumerian grammar. This way, the number ten (U, *ešeret*) is encoded in IMIN which is consequently associated with *kiššatu*, "totality". This kind of speculative interpretation seems to be rooted in the Old Babylonian period (ca. 2000–1500 BC), and it is typical of commentary texts dating to the first millennium BC.

102 The fragment K 2054 (CT 18, 29–30) is part of a series entitled *šarru* according to the colophon. It is similar to what has been labelled "group vocabularies", which list groups of semantically and phonetically related entries (Veldhuis 2014: 361–363, fn. 879; Cavigneaux 1983: 638–639).

(written *ki-ši*) could be based on the first syllable of *kiš-šá-tu*, whereas IMIN = Uruk (written ú-ru-uk) on the logogram for "ten" (U).[103]

Moreover, several fragments of lexical lists dating to the Middle Babylonian period (ca. 1400–1100 BC) preserve the use of IMIN as an attribute for LUGAL,[104] "king", when it is semantically connected to the idea of the "king of all the countries" (Veldhuis 2014: 253).[105] One example of this is given in a Kassite fragment from Babylon belonging to the list of professions labelled Lú = *ša*, and quoted below:

11. [luga]l <<LÁ-IGI>> u$_5$	[Kin]g of all
12. lugal ninnu	King of fifty
13. lugal imin	King of seven
14. lugal kur$^?$	King of the countrie(s)$^?$
15. lugal kiš$^?$	King of the world$^?$ (i.e. totality)

(VS 24, 15 rev. 11–15; Bartelmus 2016: 369)

Mesopotamians were certainly fond of the number seven. Regardless of lexical lists, the number seven, as a "mystical" number, qualifies groups of elements in every cultural context. In the following overview, the most relevant instances in which the number seven has its attributive usage are mentioned and commented on.

– As early as the third millennium BC, the number seven seems to have had a mystic relevance. One of the earliest Sumerian literary texts from Abu Salabikh, although fragmentary, mentions Ašnan/Ezina, a goddess of grain and pulses, being pregnant for seven months with seven children (OIP 99, 283 obv. ii 10–16, see Alster 1976: 124–125). Colophons and dedicatory inscriptions to Nisaba are found on lexical prisms and cylinders dating to the late third millennium BC (Veldhuis 2014: 68–71). In these prisms, Nisaba is described as the woman with "seven ears", meaning that she possessed an exceptionally powerful wisdom:

  {^{d}}nisaba / munus zi munus saga / munus mul-mul-e / ki-ág an-na / ḫi-li é kur-kur-ra / géštug imin

  Nisaba, / true woman, beautiful woman, / starry woman, / beloved of the sky, / luxuriant one of the temple of all the lands, / with seven ears.
  (AO 337 surface h 1–6, see CDLI n. P481014)[106]

---

103 Similar patterns are found in other contexts as well, for instance in cryptographic spelling where the writing of a word mirrored its sound. See Maul (1999: 6), who noted that the Akkadian word for "mirror", *mušālu*, was associated with Sumerian GE$_6$.URU. The latter is composed by GE$_6$, "night", *mūšu* in Akkadian, and URU, "city", *ālu* in Akkadian. The meaning of the Sumerian GE$_6$.URU reflects the reading of the two signs in Akkadian (*mūšu* + *ālu* = *mušālu*).

104 It must be noted that u-mu-un in the Sumerian Emesal dialect is equated to LUGAL, *šarru*, "king", in the vocabulary Emesal II (MSL 4: 13 ll. 8–9).

105 See, e.g., MSL 12: 94 ll. 46–56.

106 https://cdli.ucla.edu/search/search_results.php?SearchMode=Text&ObjectID=P481014 accessed

- The number seven was used to define divine hierarchies. For instance, seven are the major gods of the pantheon (see 2.3.2.); seven are also the gods who determine the fates and have an astral representation: Nanna/Sin is the Moon, Utu/Šamaš is the Sun, Inanna/Ištar is Venus, Marduk is Jupiter, Nergal is Mars, Adad is Saturn and Marduk, Nabu or Ninurta can be Mercury.[107] As will be argued in the next chapter (see 3.2.5.1.), the gods who belong to this group are not always the same, as there are usually shifts and changes according to different panthea during time. The number seven is also an attribute of the Igigi, the group of gods sometimes equated with the Anunnaki. Both Anunnaki and Igigi are groups of the highest gods in the Mesopotamian panthea, sometimes designated as underworld gods (Anunnaki), opposed or equated to the gods of heaven (Igigi).[108] The word for Igigi, ^{d}*i-gi-gi*, can be written ^{d}*í-gì-gì*: the latter writing might be based on the number seven, as it is composed by the signs IÁ (5) + DIŠ (1) + DIŠ (1); hence, the writing could be also interpreted as ^{d}Í.MIN or Í.DIŠ.DIŠ (Jacobsen 1977: 116–117; see also 3.2.5.1.).
- As already mentioned (see 2.3.2.), there are the scattered divine heptads of the Mesopotamian religion. The most relevant ones are the warlike and demonic Sebettu (see 3.1.2.1.), and the seven heroes of Gilgameš (see 3.1.1.1.). Primordial creatures (i.e. monsters, demons, and sages) are often grouped in heptads (see 3.1.3.). The identities of these creatures find their roots in legends: for instance, the seven sages or *apkallū*, the depositary of the primordial wisdom, were associated with seven antediluvian kings, who lived in a mythical time (Annus 2016: 9–13). These are liminal creatures, who act as a homogeneous group, without having a proper and defined identity.
- Not only gods and demons are seven in number: the so-called archive of mystic heptads (KAR 142, see Pongratz-Leisten 1994: 221–227), dating to the Neo-Assyrian period (ca. 911–612 BC) lists several heptads of buildings, gates, and streets.
- Traditionally, the ziqqurrat has a sevenfold structure like the levels of the cosmos: two examples are the Etemenanki in Babylon or the Eninnu of Gudea of Lagaš (Annus 2008: 7–14).
- Concerning the meaning of seven as a literary device, Konstantopoulos (2015: 16) underlined how this number works as a spatial and temporal marker. The sevenfold structure is very well emphasised in mythological texts of ancient Mesopotamia: lists of seven elements, in which the seventh is the last and conclusive one, are a poetic device for a climax. For instance, in the Standard Babylonian epic of Gilgameš, the sevenfold structure symbolises the entirety of the hero's journey (Konstantopoulos 2015: 16–18).[109] Likewise, in the Akkadian myth *Atra-ḫasīs*, seven are the days of the flood (Lambert-Millard 1968: 96–97 iv 24–25). In the Sumerian poem "Inanna's

18.06.2021.

107 For an overview of the equations between gods and planets, see Brown (2000: 54–74).

108 For the organisation of the pantheon (Anunnaki and Igigi) according to *Enūma eliš*, see Lambert (2013: 193–196).

109 E.g. the six days and seven nights Gilgameš slept when visiting Ūta-napišti is proven by the number of loaves of bread the latter baked each day (George 2003: 716–719 ll. 223–230).

Descent to the Netherworld", Inanna passes through seven gates to reach the core of the netherworld (Sladek 1974: 243–245 ll. 42–62). Furthermore, a similar pattern was suggested by Kapelrud (1968), in commenting on the use of the seven on a qualitative level in the Ugaritic mythology. While he suggested that the number seven is related to the fate, being it good or bad, he also added that seven "was the number in which everything concentrated itself fully. Saying seven was like saying 'maximum'" (Kapelrud 1968: 497).

– The use of seven is not exclusive for groups of elements, as it also recurs as a mystical number in rituals and repetitive apotropaic formulas of prayers and incantations. For instance, in the ritual *bīt rimki*, "house of ablution", the king should walk through "seven houses" made of reed, to send the evil away (Ambos 2013a: 42). Moreover, anti-witchcraft rituals are usually recited seven times, and so are the ritual acts enacted, as shown in the following example:

> IGI [d]*šá-maš* IMIN-*šú tu-šad-bab-šu-ma* IMIN-*šú* KI NU NITA IMIN-*šú* KI NU MUNUS / TÚG.SÍK-*šú ta-bat-taq*
>
> You make him say (the formula) seven times before Šamaš, and seven times with the figurine of a man (and) seven times with the figurine of a woman / you cut his hem.
> (KAL 2, 34 rev. iv 5'–6'; Abusch-Schwemer 2011: 134 C rev. 23'–24')

The idea of the seven as "totality" or "completeness" was also part of less traditional, and crystallised contexts like mythology or ritual magic. One example is the following idiomatic expression attested in one of the so-called astronomical and astrological report to a Neo-Assyrian king:

> *pu-*⸢*tú*⸣ [*a*]-*di** IMIN-*šú na-šá-ka*
>
> I guarantee (lit. "the front is raised up to you") as long as seven times.
> (SAA 8, 447 rev. 1)

The number seven can also easily have a negative meaning. For instance, Livingstone (2013: 258) noted how in hemerologies dating to the Neo-Assyrian period, a day multiple of seven is considered an UD.ḪUL.GÁL, *ūmu lemnu*, "evil day".

– The Pleiades are a group of seven very bright stars, which resembles the idea of "completeness" par excellence, when observed with the naked eye (see 2.1., 2.3.3.). Nevertheless, despite the importance of the number seven in Mesopotamia, it cannot be excluded that the Pleiades were not always perceived as seven in number. Not by chance, modern historians of astronomy still argue on what was the actual observable number of the Pleiades in the past (Rappenglück 2008: 14–15; Kelley-Milone 2005: 141–143). Considering the relevance of the number seven as an attribute in every context of reality, it seems likely that the seven was associated with the Pleiades as a fixed rule.

This overview of the uses of the number seven as an attribute adds more elements to the already complex background for the Pleiades in Mesopotamia. Regardless of the different fields of application of the number seven, the common denominator seems to be the meaning of seven as "completeness", or even exceeding from completeness. Under these circumstances, the association between the Pleiades and the number seven can be thought of as an endured tradition, and not as a reference to the exact number of stars of the Pleiades. This association of ideas was likely suggested by the appearance of the asterism itself: a cluster whose most brilliant stars are approximately seven, and very well resemble the meaning of seven as an attribute.

2.5. Conclusion for the Nature of the Pleiades in Their Context

This chapter presented the Pleiades and how they were perceived in their context, in order to introduce the preliminary concepts necessary to deal with the texts discussed in the following chapters. More specifically, it discusses the relationship between celestial bodies and gods in Mesopotamian culture, and the names, iconography, and number of the Pleiades. What emerged from the first part of the chapter (see 2.2.–2.2.3.3.) is that the binomial god-star comes from an inherent interchangeability in the taxonomy and ontology. A look at the theories about metaphors according to cognitivists and semiologists reinforced the idea that this interchangeability should not be considered a problem, but a feature of Mesopotamian culture.

First, a unique characteristic of Mesopotamian culture was the coexistence of two different languages and cultural systems, Sumerian and Akkadian: Sumerograms, such as those for "star" (see 2.2.1.), or "Pleiades" (see 2.3.1.) had different determinatives, readings, and synonyms, which covered various semantic fields, and which were used to build analogies and association in the textual sources. The Mesopotamian scribes skilfully played with signs, words, and meanings: this way of reasoning is, for example, also witnessed in the word for "seven", and its attributive usages as "completeness" or "totality" (see 2.4.–2.4.2.).

Second, a look at the Mesopotamian lexical repertoire was the catalyst to go beyond the cliché of Pan-Babylonism in trying to define the perception of celestial bodies in Mesopotamia, a topic that has been sadly a taboo for Assyriologists for too long (see 2.2.3.). It is safe to assume that there were several ways in which Mesopotamians perceived the relationship between gods and celestial bodies. Overtaking the idea of an Ancient Near Eastern "astral religion", and Frankfort's "mythopoetic thought", Rochberg (2009; 2010: 317–338; 2011) proposed three new perspectives according to which this relationship can be understood, and which are consistent with the textual evidence: the gods can be considered as stars (see 2.2.3.1. § 1), as it is witnessed especially in the earlier stages of mythology; the celestial bodies can be the representation of the gods, or their image in the sky, so the celestial phenomena represent the will and actions of the gods (see 2.2.3.1. § 2); the celestial bodies and their phenomena can also be personified (see 2.2.3.1. § 3). The three perspectives on celestial bodies point towards a conceptual distinction at least between what is divine and human, and what is divine and astral, or non-human. Such a statement sheds light on how much the cosmos and the divine agency were multifaceted in Mesopotamia (see 2.2.3.2.–2.2.3.3.).

In the second part of this chapter (see 2.3.–2.4.2.), the Pleiades (MUL.MUL) have been framed within the Mesopotamian scribal cultural landscape, with its unique modality of building knowledge. From the lexical sources, we learn that the Sumerian MUL.MUL, lit. "stars", or "to constantly shine", was associated with the Akkadian *zappu*, "Bristle", the name for the Pleiades as an asterism within Taurus, at least from the Old Babylonian period (ca. 2000–1500 BC) onwards. Likely by the end of the second millennium BC, the Pleiades were regarded as a group of seven stars as a fixed tradition. First, they are seven stars because of their appearance – they are roughly seven stars when observed with naked eye. Second, the Pleiades are seven stars because of the analogy with the number seven as an attribute for "completeness" or "totality". Like many other celestial bodies (e.g. the Moon, the Sun, or Venus), they are associated with one or more fixed groups of gods, the divine heptads named ^{d}IMIN.BI, "divine Seven", or "Sebettu". The origin and identity of the divine heptads are not always easily detectable, and they often depend on the context, as it will be discussed in the following chapters.

Chapter 3

The Heptads and the Pleiades in Mesopotamian Literature

The role of the divine heptads and the Pleiades in Mesopotamian literary texts is discussed in this chapter with the aim to show if there is an association between the heptads and the Pleiades, and what kind of association it may be. The first part of the chapter (see 3.1.–3.1.3.) concerns myths featuring an astral component, to clarify whether they are actually associated with the Pleiades. The texts analysed are written in both Sumerian (see 3.1.1.–3.1.1.3.1.) and Akkadian (see 3.1.2.–3.1.2.2.2.) language; hence, they belong to different traditions within circa two millennia. The second part of the chapter (see 3.2.–3.2.7.) discusses the Akkadian magic-religious texts (see prayers in 3.2.1.–3.2.4., rituals in 3.2.5., and royal inscriptions in 3.2.7.) where the Pleiades are mentioned. These texts date to the first millennium BC, and the great majority of them were found in Ashurbanipal's library in Nineveh. Altogether, the two parts of this chapter represent two of the perspectives on celestial bodies proposed by Rochberg (2009; 2010: 317–338; 2011; see 2.2.3.1. § 1–2): first, the gods as celestial bodies in the myths, and second, the celestial bodies as the representation of the gods in prayers and rituals.

Generally speaking, understanding the Mesopotamian literary texts beyond the historical and philological approach can be challenging. Since the end of the nineteenth century, many scholars have attempted to find astral motifs behind the structure of Sumerian and Akkadian myths, because they are rich in celestial symbolics. That is what has been defined as astralisation, or the outcome of an assumed "astral religion". This argument will be reconsidered, but only if the trap of Pan-Babylonism (see 2.2.3.) is avoided. To achieve this purpose, the approach adopted for the literary texts in this chapter will be both philological and intertextual: epithets, idioms, or recurrent lines will be treated as parameters for intertextuality. The sources will be arranged chronologically to highlight patterns and differences among them.

3.1. Gods as Celestial Bodies: Heptads in Myths with Astral Components

As celestial bodies with a divine counterpart, the Pleiades are involved in the complexity of the binomial god-star discussed in chapter 2. The Pleiades (MUL.MUL or *zappu*, "Bristle") are associated with one or more groups of seven (IMIN, *sebe*) divine entities. As discussed in 2.4.2., the heptads and the Pleiades share the "mystical" meaning of the number seven as "completeness", or "totality", following an endured tradition. The Mesopotamian mythology – built on different local traditions – is full of such scattered heptads, who have the same name but different features: the God List *An* = *Anum* (Litke 1998: 211–213 ll. 149–184) attests up to six different heptads, and more in other types of sources (see 2.3.2.). Given the nature of cuneiform knowledge, often based on analogies, or on the so-called "orthographic-semantic method" (Rochberg 2016: 92; see 2.2.3.3.), the line between the divine heptads and the Pleiades appears blurred whenever a heptad, in a nocturnal or astral context, is mentioned in literary texts. Therefore, the goal of this first part of this chapter (see 3.1.–3.1.3.) will be to underline whether an analogy between the Pleiades and a heptad occurred, or if the number seven only acted as a literary device, an attribute to describe a group of individuals.

3.1.1. Sumerian Sources

Three Sumerian compositions will be here discussed: "Gilgameš and Ḫuwawa" version A and B (see 3.1.1.1.), the "Hymn of Ḫendursaĝa" (see 3.1.1.2.), and "Lugalbanda in the Wilderness" (see 3.1.1.3.). Whenever needed, the sources are compared to other Sumerian and Akkadian compositions (e.g. see 3.1.1.3.1.).

3.1.1.1. "Gilgameš and Ḫuwawa" A and B: the Seven Warriors

The first heptad encountered in the God List *An = Anum* (VI 150, see Litke 1998: 211 l. 150) is the "seven heroes" (UR.SAG ^{d}IMIN.BI), known from the Sumerian composition "Gilgameš and Ḫuwawa" (see 2.3.2.). The composition "Gilgameš and Ḫuwawa" is the prototype of the fourth and fifth tablet of the Standard Babylonian epic of Gilgameš. Whereas "Gilgameš and Ḫuwawa" is written in Sumerian and known from copies produced during the Old Babylonian period (ca. 2000–1500 BC), the Standard Babylonian Epic is an Akkadian composition reconstructed through textual evidence written in Standard Babylonian (i.e. literary Babylonian), and dating to the second half of the second millennium BC onwards (George 2003: 9–11, 28–33). In the Standard Babylonian Epic, the two main characters, Gilgameš and his companion Enkidu, fight and win against the antagonist monster Ḫuwawa when travelling through the cedar forest. However, in "Gilgameš and Ḫuwawa", during the journey in the cedar forest the Sun god Utu grants protection to Gilgameš and gives him seven heroes (ur-saĝ) with hybrid and starlike features, to help him on his way.

There are two versions of the composition "Gilgameš and Ḫuwawa", version A (Edzard 1990; 1991) and B (Edzard 1993).[110] Delnero (2006) collated different copies of the text from several Mesopotamian cities, and established the following composite edition for version A:

36. ur-saĝ dumu ama AŠ imin-me-eš
37. diš-àm šeš-gal-bi piriĝ-ĝá umbin ḫu-rí-in-na
38. min-kam-ma muš šà-tùr ka [...] KU šu ⸢UŠ⸣
39. eš$_5$-kam-ma muš ušumgal ⸢muš⸣ [...] ⸢x⸣ RU
40. limmu$_5$-kam-ma izi bar$_7$-bar$_7$ [...] ⸢ku$_4$⸣-ra
41. iá-kam-ma muš-saĝ-kal šà gi$_4$-a ⸢ub⸣ KA ⸢x⸣
42. àš-kam-ma a-ĝi$_6$ a gul-gul-dam kur-ra gaba ra-ra

110 The editions of both versions are also available online at the website ETCSL, 1.8.1.5. and 1.8.1.5.1.

43. imin [...] ⸢nim⸣ ĝír-ĝír-re lú nu-⸢da-gur⸣-dè
44. imin-⸢bi-e⸣-ne mu-⸢un-na-ra-an⸣-šúm
45. má-ùr-má-ùr-ḫur-saĝ-ĝá-ke$_4$ e-ne bi-in-tu-mu

36. They were seven heroes, sons of one mother,
37. the first, their eldest, has paws of a lion and claws of an eagle,
38. the second is a snake that [opens] the mouth [...]... ,
39. the third a snake-lion-dragon hurling [...]... ,
40. the fourth spat fire [...]... ,
41. the fifth a *sagkal*-serpent... ,
42. the sixth, a spring that strikes the mountain on the chest,
43. the seventh [...] lights up like a lightning no one could avoid.
44. These seven, he (i.e. Utu) gave to him (i.e. Gilgameš),
45. through the mountain passes he will lead them.
(Delnero 2006: 2416–2419 ll. 36, 37a$_1$–37g$_1$, 38; for the translation, see George 1999: 151–152 ll. 36–45)

Version A has a one-by-one description of the seven heroes (i.e. lion-eagle, snake, snake-lion-dragon, a fire-breathing creature, *sagkal*-serpent, a spring, a creature that lights up) who are hybrids and who are supposed to escort Gilgameš "through the mountain passes" (l. 45). In the text of version B, only the descriptions of the last two of the seven heroes are preserved; yet, version B has four additional lines where, as George (1999: 163 ll. 45–46) noted, the heroes are described as both terrestrial and astral entities:

45. e-ne-ne an-na mul-la-me-eš / ki-a ḫar-ra-an zu-me-eš
46. an-na mul-la ⸢x x x⸣ íl-la-me-eš
47. ki-a-kaskal-aratta^{k}[^{i} zu-me-eš]
48. dam-gàr-ra-[gin$_7$] / ĝìri-bal zu-me-eš
49. tu^{mušen}-gin$_7$ ab-lá-kur-ra zu-me-eš
50. má-ùr-má-ùr-ḫur-saĝ-ĝá-ka ḫé-mu-e-ni-túm-túm-mu-ne

45. In the sky they are bright, / on earth they know the paths,
46. in the sky they are stars... they are high,
47. on earth [they know] the road to Aratta,
48. [like] merchants / they know the pathways,
49. like pigeons, they know the caves in the mountains.
50. Through the mountain passes they will lead us.
(Edzard 1993: 21 ll. 45–50)

Version B diverges from version A only for five lines, but these five lines are rich in meaning. The scribe arranged the verses in a dichotomous scheme which underlines the double aspect of the seven heroes, who belong to both the terrestrial and astral realm (i.e. "In the sky" versus "on earth", ll. 45–47; "[like] merchants" versus "like pigeons", ll. 48–49).

In an article, Alster (1985: 225–226) noted a case of intertextuality in the "Song of Inanna and Dumuzi". The "Song of Inanna and Dumuzi" is an *eršemma* song dating to the Old Babylonian period (ca. 2000–1500 BC),[111] where Geštinanna, Dumuzi's sister, discovers the death of her brother, and lamentations for the mourning are performed by seven singers of Uruk. The hymn has one line that is a close parallel of "Gilgameš and Ḫuwawa" version B:

> e-ne-ne an-na mul zu-me-eš ki-a ḫar-ra-an zu-me-eš / an-na mul zu múru íl-la-me-eš an-na daĝal-ba nesaĝ íl-la-me-eš
>
> They know (i.e. the seven singers of Uruk) the star(s) in the sky, on earth they know the paths. / The ones that know the stars in the sky are carrying the *middle.*[112]
> (Alster 1985: 223 l. 27–28; Peterson 2019: 63)

In "Gilgameš and Ḫuwawa" version B, the heroes are both terrestrial and astral guides of Gilgameš through the mountains; but in the "Song of Inanna and Dumuzi", the seven singers of Uruk only know the stars and the paths on earth: they are depositories of ancient wisdom, like the seven sages (*apkallū* or *ummânū*).[113] The seven sages are hybrids who lived on earth in a mythical time before the deluge, associated with seven antediluvian kings from the so-called Sumerian King List (Finkelstein 1963; Wiggermann 2011: 305–307). The heroes and the singers are clearly different entities with different origins, yet they have the number seven and the knowledge of stars and paths in common. What we have here is a case of intertextuality: "Gilgameš and Ḫuwawa" version B and the "Song of Inanna and Dumuzi" parallel each other, and this shows that having the "knowledge of the stars and the paths (on earth)" means having a far-ranging ancestral ability to lead.

It must be noted that it "is unique in Sumerian literature" that "the stars were useful in finding the way over the mountains" (Alster 1985: 225). It cannot be excluded that the role of the guide came from the use of stars as tools for celestial orientation (Bilić 2007). Indeed, there are other instances in which the stars seem to be used for orientation: in the Sumerian epic "Enmerkar and the lord of Aratta", the legendary king Enmerkar of Uruk sends a messenger to Aratta, to establish a trade with that city; the messenger "at night journeyed by starlight",[114] meaning that the starlight allowed travellers to see the path at night. Ultimately, the Pleiades themselves were likely a tool for orientation. As previously mentioned (see 2.1.), they are observable in both the eastern and western sky. We found the

111 The *eršemma* songs are identified by their rubrics. They are lamentations written in the emesal dialect, and many of them are connected to Inanna (Cohen 1981).

112 Alster (1985: 233) and ETCSL, 4.08.10 read the line differently: an-na daĝal-ba nesaĝ íl-la-me-eš, "In the broad sky, they carried the *first-fruit offering(s)*". For a discussion on the meaning of nisaĝ, *nisannu*, as "first-fruit offerings" in this context, see Alster (1985: 225–226).

113 These primaeval beings were created by Ea and sent from heaven to earth to start civilisation. For their aetiological myth, see Reiner (1961). They also appear in the Standard Babylonian epic of Gilgameš I 21 (George 2003: 538): IMIN *mun-tal-ku*, "seven sages". For references for the alternation of the terms NUN.ME^{meš}, *apkallū* ("sage"), and *ummânū* ("craftsman") for the seven sages see Reiner (1961: 8–9).

114 gi$_6$-ù-na-ka mul-àm im-ĝin (Vanstiphout 2003: 64 l. 161).

Pleiades as markers of the compass point east, according to the Akkadian astral composition MUL.APIN:

> ^{mul}ŠU.GI *u* MUL.MUL *ina* ZI IM.KUR.RA GUB^{meš}-*z*[*u*]
>
> The Old Man and the Pleiades stan[d] where the east wind rises.
> (MUL.APIN II i 70, see Hunger-Steele 2019: 83–84)

MUL.APIN was composed later than the mythological compositions discussed so far, between the end of the second and the beginning of the first millennium BC (see 4.1.1.). One should assume that MUL.APIN attest to a much older tradition according to which the Pleiades were an orientation tool, a tradition that would lie behind the origin of the seven heroes. Nevertheless, even though the seven heroes from "Gilgameš and Ḫuwawa" have starlike features and guided Gilgameš at night, an analogy between the Pleiades as a tool for orientation and the seven heroes shall remain a hypothesis, since a reference to seven stars is never explicitly remarked.

3.1.1.2. The "Hymn of Ḫendursaĝa": the Seven Heralds

Ḫendursaĝa (^{d}PA-saĝ-ĝá) is a god attested in a pre-eminent role for the city of Lagaš (Edzard 1972–1975: 324). He is identified with the god Išum, the "herald of the street", from the Old Babylonian period (ca. 2000–1500 BC) onwards (George 2015: 1–4). Ḫendursaĝa/Išum is connected to carrying the fire, the light, or a torch during the night against evil. He is the main character of the Sumerian hymn "Ḫendursaĝa A":[115] he is exalted as an infernal god, protector of people who revere their personal god, and as a counsellor of the Sun god Utu. In his duties, Ḫendursaĝa is assisted by seven heralds (niĝir, *nāgiru*)[116] described with hybrid features. The modality of description recalls the above-mentioned seven heroes of "Gilgameš and Ḫuwawa" version A (see 3.1.1.1.):

78. imin-ba diš ka$_5$-a-àm kun im-ùr-ùr-re
79. min-kam-ma ur-gir$_{15}$-gin$_7$$^?$ si-im-si-i[m] ⸢ì⸣-AK-e$^!$
80. eš$_5$-kam-ma ^{u2}⸢uga$^{mušen?}$-gin$_7$ za⸣-na gug$^!$ im-kul-e
81. limmu$_5$-kam-ma ⸢ti$_8$⸣^{mušen} maḫ ad$_6$ gu$_7$-a-gin$_7$ ka ì-ša-an-ša-ša {x}
82. iá-kam-ma ⸢ur-bar⸣-ra nu-⸢me-a⸣ sila$_4$ ĝi$_6$-⸢ga⸣ ì-šub
83. àš-kam-[ma ^{d}nin-imma$_x$^{mu}]^{šen}-gin$_7$ eri [u$_3$$^?$] ⸢ku-a-gin$_7$$^?$ mi-ni-ib-ra-ra
84. [imin-kam-ma kúšu^{ku6}-àm... i]m-BU-BU-BU

78. The first of the seven is a fox, it drags its tail,
79. the second sniffs like a dog,

---

115 The *editio princeps* of the composition was published by Edzard and Wilcke (1976). For the latest edition, see Attinger-Krebernik (2005); see also online at the website ETCLS, 4.06.1.

116 The translation "herald", though conventional, has been considered misleading by George (2015: 3 fn. 12) because the function of the niĝir, Akkadian *nāgiru*, is mainly that of a night watchman.

80. the third pecks larvae like a raven,
81. the fourth overwhelms like a huge carrion-devouring vulture,
82. the fifth, despite not being a wolf, attacks black lambs,
83. the sixth, like [an *owl*], hoots against it after the city fell asleep,
84. [the seventh is a *crocodile*], it... the *waves*.
(Attinger-Krebernik 2005: 41–42 ll. 78–84)

However, the individual descriptions of the seven heralds (i.e. fox, dog, raven, vulture, *wolf*, *owl*, *crocodile*) are different from the seven heroes (i.e. lion-eagle, snake, snake-lion-dragon, a fire-breathing creature, *sagkal*-serpent, a spring, a creature that lights up). The seven heralds are theriomorphic creatures with features of wild animals, whereas the seven heroes are monstrous hybrids. These descriptions of theriomorphic or hybrid individuals seem to be a literary device for climax, especially because the number seven is involved in its meaning of "totality", or "completeness" (see 2.4.1.–2.4.2.).

Attinger and Krebernik (2005: 31–32) underlined that the seven heralds are related only to death, and assist the god in his infernal duties with a psychopomp function, by guiding people in their ultimate journey to the netherworld. Hence, their purpose is clearly different from the seven heroes of "Gilgameš and Ḫuwawa", though both heptads are guides. The seven heralds might have an implicit connection with the night, because they are associated with Ḫendursaĝa/Išum, who is the herald of the streets at night. Nevertheless, regardless of that, the seven heralds do not have any association with the stars in general, or the Pleiades.[117]

3.1.1.3. "Lugalbanda in the Wilderness": the Seven Weapons of Inanna

"Lugalbanda in the Wilderness" (or "Lugalbanda in Ḫurrumkurra") is a Sumerian myth in which the main character Lugalbanda is abandoned in a cave in the mountains for three days and three nights.[118] Lugalbanda is a mythological hero whose name is listed in the Sumerian King List as the antediluvian ruler of Uruk (Jacobsen 1939: 88–89); owing to that, he enjoyed a divine status. His story tells us about various stages of purification during his journey towards the city of Aratta. Indeed, in the myth, Lugalbanda falls sick in a cave, and prays to the stars to recover. The purification, or the process of healing, is achieved through the exposition of Lugalbanda to the starry sky:[119] his journey is a metaphorical battle between the light and the darkness, under the presence of the stars.

The sequence of the prayers that Lugalbanda performs to the setting Sun, Venus, the Moon, and the rising Sun mirrors the mechanism of celestial bodies, moving in the sky

117 For a specific study on the seven heralds in comparison with demons and the Sebettu, see Konstantopoulos (2015: 95–107), and Verderame (2017a).

118 The main edition for the first part of the myth is by Wilcke (1969), whereas a comprehensive edition can be found in Vanstiphout (2003: 97–131). The list of manuscripts used for the composite texts by Vanstiphout is in Vanstiphout (1995). The edition of the myth is also available online at the website ETCSL, 1.8.2.1.

119 Wee (2014a), following earlier suggestions of Reiner (1995), included the narrative of Lugalbanda in the broader concept of celestial healing or "astral irradiation" of medicine (see 6.2.2.).

from dusk until dawn. In the last section of the myth, a battle starts at midnight, and seven "torches of battle" of Inanna/Venus are involved:[120]

> 470. lú šag$_4$ ^{d}inanna sag$_9$-ge-me-eš mè ba-súg-ge-eš
> 471. izi-ĝar mè imin-me-eš nu-ga-mu-rib-ba-me-eš
> 472. an sig$_7$[121]-ga men saĝ ìl-la-gin$_7$ ul-la ba-an-súg-ge-[eš]
> 473. saĝ-ki-ne-ne igi-ne-ne an-usán sig$_7$-ga-me-eš
>
> 470. They, who are favoured by Inanna's heart, who stand in battle,
> 471. they are the seven torches of the battle, ... who are mighty.
> 472. In the pale sky they stand joyfully while she wears the crown,
> 473. their foreheads and their eyes enlighten the evening.[122]
> (Vanstiphout 2003: 128 ll. 470–473)

Further in the myth, these "seven torches of battle" (izi-ĝar mè imin-me-eš) are called "pure weapon" (šíta kù) of Inanna:

> šíta kù mul-mul zag an ki-šè mu-un-ne-dè-ĝál
>
> The pure weapon, the Pleiades (or: the stars, starry), she carries to the confines of the sky and the earth.
> (Vanstiphout 2003: 130 l. 482)

This line preserves the signs mul-mul, interpreted by Vanstiphout as an attribute of the pure weapon, and not as the "Pleiades". Nevertheless, given the reference to the number seven in the preceding lines (l. 471), mul-mul could also be the earliest mention of the Pleiades as an asterism, pictured as a weapon, or seven torches (Wiggermann 2011a: 463).

3.1.1.3.1. Inanna as Venus and her Relationship with the Pleiades

From the above-mentioned passages of "Lugalbanda in the Wilderness" (see 3.1.1.3.), a particular relevance of Inanna/Ištar as a celestial body with "seven torches" in the sky

---

120 A few lines before, the torches of battle are reduplicated and called "fourteen" (Vanstiphout 2003: 128 l. 464). The reason that they are called first fourteen and then seven is not clear, but the multiple of seven might be a standardised and redundant formula to indicate a surplus of something, as the number seven is meant as "totality".

121 For references to the meaning of sig$_7$, "pale", and "yellow/green", see Thavapalan (2020: 27, 65–67).

122 The fact that the seven torches of the battle have a forehead (saĝ-ki, *pūtu*), eye(s) (igi, *īnu*), or they are a weapon (šíta, *kakku*) illuminating the sky would attest not only to the use of a metaphorical language and personification, but also to an embryonic concept of corporeality, a concept which is especially being developed in astrological texts (e.g. see 6.2.2.). More specifically, "corporeality is understood in a literal sense as pertaining to bodies of animate beings or inanimate objects with a spacial extension, perceived or imagined, in at least two dimensions." (Ossendrijver 2016: 143). For further bibliography and references to corporeality, see Ossendrijver (2016) and Schreiber (2020: 130 fn. 70).

emerges, so she now deserves further attention. Inanna/Ištar is the planet Venus, called ^{mul}NIN.SI$_4$.AN.NA in Sumerian,[123] and ^{mul}*dilbat* in Akkadian,[124] and she is considered an evening and a morning star. Her actions in several myths may metaphorically reflect the celestial movements of Venus: this has been noted, for instance, in the poem "Inanna and Šukaletuda".[125] In this poem, a travelling Inanna sleeps in the garden of Šukaletuda, who has intercourse with her while she is sleeping. A furious Inanna sends down a series of plagues seeking revenge against Šukaletuda. In the underworld, Enki helps Inanna in her purpose and, after travelling across the sky in the form of a rainbow[126], she finds and kills Šukaletuda. Cooper (2001: 142–144) and later Cooley (2013: 164–173) suggested that the poem has an astral motif: the description of Inanna's movements, i.e. her first appearance to Šukaletuda, her rest and rape, her search and plague against the land, the visit to Enki, the descent to the underworld, and the miraculous crossing in the sky, all match with the synodic cycle of the planet. Venus has a synodic period of circa 584 days, i.e. the planet takes circa 584 days to be seen from earth in the same starting position with respect to the Sun. During this period, she disappears from the east to be visible again in the west after roughly 50 days (i.e. superior conjunction with the Sun). She is then visible for around 8 months as an evening star, until she approximately passes between the earth and the Sun (i.e. inferior conjunction) and disappears for approximately 8 days. Finally, she is visible again in the east for 8 months circa as a morning star (Reiner-Pingree 1975: 15). Thus, Venus is visible as an evening star and a morning star for 8 months each. During her cycle, Venus is first visible in the east, then she is invisible for 80 days in conjunction with the Sun, and at the end she becomes visible in the west and disappears in the east. If that suggestion about the motif of "Inanna and Šukaletuda" is correct, the ancient author probably believed that the period of western visibility of Venus was due to her extraordinary ability to cross the sky. According to Volk (1995: 179 fn. 842) the poem undoubtedly has something to do with Venus as a star, and how important the planetary motion was as a seasonal marker for a successful fieldwork; nevertheless, planetary motion cannot be used as a seasonal marker because they can occur in any season. Therefore, it is not possible to say with certainty if there was any close relationship between the myth and Venus.

---

123 The first attestations of Ninsianna date to the Ur III period (ca. 2100–2000 BC) but the name is mainly attested in the Old Babylonian period (ca. 2000–1500 BC) (e.g. the hymn to Ninsianna for Idin-Dagan, see Reisman 1973). It seems that Ninsianna was syncretised with Inanna at the beginning of the second millennium BC, even though the name survived in the copies of *Enūma Anu Enlil* tablet 63 dating to the Neo-Babylonian period (ca. 626–484 BC), the so-called Venus Tablet of Ammiṣaduqa, based on observations dating to the Old Babylonian period (ca. 2000–1500 BC) (Reiner-Pingree 1975).

124 Inanna is referred to as the planet Venus, at least already at the beginning of the second millennium BC (Reisman 1973: 197 l. 133; Szarzyńska 1993: 8).

125 For the edition, see Volk (1995).

126 [n]í-te-a-ni ^{d}tir-an-na-gim an-na ba-an-gi$_{16}$-ib, "She lays herself like a rainbow in the sky" (Volk 1995: 122 l. 252).

Another example of catasterism for Inanna is found in "Inanna's Descent to the Netherworld",[127] perhaps an aetiology for Venus during her revolutionary period (Wilcke 1980: 83). Likewise, the Akkadian fragment AO 6035, known as the "Ištar-Louvre" tablet, probably composed in the Old or Middle Babylonian period (ca. 2000–1500/1400–1100 BC), refers to Venus as a guide-star for travellers and caravans:

44'. [*e*$^{?}$]-*ma ur-ḫa-am še-na e-*⸢*eṭ*⸣-*la i-la-ka ša-di-a*
45'. ⸢*e*$^{?}$⸣-*li-iš mu-ut-ti-ki uz-na-aš-nu iš-ku-nu e-li**-*ik i-*{*si*}-*su*$^{!}$
46'. ⸢*t*⸣*a-ma-ag-ri a*$^{!}$-*pu-uz-ri-im ta-ḫu-zi e-mu-uq li*$^{!}$*-*le*$^{!}$*-*en*

44'. Wherever$^{?}$ two young men were going on the path through the mountains,
45'. they placed their attention to you above$^{?}$, and they called for you.
46'. You were favourable, you took (them) into a shelter, power of the evening!
(Streck-Wasserman 2018: 27)[128]

Inanna/Ištar also has a double aspect: she is the goddess of love and war, and she is sometimes described and depicted with a beard. Her androgynous appearance alludes not only to her masculine and warlike aspect but also to her synodic cycle. Indeed, as an evening star, Venus was male,[129] and a bearded Venus is often mentioned in the protases of celestial omens. In omens, the beard (SU$_6$, *ziqnu*) of Venus is a celestial phenomenon, once explained as a conjunction between Venus and the Pleiades:

[DIŠ ^{d}]*dil-bat ina* ^{iti}BÁR *nap-ḫat* SU$_6$-*u* SU$_6$-*át* ⸢DINGIR^{meš}⸣ [ḪÉ].GÁL *ina* KUR *i-tab-ba-ku* SI.SÁ BURU$_{14}$ *na-pa-áš* ^{d}NIDABA *ur-ru-uk u*$_4$-*mì* NUN MUL.MUL ⸢12 IGI$^{?}$⸣-*šá* KIMIN *ina* Á-*šá* GUB-*zu-ma šá ina* ^{d}UTU.È SA$_5$-[*á*]*t u ba-ʾa-lat*

[If] Venus in Nisannu (i.e. Month I) rises (and) has a beard, the gods will pour out the [abun]dance in the country, (there will be) prosperity of the harvest, expansion of Nisaba (i.e. grain), prolongation of the days of the prince. (It means that) the 12 Pleiades (in) front$^{?}$,[130] ditto (i.e. the 12 Pleiades) stand at her side. (It means) that in the east (lit. sunrise) she is re[d] and she shines brightly.
(Reiner-Pingree 1998: 150 rev. 1)

127 For the edition, see Sladek (1974) and Heimpel (1982).

128 These lines might be an allusion to "Gilgameš and Ḫuwawa" travelling through the mountains (see 3.1.1.1.) (Streck-Wasserman 2018: 34).

129 For instance, in Ashurbanipal's hymn to Ištar of Nineveh: *a-ki* AN.ŠÁR *ziq-ni zaq-nat nam-ri-ri ḫal-*⸢*pat*⸣ [x x x], "Like Aššur, she has a beard, she is clad in brilliance [...]" (SAA 3, 7 obv. 6); see also Groneberg (1986).

130 Reiner and Pingree (1998: 150 rev. 1) restored the first half of K 137 rev. 6' as MUL.MUL [*ina*] ⸢IGI$^{?}$⸣-*šá* KIMIN *ina* Á-*šá* GUB-*zu-ma*, based on a similar entry (Reiner-Pingree 1998: 215, 24): DIŠ ^{mul}*dil-bat ziq-na zaq-na-at* MUL.MUL *ina* Á-*šà* : *ina* IGI-*šá* GUB^{me}-*zu-ma*, "If Venus has a beard, the Pleiades at her side (– var.) in front of her stand". Jeanette C. Fincke pointed out to the author that the signs that Reiner and Pingree restored as [*ina*] ⸢IGI$^{?}$⸣ are rather ⸢12⸣. For a further discussion on the meaning of the 12 Pleiades, see 4.2.2. § 2c and 5.1.8.4. § EAE 59–61.

The beard of Venus in this long entry from the series *Enūma Anu Enlil* (see 5.1.8.4. § EAE 59–61) is explained either with a conjunction with the Pleiades, or with a particular brightness and redness of the planet. To some extent, the omen uses a metaphorical language (further discussed in 5.1.8.4.); most importantly, it leads to the third perspective on celestial bodies discussed in the previous chapter (see 2.2.3.1. § 3), which is the personification of celestial bodies. In other words, "the scribes sought to explain the anthropomorphic image of a bearded goddess in terms of an optical description" (Rochberg 2004: 172). It is difficult to assess whether the beard of Venus, associated with the Pleiades, might be a later reference, or an allusion to the seven torches of Inanna in "Lugalbanda in the Wilderness" (see 3.1.1.3.). In the Sumerian composition, the "seven torches of battle" (izi-ĝar mè imin-me-eš) or "pure weapon" (šíta kù) are tightly related to Inanna as Venus, but their relationship is only known from a few attestations. Weapons are mentioned in other mythological contexts related to Inanna: for instance, "seven weapons" (šíta imin-e) belong to Inanna in one of the Sumerian temple hymns collected by the priestess Enḫeduanna,[131] and in the Sumerian mythological composition "Inanna and Ebiḫ".[132] Nevertheless, they are never called "starry", or associated with the "Pleiades" (mul-mul).[133]

To sum up, "Lugalbanda in the Wilderness" shows Inanna side by side with "seven torches of the battle" (izi-ĝar mè imin-me-eš), or a "pure weapon" (šíta kù), called mul-mul, lit. "stars", "starry", or "Pleiades". This association portrays an anthropomorphic and, implicitly, a warlike and masculine aspect of Inanna/Ištar because of the weapons and because she is Venus as an evening star; the torches of battle, then, can be a descriptive metaphor for seven stars next to her in the night sky. However, there is an inherent difficulty in determining whether these seven stars are the Pleiades or not. First, mul-mul can be either an attribute for being "starry", or the name of the Pleiades. Second, the number seven can either indicate a sky full of stars (i.e. seven as "totality") or the actual number of the stars. While we are left with the impossibility of determining a "secure" identity for the seven torches, the omens dating to the first millennium BC show that when Venus is seen surrounded by other stars, i.e. the Pleiades, she is perceived in the sky with a "beard", a reference to her warlike aspect. Hypothetically, the omens testify only a possible, later, playful analogy between the seven torches of Inanna and the beard of Venus, as both metaphors for the Pleiades.

---

131 See Sjöberg-Bergman-Gragg (1969: 47 l. 514).

132 See Attinger (1998: 170, 171 l. 56).

133 Only in a Sumerian Temple Hymn to the Eanna of Uruk, the temple of Inanna, is there a mention of seven fires in a nocturnal context: é-an-na é-ub-imin izi-imin ĝe$_6$-ù-na íl-la ka an-né si imin-e igi-ĝál, "Eanna, the house of seven corners, which lifts the seven fires at night, which checks the seven *desires* (lit. words which fill the sky)" (Sjöberg-Bergman-Gragg 1969: 29 ll. 201–202). For readings of ka an-né si, lit. "the word which fills the heaven", see Sjöberg Bergman-Gragg (1969: 92–93). See also ka an.né [s]i = *iš-ta-ra-an*, lit. "the two goddesses" (CAD I/J 274b 4).

3.1.2. Akkadian Sources

In this subchapter, Akkadian compositions will be discussed: the poem of "Erra and Išum" (see 3.1.2.1.–3.1.2.1.5.), and the UDUG.ḪUL incantations (see 3.1.2.2.–3.1.2.2.2.). The UDUG.ḪUL incantations, even though rooted in the third millennium BC, are here discussed in their Akkadian version, with focus on UDUG.ḪUL tablet 16, an additional Akkadian composition dating to the first millennium BC. Whenever needed, the Akkadian sources are compared to other Sumerian or Akkadian compositions (e.g. see 3.1.2.1.3., 3.1.2.1.5., 3.1.2.2.2.).

3.1.2.1. The Poem of "Erra and Išum": the Sebettu

The Akkadian poem of "Erra and Išum" is part of the traditional mythological narrative poetry of Babylonia.[134] The dating of the text is controversial, though it is almost surely to be ascribed to the first millennium BC.[135] It is a narrative dialogue between three protagonists, Erra, the god of plague and pestilence, Išum, his vizier, and Marduk, the king of the gods. The god Erra, who has a Semitic origin, is associated with Nergal, the god of the underworld, since the Old Babylonian period (ca. 2000–1500 BC) (Wiggermann 1998–2001: 217–218), whereas Išum is associated with the Sumerian god Ḫendursaĝa (see 3.1.1.2.). In this poem, the Sebettu (^{d}IMIN.BI) appear as seven warlike gods, helpers of Erra in his raging mission.

3.1.2.1.1. The Origins of the Sebettu

The beginning of the poem talks about a sleepy Erra, convinced by Išum to start a war against humankind with the help of the Sebettu (^{d}IMIN.BI). The Sebettu are introduced as warriors created by Anu (Erra I 31–40), and given to Erra as companions, to help him with his mission:

24. *i-lit-ta-šú-nu a-ḫa-at-ma ma-lu-u* ⸢*pul*⸣*-ḫa-a-ti*
(...)
31. *iš-ten i-šak-ka-na ṭ*[*è*]*-e-ma*
32. *e-ma* [*ta-a*]*n-di-ru-ma* ⸢*ta*$^{!}$⸣*-tal-ku ma-ḫi-ra* ⸢*e*⸣ [*t*]*ar-ši*
33. *i-qab-bi ana šá-né-e* GIM ^{d}GIŠ.BAR *ku-bu-um-ma ḫu-muṭ* GIM [*n*]*ab-li*
34. *i-t*[*a-mi*] *ana šal-ši zi-im la-bi lu šak-na-*[*t*]*a-ma a-mir-ka liš-ḫar-miṭ*
35. *i-qab-bi ana re-bi-i a-na na-še-e* ^{giš}TUKUL^{meš}*-ka ez-zu-ti* KUR*-ú li-*⸢*tab*⸣*-bit*
36. *a-na ḫa-an-ši iq-ta-bi* GIM IM *zi-i*[*q*]*-ma kip-pa-ta ḫi-i-ṭa*

---

134 Gössmann (1955) and Cagni (1969, 1977) were the first to collect and edit the poem of "Erra and Išum"; for the second tablet of the composition, see Al-Rawi-Black (1989); for a recent translation see Foster (2005: 880–911). In the present study, all the transliterations are taken from Cagni's edition (1969) for the first tablet, and from Al-Rawi-Black's edition (1989) for the second one.

135 See George (2013: 47) and Konstantopoulos (2015: 162–163) for an overview of the different suggestions of the date of composition.

37. *šeš-šá um-ta-ʾi-ir e-liš u šap-liš ba-ʾa-ma la ta-gam-mul mam-ma*
38. *se-ba-a i-mat ba-áš-me i-ṣe-en-šú-ma šum-qí-ta* ZI-*ta*
39. *ul-tu ši-mat* ᵈIMIN.BI *nap-ḫar-šú-nu i-ši-mu* ᵈ*a-num*
40. *id-din-šu-nu-ti-ma ana* ᵈ*èr-ra qar-rad* DINGIRᵐᵉˢ̌ *lil-li-ku i-da-ka*

24. Their descent is strange and they are covered with terror.
(...)
31. He (i.e. Anu) summoned the first:
32. "Rage everywhere you go and have no rival."
33. He said to the second: "Burn like fire and glow like a flame."
34. He ad[jured] to the third: "Look like a lion, may it (i.e. your appearance) destroy who looks at you."
35. He said to the fourth: "May the mountain collapse when you lift your fierce arms."
36. To the fifth he said: "Blow like the wind and explore the totality of earth."
37. He ordered the sixth: "Go up and down and spare no one."
38. He loaded the seventh with a snake's poison: "Kill (every) life."
39. After Anu had determined the fates of all the Sebettu,
40. he gave them to Erra, warrior of the gods, (he said:) "May they go at your side."
(Cagni 1969: 60, 62 ll. 24, 31–40)

The origin of the Sebettu, whose fate is determined by Anu, is unique compared to other heptads so discussed earlier (see 3.1.1.1.–3.1.1.3.). That is perhaps recalled in the astral science of Mesopotamia: the Pleiades are stars in the path of the god Anu (*šu-ut* ᵈ*a-nim*) in the night sky according to MUL.APIN I i 44 (see 4.1.1., 4.1.3.2.). Despite this, within the poem, the origin of the Sebettu is demonic, and never celestial: the Sebettu are secondary characters in the poem, and their descriptions (i.e. raging creature, burning creature, *lion*, creature that destroys the mountain, windy creature, destructive creature, *snake*) are reminiscent of the earlier traditions of the seven heroes (i.e. fox, dog, raven, vulture, *wolf*, *owl*, *crocodile*) in "Gilgameš and Ḫuwawa", and the seven heralds (i.e. lion-eagle, snake, snake-lion-dragon, a fire-breathing creature, *sagkal*-serpent, a spring, a creature that lights up) in the "Hymn of Ḫendursaĝa" (see 3.1.1.1.–3.1.1.2.). At this point, it is safe to assess that a description of seven beings is used as a literary device to create a kind of climax when introducing theriomorphic, hybrid, demonic entities, or awe-inspiring characters. It cannot be excluded, however, that the descriptions of the Sebettu, as Bottéro (1978: 160) first noted, are Akkadian renderings of their names. This kind of speculative interpretation (see fn. 101) of sacred names has been studied and called "encoding" by Bennett (2021: 384–391): the Akkadian description would derive from an interpretation of the Sumerian name of the Sebettu. This very particular practice was already observed in *Enūma eliš* I, and in the Standard Babylonian hymn to Gula, the healing goddess, reconstructed and edited by Bennett (2021: 159–335). Nevertheless, it is not possible to verify whether the poem of "Erra and Išum" contains the encoded names of the Sebettu, since we do not know them. The section of the God List *An* = *Anum* (VI 149–184) dedicated to the Sebettu is fragmentary and their names are poorly preserved.

3.1.2.1.2. The Purpose of the Sebettu

Throughout the first tablet of the poem of "Erra and Išum", the Sebettu are said to be especially devoted to destroying the livestock (Erra I 43, 74, 77, 83, 85) protected by Šakkan, the pastoral god of animals (Wiggermann 2013):

> *ṣal-mat* SAG *a-na šu-mut-ti šum-qu-tu bu-ul* ^{d}GÌR
>
> To kill mankind, to destroy the herd of Šakkan.
> (Cagni 1969: 62 l. 43)

On an intertextual level, the purpose of destroying the herds is mirrored in apodoses of celestial omens when the Pleiades (MUL.MUL) are addressed in the protases. This association is indicated in the assumed *Enūma Anu Enlil* tablet 50:

> DIŠ ^{d}IMIN.BI *a-na* GU$_7$*-ti bu-lì* (...) ^{mul}UDU.IDIM *ana* MUL.MUL KUR-*ud* ^{d}IMIN.BI KUR GU$_7^{meš}$
>
> ¶ The Sebettu (are) for the consumption of the herd: (...) (If) a planet reaches the Pleiades, the Sebettu will devour the country.
> (Reiner-Pingree 1981: 48 VI 2, 2b)

In these entries of *Enūma Anu Enlil* tablet 50, the Pleiades are associated with the Sebettu, hence a phenomenon of the Pleiades has the destruction of the livestock by the Sebettu as a prediction. To this topic a further discussion, which goes beyond the purpose of the present subchapter, will be dedicated in chapter 5 (see 5.1.8.4. § EAE 50–51 and 5.3.1.).

3.1.2.1.3. The Knowledge of the Mountain Paths

Always at the beginning of the poem of "Erra and Išum" (Erra I 87), the Sebettu desire to go to war together with Erra. One of the reasons is that they "are going rusty" (George 2013: 53):

> *ù né-e-nu mu-*⸢*de*⸣*-e né-reb* KUR-*e nim-ta-á*[*š-ši*$^?$ *ḫa*]*r-ra-nu*[136]
>
> And we, who know the pass of the mountain, we used to for[get$^?$ the p]ath.
> (Cagni 1969: 66 l. 87)

136 Lambert (1957–1958: 401) suggested restoring *nim-ta-*⸢*al*⸣*-*[*lik*], "we kept traver[sing]", a Gtn-stem preterite from *alāku*, "to go", based on similar forms in the epic of Gilgameš; Cagni (1969: 175) followed the reading suggested by von Soden (Falkenstein 1959: 204), that is *nimtaššī*, Gtn-stem preterite of *mašû*, "to forget". This interpretation would be consistent with the general meaning of the passage of the poem (Erra I 82–91), which expresses the impossibility of the Sebettu to go to war. See also Frankena (1959–1962: 47) and Foster (2005: 885 l. 87).

There is a case of intertextuality here: the line of "Erra and Išum" quoted above parallels "Gilgameš and Ḫuwawa" version B and the "Song of Inanna and Dumuzi" (see 3.1.1.1.).[137] Having the knowledge of the paths means to have an ancestral knowledge that the Sebettu lost. These statements, and the presence of the UR.SAG ᵈIMIN.BI, the "seven heroes", in the God List *An = Anum* (VI 150) just before the DINGIR^meš ᵈIMIN.BI, the "seven gods" (*An = Anum* VI 151) (Litke 1998: 211–212), point towards considering the seven heroes as predecessors of the Sebettu.

3.1.2.1.4. Understanding the Poem of "Erra and Išum"

The poem of "Erra and Išum" is a complex composition that has been interpreted by modern scholars according to a multiplicity of perspectives. This poem shows a widespread usage of word plays and hermeneutics indeed (Noegel 2011; Ponchia 2013–2014). The composition is also unique in its polyfunctional aspect, and "it is the power of language itself which is the constitutive element of the poem" (Machinist-Sasson 1983: 226).

According to the interpretation of Machinist (2005) and George (2013), the poem of "Erra and Išum" could be understood as a critique of war, a subversion of the cosmic order. In the story, the war turns upside down the rules of cosmos established by Marduk in *Enūma eliš* VI (Lambert 2013: 108–121): Marduk leaves his throne as the king of the gods (Erra IIIc 49, see Cagni 1969: 98); the cosmic bonds are destroyed by Erra (Erra IV 2, see Cagni 1969: 104); the seven sages are sent away by Marduk into the abyss of Ea, the god of wisdom, because they were not needed anymore:

> 147. *um-ma-a-[n]i šu-nu-ti a-na* AB:ZU *ú-še-rid-ma e-la-šú-nu ul aq-bi*
> (...)
> 162. *a-li* IMIN [N]UN.ME^meš *ap-s[i] pu-ra-di eb-bu-te*
>
> 147. I sent these craftsmen down to the Apsû, I did not tell them to come up.
> (...)
> 162. Where are the seven [sa]ges of the Apsû, the pure fishes?
> (Cagni 1969: 74, 76 ll. 147, 162)

The seven sages lived on the earth before the deluge; still, at the mythical time of "Erra and Išum", the sages had already accomplished their purpose long ago, so they were useless (Annus 2016: 9–13). The war has subverted the cosmic rules created by Marduk, and the seven sages are replaced by the warlike and demonic heptad, the Sebettu.

Another interpretation of "Erra and Išum" is as a metaphor for astral motifs. Brown (2000: 256–257) was the first to stress how the scribe of the poem applied not only a basic knowledge of the sky but also concepts, terms, and working principles of celestial

137 "Gilgameš and Ḫuwawa": ki-a ḫar-ra-an zu-me-eš, "on earth they know the paths"; má-ùr-má-ùr-ḫur-saĝ-ĝá-ka ḫé-mu-e-ni-túm-túm-mu-ne, "through the mountain passes they will lead us (?)" (Edzard 1993: 21 ll. 45, 50); "Song of Inanna and Dumuzi": ki-a ḫar-ra-an zu-me-eš, "on earth they know the paths" (Alster 1985: 223 l. 27).

divination to describe the actions of the characters in the plot. He assumed that Erra is the planet Mars, Marduk is Jupiter, and the Sebettu are the Pleiades. Marduk, when leaving his throne in favour of Erra, can be compared to the absence of Jupiter and the presence of Mars, the latter being a negative ominous sign in celestial divination.[138] The interpretation of Brown is corroborated by one of the wills of Erra in the narrative (Erra IV 124):

> *šá* ^{d}*šul-pa-è-a šá-ru-ru-šu lu-šam-qit-ma* MUL^{meš} *šá-ma-m*[*i*] *lu-šam-sik*
>
> Let me defeat the brilliance of Šulpaea (i.e. Jupiter)[139] and remove the stars of the sk[y].
> (Cagni 1969: 116 l. 124)

Cooley (2008) compared the technical language of celestial divination (e.g. *Enūma Anu Enlil* and MUL.APIN) with the poem, and suggested that it could be considered the ultimate expression of the use of celestial divination. Indeed, "divine royals use divination to decode each other's movements and intentions and to predict the final results of conflicts" (Cooley 2013: 179). That is one of the main purposes of celestial divination, and it would be consistent with the idea of disseminating references to celestial divination in a poem about war. As a matter of fact, astral references are explicit in the poem. It suffices to mention a few other passages; for instance, when Marduk leaves his throne and realises that something is happening in the sky (Erra II iii 6'–7', 10'):

> 6'. *ina* MUL *šá-ma-mi* ^{mul}KA$_5$.⸢A⸣ [...]
> 7'. *um-mul-ma ana šá-a-šú šá-ru-ru* x[...]
> (...)
> 10'. *kak-kab* ^{d}*èr-ra um-mu-lu šá-ru-ri i-na-áš-ši-ma*$^{!}$ *ka-tim* ^{d}*a-nu-ni-ti*
>
> 6'. Among the star(s) of the sky, the Fox-star [...]
> 7'. scintillates and towards him the brilliance ...[...].
> (...)
> 10'. The star of Erra scintillates, it has a brilliant sheen, and it covers Anunitu.
> (Al-Rawi and Black 1989: 118 ll. 6'–7', 10')

Regardless of the use of technical terms borrowed from celestial divination,[140] the Fox (^{mul}KA$_5$.A, *šēlebu*) mentioned in the lines given above is known as one of the names of the planet Mars in the omens (Brown 2000: 62), and in the Great Star List (Koch-Westenholz

138 See SIT 6: 40: DIŠ ^{mul}*ṣal-bat-a-nu ú-tan-na-at-ma* SIG$_5$ GUR$_4$-*ma a-ḫi-tú*, "If Mars is faint, it is good; (if) it becomes bright, (there will be) misfortune".

139 Šulpaea, as a name for Jupiter (i.e. the Marduk-star), stands for the planet when it rises heliacally (Brown 2000: 58), e.g. $^{[m]ul.d}$AMAR.UTU *ina* IGI.LÁ-*šú* ^{d}ŠUL.PA.È, "The Marduk-star, when it appears (is) Šulpaea" (SAA 8, 147 obv. 7).

140 The terms *wamālu*, "to scintillate" (CAD U/W 401), and *šarūru*, "brilliance" (CAD Š2 140–143) in l. 7' are widely used in the context of celestial omens, see K 8744 l. 10' (see Commentary EAE 53: 23) and 5.1.4. commentary to the text of entries 37–43, p. 166.

1995: 190 l. 93).[141] Al-Rawi and Black (1989: 112–113) were the first to notice this association, suggesting that Anunitu could be also mentioned in her astral form. Thus, there would be the possibility that the characters of the poem can be all associated with celestial bodies.

The end of the poem shows even a further bond between the story and the art of exorcists – the same exorcists who likely dealt with celestial divination. The last lines (Erra V 57–58) seal the apotropaic function of the entire poem against the plagues:[142]

> 57. *ina* É *a-šar tup-pu šá-a-šú šak-nu* ^{d}*èr-ra li-gug-ma liš-gi-šú* ^{d}IMIN.BI
> 58. *pa-tar šip-ṭi ul i-ṭe-ḫi-šu-ma šá-lim-tu šak-na-as-su*

> 57. In the house where this tablet is placed, may Erra be angry and may the Sebettu slaughter.
> 58. The sword of the punishment will not come close to it, health will be set for it.
> (Cagni 1969: 128 ll. 57–58)

It is evident that this poem can be interpreted based on different perspectives, which is what makes it such an interesting composition. Every point of view (i.e. political, astral, or hermeneutical) contributes to build the idea of "Erra and Išum" as an "erudit" poem, in which every element is tied by analogy to something else. Under these circumstances, one can suggest that the Sebettu in the poem probably aroused several associations of ideas for an ancient reader: for instance, the Pleiades and their relationship to the livestock in omens. While this is true for omens, the poem does not strictly show any astral aspect of the Sebettu, as they are only described as warlike and demonic.

3.1.2.1.5. The Sebettu, the Earlier Heptads, and the Pleiades

A final remark should be made on the Sebettu compared to the other heptads so far discussed. The Sebettu are secondary characters described in their hybrid and warlike aspect. The seven heroes, who show in the Sumerian composition "Gilgameš and Ḫuwawa" an astral feature in guiding the protagonist through the mountain passes, are likely a sort of forerunner to the Sebettu who have instead forgotten how to cross them (see 3.1.2.1.3.). Indeed, in the Akkadian poem of "Erra and Išum", the Sebettu are never mentioned as stars or starlike, as can be witnessed with the god Erra (i.e. the Fox star) and Marduk (i.e. Jupiter). The poem shows a "demonisation", more than astralisation, of the Sebettu. The "apparent" demonisation of the Sebettu in the poem, despite their status as higher gods in the first millennium BC, is "probably linked to the absence of a (substantial) cult in Babylonia during the period of its composition" (Wiggermann 2011a: 461–462). We only

141 Beaulieu-Frahm-Horowitz-Steele 2018: 37 D iii 19: ^{mul}⌜KA$_5$.A ^{d}*ér-ra*⌝ *gaš*$^{!}$(BE)-*ri* D[INGIR^{meš}], "The Fox, Erra, strong among the g[ods]".

142 Several lines from the poem of "Erra and Išum" (e.g. Erra IIId 3–15, see Cagni 1969: 100–103) were inscribed on an amulet to protect the household; other amulets also address Marduk, Erra, Išum and the Sebettu for protection (Reiner 1960).

learn of an association between the Sebettu and the Pleiades in the prayers (see 3.2.1.5.) and celestial omens (see 3.1.2.1.2. and 5.3.1.), which are rich in details about this analogy, but never from "Erra and Išum".

3.1.2.2. UDUG.ḪUL Incantations: the Seven Demons

This bilingual incantation series has its roots in the third millennium BC, and it is known as UDUG.ḪUL, *utukkū lemnūtu* in Akkadian, literally "evil utukku-demons" (in the following abbreviated UḪ). By the first millennium BC, the series evolved in a 16-tablets composition.[143] Even though UḪ is not narrative poetry, this series of incantations still contains long mythological narratives that reveal aetiological motifs for demons and illnesses portended by demons. The purpose of the overall collection of incantations is to fight these ill-portending demons, in order to heal a sick patient. The exorcists performed the rituals using a therapeutic magic, similar to many other types of Mesopotamian incantations, such as the series of incantations and rituals *Maqlû* and *Šurpu* (Geller 2016: 4).[144]

Within the composition, UḪ tablet 5 (Geller 2016: 174–216) has a long list of several ill-portending groups of seven demons. The exorcist should invoke the gods of the pantheon in order to counteract the demons and their powers, which are:

- Atmospheric heptads (e.g. UḪ 5 77: *u*$_4$*-mu šá* ḪUL-*tì*, "Evil storms")
- Chthonian heptads (e.g. UḪ 5 84: *se-bet* DINGIR^{meš} *ma-a-ti ra-pa-áš-ti*, "Seven gods of the wide netherworld")
- Celestial heptads (e.g. UḪ 5 83: *se-bet* DINGIR^{meš} AN-*e rap-šú-ti*, "Seven gods of the wide sky")
- Heptads bringer of diseases (e.g. UḪ 5 87 *se-bet* DINGIR^{meš} *lem-nu-tu*$_4$, "Seven evil gods"; UḪ 5 88: *se-bet la-maš-tu*$_4$ *lem-nu-tu*$_4$, "Seven evil Lamaštu"; UḪ 5 89 *se-bet la-ba-ṣi li-ʾi-bu lem-nu-tu*$_4$, "Seven Labaṣu and *liʾibu*-disease demons")
- Sexless heptads (e.g. UḪ 5 171: *ul zi-ik-ka-ri šú-nu ul sin-niš-a-tú šú-nu*, "They are not male, they are not female")

The incantations of UḪ have a dichotomic aspect in their apotropaic function: hybrid entities can be good or bad. If the heptads are ill-portending, they also have a counterpart, heptads of helper-spirits who can fight the evil of demons. The incantations also cater for the well-attested prophylactic use of specific items in rituals, such as ingredients or figurines. Indeed, it has been noted that UḪ 12 (120–139; Geller 2016: 423–425) partly duplicates the *bīt mēseri*,[145] the ritual of protection of houses against evil, which involves

---

143 An extensive edition was published by Geller (2016). In the present study, the focus is on the content of the UḪ series; hence, only the Akkadian version of the relevant lines is quoted since it does not have specific or noteworthy differences in contents against the Sumerian version.

144 For the editions of *Maqlû* and *Šurpu*, see Abusch (2015) and Reiner (1958), respectively.

145 See Wiggermann (1992: 105–130) for an overview and comments of the edited and unedited sources belonging to the *bīt mēseri* ritual.

the use of groups of seven apotropaic figurines to avert the fate of a patient (Wiggermann 1992: 113–114; Geller 2016: 16).

Konstantopoulos (2015: 244–247) rightly argued that not all of these heptads are associated with the warlike and demonic Sebettu. Only the "seven sexless demons" aspect, that is eventually recalled in the forerunners of UḪ,[146] might be ascribed to the tradition of the Sebettu. All these heptads are not associated with the Pleiades either: they are demons or "unrepresentable not-yet beings" (Wiggermann 2011: 309). Thus, only the number seven is relevant in this context because it has an apotropaic function, and it is used as a literary device to define various shades of demonic evil and their actions within the UḪ series.[147]

3.1.2.2.1. The Lunar Eclipse Myth as Part of UDUG.ḪUL

Contrary to the incantations in UḪ 5 (see 3.1.2.2.), UḪ 16 is not attested in the second millennium BC sources, but dates to the first millennium BC. It seems an independent composition about an eclipse of the Moon god – the worst ominous sign for the king and his kingship – brought by seven demons.[148] When a lunar (or a solar) eclipse was expected, there usually was the enactment of "substitute king" (*šar pūḫi*) ritual: the king was supposed to be replaced by someone else for a certain period of time, during which the evil portended by the eclipse could have hit the reign.[149] The evil was then transferred to the substitute king, and by his death, and enacting prayers of reconciliation, the real king was enthroned again (Parpola 1983: xxii–xxxii).

In contrast with the other heptads, in UḪ 16 (ll. 5–12) seven "butting storm-demons" (*umū muttakpūtu*) are described one by one in their hybrid and theriomorphic aspect:

5. *ina se-bet-ti-šú-nu iš-ten šu-ú-tu*$_4$ *ez-ze-tùm-ma*
6. *šá-nu-ú u-šum-gal-lu šá pi-i-šú pe-tu-ú man-ma la i-ʾi-ir-ru-šú*
7. *šal-šu nim-ru ez-zu šá pi-i-ru i-ba-ʾa-a*
8. *re-bu-ú šib-bu gal-ti* ⸢*šu*⸣-[*ú*]
9. *ḫa-áš-šá lab-bi na-ad-ri šá ana* EGIR-*šú ne-ʾe-a l*[*a i-le-ʾi*]
10. *šeš-*[*šu a-gu*]*-ú* ⸢*té*⸣*-bu-ú šá ana* DINGIR *u* LUGA[L...]
11. ⸢*se-bu*⸣*-ú me-ḫu-ú šá-a-ru lem-nu šá* ⸢*gi-mil-li*⸣ *t*[*ur-ru*]
12. *se-bet-ti-šú-nu mar* ⸢*šip-ri*⸣ *šá* ^{d}*a-nim šar-ri šú-nu*

5. Among these seven, the first is the furious south wind,
6. and the second is a lion-dragon whose mouth is open, no one will approach.
7. The third is a furious panther which passes an elephant.

146 The forerunners of UDUG.ḪUL were edited by Geller (1985) and commented by Konstantopoulos (2015: 90–95) regarding the features strictly related to the Sebettu, mainly their warlike aspect.

147 "Suspended between the Sky and the Netherworld, demons are allowed to move between these spheres and to act in the in-between area, the Earth" (Capomacchia-Verderame 2011: 296).

148 See Hall (1985: 878–899) for an overview of the Moon god as protector of kingship and the well-being of the country (i.e. fertility) in the Ur III period (ca. 2100–2000 BC), and Hätinen (2021: 229–247) for the Moon god in the first millennium BC.

149 For the substitute king ritual, see Lambert (1957–1958a; 1959–1960) and Parpola (1983: xxii-xxxii).

8. The fourth is an angry snake.
9. The fifth is a raging lion which (no one) [is able] to turn back.
10. The sixth is a rising flood which [overwhelms][150] both god and kin[g].
11. The seventh is a storm, an evil wind which takes revenge.
12. The seven of them are the messengers of the king Anu.[151]
(Geller 2016: 502–504 ll. 5–12)

The modality of description of the seven demons (i.e. south wind, lion-dragon, panther, snake, lion, flood, storm) is, again, reminiscent of "Gilgameš and Ḫuwawa" (see 3.1.1.1.), the "Hymn of Ḫendursaĝa" (see 3.1.1.2.), and "Erra and Išum" (see 3.1.2.1.), and it works as a literary device. However, the actions of the seven demons in UḪ 16 are unique, because they bring an occultation of the Moon god by encircling his front, so that he becomes invisible to the people:

ina ma-ḫar ^{d}NANNA-*ri* ^{d}30 *ez-zi-iš il-ta-nam-mu-u*

In front of the divine luminary of the Moon they kept circling furiously.
(Geller 2016: 508 l. 30)

In light of their purpose, the seven demons are also described as atmospheric agents (i.e. clouds, breezes, winds, rains), helpers, or weapons, of Adad, the storm god:

15. *er-pe-tu*$_4$ *šá-pi-tu*$_4$ *šá ina* AN-*e da-um-ma-ta i-šak-ka-nu šú-nu*
16. *zi-iq šá-a-ri te-bu-tu*$_4$ *šá ina u*$_4$*-mi nam-ri e-ṭu-ta i-šak-ka-nu šú-nu*
17. *it-ti im-ḫul-li šá-a-ri lem-ni i-sur-ru šú-nu*
18. *ri-ḫi-iṣ-ti* ^{d}IŠKUR *te-šu-ú qar-du-*⸢*te*⸣ *šú-nu*
19. *ina i-mit-ti* ^{d}IŠKUR *il-l*[*a-ku*]

15. They are a thick cloud which makes the gloom in the sky.
16. They are the breeze of the rising winds which makes the darkness on a bright day.
17. They circle with the destructive wind, the evil (wind).
18. They are the inundation of Adad, a warlike confusion.
19. [They wa]lk on the right side of Adad.
(Geller 2016: 504–505 ll. 15–19)

The aetiology of the lunar eclipse in UḪ 16 is at odds with the lunar eclipse as described in the series *Enūma Anu Enlil*. In celestial omens, it is the Moon god who decides the eclipse, or the Moon is said to set or grieve in distress (see 2.2.3.1. § 3). The eclipse of the Moon as an ominous sign is usually described through a metaphorical language, but never through mythological metaphors. Under these premises, several scholars have suggested different

150 The meaning is restored from parallels written in Sumerian: àš-kam-ma [a-gi$_6$]-⸢a⸣ zi-ga dingir lugal-la-šè ḫul ⸢mu⸣-u[n-...] (Geller 2016: 504 l. 10).

151 In the poem of "Erra and Išum" (Erra I 39–40), the Sebettu are also created by Anu (see 3.1.2.1.1.).

interpretations for the lunar eclipse myth in UḪ 16. For instance, Lawson (2012: 225–226), and later Wee (2014: 66) suggested that UḪ 16 is a royal apology, a different tendency in respect of *Enūma Anu Enlil* and the ritual of the substitute king, where the eclipse is never caused by external forces. Nonetheless, the closing paragraph of *Enūma Anu Enlil* tablet 22[152] refers to a *tamītu* text, "oracle-query",[153] where the eclipse of the Moon is brought about by "seven *gallû* demons" (GAL$_5$.LÁ^{meš}):

1. *ta-mi-a-tum an-na-tum e-nu-ma* ^{d}30 *mit-lu-uk-ta* GAR-*nu* DINGIR^{meš} *šá* AN *u* KI
2. *ep-šet a-me-lu-ti ṭu-pu-ul-šú-nu i-ši-im-ma*
3. AN.TA.LÙ *ri-iḫ-ṣu mur-ṣu mu-tum* GAL$_5$.LÁ^{meš} GAL^{meš} ^{d}IMIN.BI
4. *ma-ḫar* ^{d}30 *it-ta-nap-ri-ku*

1. These (are) the queries: when the Moon establishes the advice, determines (a fate for) the gods of the sky and the earth,
2. the deeds of humankind (and) their disgraces.
3. Eclipse, devastation, illness, death, the great *gallû*-demons, the divine Seven,
4. keep standing in front of the Moon.

(Rochberg-Halton 1988: 269–270 II § XII concluding paragraph 1–4)

In these entries, the seven *gallû*-demons are associated with several catastrophes, yet their identity is unclear. Together with the *rābiṣu*-demon, "bailiff" or "deputy", the *gallû*-demons should represent the evil counterpart of a corrupt official in the human world (Geller 2016: 28). The latter, always seven in number, are the same group of chthonian entities who threaten Dumuzi and accompany Inanna in "Inanna's Descent to the Netherworld" (Sladek 1974: 147 ll. 350–351). In UḪ 5 (129), these seven *gallû*-demons are separated from the preceding groups of seven demons, and were thought to harass a patient, causing depression or anxiety (Geller 2016: 202 fn. 129). Whereas in UḪ 16, they cause the occultation of the Moon god, the evillest omen for the king.

Wee (2014: 64) also compared the above-mentioned *tamītu* text from *Enūma Anu Enlil* tablet 22 and UḪ 16 to three ritual texts from Uruk (Linssen 2004: 306–320), dating to the Hellenistic period (ca. 323–63 BC). In one of these texts (BM 134701), which describes a royal ritual, the lunar eclipse is caused by seven entities who cover and dethrone the Moon god:

10'. [... DUMU] *šip-ri* IMIN^{meš}-*šú-nu šá ina* AN-*e rap-šu-tu ina šu-bat* ^{d}60 *šar-ri ra-biš* ⸢x⸣ [...]
(...)
12'. [*ina ma*]-*ḫar* ^{d}NANNA-*ri* {x} ^{d}30 *ez-*{x}*-zi-iš il-ta-n*[*am-mu-u*...]

10'. [... the me]ssengers, these seven, which in great manner... in the wide sky, in the dwelling of Anu, the king... [...]

152 For the edition, see Rochberg-Halton (1988: 251–272).
153 For the *tamītu* texts, see Lambert (2007).

(...)
12'. [In fr]ont of the divine luminary of the Moon, they kept cir[cling] furiously [...].
(Linssen 2004: 309 rev. 10', 12')

It seems that UḪ 16, the *tamītu* text and the ritual texts from Uruk display a tradition in which seven *gallû*-demons bring about the occultation of the Moon. While the substitute king ritual implies a guilt to be expelled by the king by hiding and praying, in this parallel tradition the role played by the guilt is absent, because the eclipse is brought by external forces, or demons. The image described in the *tamītu* text could integrate the one of UḪ 16, but this explanation "still fails to provide a satisfactory synthesis of the lunar eclipse myth and the *Enūma Anu Enlil* omen tradition" (Wee 2014: 63). Indeed, the omens about lunar eclipses never mention the *gallû*-demons, or even an occultation of the Moon brought by seven demons. Nevertheless, Rochberg (2018a) rightly argued that there were different views and responses to the lunar eclipse scattered across Mesopotamia. Even though different traditions seem to contradict each other, they can just reflect different "perspectives on the theological 'problem of evil'" (Rochberg 2018a: 311). From this point of view, the demons in UḪ 5 and UḪ 16 could be the same *gallû*-demons of the *tamītu* text in *Enūma Anu Enlil* tablet 22.

3.1.2.2.2. The Pleiades as Seven Demons in UDUG.ḪUL 16?

In the preceding subchapters (see 3.1.2.2.–3.1.2.2.1.), it has been noted that in UḪ 5 there is a long list of seven demons with different features, but they are not related to the Sebettu nor to the Pleiades.[154] An assumed astral aspect of the seven demons in UḪ lies only in the background. The seven demons have a prominent role only in the narrative of UḪ 16, which is a later composition dating to the first millennium BC, whereas the rest of the composition mainly dates to the second millennium BC. Only in UḪ 16, a more atmospheric aspect of the heptad appears, i.e. demons in the shape of clouds and winds of Adad, causing the occultation of the Moon. In one omen from *Enūma Anu Enlil* tablet 46,[155] the divine Seven in the shape of seven stars meet Adad. They are said to "descend to the earth" and cause terrible consequences (see 5.1.8.3.):

DIŠ ^{d}IŠKUR *ina* MÚRU ^{d}IMIN.BI GÙ-*šú* ŠUB-*ma* MUL^{meš}-*šú-nu se-bet-ti-šú-nu ana* KI-*tì it-tab-ku-ni* ^{d}IŠKUR *šá* KUR : *šá* A.AB.BA *ḫar-gal-li-šá* ḪE.ḪE ^{giš}ŠINIG [...] ⸢É⸣^{meš} DINGIR^{meš} *a-šá-re-du-tu*$_4$ *ú-nap-pa-aṣ* URU KI URU É KI É ŠEŠ KI ŠEŠ-*šú* LÚ KI LÚ *nam-ga-ru* KI ÍD *a-tap-pu* KI *nam-ga-ri-šú* PA$_5$ KI *a-tap-pi-šá* KÚR^{meš}-*ma* 55 MU^{meš} LÚ UZU LÚ GU$_7$ LÚ KUŠ LÚ MU$_4$.MU$_4$-*aš*

If Adad thunders in(to) the middle of the divine Seven and their stars, their seven, are poured over the earth, Adad the locks of the country – (var.) of the sea will mix up he

154 Wiggermann (2011a: 461) pointed that out: "the association of the Sebettu with the netherworld and death does not necessarily identify them with the seven Evil Demons (UDUG.ḪUL)."
155 For the main edition, see Gehlken (2012: 76–121).

will smash the tamarisk of [...] the temples of the gods of the highest rank; a city to a city, a household to a household, a brother to his brother, a man to a man, an irrigation canal to a river, a small branch of an irrigation canal to its irrigation canal, an irrigation ditch to its small branch of an irrigation canal will show hostilities, for 55 years a man will eat flesh of a man, a man will dress in the skin of a man.
(Gehlken 2012: 80 obv. 1)

In the protasis of this omen, the seven stars (MUL^meš^-*šú-nu se-bet-ti-šú-nu*) of the divine Seven (^d^IMIN.BI) mean the Pleiades. The phenomenon described, the stars poured over the earth, could be a metaphor for the beginning of a violent storm around the Pleiades in the night sky. The apodosis is long and detailed, very reminiscent of the consequences of a demonic attack, which usually subverts the regular cosmic order by threatening and destroying.[156] Therefore, it might be suggested that the divine Seven in the protasis are reminiscent of the seven *gallû*-demons of UḪ 16. This hypothesis remains speculative, though we are aware that the scribes of the series *Enūma Anu Enlil* were also *āšipū* ("exorcists"), so they were familiar with both divination and incantations (Koch 2015: 20–21). It is undoubtedly problematic for a modern reader to trace back the origin of eventual associations of ideas through the intertextual analysis of different genres of texts. Yet, it cannot be excluded that certain analogies – like the Pleiades and the seven demons – were maybe implicit, if not obvious, for both the compilers of omens and of incantations.

3.1.3. Summary of the Nature of the Heptads and the Pleiades

From the mythological compositions discussed so far (see 3.1.1.1.–3.1.2.2.2.) a consistent variety of heptads with astral features, or in astral contexts, emerges. A summary about the role of all these heptads is, at this point, necessary: the aim is to clarify until which instance one can talk about an analogy between the terrestrial and the astral aspect of the heptads in myths, or about a recurrent use of the description of seven entities as literary device.

Starting from the narrative structures of the myths, all heptads are helpers, i.e. secondary characters at the service of a primary character:

- Seven heroes help Gilgameš in his journey (see 3.1.1.1.)
- Seven heralds help Ḫendursaĝa in his psychopomp duties (see 3.1.1.2.)
- Seven weapons help Inanna during the battle in the sky (see 3.1.1.3.)
- The Sebettu help Erra in his mission of war (see 3.1.2.1.)
- Seven demons help Adad in obscuring the Moon god Sin (see 3.1.2.2.)

In the Sumerian composition "Gilgameš and Ḫuwawa", seven heroes guide Gilgameš at night through the mountains towards Aratta.[157] It could be argued that the ancestral ability

156 E.g. see Wiggermann (1992: 151–152). For references to cannibalism in Mesopotamia, see Oppenheim (1956: 270–271), Edzard (1980), and Wilhelm (1997).

157 ki-a ḫar-ra-an zu-me-eš, "on earth they know the paths"; má-ùr-má-ùr-ḫur-saĝ-ĝá-ka ḫé-mu-e-ni-túm-túm-mu-ne, "through the mountain passes they will lead us (?)" (Edzard 1993: 21 ll. 45, 50).

to escort an individual at night is to be ascribed to an original use of stars as a tool for orientation and, therefore, that this feature of the seven heroes comes from their association with the stars. Through intertextuality, we learn that the knowledge of the stars and the mountain paths also means possessing ancestral wisdom, whose depositories were the seven singers of Uruk, or the seven sages, in the "Song of Inanna and Dumuzi".[158] In the poem of "Erra and Išum", one witnesses the opposite situation: the Sebettu forgot the knowledge of the mountain paths.[159] Through this statement, the poem of "Erra and Išum" refers to "Gilgameš and Ḫuwawa", and to the seven heroes as a sort of precursors for the later Sebettu.

The seven heroes, the seven heralds, the Sebettu, and the seven demons of Adad are described one-by-one as hybrids, but their identities do not match. In the "Hymn of Ḫendursaĝa", seven heralds are essentially described as animals;[160] in "Erra and Išum", the Sebettu are "demonised" gods because they have a blurred origin and forgot the knowledge of the mountain paths of their forerunners, the seven heroes; in UḪ 16, seven demons are described as atmospheric events because they are weapons of Adad and cause a lunar eclipse, the worst threat for the king. The descriptions of all these heptads are relevant for the attributive use of the number seven, because they all act as the same literary device: a list of seven beings is a tool for climax,[161] a tool to describe particularly awe-inspiring entities. This last statement would be confirmed by a Sumerian hymn to Numušda, deity of the city of Kazallu and son of the Moon god Sin (Cavigneaux-Krebernik 1998–2001). In the hymn, dated to the Old Babylonian period, the god is described as a lion (piriĝ), a snake (muš), and a snake-lion-dragon (ušumgal) (Sjöberg 1973: 108 ll. 11–13),[162] in the style of the descriptions in "Gilgameš and Ḫuwawa" version A ll. 37–39 (see 3.1.1.1.).

The theriomorphic or hybrid heptads, from the earliest to the latest attestations, are either benevolent or malevolent in their actions. Even though they were sometimes determined with the logogram DINGIR, *ilu*, "god",[163] they can all be formally ascribed to the flexible characteristics of the Mesopotamian pandemonium. Wiggermann (2011) divided the huge, sometimes, unclear, number of demons of Mesopotamia into various categories; according to him, hybridity defines two groups of entities:

---

158 ki-a ḫar-ra-an zu-me-eš, "on earth they know the paths" (Alster 1985: 223 l. 27).

159 *ù né-e-nu mu-⸢de⸣-e né-reb* KUR-*e nim-ta-á*[*š*?-*ši*? *ḫa*]*r-ra-nu*, "And we, who know the pass of the mountain, we used to for[get? the p]ath" (Cagni 1969: 66 l. 87).

160 The imagery portrayed by the "Hymn of Ḫendursaĝa" is the most discordant, compared to the others. The myth has, indeed, a different tradition, which can be ascribed to Lagaš (Konstantopoulos 2015: 95–107). Verderame (2017a) suggested comparing the seven heralds of Ḫendursaĝa to the names of the planet Mars. For instance, according to the Great Star List (Koch-Westenholz 1995: 190 ll. 93, 96, 97, 105), the Fox, the Wolf, the Eagle, and the Raven can be substitute names for Mars.

161 As Cooley (2013: 90–91 fn. 7) stated, the ancient authors drew probably from the same literary *topos* of hybrids in Mesopotamian literature.

162 See also online at the website ETCSL, 2.6.7.1.

163 E.g. Wiggermann (2011a: 462–463) rightly noted that the seven heroes are presented as gods in the list *An* = *Anum* VI 150 (dur-⸢sag⸣-[imin-bi]), i.e. with the determinative DINGIR (Litke 1998: 211 l. 150).

- Monsters, who fight with brute force. To this group belong all demons, such as the atmospheric ones. The seven heroes, the seven heralds, the seven weapons of Inanna, and the seven *gallû*-demons from UḪ 16 are part of this category.
- Sages, who fight evil with wisdom and purifying rituals. The seven sages belong to this group (see 3.1.1.1.).

Malevolence and benevolence seem to be inherent features of monsters, demons, and sages. The UḪ's incantations, for instance, already show in sources dating to the Old Babylonian period (ca. 2000–1500 BC) a dichotomous and apotropaic aspect of the demons because they can be protective spirits as well (Wiggermann 2011: 305). In this framework, the Sebettu seem at first sight to constitute an exception. They reached the status of higher gods during the first millennium BC, even though their origin is almost clearly monstrous, as their forerunners would have been the seven heroes of "Gilgameš and Ḫuwawa" (see 3.1.1.1.). The reason for this exception was investigated by Konstantopoulos (2015) in her PhD dissertation. The Sebettu seem to have reached the status of "great gods" mainly because of the Neo-Assyrian aggressive propaganda: they were demons who became awe-inspiring gods at the service of the kings, yet they were never, in practice, treated as such (Konstantopoulos 2015: 312–317). Regardless of the Sebettu in Neo-Assyrian royal inscriptions and treatises (see 3.2.7.), they were never described as great gods, like Marduk, Sin, or Šamaš, nor did they have a cult of their own. In the poem of "Erra and Išum" (see 3.1.2.1.) their role is also secondary, and their aspect is demonic. Thus, despite their official insertion in the primary pantheon during the Neo-Assyrian period (ca. 911–612 BC), the Sebettu must be considered "one expression of the overall flexibility inherent in demonic beings" (Konstantopoulos 2015: 312).

How do the Pleiades fit into the complex system of the Mesopotamian pantheon and pandemonium? In mythology, astralisation lies in the background. Only two myths attest heptads with both terrestrial and astral components: the Sumerian myths "Gilgameš and Ḫuwawa" version B (see 3.1.1.1.) and "Lugalbanda in the Wilderness" (see 3.1.1.3.). An astral metaphor lies behind these heptads, hence they fit the perspective according to which the gods, or other entities, can be considered starlike (i.e. astralisation) (see 2.2.3.1. § 1). In "Gilgameš and Ḫuwawa", the ability of the seven heroes to guide at night could derive from an original use of all stars as a tool for orientation. In "Lugalbanda in the Wilderness", the seven torches (izi-ĝar mè imin-me-eš), or pure weapon (šíta kù) of the warlike Inanna/Venus could be "starry", or "the Pleiades" (mul-mul). Hypothetically, the origin of this catasterism could be sought in a conjunction of the Pleiades and Venus, a conjunction once described by a celestial omen dating to the first millennium BC, in which the "beard" of Venus, i.e. when she is male and warlike, is a metaphor for the Pleiades (see 3.1.1.3.1.). On the contrary, in Akkadian mythology, the heptads are subjected to demonisation, rather than to astralisation, as it is the case of the Sebettu and the seven demons (see 3.1.2.–3.1.2.2.2.).

3.2. Celestial Bodies as Gods: Cult, Prayers and Rituals

The prayers and rituals discussed in the second part of this chapter are written in Akkadian and date to the first millennium BC from Nineveh. These sources belong to the multiple fields of the exorcists' expertise, so they testify to the use and the relevance of stars in magic, medicine, and apotropaia. On account of their polyfunctional nature, prayers and rituals are useful tools to understand the perception of gods and celestial bodies in Mesopotamian culture. For instance, Maul (1994: 46–47), in his book *Zukunftsbewältigung*, stresses the role of the stars as messengers between people and gods. According to Reiner (1995: 15–24), stars have a dual role because they can transform ordinary objects into magical ones for divination and medicine, and they are mediators between humankind and the gods. The latter function has the purpose to protect the king or the throne against evil and black magic: for example, prayers to stars feature in royal ritual units, such as the substitute king ritual (e.g. SAA 10, 240). Prayers and rituals testify overall a role of celestial bodies as counterparts of the gods, and thus they are consistent with the second perspective on celestial bodies in Mesopotamia (see 2.2.3.1. § 2). Therefore, the purpose of this part of the chapter will be to examine the role of the Pleiades in prayers and rituals, both as stars and whenever associated with a heptad; for the latter, the goal will be to identify a possible origin of such association.

Regarding a practical cult of celestial bodies in Mesopotamia, including the Pleiades, there is not much preserved. The textual evidence shows that there were places of worship dedicated to astronomical observations, or stars in general.[164] For instance, the ziqqurrats were likely astronomically oriented and used for astronomical observations, as deduced from the mythology of the Moon god Nanna at Ur, and the Sun god Utu at Larsa (Nadali-Polcaro 2016). The Moon, the Sun, and Venus, whose cults are related to the main gods of the panthea (Suen/Sin, Utu/Šamaš, Inanna/Ištar), were worshipped in complex religious buildings within the cities, but the astral aspects of the gods within the places of worship is difficult to detect, regardless of a few instances.[165]

Based on the textual evidence at our disposal, the Pleiades, like the majority of celestial bodies in Mesopotamia, seem to be worshipped mainly during the first millennium BC. They are part of extensive Mesopotamian ritual enactments, such as the *bīt rimki*, "house of ablution", *bīt salā' mê*, "house of water sprinkling", the series of rituals *Maqlû*, "Burning", and of smaller apotropaic rituals for incantations, which will be discussed in further sub-chapters (see 3.2.1.–3.2.5.). The ritual acts, of which several stars were part, were usually enacted on an important day for a calendric or astronomical reason, but in a rather generic setting. For instance, in *bīt salā' mê* ritual (see 3.2.1.1.), the diviners enacted the rituals from dusk until dawn, in an open-air space:

[*ana* IGI] ^{d}IMIN.BI NÍG.NA ^{šim}LI GAR-*an* KAŠ.SAG BAL-*qí*

164 E.g. see George (1993: 128 n. 817 and 820): é.mul, "the house of the star", is a sanctuary of Nabu, and é.mul.mul, "the house of the stars", is a temple of Nisaba in Ereš.

165 E.g. é.an.da.sá.a, "the temple of Venus (i.e. Ištar-of-the-stars)" in Babylon (George 1993: 34 n. 434, 67 n. 63–64).

[In front of] the divine Seven (i.e. the Pleiades) you set up a censer of juniper (and) pour out first-class beer.
(Ambos 2013: 188 l. 62)

There are also mentions of offerings to the Pleiades addressed as seven gods in connection with a terrace as a ritual place. This is witnessed in a ritual for the Lady-of-the-Mountain, linked to the cultic calendar of the Ištar temple in Aššur (Parpola 2017: xxii):

11. [*ina*] ⸢UGU⸣ *tam-le-e ina* IGI ^{d}IMIN.BI ⸢MUN⸣ *i*-[*kar-ra*]-⸢*ár*⸣
(...)
18. ⸢^{lú}⸣SANGA [*šá*] ⸢^{d}⸣*aš-šur* A^{meš} ŠU^{II}
19. [*a*]-*na* ^{d}⸢GAŠAN⸣ [KUR]-⸢*e a-na*⸣ LUGAL *ú-qar-rab*
20. UZU.TI [*ina*] ⸢UGU *tam*⸣-*le-e ina* IGI ^{d}IMIN.BI GAR-*an*

11. He s[trew]s salt [on] the terrace before the divine Seven (i.e. Pleiades).
(...)
18. The priest [of] Aššur offers water by hand
19. [t]o the Lady-of-the-[Mountai]n and to the king,
20. (and) places a rib [on] the terrace before the divine Seven (i.e. Pleiades).
(SAA 20, 21 obv. 11, 18–20)

While in texts dating to the first millennium BC the Pleiades are mainly part of bigger ritual enactments, the only attestation for them in an earlier cultic context is found in a record of sheep offerings to the stars from Tuttul, dating to the Old Babylonian period (ca. 2000–1500 BC):

2 UDU 6 ^{giš}BANŠUR$^{\text{ḫi.a}}$ / *a-na* ^{mul}*za-ap-pí-im* / ^{mul}*ba-li*-[*im*]

Two sheep, six offering tables / for the Bristle / (and) Bālu.
(Krebernik 2001: 146 obv. 6–8)

The text mentions an offering of two sheep and six tables to *zappu*, "Bristle", followed by Bālu, a name for Mars[166] or Orion[167] that is otherwise found only in texts dating to the first millennium BC. The text from Tuttul is among the earliest in which *zappu*, "Bristle", is mentioned. Only one other text dating to that period, one *ikribu* prayer to the gods of the night (AO 6769), mentions the Pleiades as *zappu*.[168]

There is very little evidence of an individual cult for the Pleiades as an individual asterism. Most of the textual evidence dating to the first millennium BC mentions temples

166 See the lexical list Murgud (MSL 11: 40 l. 31).

167 See the astrolabe fragments dating to the Neo-Assyrian period (ca. 911–612 BC) (Horowitz 2014: 230 B rev. 4; 231 C 6').

168 The prayer to the god of the night is discussed in 3.2.2. For a bibliography of the prayer and the most updated transliteration and translation, see online at the website SEAL, see SEAL n. 7491 (https://seal.huji.ac.il/node/7491 accessed 05.03.2021).

or streets of seven gods or the Sebettu (George 1992: 69 l. 78, 365–366; Grayson 1991: 380 Ashurnasirpal II 131), whereas in earlier periods the attestations are scarce (Wiggermann 2011a: 464–465). The name of the celestial body (MUL.MUL or *zappu*) is not mentioned, so it is not possible to state if an astral-related background was intended in those instances.[169] Only a few names for places of worship, attested in the form ki DN, "place of DN", are possibly related to an open-air cult. These open-air sanctuaries date to the Ur III period (ca. 2110–2000 BC) from Sumer, and are dedicated to deities with astral or atmospheric features (e.g. Anu, Sin, Šamaš, Adad and Inanna), and for seven gods as well (Richter 2004: 241–245). There are two instances about the Pleiades or seven gods in open air cult places from Sumer: the first textual evidence from Puzriš-Dagan/Tell Drehem has ki dingir-imin, "the place of the divine Seven" (Yildiz-Gomi 1988: 21 n. 767 obv. ii 4); the second evidence, from Nippur, has é dingir-imin-bi, "the temple of the divine Seven" (Robertson 1981: 232, CBS 7651 obv. 3). If this interpretation proves to be correct, then a sanctuary of seven gods as the Pleiades would probably be the first attestation for an individual place of worship for this asterism (Wiggermann 2011a: 464).

3.2.1 *Šu'illa* Prayers

In Akkadian there is no general term for "prayer", yet many different words refer to specific kinds of prayers.[170] Only the rubrics, at the end of the prayers, help the modern reader to identify their purpose, more than the genre.[171] The term *šu'illa* is a loanword from the Sumerian ŠU.ÍL.LA, which means "hand-lifting", and refers to the gesture of raising the hand, a formal gesture to greet an authority, or to what is intended by a *šu'illa*, i.e. the act of calling for a divine intervention. In other words, they are prayers, whose wording can fit various purposes. A *šu'illa* is usually considered part of the *Gebetsbeschwörungen*, "incantation-prayers",[172] but it holds its specific rubric: KA.INIM.MA ŠU.ÍL.LA, "word of the hand-lifting".

The *šu'illa* prayers are preserved almost exclusively on tablets dating to the first millennium BC, from Assyria and Babylonia (i.e. cities of Sippar, Ur, and Uruk).[173] They are dedicated not only to all the higher gods of the individual panthea, but also to celestial bodies.[174] Generally speaking, these prayers are formulaic and rarely specific in their wording. They show adaptability in multiple respects because they can be enacted for different occasions: for instance, one prayer to the Pleiades, the prayer entitled "Zappu 1"

169 For the attestations of the warlike and demonic Sebettu in cultic texts, see Konstantopoulos (2015: 369–372).

170 E.g. *ikribu*, "dedication" (CAD I 62–66), or *teslītu*, "appeal" (CAD T 369–371).

171 Regarding the issue of applying a genre to prayers and incantations in Mesopotamia, see 1.4.2.

172 These incantation-prayers are traditionally *namburbi*s, i.e. incantations to avert inauspicious events, the *šu'illa*s, and all the prayers to free an individual from evil.

173 The first comprehensive and in-depth critical study on the features of *šu'illa* prayers was conducted by Mayer (1976). See also Frechette (2012) and Lenzi (2011).

174 A list of all the best-preserved *šu'illa*s can be found in Mayer (1976: 378–437), but also in Frechette (2012: 252–275) and online at the website *Corpus of Akkadian Shuila Prayers Online* (http://shuilas.org/index.html accessed 25.01.2023).

(see 3.2.1.1.), is part of the *bīt salā' mê* ritual, but its incipit is also found on a fragment belonging to a different collection of *šu'illa* prayers, suitable for other purposes.[175]

There are in total six known *šu'illa* prayers addressed to the Pleiades. Each will be individually presented in transliteration and translation (see 3.2.1.1.–3.2.1.6.) and named using the titles given by Mayer (1976: 431–432), as shown in Table 1. Within the texts and the rubrics, the prayers are addressed to the Pleiades (MUL.MUL or *zappu*, "Bristle"), or to the divine Seven or Sebettu (^{d}IMIN.BI). The epithets in the prayers refer to both an anthropomorphic and an astral aspect of the addressee at the same time, regardless of the name written in the rubric. The names ^{d}IMIN.BI, MUL.MUL or *zappu* alternate, and anthropomorphic and astral features also fuse together.[176] More specifically, all the prayers refer to the stars, to war, to demonic features, and also to divine judging and fates. Therefore, the ultimate goal of the following subchapters will be to comment, whenever possible, on the nature of the symbolism emerging from the content of prayers.

Prayer	Main edition	Incipit	Addressee in the rubric
Zappu 1	Ambos 2013: 200 ll. 68–76'; Jiménez 2014: 108–109	*attunu zappu šarḫūtu ša mušīti*, "You (are) the Pleiades, splendid of the night".	MUL.MUL, "Pleiades".
Zappu 2	Abusch-Schwemer-Luukko-van Buylaere 2020: 231–232	Unknown	MUL.MUL, "Pleiades".
Zappu 3	AGH: 149–151; BMS 52 (only incipit)	*šar ilī gašrūti ša napḫar māti šūpû* ^{d}IMIN.BI *attunuma*, "The king(s) of the mighty gods, of the totality of the country, brilliant, you are the Sebettu!".	Unknown
Zappu 4	Mayer 1976: 534–535	Unknown	^{d}IMIN.BI, "Sebettu".
Zappu 5	STT 69	[... *gaš*]*rūti dannūti*, "[...mi]ghty, strong".	^{d}*zappu*, "Bristle".
Zappu 6	Ambos 2013: 198 l. 63 (only incipit)	[...] *zappu ilū rabûtu*, "Pleiades, the great gods".	MUL.MUL, "Pleiades".

Table 1. The incipits of the *šu'illa* prayers to the Pleiades.

175 For an overview of the *šu'illa* prayers used in rituals, see Frechette (2012: 165–224).

176 This trend is also witnessed in the famous acrostic hymn to Marduk by the king Ashurbanipal. Marduk is called with his celestial name (^{d}SAG.ME.GAR, "Jupiter"), and among the numerous stars mentioned in the hymn, the Pleiades appear by the name ^{d}IMIN.BI, "divine Seven", and not MUL.MUL (SAA 3, 2 rev. 4).

3.2.1.1. Zappu 1 (Addressed to MUL.MUL)

The prayer "Zappu 1" was edited by Ambos (2013: 200 ll. 68–76') and then restored by Jiménez (2014: 108–109). The prayer was part of *bīt salāʾ mê* ritual, lit. "house of water sprinkling". This was a royal ritual in which Akkadian *šuʾilla*s functioned as simple elements within a complex enactment.[177] The ritual took place just before the autumnal equinox during the second New Year Festival of Babylon, specifically from the fourth to the eighth of Tešritu (i.e. Month VII). It was also carried out in connection with the substitute king ritual, to reinstate the actual king after the death of a substitute king.

In the cycle of prayers to be recited at night, i.e. in the eighth section of *bīt salāʾ mê*, there is a prayer to the Pleiades, addressed with the name MUL.MUL (Ambos 2013: 164 l. x+24'):

22. ÉN *at-tu-nu* MUL.MUL *šar-ḫu-tu*$_4$ *šá mu-*[*š*]*i-t*[*i*]
23. *nam-ru-ti šá* DINGIR^{meš} GAL^{meš} M[U]L.MU[L]
24. *a-na ḫul-lu-qu lem-nu-ti ib-nu-ku-nu-ši* ^{d}*a-num* : *ina šá-ma-me* ⸢MU⸣-*ku-nu* ^{d}IMIN.BI M[UL.MUL$^{?}$]
25. [*za-ʾa-n*]*u ki-li-lu* ^{na4}MUŠ.GÍR *ra-ki-*⸢*su*⸣ *me-sír-r*[*i*]
26. [x x x] x-*su-ti šá til-le-e mu-šam-qí-tu*$_4$ *bu-l*[*ì*]
27. [*mu-pa-áš-š*]*i-ḫu* EDIN *da-li-ḫu* A.AB.BA^{me}[^{š}]
28. [x x x]-⸢x x-x-*tum*⸣ ⸢*gaš*-ru*-tu*$_4$⸣ DUMU ^{d}*a-nim*
29. [x x x x x x x x]x-*ku-nu-ši*
30. [(x x) *a-na da-ra-a-ti dà*]-⸢*lí*⸣-*lí-ku-nu lud-lu*[*l*]
31. ⸢*ki*⸣-*m*[*a* A^{meš} ÍD *e*]⸢*š*-šu*-ti*⸣* ⸢*iṭ-ru-du la-bi*⸣-*ru-ti* TU$_6$.É[N]

---

32. KA.INIM.MA ŠU.ÍL.LÁ MUL.MUL.KAM

---

22. Incantation: You (are) the Pleiades, splendid of the n[i]gh[t],
23. bright among the great gods, P[l]eiade[s].
24. Anu created you to destroy the evil, in the sky your name is Sebettu, the P[leiades$^{?}$].
25. [Decora]ted with the headdress of serpentine, harnessed with the belt,
26. [...]... of the weaponry, who consume the her[d],
27. [who ca]lm the steppe, who roil the seas,
28. [...]... powerful sons of Anu,
29. [...]... to you.
30. May I proclaim your [p]raises [forever]!
31. Lik[e] the fresh [water of a river] expelled the old things. Incantati[on].

---

177 For the edition and study of the ritual, see Ambos (2013: 155–173, 198–224).

32. Wording of a *šu'illa*-prayer to the Pleiades

(Jiménez 2014: 108 rev. 22–32)

In this prayer, the Pleiades (MUL.MUL) are called Sebettu (^{d}IMIN.BI), sons of Anu (*mār anim*), but they are also called *namrūtu*, "brilliant", and decorated with "serpentine" (^{na4}MUŠ.GÍR, *muššāri*). The last feature may fit the traditional belief that heaven is made of precious stones (Horowitz 1998: 263). The Pleiades or Sebettu are also "fastened with the belt" (*rākisu mesirri*), probably a reference to the Pleiades seen as the "Bristle" (*zappu*) of the Bull of Heaven (^{mul}GU$_{4}$.AN.NA), or Taurus, which was perceived as one zodiacal sign together with the Pleiades in later zodiological literature (see 6.1.1.).

On the whole, this prayer recalls both astral (i.e. Pleiades) and warlike (i.e. Sebettu) features. As rightly noted by Jiménez (2014: 109), the prayer also shows intertextuality, and it has a direct correspondence with the poem of "Erra and Išum": *mūšamqitū būli*, "who consume the herd", refers to the propensity of the Sebettu to "destroy the herd of Šakkan" (*šumqutu būl Šakkan*, see 3.1.2.1.2.).

3.2.1.2. Zappu 2 (Addressed to MUL.MUL)

The prayer "Zappu 2" is preserved in a broken fragment (K 8808 rev. 1'–8', see BMS 47) with the rubric [DUB.x.KAM] ⌜ÉN⌝ ŠU.ÍL.LA.KAM-*ni*, "[Tablet x of] Incantation of the *šu'illa* prayers". This fragment is part of a collection of *šu'illa*s and not of a specific ritual. On the reverse of the fragment (rev. 8'), right after the prayer "Zappu 2", there is part of the incipit of the prayer "Zappu 3" (see 3.2.1.3.).

1'. [... *kišpī ru-h*]*e-e ru-s*[*e-e*]
2'. [*upšāšê*] *lem-nu-*[*ti*]
3'. [(*zīra*) *dibalâ* ZI.KU$_{5}$.R]U.D[A-*a*] KA.DAB.⌜BÉ⌝.DA DÙ-[*ni*]
4'. [x x x x x x (x) *d*]*an-*⌜*na-tu*⌝*-nu* TI.LA BA-*a-ni*
5'. [*narbîkunu lu-šá-p*]*i* ⌜*dà*⌝*-lí-lí-ku-nu lud-lul*

6'. [KA.INIM.MA ŠU.Í]L.LÁ MUL.MUL.KE$_{4}$

7'. [DÙ.DÙ.BI *lu ina* KÉŠ *l*]*u ina* NÍG.NA DÙ-*uš*

8'. [ÉN LUGAL DINGIR^{me}]^{š} *gaš-ru-ú-t*[*i*]

1'. [... witchcraft, m]agic, sorce[ry],
2'. ev[il machinations],
3'. [(hate-magic), distortion-of-justice magic, cutt]ing-[of-the-throat magic] (and) aphasia performed [against me].
4'. [...] you are [st]rong, grant me life!
5'. [Your greatness may I glorif]y, may I proclaim your praises!

6'. [Wording of a *šu'i*]*lla*-prayer to the Pleiades

7’. [Its ritual:] you perform (this) [in a ritual arrangement o]r in a censer.

8’. [Incantation: “The king(s)] of the mighty [god]s”.
(Abusch-Schwemer-Luukko-van Buylaere 2020: 231 ll. 1’–8’)

The preserved part of the prayer quoted above, which corresponds to the closing lines of the text, is unfortunately too generic to allow an analysis for the perception of the Pleiades. Therefore, one can only formulate hypotheses. First, one hint is given by the rubric of the prayer, which is addressed directly to the Pleiades (MUL.MUL). Second, the preserved part has a reference to *kadibbidû* (“aphasia”, or “seizing-of-the-mouth”) and *zikurudû* (“the cutting-of-the-throat”) magic: these special types of witchcraft are performed before the stars,[178] as shown in the following quotation.

8. [DIŠ NA IGI^{meš}-*š*]*ú iṣ-ṣa-nun-du* ⸢GÌR^{II}-*šú* ⸢*it-te*⸣-*n*[*e*]*n-ṣi-la*
9. [Á$^{II?}$-*šú*] *i-šam-ma-ma-šú* ⸢UZU^{meš}-*šú*⸣ *i*[*k-t*]*a*-⸢*na-su-u*⸣
10. [GU$_7$] *u* NAG-[*m*]*a* UGU-⸢*šú*⸣ NU GUB-*za* x x x-*su-ma*
11. [x x S]ÌG-*iṣ* SU DÙ.A.⸢BI⸣ *ú*-⸢*zaq-qat-su*⸣
12. [*ana* NA BI *n*]*a*-⸢*áš*⸣-*pa-rat* ⸢ZI.KU$_5$.RU⸣.<DA> IGI MU[L.MU]L DÙ-⸢*šú*⸣ ÚŠ$^!$ (KÚR)

8. [If a man] has [ver]tigo, his feet are constantly contracted,
9. [his arms$^?$] are paralysed, his flesh is constan[tly b]ent,
10. [he eats] and drinks, but he does not retain (the food), ... him and
11. [... is af]fected, his entire body hurts,
12. [against that man, m]essages of cutting-of-the-throat magic have been performed before the Pl[eiade]s; he will die.

(Abusch-Schwemer 2011: 436 obv. i 8–12).

These lines belong to the “Diagnostic Handbook”, also referred to as the diagnostic-prognostic omen series SA.GIG, *Sakikkû* (STT 89, see Abusch-Schwemer 2011: 434–443). The series provides a symptom description, diagnosis, and prognosis of witchcraft-induced illnesses. As shown in the lines quoted above, the Pleiades are among the stars invoked to send the *zikurudû*, “cutting-of-the-throat”, magic, so this could constitute a parallel with the fragmentary prayer “Zappu 2”.

3.2.1.3. Zappu 3 (Unclear Address)

The prayer “Zappu 3” is partially preserved in three fragments of collections of *šu’illa*s, among which a prayer to the True Shepherd of Anu (^{mul}SIPA.ZI.AN.NA, *šitaddaru*) is

---

178 E.g. before the Wagon (^{mul}MAR.GÍD.DA, *ereqqu*), see Abusch-Schwemer (2011: 399–406).

included,[179] likely because these prayers were part of a collection of *šu'illas* to the stars. The prayer "Zappu 3" is presented in a score transliteration and translation:

List of Sources

Source:	A
Siglum:	K 6395 + K 10138
Edition:	BMS 52; AGH 148–151
CDLI n.:	P396503 (with photograph)

Source:	B
Siglum:	VAT 11126
Edition:	KAR 258; AGH 150–151
CDLI n.:	P369223 (without photograph)

Source:	C
Siglum:	K 8808
Edition:	BMS 47; Abusch-Schwemer-Luukko-van Buylaere 2020: 231–232
CDLI n.:	P397768 (with photograph)

Text

1.	**ÉN LUGAL DINGIR^{meš} *gaš-ru-ú-ti***	
A rev. 5'	ÉN LUGAL DINGIR^{meš} *gaš-ru-ú-ti*	
B 5'	ÉN L[UGAL DINGI]R^{meš} *g*[*aš*]-*r*[*u-ú-ti*]	
C rev. 8	[ÉN LUGAL DINGIR] ⸢^{meš}⸣ *gaš-ru-ú-*⸢*ti*⸣	

Incantation: The king(s) of the mighty gods

2.	***šá nap-ḫar ma-a-ti*** **(**var.: **KUR)** ***šu-pu-u***	
A rev. 5'	*šá nap-ḫar ma-a-ti*	⸢*šu*⸣-*pu-u*
B 6'	⸢*ša*⸣ *na*[*p-ḫa*]*r* KUR	*š*[*ú*$^{?}$-*pu-u*]

of the totality of the country, brilliant,

3.	**^{d}IMIN.B[I *a*]*t*-[*tu-nu-ma*]**
B 7'	^{d}IMIN.B[I *a*]*t*-[*tu-nu-ma*]

you are the Sebettu!

179 The prayer is entitled "Sipazianna 3" in Mayer (1976: 431).

4. **NIGIN-*ku*-[*nu-ši*...]**
B 8' ⌜NIGIN-*ku*⌝-[*nu-ši*...] ⌜x⌝ [...]

I turned back to y[ou...].

5. **mu šu$^{?}$ ši [...]**
B 9' mu šu$^{?}$ ši [...]

... ... [...].

6. **[Á]Š$^{?}$-*ti* [...]**
B 10' [Á]Š$^{?}$-*ti* [...]
[the c]urse$^{?}$ [...].

7. **x ú [...]**
B 11' ⌜x⌝ ú [...]

... ... [...].

B ———————————————

8. **[K]A.INIM.[MA ŠU.ÍL.LÁ...]**
B 12' [K]A.INIM.[MA ŠU.ÍL.LÁ...]

[Wo]rding o[f a *šu'illa*-prayer...]

B (remainder is missing)

This prayer is highly fragmentary: only the attribute *šūpû*, "brilliant" may be a reference to the Pleiades, whereas the apposition *šarrū ilī gašrūti*, "king(s) of the mighty gods", could be a reference to the Sebettu (^{d}IMIN.BI).

3.2.1.4. Zappu 4 (Addressed to ^{d}IMIN.BI)

The fragment Sm 1025 preserves only the last part of the prayer "Zappu 4", including the rubric, which is addressed to the divine Seven (^{d}IMIN.BI).

Transliteration

Obverse
1'. [x x x x] ⌜SUKKAL KI-*ti li-iṣ*$^{?}$*-bat*$^{?}$*-su-nu*⌝*-ti*
2'. [Ì$^{?}$.D]U$_{8}$$^{?}$.⌜GAL⌝ *gaš-ru* ^{d}NE.DU$_{8}$ KÁ^{meš}-*šú-nu li-dil*
3'. [*it-t*]*i* IM *lu ṣa-an-du*
4'. [*it-t*]*i me-ḫe-e lu rak-su*
5'. [SUD$^{?}$]^{meš}-*ma a-a* GUR^{meš}-*ni*

6’. [BAD$^{?}$]^{meš}-*ma a-a* NIGIN^{meš}-*ni*
7’. [*ina q*]*í-bit* DINGIR-*ti-ku-nu* GAL-*ti šá* NU KÚR-*ru*
8’. [*ina q*]*í-bit* ^{d}60 ^{d}UTU ^{d}ASAR.LÚ.ḪI TU$_6$.ÉN

9’. [KA].INIM.MA ŠU.ÍL.LÁ ^{d}IMIN.BI.KE$_4$

10’. [GI]M *an-na-a* DÙ-*šú tuš-tam-šal-ma ina* MÚRU *rik-si šú-nu-t*[*i*]
11’. [*t*]*a-za-za* ŠU-*ka* ÍL-⸢*ma*⸣
12’. [É]N *al-si-ku-nu-ši* EN^{meš}-*e* GAL$^{⸢\text{meš}⸣}$
13’. [*ana*] IGI DINGIR.DINGIR GAL.GAL 3-*šú* ⸢ŠID⸣-[*nu*]

(end of the fragment; reverse is lost)

Translation

Obverse
1’. May [...] the vizier of the netherworld catch$^{?}$ them.
2’. May the mighty chief [gate]keeper, Nedu, lock their doors.
3’. May they be harnessed [wi]th the wind.
4’. May they be tied up [wi]th a violent storm.
5’. [May] they [go away$^{?}$], and not return to me.
6’. [May] they [depart$^{?}$], and not turn back to me.
7’. [At the com]mand of your great divinity which cannot be changed.
8’. [At the com]mand of Anu, Šamaš, Asalluḫi, incantation formula.

9’. [Wo]rding of a *šu’illa*-prayer of the divine Seven.

10’. [On]ce you have done that, you do it again and you stand in the middle of these ritual arrangements,
11’. [y]ou raise up your hand and
12’. you recite the [pra]yer “I sought you, great lords”
13’. three times [be]fore the great gods.

(end of the fragment; reverse is lost)
(Sm 1025, see Mayer 1976: 534–535)

According to Mayer (1976: 14), prayers like “Zappu 4” strongly distinguish themselves in content from most prayers labelled as *šu’illa*. This is true because of the presence, in obv. 8’, of the “incantation formula” (TU$_6$.ÉN) to Anu, Šamaš, and Asalluḫi, a designation that ascribes the prayer to an apotropaic purpose, more common to *namburbi*s rather than *šu’illa*s.[180] The prayer “Zappu 4” was probably meant to obtain the intervention of (seven)

180 Asalluḫi, the god of magic, is mainly addressed in the context of exorcism, to dispel the evil of ominous signs. In *namburbi*s, the gods Ea, Šamaš and Asalluḫi are always addressed to avert the fate

great gods more than of the Sebettu, as also implied by the ritual instructions at the end (i.e. *ana* IGI DINGIR.DINGIR GAL.GAL 3-*šú* ŠID-*nu*, “you recite three times before the great gods”). The text could be interpreted as “against” the violence of seven demons, who must be “tied up” (*lū raksū*) or “harnessed” (*lū ṣandū*) and sent back to the netherworld. That is very reminiscent of incantations against demons (Jiménez 2018: 327; see 3.1.2.2.): indeed, to achieve this purpose, the prayer addresses a health-portending counterpart of such demons, that is, (seven) great gods.

3.2.1.5. Zappu 5 (Addressed to ^{d}*zappu*)

The prayer “Zappu 5” is preserved in a manuscript from Sultantepe (STT 69) dating to the Neo-Assyrian period (ca. 911–612 BC). It has the typical subscript of the *šu’illa* rubric, together with ritual instructions to be performed at night by an incantation priest (^{lú}MAŠ.MAŠ).

Transliteration

Obverse
1. [ÉN... *gaš*$^{?}$]- ⸢*ru*$^{?}$⸣-*ti dan*- ⸢*nu*$^{?}$-*ti*⸣
2. [...]-*ti al-lal*- ⸢*li*⸣ DINGIR^{meš} *šá*-⸢*qu-ti*⸣
3. [...] *na-mir-ti ra-’i-mu ba-am-ma-a-ti*
4. [*mut-tal-li-ku sa-an*]-⸢*ga*⸣-*a-ni mu-šad-di-ḫu ṣu*-⸢*ṣe*⸣-*e*
5. [*mu*$^{?}$-*šal*$^{?}$]- ⸢*li*$^{?}$-*mu*$^{?}$ ⸣ ZI-*ti* ⸢NAM$^{?}$⸣ TI.LA *at*-⸢*tu*⸣-*nu-ma* ⸢*ta-šim*⸣-*ma*
6. [... TI].⸢LA *at-tu-nu-ma tu*-⸢*uṣ*⸣-*ṣa-ra* EŠ.BAR TI.LA *at-tu*-⸢*nu*⸣-[*ma*] KUD-[*sa*$^{?}$]
7. [... TI].LA ⸢*ṣi*⸣-*it* KA-*ku-nu šá*-⸢*la-mu*⸣ *e-piš* KA-*ku-nu* TI.⸢LA-*ma*⸣
8. [*ana-ku* NENNI] A NENNI ⸢*šag-šu* ARAD *pa*⸣-*liḫ-ku-nu*
9. [...] SAG.KI SAG.DU[...] ⸢ŠU⸣.MIN *u* ⸢GÌR.MIN⸣
10. [...] ⸢*šim*⸣-*ma-ti* ⸢*ri*⸣-[*mu-ti*...] x ra x x
11. [...]^{meš}-*ni* ⸢*ul*⸣ [...] sag
12. [...] x AN.NA *i-ba*-x [...] a a x
13. [...] x x x 5 x [...]
14. x lu ba ni x [...]
15. ⸢GIG⸣ SAG.KI-[*su*$^{?}$...]
16. ⸢GIG⸣ UZU^{meš}-⸢*šu*$^{?}$ ⸣ [...]
17. x da ma x [...]
18. x x x lu x [...]
19. [...] di [...]
(end of the obverse)

Reverse
1. [...] ri ⸢li⸣ x [...] x [...] x x

of a negative sign. According to Maul (1994: 48–57), the order of the gods reflects the arrangement of altars during the rituals, to mimic a trial in which Ea and Asalluḫi assist the judge god Šamaš.

2. ⸢*nar*⸣*-bi-*⸢*ku*⸣*-nu lu-šá-pi dà-lí-lí-ku-nu ana** ⸢*u*$_4$⸣-[*me*...] <<⸢A$^?$⸣>> *ṣa-a-ti* ⸢*lud-lul*⸣

---

3. ⸢KA.INIM⸣.MA ŠU.ÍL.LÁ ^{d}⸢*zap-pu*⸣.KE$_4$

---

4. DÙ.DÙ.BI *lu ina ši-im-tan lu-ú ina qid-da-at u*$_4$*-mi*
5. *ina* EDIN *ina ba-li-ti* KI ⸢GÌR⸣ KU$_5$-⸢*at*⸣ x [...] x x x
6. GU.DU$_8$ GUB-*an* NÍG.NA^{šim} ⸢LI MÚ KAŠ.SAG⸣ *u* ⸢GEŠTIN$^?$ BAL-*qí*⸣
7. ⸢^{udu}⸣SISKUR *eb-ba* BAL-*qí* ^{uzu}ZÀ ^{uzu}ME.ḪÉ *u* ^{uzu}⸢KA.IZI *tu-ṭaḫ-ḫa*⸣
8. [x] *kup-pi-né-e-ti* ^{še}*bu-ṭu-ut-ti eb-bé-ti ina* Ì.GIŠ LÀL
9. [Ì].NUN *ta-mar-ras kup-pi-né-e-ti tu-kap-*⸢*pat*⸣
10. ⸢*lìb*$^?$⸣*-bi tu-šá-saḫ* KI.SIG$_{10}$.GA *ta-kás-sip* 7 ma$^?$ qi$^?$ [...]
11. [...] x e GEŠTIN^{meš} BAL-*qí* ÉN *an-ni-tu*$_4$ x [...]
12. [ŠID]-*ma* ^{lú}GIG BI ⸢TI$^?$⸣

---

13. [...]x PAB$^?$ ^{md}AMAR.UTU.DUB.NUMUN ^{lú}MAŠ.MAŠ KÀ.DINGIR.RA⸢^{ki}⸣
(after a few lines of blank space, end of the reverse)

Translation

Obverse
1. [Incantation: ... mi]ghty$^?$, strong$^?$ [...],
2. powerful [...]..., elevated gods,
3. [...] of brightness, lovers of the open country,
4. [who roam on the mo]untain passes, who march through the swamps.
5. [Who gu]ard$^?$ the life, you determine the fate of life,
6. you draw up the [...] of life, you decide the verdict of life.
7. [...] is life, the utterance of your mouths is peace (and) your command is life.
8. [I, so-and-so], son of so-and-so, the afflicted, the slave who serves you,
9. [...], forehead, head, [...] hands and feet,
10. [...] *šimmatu*-paralysis (and) *rim*[*ûtu*-paralysis...] ...
11–14. (too broken for translation)
15. A patient, [his$^?$] forehead [...],
16. A patient, his$^?$ flesh [...].
17–19. (too broken for translation)
(end of the obverse)

Reverse
1. (too broken for translation)
2. May I glorify your greatness, may I proclaim your praises forever!

---

3. Wording of a *šuʾilla*-prayer to the Bristle.

---

4. Its ritual: during the evening or late afternoon
5. in the steppe, in the desert, where the access is blocked [...]

6. you set up an offering table, light up a censer of juniper (and) pour out first class-beer and wine.
7. Present a pure sheep offering; offer the shoulder, the fatty tissue, and the roast meat.
8. Mix ... pellets of pure turpentine in oil, honey,
9. butter (and) roll the pellets.
10. You cut out the heart[?] (and) make a funerary offering. Seven... [...]
11. [...]... you pour out wines. This incantation... [...]
12. you [recite], and that patient will recover[?].

---

13. [...]... Marduk-šapik-zeri, incantation priest of Babylon.
(after a few lines of blank space, end of the reverse)

(STT 69, see http://oracc.org/cams/gkab/P338387 accessed 05.03.2021).

This prayer is addressed to the Pleiades called "Bristle" ([d]*zappu*), and it is followed by a ritual (rev. 4–12) to perform in the evening in an open space in the desert; the setting may be related to the presence of the asterism itself during the ritual. The ritual is meant to ward against at least two illnesses, the *šimmatu*-paralysis and numbness (*rimûtu*) (obv. 10).

The epithets in the prayer shed light on traces of intertextuality. First, the prayer mentions the ability of the Bristle (*zappu*) to "roam on mountain passes" (obv. 4). The propensity to travel and the knowledge of the mountain passes are features of the seven heroes in the Sumerian composition "Gilgameš and Ḫuwawa" (see 3.1.1.1.);[181] the same knowledge was forgotten by the Sebettu in the Akkadian poem of "Erra and Išum" (see 3.1.2.1.3.).[182] Second, the epithets (obv. 5–6) NAM TI.LA *at-tu-nu-ma ta-šim-ma*, "you determine the fate of life", and EŠ.BAR TI.LA *at-tu-nu-ma* KUD, "you decide the verdict of life", refer to abilities untypical of the Sebettu or the seven heroes. One could argue that these epithets are more similar to the seven singers of Uruk, or the seven sages from the Sumerian "Song of Inanna and Dumuzi" (see 3.1.1.1.); the two epithets may come from different traditions unknown to us, or they may simply belong to a typical aulic and repetitive language of prayers.[183] However, the feature of the Pleiades as "determiners" is not unique and it requires more attention as it illuminates a parallelism with other prayers (e.g. see 3.2.2.), or celestial omens (see 5.1.8.4. § 50–51, 5.3.2.); therefore, this topic will be further discussed at length in 3.2.5.1.

181 ki-a ḫar-ra-an zu-me-eš, "on earth they know the paths"; má-ùr-má-ùr-ḫur-saĝ-ĝá-ka ḫé-mu-e-ni-túm-túm-mu-ne, "through the mountain passes they will lead us (?)" (Edzard 1993: 21 ll. 45, 50).

182 *ù né-e-nu mu-*⸢*de*⸣*-e né-reb* KUR*-e nim-ta-á*[*š*[?]*-ši*[?] *ḫa*]*r-ra-nu*, "And we, who know the pass of the mountain, we used to for[get[?] the p]ath" (Cagni 1969: 66 l. 87).

183 The language and the style of prayers, the so-called "hymno-epic dialect", has been investigated by von Soden (1931; 1933), and his study was reprised and expanded by Pohl (2022) with a focus on Old Babylonian sources.

3.2.1.6. Zappu 6 (Addressed to MUL.MUL)

Only the incipit of the prayer "Zappu 6" is known as part of the *bīt rimki* ritual, lit. "house of ablution". The purpose of this ritual was to cleanse the king and his household from the evil portended by a lunar eclipse, or from witchcraft (Ambos 2013: 96–99).[184] In the prayer "Zappu 6", the Pleiades (MUL.MUL) are directly addressed as "great gods", and not as the divine Seven (dIMIN.BI) like in "Zappu 4" (see 3.2.1.4.):

> [ÉN (...)] MUL.MUL DINGIRmeš GALmeš 3-*šú* ŠID-*nu*
>
> [The incantation (...)] Pleiades, the great gods, you recite three times.
> (Ambos 2013: 189 l. 63)

A letter from the chief exorcist Marduk-šākin-šumī to Esarhaddon (ca. 680–669 BC) reports the enactment of "Zappu 6" among a lengthy sequence of *šuʾilla*s, carried out for the ritual of the substitute king (K 602 obv. 5–6, see SAA 10, 240).

3.2.2. The Gods of the Night

The famous prayer to the gods of the night (*ilū mušīti*)[185] is an *ikribu*, a type of prayers "assumed to designate a special category of prayer accompanying a nocturnal consultation by the diviner" (Reiner 1995: 76). It is preserved in copies dating to the Old Babylonian period (ca. 2000–1500 BC) onwards, and there is also one version from the Hittite capital Ḫattuša (Boghazkoy).[186] The prayer evokes the quiet night when rituals and divination are performed, and the gods are supposed to assist the diviner (*bārû*, "seer") during his performing act (Koch 2015: 72–75). The purpose of the prayer to the gods of the night is to obtain a propitious sign from the extispicy ritual by addressing several gods (Šamaš, Sin, Adad, Ištar, Ea, Girra and Erra) and celestial bodies.[187] In such a context, stars and planets are considered divine agents and not mere mediators, because they have an active role, together with the anthropomorphic gods, in determining the outcome of the extispicy (Rochberg 2009: 75–76).[188]

---

184 For a study on the ritual, see Læssøe (1955); see Frechette (2012: 176–180) for *šuʾilla*s within the *bīt rimki*. An updated edition of the ritual tablet of *bīt rimki* is given in Ambos (2013: 188–189).

185 For the partiture and the references to all the studies and editions, see online at SEAL n. 7491 (https://seal.huji.ac.il/node/7491 accessed 05.03.2021).

186 The last edition of the Hittite prayer to the gods of the night (KUB 4, 47) is given in Mayer (2018). Brown (2000: 251) stressed that this fragment lists 17 celestial bodies, among which the first five are the planets: this could be a *terminus ante quem* for the identification of Saturn and Mercury. The prayer is also included in a Hittite ritual against depression to be performed at night (Beckman 2007).

187 According to Oelsner and Horowitz (1997–1998: 182–183), the prayer to the gods of the night could represent the earliest attestation of the so-called 10-stars list's tradition, in which one star corresponds to one month of the ideal Mesopotamian calendar (see 4.2.1.).

188 See Fincke (2009) for a discussion on the relationship between the prayer to the gods of the night and nocturnal extispicy.

The following list gives the sequence of the stars named in the prayer to the gods of the night according to its best-preserved fragments, AO 6769 (Dossin 1935) and Erm. 15642 (Horowitz 2000: 195–198), both dating to the Old Babylonian period:

Bow (*qaštum*, ^{mul}PAN)
Yoke (*nīrum*, ^{mul}ŠUDUN; also *elamātum*[189])
Pleiades (*zappum*, MUL.MUL)
True Shepherd of Anu (*šitaddarum*, ^{mul}SIPA.ZI.AN.NA)
Dragon (*mušḫuššum*, ^{mul}MUŠ)
Wagon (*ereqqum*, ^{mul}MAR.GÍD.DA)
She-Goat (*enzum*, ^{mul}ÙZ)
Bison (*kusarikkum*)
Horned-Serpent (*bašmum*)

The Pleiades are named among the gods of the night in AO 6769 (obv. 18: *za-*⸢*ap*⸣*-pu*),[190] and that may constitute another *terminus ante quem* for the identification of *zappu*, “Bristle”, as the name of the Pleiades in Mesopotamia, together with a previously mentioned record from Tuttul dating to the Old Babylonian period (ca. 2000–1500 BC) (see 3.2.).

The gods of the night are also evoked in the context of other rituals. For instance, they appear in the series of rituals *Maqlû*, “Burning” (Abusch 2015: 17–18 ll. 1–36), with the purpose of helping the patient not be attacked by witchcraft at night; or they appear in *namburbi*s, i.e. incantations to avert inauspicious events, as intermediaries between the earth and the sky (Maul 1994: 45–46).[191] One *namburbi* addressed to the gods of the night has a prayer that describes a particular situation to observe in the sky at night, when the incantation was recited:

9. ÉN.É.NU.RU *i-zi-za-nim-ma* DINGIR^{meš} *mu-ši-tim*
10. *a-na ṣi-it pi-ia ú-taq-qá-a* DINGIR^{meš} EN NA[M^{meš}]
11. ^{d}*a-nu* ^{d}*en-lil* ^{d}*é-a ù kal* DINGIR^{meš} GAL[^{meš}]
12. *a-ša-as-si-ki* ^{mul}*dil-bat be-let qab-la-a-*[*ti*]
13. *a-ša-as-si-ki mu-ši-tum kal-la-tum* ^{d}*a-*[*nim*]
14. MUL.MUL *ina* ZA[G]$^{!}$.MU ^{mul}ÉLLAG *ina* GÙB.[MU *i-zi-za*]

9. Enuru incantation: stand by me gods of the night!
10. Heed my utterances, gods, lords of fa[tes],
11. Anu, Enlil, Ea and all the great gods!
12. I call you, Venus, lady of battl[es],

189 The astronomical identification of *elamātum*, as well as *bašmum* and *kusarikkum*, is uncertain (Walker 1983: 146–147).

190 For *zappu*, see also K 2315+ obv. 17 (Oppenheim 1959: 283). For MUL.MUL, see KUB 4, 47 rev. 44 (Mayer 2018: 270) and K 10659 l. 11 (Mayer 1976: 533).

191 E.g. see the *namburbi* against chariot accidents (Maul 1994: 393, 398 l. 31’).

13. I call you, Night, bride of A[nu],
14. Pleiades, [stand] at my righ[t], Kidney at [my] left.
(Maul 1994: 422 obv. 9–14 source A)

These few lines are steeped in analogies. First, the prayer underlines immediately that the gods of the night are the "gods, lords of fates" (DINGIR^{meš} EN NAM^{meš}) because they dispel their decisions through the outcomes of extispicy.[192] The Kidney (here written ^{mul}ÉLLAG, *kalītu*, lit. "Kidney-star") is a reference to one of the organs inspected during the extispicy ritual. Its star can be used in celestial omens as a substitute name for Mars (Koch-Westenholz 1995: 190 l. 106), and two sources of the standard recension of Ura = *ḫubullu* 22 give the word *kalītu*, "kidney", as one of the readings for MUL.MUL (see 2.3.1.). Furthermore, the Moon during the 7th day of its cycle is sometimes called "kidney" (Hätinen 2021: 102–103; see 3.2.4.). Ultimately, in this *namburbi*, the Pleiades and the Kidney represent the two leading divinatory practices of ancient Mesopotamia, that is the celestial divination and the extispicy, respectively performed by the exorcist (*ašīpu*) and the seer (*bārû*) (Koch 2015: 20–23).

Additionally, in one of the copies of this *namburbi* (VAT 8240 rev. 40, see Maul 1994: 428 source A), a ritual was enacted against the *katarru* fungus.[193] The ritual against the *katarru* prescribes nocturnal offerings to the Pleiades (MUL.MUL), and an invocation to the "divine Seven, the great gods" (^{d}IMIN.BI DINGIR^{meš} GAL^{meš}):

ina IGI MUL.MUL KUD-*is-ma mu-uḫ-ra* ^{d}IMIN.BI DINGIR^{meš} GAL^{meš}

Before the Pleiades, he renders "accept, divine Seven, the great gods."
(Maul 1994: 360, 63)

This line endorses an association between the Pleiades and the "divine Seven, the great gods", a group of unidentified great gods, different from the demonic Sebettu (^{d}IMIN.BI), perhaps the major gods of night, gods who have an astral counterpart, and who carry out the role of judges or determiners in heaven, as much as on earth (see 2.3.2., 2.4.2., 3.2.1.5. and further 3.2.5.1.).

3.2.3. Love Incantations

The prayer entitled by Mayer (1976: 432) "Zappu-Muštarīlu 1", i.e. "Pleiades and Mercury", is on a tablet from Aššur (VAT 8251, see KAR 69) dating to the Neo-Assyrian period (ca. 911–612 BC). The rubric of the prayer, ŠÀ.ZI.GA, "rising of the heart", shows that it is a love incantation, a type of incantation that dates to the first millennium BC. The

192 The predictions of omens are, indeed, the will of the gods, their "decisions" (EŠ.BAR, *purussû*) (see 5.1.).

193 The *katarru* is a fungus which grows on the walls of houses. It received particular attention in terrestrial omens (*Šumma ālu* tablet 12, see Freedman 1998: 191–205), likely because it was a threat to human health (Maul 1994: 354–355; Fincke 2018: 208).

purpose of love incantations is usually to make a woman fall in love with a man, or to enhance the sexual potency of men.[194] The prayer "Zappu-Muštarīlu 1" is addressed to the Pleiades – written ^{d}MUL.^{d}MUL – and to Mercury (^{d}GU$_{4}$.UD), in order to make a woman change her mind:

7. [É]N ^{d}MUL.^{d}MUL ^{d}⸢GU$_{4}$.UD⸣
8. [*at*]-*tu-nu* MUL^{meš} ⸢*šá še*$^{!}$-*re-tì*⸣
9. [*šá*] ^{d}*en-líl* ⸢*ib-nu-ku-nu*⸣-[*ši*]
10. [*er*]-*šu*$^{!}$(MA) ^{d}*nu-dím-mud ul-ta*$^{!}$(RU)-*mi*-⸢*ku-nu*⸣-*š*[*i*][195]
11. [*a*]-*šap-par-ku-nu-ši a*-⸢*na* NENNI⸣ A ⸢NENNI⸣
12. ⸢*ša*⸣ *šab-sa-tu* UGU.⸢MU⸣
13. [*l*]*a i-ba-áš-šu*-⸢*ma*⸣ *ina lìb-bi*-[*šá*]
14. ⸢*lid*⸣-*di*-⸢*in*$^{?}$⸣-*an-ni lit-ta-tap-ra-ar*
15. ⸢*ur*⸣-*ra ù* GE$_{6}$ *da-ba-bi lid-bu-ub*
16. *ina qí-bit*$^{!}$(KID) *iq-bu-ú* ^{d}ZÍB ^{d}*iš*-⸢*tar* ÉN$^{?}$⸣

---

17. KÌD.KÌD.BI Ì.UDU ⸢ÉLLAG⸣ UDU BABBAR *ša* ⸢GÙB⸣ TI-*qí* NU-*šá* DÙ-*uš*
18. MU-*šá ina* MAŠ.SÌLA GÙB-*šá* SAR NÍG.NA ^{šim}LI *ana* IGI ^{d}15 GAR-*an*
19. KAŠ BAL-*qí* ÉN IMIN-*šú ana* UGU ⸢ŠID⸣-*n*[*u*] ⸢x x⸣ DU$_{11}$.DU$_{11}$-*ma* ⸢GIN.NA$^{?}$⸣

---

7. [Incan]tation: Pleiades, Mercury,
8. [y]ou are the stars of the morning,
9. Enlil created [you],
10. [the wi]se Nudimmud (i.e. Enki) surrounded yo[u].
11. [I] send you to so-and-so son of so-and-so,
12. because she is angry with me.
13. He i[s n]ot in [her] heart,
14. let her give (it)$^{?}$ to me, may she run aimlessly,
15. day and night may she speak (about) me,
16. At the command (which) the capable, Ištar, said. Incantation$^{?}$.

---

17. Its ritual: you take the fat of the left kidney of a white sheep, you make her figurine,
18. you write her name on her left shoulder, you set up a censer of juniper before Ištar,
19. you pour out beer (and) you recite the prayer seven times before (it) ... she will speak and she will come$^{?}$.

(Biggs 1967: 74 obv. 7–19)

194 For the edition and study of ŠÀ.ZI.GAs, see Biggs (1967).

195 Obv. 9–10 quotes the prayer to the gods of the night; for the reading of these lines, see Mayer (2018: 269).

The prayer is followed by another short incantation (obv. 20'–rev. 1), and one short bilingual incantation to the Pleiades (rev. 2–6) to appease the anger of a woman. The Pleiades are here written in a late pseudo-Sumerian form, ^{d}ZA.BA, for *zappu*, "Bristle":[196]

2. [ÉN ^{d}ZA].⌜BA⌝.KÙ.GA ^{d}ZA.BA.KÙ.GA
3. [AN.D]A.GUB.BA AN.DA.GUB.BA
4. [x x]x MA DÍM.E.DÈ
5. ⌜*an-na*⌝-*ni-tu-ú-a* DUMU.MUNUS *an-na-ni-tú-ú-a*
6. *i-tam-gu-ug* GIM ANŠE-*ma i-tal-su-ma*[197] *ana muḫ-ḫi-ia*

2. [Incantation:] Brilliant [Ple]iades, brilliant Pleiades
3. who are stationed in [the sky], who are stationed in the sky,
4. [...]... he will make.
5. (For) so-and-so, daughter of so-and-so,
6. keep being tense like a donkey! *Keep on running* on me!

(Biggs 1967: 76 rev. 2–6)

The reason that the Pleiades and Mercury are mentioned in these prayers is unclear, but it is not unusual for love incantations to evoke the stars. When the petitioner asks the gods to intercede for him in a nocturnal love context, the stars can act as mediators. Indeed, the ritual enactment of the first prayer quoted above should have been performed before Ištar, probably at night: the rituals to appease the anger of a woman are usually performed before *Ištar-kakkabī* (^{d}15 MUL^{meš}), "Ištar-of-the-Stars", which is a name for the planet Venus (Biggs 1967: 27–30). Regardless of the specific and unclear significance of the Pleiades and Mercury within such a context, love incantations cast light upon the role of the Pleiades as mediators, a role attested in other prayers as well (e.g. see 3.2.1.4. and 3.2.2.).

3.2.4. Polyfunctional Incantations

The cryptic incantation *anāku nubattu aḫāt Marduk*, "I am the vigil, the sister of Marduk", attests the polyfunctionality of the Mesopotamian incantations, but also to celestial bodies being used in the art of the exorcist. The prayer is preserved in copies dating to the Neo-Assyrian (911–612 BC) and Late Babylonian period (ca. 484–30 BC), and belonging to different contexts (e.g. incantations against adversaries, fever, salivation, curses, etc.).[198] The text of the prayer given below is taken from one of the best-preserved fragments:

---

196 A "pseudo" translation (or the pseudo-Sumerian) refers to a tendency of Mesopotamian scholarship to use – expecially in religious or mythological literary texts – highly learned Sumerian, i.e. writings that do not adhere to the Sumerian grammar or syntax but rather reflect the Akkadian scribes' take on the language. For this topic, strictly related to the issues of the Sumerian-Akkadian bilingualism, see e.g. Hallo (1996), Briquel-Chatonnet (1996), Foster (2005: 44–45), and Woods (2006).

197 The form *i-tal-su-ma* is unclear and is not attested elsewhere. Biggs (1967: 77) interpreted it as an Ntn-stem third person plural imperative of *lasāmu*, "to run", referring to the Pleiades to whom the incantation is addressed.

198 The latest edition of the prayer can be found in Abusch-Schwemer (2016: 48–63), but the prayer was

15. ÉN *ana-ku* [*n*]*u-bat-*⸢*tum*⸣ *a-ḫat* ^{d}AMAR.UTU
16. ^{d}*za-ap-pi i-ra-an-ni* ^{d}*bal-*⸢*lum ú*⸣*-li-dan-ni*
17. ^{d}LÚ.ḪUŠ ⸢*ana*⸣ *li-qu-ti il-qa-an-ni*
18. ⸢*ana-áš*⸣*-ši* ŠU.SI.MEŠ.MU *ina bi-rit* ^{d}*za-ap-pi u* ^{d}⸢*bal-lum*⸣ *ú-šeš-šeb*
19. ⸢*ú*⸣*-še-*⸢*eš*⸣*-šeb ina* IGI.MU ^{d}15 GAŠAN GAL-*tum a-pi-lat* ⸢*ki-mu-ú-a*⸣
20. ⸢ŠEŠ ^{d}⸣AMAR.UTU *um-m*[*i*] UD.15.KAM AD-*a* u_4*-mu*[199]
21. KI-*ia-a-ma lip-šu-ru ka-la* [*ta-m*]*a-a-ti*
22. *ma-miti šá at-tem-mu-ú la ú-qar-ra-ab re-mé-nu-ú* ^{d}⸢AMAR.UTU⸣ TU$_6$.ÉN

15. Incantation: I am the [v]igil, the sister of Marduk,
16. the Pleiades conceived me, Bālu gave birth to me,
17. Luḫušû took me as adoptive child.
18. I raised my fingers, I placed (them) between the Pleiades and Bālu.
19. I placed before me Ištar, the great lady, who answers for me.
20. (My) brother is Marduk, m[y] mother is the 15th day, my father is the (1st) day (of the month).
21. May all the [sea]s undo (the evil) with me!
22. May the merciful Marduk not allow near (me) the oath that I have spoken! Incantation formula.

(Abusch-Schwemer 2016: 52–54 ll. 15–22 source n)

The meaning of this prayer is complex, and the analysis provided by Abusch and Schwemer (2016: 62–63) is the most agreeable so far. According to the different rubrics of the prayer, it was used against various types of witchcraft, illnesses (*di'u* and *diliptu* disease), curses (NAM.ÉRIM.BÚR.RU.DA, "for undoing curses", *māmītu*, "curse", "oath") and for healing the kidney (*kalīti šumēli*, "left kidney"). The *nubattu*, "vigil", is the personified vigil of the seventh day of the month. That period of the lunar cycle is called *kalītu*, "kidney", because the shape of the Moon resembles a kidney (Hätinen 2021: 102–103). The vigil of the seventh day is considered a favourable time for the performance of rituals, and it is associated with Ea and the month Nisannu (i.e. Month I), whereas the vigil of the third day of Nisannu is associated with Marduk ("My brother is Marduk").[200]

Line 20 of the prayer ("my mother is the 15th day, my father is the 1st day of the month") refers to the subdivision of the lunar cycle in a lunar month of twenty-eight days (Stol 1992: 253). In addition to that, the Pleiades, Bālu and Luḫušû, among which Ištar/Venus is rising, are said to be the parents of the vigil. While Luḫušû is a name of

already studied and commented on by Stol (1992), Livingstone (1999: 136–137), and Koch (2003). See also Bácskay (2015) for the use of the prayer in magical-medical prescriptions. More recently, a duplicate was published in KAL 10, 53.

199 Instead of AD-*a* u_4*-mu*, "my father is the (1st) day (of the month)", in other sources there is AD ITI, "my father is (the 1st day of) the month" (Abusch-Schwemer 2016: 53 l. 20 source J).

200 DIŠ UD.3.KÁM *nu-ba-tú ša* ^{d}MES, "The 3rd day (of Nisannu, i.e. Month I), the vigil of Marduk"; [DIŠ UD.7].KÁM *nu-bat-tú ša* ^{d}*é-a*, "[The 7]th [day] (of Nisannu), the vigil of Ea" (Livingstone 2013: 108 obv. i 21; 109 obv. i 48).

Nergal (Röllig 1987–1990: 159), Bālu is usually a name of Mars,[201] even though Koch (2003: 94) suggested to identify Bālu with Aldebaran, and Horowitz (2014: 234) suggested him to be Orion.[202] In line 18, the vigil puts her finger between the Pleiades and Bālu ("I raise my fingers, I let them rest between the Pleiades and Bālu"), which sounds like a rule of thumb for the measurement of the distance between two celestial bodies (Koch 2003: 90). Regardless of the identification of the exact stars involved, the prayer is apotropaic, and it describes the nocturnal sky in one particular night, the night of the seventh day of Nisannu. The beginning of Nisannu is also the moment when the Pleiades are in balance with the Moon, a phenomenon which determined the length of the Mesopotamian year (see 4.2.2.).

The performative acts and the offerings related to the prayer "I am the vigil, the sister of Marduk" could be older than the first millennium BC. Indeed, a record from Tuttul testifies that offerings to *zappu*, Pleiades ([mul]*za-ap-pí-im*) and to Bālu ([mul]*ba-li-*[*im*]) were already being performed in the Old Babylonian period (ca. 2000–1500 BC) (see 3.2.). The prayer is also used for therapies of the left kidney in two copies:[203] the association of the Pleiades with the kidney is perhaps unusual, but not unique, as it exists in the lexical lists (see 2.3.1) and in one *namburbi* to the gods of the night (see 3.2.2.).

One last remark about the influence of celestial bodies on mundane events and even medical treatments is needed. The various uses of prayers like "I am the vigil, the sister of Marduk" could be precursors of astrological practices. For instance, in cuneiform archives from Uruk dating to the Late Babylonian period (ca. 484–30 BC), there are many medical texts, as well as unique medical commentaries, in which celestial bodies were thought to influence both the fate of a patient and the effectiveness of a medicament. In one commentary dating to that period, the kidney was considered to be governed by the planet Mars,[204] a statement consistent with the use of a prayer like "I am the vigil, the sister of Marduk" to heal the kidney. The idea that Mars pertained to the kidney even testifies to precursors of planetary *melothesia*, a concept well attested only in Greek astrology (Reiner 1993; Geller 2014). However, during the Late Babylonian period, the influence of the celestial bodies on earth depended heavily on the zodiac as a conceptual and technical framework; this framework was absent at the time that "I am the vigil, the sister of Marduk" was probably composed. Therefore, the texts dating to the Late Babylonian period (ca. 484–30 BC) will be discussed in chapter 6 (see 6.2.2.), even though they likely attest to a *fil rouge* with the previous tradition.

201 See the lexical list Murgud (MSL 11: 40 l. 31).

202 Bālu is a name for Orion ([mul]SIPA.ZI.AN.NA, "True Shepherd of Anu") in two astrolabe fragments dating to the Neo-Assyrian period (ca. 911–612 BC) (Sm 1113 l. 6' // Sm 162 rev. 4, see Horowitz 2014: 230–231). The same is found in a mathematical-astronomical problem (HS 245 and parallels) dating to the Old Babylonian period (ca. 2000–1500 BC), see Oelsner (2005–2006: 119–122).

203 Abusch-Schwemer 2016: 54 J rev. 8, L rev. iv[?] 10.

204 DIŠ NA ÉLLAG-*su* KÚ-*šú* [d]U.GUR *šá* E-*u* / [mul]ÉLLAG : [d]*ṣal-bat-a-nu* "If a man's kidney hurts him (the disease comes from) Nergal. / The Kidney (is) Mars" (Civil 1974: 337 ll. 20–21).

3.2.5. Rituals

As showed in the previous subchapters (see 3.2.1.–3.2.4.), the prayers addressed to the Pleiades were recited within the context of big ritual units, i.e. *bīt rimki*, *bīt salā' mê*, but they were also related to ritual instructions of *namburbi*s and anti-witchcraft incantations. However, the Pleiades also appear in more specific and less-known ritual texts. For instance, a ritual quoted in SAA 20, 32 (KAR 141) refers to the duties of the chief singer of Aššur during the ritual for the "Daughter-of-the-river" (DUMU.MUNUS.ÍD), an apotropaic ritual for the purification from sins. During the afternoon, a priest would go to a river, which symbolises the entrance to the underworld; he would then set up a tent on the bank of the river and gather seven divine judges (^{d}DI.KUD^{meš}): Šamaš, Ištar, Enlil, Aššur, Ea, and two others whose names are not preserved. The sinner would pour water and wine into the river, while simultaneously offerings and cultual songs would be enacted before the gods, like the one mentioned in the following line:

> ⸢MUL⸣.MUL ⸢*aš*⸣-*ṭu-ma mu-šim-*<*mu*> ⸢*i*⸣-*za*-⸢*mur* x⸣ *ú*-[*še*]-*rab*
>
> He b[ri]ngs in ... (and) he sings: "the Pleiades are fierce and determiners (of fate)." (Ebeling 1931: 86–90; SAA 20, 32 rev. 11)

The title of the song quoted in this line is addressed to the Pleiades (MUL.MUL) as "determiners (of fate)", a statement pointing again towards the identification of the Pleiades as divine judges (see previously 3.2.1.5. and 3.2.2.). This aspect will be discussed in more detail in the subchapter given below (see 3.2.5.1.).

3.2.5.1. The Seven Gods Appointed to the Fates

The epithet of the Pleiades as "determiners (of fates)" in the ritual for the "Daughter-of-the-river" (see 3.2.5.) needs further scrutiny. As already noted, for instance, in the prayer "Zappu 5" (see 3.2.1.5.), the Pleiades are sometimes called "determiners of fates"; yet the role of "gods of the fates" is shared, on a broader level, by all the gods of the night because of their ability to influence the outcome of divination (see 3.2.2.). Thus, clarifying the identity of the "determiners of fates" in relation to the Pleiades appears problematic. A first group of "gods who decide the fates, who are seven" (dingir nam tar-ra imin-na-ne-ne) are known from the Sumerian composition "Enlil and Ninlil", a composition dating to the second millennium BC.[205] According to it, the pantheon was conceived as an assembly consisting of the king, seven decrees of fates, and fifty gods (dingir gal-gal ninnu-ne-ne) (Behrens 1978: 27 l. 56, 57). Second, the Sumerian *balag* lamentations, that preserves mythological hints dating to the Old Babylonian period (ca. 2000–1500 BC), also mention "seven gods of fates" together with the Anunnaki and the Igigi (Cohen 1988: 241, c+353; 310, c+227). Third, seven gods of fates are mentioned in the later Akkadian epic of creation *Enūma eliš*:

---

205 For the edition, see Behrens (1978).

80. DINGIR.DINGIR GAL.GAL *ḫa-am-šat-su-nu ú-ši-bu-ma*
81. DINGIR.DINGIR NAM^meš IMIN-*šú-nu a-na* EŠ.BAR *uk-tin-nu*

80. The fifty great gods sat down, and
81. the gods of fates, these seven, were assigned to (make) the decision (lit. verdict).
(Lambert 2013: 114 ll. 80–81)

Considering the three examples mentioned above, the seven gods appointed to the fates are in contexts where a cosmogony is involved. In *Enūma eliš* VI 80–81, the numbers seven and fifty represent both a numerical and cosmogonic hierarchy. More specifically, the Akkadian epic describes the divine order "in terms of a political metaphor of hierarchical cosmic rule" (Rochberg 1999a: 56), according to which Marduk is like the "king" followed by seven gods of fates, the fifty Anunnaki, and three hundred Igigi (Lambert 2013: 114 l. 69, 193–196).

Though the Igigi are three hundred in *Enūma eliš*, the point made by Jacobsen (1977: 116–117) about their etymology is interesting: in the Akkadian myth *Atra-ḫasīs* (Lambert-Millard 1968: 42 ll. 5–6) the Igigi (ᵈ*i-gi-gi*) are called "seven" (*se-be-et-tam*), and the writing for "Igigi" equals the writing for "seven" (i.e. í + gì + gì = 5 + 1+ 1 = 7).[206] Although the differences between the Anunnaki and the Igigi and the numbers associated with them may vary according to different traditions, "Igigi" was likely another designation, perhaps a pun, for the totality of the gods, with "seven" as an attribute for "totality" (see 2.4.2.).

Apart from the Anunnaki and the Igigi, the ability to determine the fate is a feature inherent to all the main gods of the Mesopotamian panthea called *mušimmū šīmti* or *šīmāti*, "determiners of fate or fates" (Lawson 1994: 40–48). It is also a feature of the gods of the night (i.e. the representation of the main gods as celestial bodies), whose purpose is to guarantee the delivery of the decisions (i.e. the fate) of the gods.[207] Therefore, modern scholars suggested that the seven gods of fates are simply the major gods of the panthea who also had a preeminent astral counterpart: Anu, Enki, Enlil, Ninhursag, Nanna, Utu, and Inanna from the Sumerian pantheon (Kramer 1963: 122–123). Or that they are the "seven" planets, which are the Moon/Sin, the Sun/Šamaš, Jupiter/Marduk, Saturn/Adad, Mercury/Nabû or Ninurta, Venus/Ištar, and Mars/Nergal (Annus 2008: 2–7). The names of the seven major gods, who are also the gods of fates, seem indeed variable, as is gained from the examples given below.

---

206 Jacobsen (1977: 116) also considered an alternative reading of the beginning of *Atra-ḫasīs*. He espouses the suggestion of von Soden (1969: 420–421 fn. 2–3), according to which the accusative *sebettam* creates an enjambment with the following line. Thus, *sebettam* is an attribute of the Igigi in *Atra-ḫasīs* I 6, and not of the Anunnaki in line 5, as interpreted by Lambert and Millard (1968: 42–43).

207 A similar "problematic" epithet of the Pleiades appears in a syncretistic *šu'illa* to Marduk: *a-lik* Á^meš^-*ka* MUL.MUL DI *kit-ti mi-šá-ri* / DINGIR^meš 15^meš, "Who go at your side are the Pleiades, the judges of truth (and) justice / (and) of the gods and the goddesses" (Oshima 2011: 387 ll. 11–12).

In the God List An = Anum (Middle Assyrian period, ca. 1400–1000 BC)

151. ^{d}IMIN.⸢BI⸣	DINGIR^{meš} ^{d}IMIN.BI	The divine Seven, seven gods	
152. ^{d}LUGAL.[...]	⸢^{d}*a-nu*⸣	The king[...],	Anu
153. ^{d}⸢LUGAL⸣.[...	^{d}*en-líl*]	The king[... ,	Enlil]
154. ^{d}[...	^{d}*é-a*]	[... ,	Ea]
155. ^{d}[...]	⸢^{d}⸣30	[...],	Sin
156. ^{d}[...	^{d}UTU]	[... ,	Šamaš]
157. ^{d}[...	^{d}IŠKUR]	[... ,	Adad]
158. [...	^{d}MAŠ]	[... ,	Ninurta]

(Litke 1998: 212 ll. 151–158)

In the god list *An = Anum* (VI 151–158), the seven gods (obv. 38' DINGIR^{meš} ^{d}IMIN.BI) are the main gods of the Akkadian panthea: Anu, Enlil, Ea, Sin, Šamaš Adad and Ninurta. The names of the gods have been restored by Litke (1998: 212 fn. 152) according to lines 186–192 of *An = Anum* VI, which give the list of the main seven Elamite gods with their Akkadian equivalents.

In royal inscriptions (Neo-Assyrian period, ca. 911–612 BC)

1. [AN.ŠÁR ^{d}*a-nu* ^{d}BAD ^{d}]*é-a* ^{d}30 [^{d}]UTU ^{d}IŠKUR ^{d}AMAR.UTU
2. [^{d}INANNA ^{d}IMIN.BI DINGIR^{me} GAL]^{me} DÙ-*šu-nu mu-šim-mu šim-ti*

1. [Aššur, Anu, Enlil], Ea, Sin, Šamaš, Adad, Marduk,
2. [Ištar, the Sebettu, the great god]s, all of them, determiners of fate.

(VAG 31 ll. 1–2, see Leichty 2011: 192 Esarhaddon 103)

In the Neo-Assyrian royal inscription VAG 31 of the king Esarhaddon (ca. 680–669 BC) at Nahr el-Kelb, the gods Aššur, Anu, Enlil, Ea, Sin, Šamaš, Adad, Marduk, Ištar, and the Sebettu, addressed as "great gods", are called "determiners of fate" (l. 2 *mu-šim-mu šim-ti*). The names of the gods have been restored according to another royal inscription of Esarhaddon in which every deity is mentioned, described through epithets, and addressed at the end as *mušimmū šīmti* (Leichty 2011: 182 Esarhaddon 98 obv. 1–11).

In the so-called archive of mystic heptads (Neo-Assyrian period, ca. 911–612 BC)

25. ^{d}*a-num*	^{d}⸢x⸣-[x x]	Anu	... [...]
26. ^{d}*en-líl*	^{d}AM[AR.UTU]	Enlil	Ma[rduk]
27. ^{d}*é-a*	^{d}*é*-[x]	Ea	E[...]
28. DINGIR.MAḪ	^{d}*a*-[*ru-ru*]	Bēlet-ilī	A[ruru]
29. ^{d}MAŠ	^{d}[x x x]	Ninurta	[...]
30. ^{d}*gu-la*	^{d}[x x x]	Gula	[...]
31. ^{d}*iš-tar*	^{d}INNIN-TIN.[TIR^{ki}]	Ištar	Ištar-of-Ba[bylon]
32. IMIN DINGIR^{meš} GAL^{meš} *ina sa*-[x x]		Seven great gods in... [...]	

33. *ta-lu-ku ša* ^{iti}⸢BÁR⸣ UD ⸢*á*$^{?}$⸣-[*ki*$^{?}$-*ti*$^{?}$] — procession of Nisannu (i.e. Month I), day of the *a*[*kītu*$^{?}$].

(Pongratz-Leisten 1994: 222–223 obv. ii 25–33)

The so-called archive of mystic heptads (KAR 142) is an explanatory work which lists many heptads of gods, stars, buildings, gates, or streets. In the lines quoted above, the gods Anu, Enlil, Ea, Bēlet-ilī, Ninurta, Gula, and Ištar are the "seven great gods", protagonists in the procession of the *akītu* festival for the beginning of the new year.

In a ritual of E-sagil in Babylon (*Neo-Assyrian period, ca. 911–612 BC*)

15. [*ina u*]*p-šu-ukkin-na-ki ki-sal* UKKIN DINGIR^{meš} *a-šar de-e-ni* [KUR *ib-b*]*ir-ru*
16. [*ina d*]*u*$_{6}$*-kù* KI NAM.TAR.TAR.E.DÈ BÁRA NAM^{meš} ^{d}ŠID.DÙ.KI.ŠÁR.RA *ina* SA[G *ina a-šá*]*-bi-šú*
17. [^{d}EN.Z]A ^{d}MAḪ.ZA ^{d}KI.ZA.ZA ^{d}AM.NA ^{d}UT.U$_{18}$.LU ^{d}EN.ZAG ^{d}[MA.R]U.TU
18. [*ina u*]*p-šu-ukkin-na re-ši-ka lil-lu-ú li-šá-ti-ru* GAL^{meš} *ma-al-k*[*u-u*]*t-ka*

15. [In U]pšukinna, the court of the assembly of the gods, the place where the judgement [of the country is dec]ided,
16. [in D]uku, where the fates are decreed, the shrine of fates, [when] Nabu [si]ts in fr[ont],
17. may [Enz]a (i.e. Anu), Maḫza (i.e. Enlil), Kizaza (i.e. Ea), Amna (i.e. Šamaš), Utaulu (i.e. Ninurta), Enzag (i.e. Nabu) and [Mar]duk
18. raise you (lit. your head) [in U]pšukinna, let them increase (and) extend your kin[gs]hip.

(Lambert 1997: 60 obv. 15–18; see also George 1992: 291)

In the ritual for the Esagil in Babylon, the gods Anu, Enlil, Ea, Šamaš, Ninurta, Nabu and Marduk are sitting "where the fates are decreed" (obv. 16' KI.NAM.TAR.TAR.E.DÈ).

In the so-called Uranology texts (*Late Babylonian period, ca. 484–30 BC*)

8'. DIŠ ^{mul}SAG.ME.GAR ^{mul}*dil-bat* ^{mul}G[U$_{4}$.UD ^{mul}GENN]A
9'. DIŠ ^{mul}UDU.IDIM ^{mul}*ṣal-bat-an-nu* ^{d}30 *u* ^{d}UTU
10'. IMIN DINGIR^{meš} DUMU^{meš} ^{d}*a-num šá ina ri-*⸢*ḫa*⸣*-tú* ^{d}60
11'. *re-ḫu-ú* ^{d}*i-gi*$_{4}$*-gi*$_{4}$ ^{d}⸢*pap-sukkal*⸣
12'. *ma-lik* ^{d}60 *se-bet-ti-šú-nu ina šu-bat* ^{d}60
13'. LUGAL *rab-biš* G[UB.Z]U-⸢*ʾu*⸣*-ma* GABA.RI
14'. *ul i-šu-*⸢*ú*⸣

8'. ¶ Jupiter, Venus, Me[rcury, Satur]n,
9'. ¶ the planet Mars, the Moon, and the Sun,
10'. the seven gods, sons of Anu, who by the seed of Anu
11'. were begotten, the Igigi (and) Papsukkal,
12'. the counsellor of Anu – the seven of them in the dwelling of Anu,

13'. the king, they s[tan]d magnificently, (and) a rival
14'. they have not.
(Beaulieu-Frahm-Horowitz-Steele 2018: 38 D iv 8'–14')

In the so-called Uranology text from Uruk, the planets Jupiter, Venus, Mercury, Saturn, Mars, with the addition of the Moon and the Sun, are called "seven gods, sons of Anu" (rev. iv 10' IMIN DINGIR^{meš} DUMU^{meš} ^{d}*a-num*). This passage is a product of the Late Babylonian tradition from Uruk, and it is primarily an exaltation of Anu's abode (Beaulieu-Frahm-Horowitz-Steele 2018: 57–58).

In essence, it seems that the number seven – meant as "totality" (see 2.4.2.) – is an attribute used for a group of gods appointed to determine the fates of humankind. Just like other heptads, the identities of these gods are not fixed, as they can belong to different local panthea; but these seven gods must be distinguished from the Sebettu (see 3.1.2.1.), who are a warlike and demonic heptad, never addressed as "determiners of fates". Keeping in mind a distinction between the Sebettu and seven gods appointed to the fates, it seems that the Pleiades can very well be associated with the divine Seven. This association is paralleled in some entries from *Enūma Anu Enlil* tablet 51 (see 5.1.8.4. § EAE 50–51). In these entries, the Pleiades are seen as seven gods who can gather in an assembly and make decisions:

25. DIŠ *ina* ^{iti}GU$_4$ MUL.MUL ^{d}IMIN.B[I DINGIR^{meš} GAL^{meš} BAD-*ma ina* UD.D]UG$_4$.GA-*šu-nu* KUR^{meš}-*ni* DINGIR^{meš} GAL^{meš} NÍGIN^{meš}-*ma*
26. GALGA KUR *ana* ^{munus}SIG$_5$ GALGA^{meš} IM[^{meš} DÙG.GA^{meš} DU^{meš} BAD-*ma ina*] *la* UD.DUG$_4$.GA-*šu-nu* KUR^{meš}-*ni* [DINGIR^{meš} GAL^{meš} NÍGIN^{meš}-*ma*]
27. GALGA KUR *ana* ^{munus}ḪUL GALGA^{meš} IM^{meš} [ḪUL]^{meš} DU^{meš} ŠÀ.ḪUL UN^{meš} GAR-*an*

25. ¶ In Ajaru (i.e. Month II), the Pleiades, the divine Seve[n, the great gods. If] they rise [at] their [specified ti]me, the great gods will assemble and
26. give good counsel to the country, [good] wind[s will blow. If] they do not rise [at] their specified time, [the great gods will assemble and]
27. give bad counsel to the country, [evil] winds will blow (and) there will be grief among the people.

(Reiner-Pingree 1981: 58 IX 13, J I 25–27 and K 20'–22'; Horowitz 2014: 174 B.2)

The predictions of the omens quoted above are built on a symbolic association between the Pleiades and the seven gods, lit. "divine Seven" (^{d}IMIN.BI), as great gods, determiners of fates, and not the warlike Sebettu from the Akkadian poem of "Erra and Išum" (see 3.1.2.1.). Therefore, the entries quoted above, the epithets in the prayer "Zappu 5" ("you determine the fate of life", "you decide the verdict of life", see 3.2.1.5.), the address in the ritual instructions against the *katarru* fungus (see 3.2.2.), and the epithet *mušimmū*, "determiners", for the Pleiades in the ritual for the Daughter-of-the-river (see 3.2.5.) are all cases of intertextuality that could be understood as analogies between the Pleiades and various traditions of seven great gods appointed to the fates.

3.2.6. Summary of the Role of the Pleiades within Prayers and Rituals

What emerges from prayers and rituals addressed to the Pleiades is the role of the stars as a whole within the Mesopotamian cultural tradition of the first millennium BC. The Pleiades are either invoked together with many other stars in large ritual units (e.g. *bīt salā' mê* or *bīt rimki*) or alone in various types of prayers; in such contexts, they can be seen either as mediators or as divine agents *tout court* (see 2.2.3.1. § 2), capable of influencing the outcome of a spell, or the extispicy (e.g. see 3.2.2., 3.2.4.).

Thanks to intertextuality, prayers and rituals shed light on the association between the Pleiades and their divine counterparts. For instance, in *šu'illa* prayers (see 3.2.1.1., 3.2.1.4.), the Pleiades are referred to with quotes from the Akkadian poem of "Erra and Išum". They are also called determiners or judges, like the seven great gods of the Mesopotamian panthea appointed of fates (see 3.2.1.5., 3.2.5.1.). The number seven acted as a real catalyst attribute in these sources, and it is the element on which the analogies between the Pleiades and the heptads are built.

3.2.7. Royal Inscriptions

The warlike Sebettu (^{d}IMIN.BI), characters from the Akkadian poem of "Erra and Išum (see 3.1.2.1.), are frequently mentioned in texts of the Neo-Assyrian period (ca. 911–612 BC), when they also begin to be called "the Sebettu, the great gods" (^{d}IMIN.BI DINGIR^{meš} GAL^{meš}). Considering the discussion about the seven gods of fates (see 3.2.5.1.), it seems that from a certain period onwards the Sebettu and seven gods of fates are addressed in the same way, even though they are two different heptads. The two heptads can be recognised according to the context and, mostly, thanks to intertextuality and allusions to other sources. That is the case of Akkadian royal inscriptions, where the Sebettu are invoked as "great gods" by Neo-Assyrian kings, yet they have nothing in common with the seven gods of fates, nor with the Pleiades.

Konstantopoulos (2015: 132–134, 142–149, 176–177) underlined that the Sebettu appeared first in one inscription of Tukulti-Ninurta I dating to the Middle Assyrian period (ca. 1400–1000 BC) in the role of warlike and demonic escorts.[208] The increasing number of the attestations of the Sebettu dating to the Neo-Assyrian period can be attributed mainly to the military propaganda of Neo-Assyrian kings: for them, the warlike Sebettu are at the service of the king, and thus they are elevated in their status as "great gods" (Konstantopoulos 2015: 215). However, they were never, in practice, treated as such, and never had a main cult (Wiggermann 2011a: 464–465).

In Neo-Assyrian royal inscriptions[209] and treatises[210], mainly of Sennacherib (ca. 704–681 BC) and Esarhaddon (ca. 680–669 BC), the Sebettu are called "great gods"

208 E.g. BM 98494 (Grayson 1987: 270 Tukulti-Ninurta I 22 l. 44).

209 Tadmor-Yamada (2011: 83 Tiglath-Pileser III 35 l. i 12; 90 Tiglath-Pileser III 37 l. 9); Grayson-Novotny (2012: 234 Sennacherib 36 obv. 14–16); Grayson-Novotny (2014: 330 Sennacherib 230 l. 2); Leichty (2011: 104 Esarhaddon 48 l. 12; 182 Esarhaddon 98 obv. 10).

210 SAA 2, 2 rev. vi 20; SAA 2, 5 rev. iv 5'; SAA 2, 6 obv. 464.

(DINGIR^{meš} GAL^{meš}), they are always described as *qardūtu*, "warlike" (CAD Q 129–131), armed with weapons, bow and arrow, and marching with the king on the military campaign. The same warlike aspect, together with the ability to travel through the mountains, is found in "Erra and Išum" (see 3.1.2.1.), and partly in the Sumerian mythological composition "Gilgameš and Ḫuwawa" (see 3.1.1.1.). A royal inscription of Shalmaneser III (858–824 BC) quotes this aspect of the Sebettu and thus illustrates again how intertextuality played a huge role in Mesopotamian literature:

1. [*ana* ^{d}IMIN].BI DINGIR^{meš} GAL^{meš} *a-li-li gít-ma-lu-tú ra-ʾi-mu-ut ṣu-ṣe-e mu-*⸢*tal*⸣*-li-ku sa-an-ga-ni ḫa-i-ṭu*
2. ⸢AN⸣-*e* KI-*ti mu-ki-nu eš-re-ti še-mu-ú ik-ri-bi* TI-*ú un-ni-ni ma-ḫi-ru tés-li-ti*
3. [*mu-šam*]-⸢*ṣu*⸣*-ú mal lìb-bi mu-šam-qi-tú za-a-a-ri re-me-nu-tú šá su-pu-šú-nu* DÙG.GA *a-ši-bu-ut*
4. [NINA.KI URU$^{?}$]-*a* EN^{meš} GAL^{meš} EN^{meš}-*a*

1. [To the Sebet]tu, the great gods, noble warriors, lovers of marshes, who roam on mountain passes, who survey
2. the sky and the earth, who maintain shrines, who heed prayers, accept petitions (and) receive requests,
3. [who ful]fil desires, who destroys enemies, compassionate, to whom it is good to pray, who dwell
4. in [Nineveh], my [city$^{?}$], the great lords, my lords.

(Grayson 1996: 153–154 Shalmaneser III 95 ll. 1–4)

This royal inscription of Shalmaneser III is more representative of a cultural propaganda, heritage of the Middle Assyrian military epic of Tukulti-Ninurta I, than anything else. The royal inscription intentionally quotes mythology: the Sebettu are still "who roam on mountain passes", like their Sumerian forerunners were able to go "through the mountain passes".[211] The inscription (l. 1) is also very reminiscent of the prayer "Zappu 5" in obv. 3–4 (see 3.2.1.5.).[212] Again, the Sebettu, even if called "great gods",[213] must be necessarily distinguished from the seven gods appointed to the fates (see 3.2.5.1.), although both heptads can be associated with the Pleiades.

211 ki-a ḫar-ra-an zu-me-eš, "on earth they know the paths"; má-ùr-má-ùr-ḫur-saĝ-ĝá-ka ḫé-mu-e-ni-túm-túm-mu-ne, "through the mountain passes they will lead us (?)" (Edzard 1993: 21 ll. 45, 50).

212 Zappu 5 obv. 3-4: *ra-ʾi-mu ba-am-ma-a-ti* [*mut-tal-li-ku sa-an*]-⸢*ga*⸣*-a-ni*, "lovers of the open country, [who roam on the mo]untain passes"; Shalmaneser III 95 l. 1: *ra-ʾi-mu-ut ṣu-ṣe-e mu-*⸢*tal*⸣*-li-ku sa-an-ga-ni*, "lovers of marshes, who roam on mountain passes".

213 As noted by Wiggermann (2011a: 462–464), sometimes the expression "great gods" (DINGIR^{meš} GAL^{meš}) simply has an appositive usage in sources dating to the first millennium BC, and that applies to the Sebettu as well.

3.3. Conclusion for the Nature of the Pleiades in Myths, Prayers, and Rituals

This chapter discussed the role of the Pleiades and the divine heptads in Mesopotamian literary texts such as myths, prayers, or rituals, looking for possible associations between them. The myths discussed in the first part (see 3.1.–3.1.3.) showed many heptads with astral features or in astral contexts: seven heroes, seven singers or sages, seven heralds, seven torches, the warlike Sebettu, and seven demons. What emerged from the myths is that not all these heptads are associated with the Pleiades. Still, a tendency towards gods seen as celestial bodies – that is, the first perspective on celestial bodies proposed by Rochberg (see 2.2.3.1. § 1) – can be traced in the Sumerian sources. First, the earliest mention of the Pleiades is possibly as seven torches or weapons of Inanna/Venus in the night sky, in "Lugalbanda in the Wilderness" (see 3.1.1.3.). These weapons, called mul.mul, lit. "stars", or "starry", or "Pleiades", are helpers of Inanna/Venus in the myth. Second, there is the heptad of starlike heroes, the seven heroes from "Gilgameš and Ḫuwawa" (see 3.1.1.1.): their ability to guide at night, as attested in the version B of the composition, probably comes from an original use of stars as a tool for orientation. Hypothetically, it cannot be excluded that an analogy with the Pleiades might have been in place in both myths. Indeed, the Pleiades are one of the brightest clusters of stars in the northern hemisphere: they were recorded in conjunction with Venus in celestial omens, and were considered markers of the east in the night sky, at least from the end of the second millennium BC onwards.

In Akkadian sources, one witnesses the opposite perspective of the Sumerian ones, that is the second perspective on celestial bodies established by Rochberg (see 2.2.3.1. § 2): the Pleiades are the counterpart, the representation of the gods in the night sky. They are increasingly perceived as the astral counterpart of seven gods, mainly the Sebettu (^{d}IMIN.BI); the latter are warlike and demonic entities who recall the earlier tradition of the seven heroes (see 3.1.2.1.). The Sebettu became part of the official Assyrian pantheon only during the Neo-Assyrian period (ca. 911–612 BC), by means of Assyrian warlike propaganda (see 3.2.7.). They were never considered, in practice, among the higher gods of the pantheon because their behaviour is more similar to that of demons than to that of gods (see 3.1.3.). Indeed, an astral representation of the Sebettu does not arise from Akkadian mythology, where one witnesses a tendency towards "demonisation". The analogy between the Pleiades and the Sebettu can rather be observed in prayers (see 3.2.–3.2.4.), and further in celestial omens (see 5.3.1.), through quotations of epithets and cases of intertextuality.

The second perspective on celestial bodies is coherent especially in the prayers and rituals dating to the first millennium BC (see 3.2.–3.2.6.). The celestial bodies are perceived as having the apotropaic function of mediators between gods and humankind, or sometimes as divine agents *tout court*. One finds intertextuality especially in *šu'illa* prayers, through straight quotes from the poem of "Erra and Išum", epithets of the warlike Sebettu, or another group of seven great gods, the divine Seven "determiners of fates". These seven gods appointed to the fates are a heptad known from the Sumerian composition "Enlil and Ninlil", or from the Akkadian composition *Enūma eliš*, but actually little is known about their individual identity (see 3.2.5.1.). These seven seem to be associated with the Pleiades by the analogy with the number seven, and a differentiation between them and the Sebettu is possible only according to the context.

The fact that the Pleiades, the Sebettu, the seven gods of fates, or even seven heroes, demons, or torches are present in Mesopotamian mythology at the same time says everything about the relevance of the number seven. The seven, which represents the meaning of "completeness" or "totality" (see 2.4.1.–2.4.2.), worked as a catalyst, especially during the first millennium BC when analogies and metaphors were particularly used by ancient scholars. Thus, the Pleiades were contemporarily associated with the Sebettu and with seven gods of fates, maybe even with seven demons (see 3.1.2.2.2.). However, the fact that a number may constitute a means of association does not imply that all the heptads were associated with the Pleiades: they should be distinguished, for instance, from the heptads of demons and sages, sometimes with astral features, sometimes not, and who are scattered throughout the mythology (see 3.1.3.). The fact that the Pleiades are, theoretically, associated with this or that heptad is a conception that is relevant for the perception of the celestial bodies in Mesopotamian culture as a whole and, most of all, to the meaning of celestial divination and omens, which will be discussed in chapter 5.

3.3.1. Reconsidering Astralisation

A final question concerns how astralisation in ancient Mesopotamia worked and how the Pleiades were associated with more than one divine being. It is essential to underline that the seven torches or pure weapon of Inanna/Venus – beyond the fact that they may be or not be a reference to the Pleiades – are astralised in "Lugalbanda in the Wilderness" because they have astral attributes (see 3.1.1.3.). We could then say that astralisation was possible for anthropomorphic entities and also divine inanimate objects. However, one should not understand astralisation as a catasterism but as part of an overall conception of the cosmos. Specifically, the prayer to the gods of the night (see 3.2.2.) and the *šuʾilla* prayers (see 3.2.1.) shed light on how multifaceted the essence of the divine agency was, according to Mesopotamian culture. Indeed, celestial bodies can have both anthropomorphic and astral attributes simultaneously; a response (i.e. fate) or an influence on the outcome of a spell or a ritual is expected from them, as well as being expected from gods. Thus, celestial bodies are divine or, at least, they can be considered divine agents.

Another phenomenon inherent in Mesopotamian cultural and religious beliefs should be separated from astralisation: the many facets to the cosmos and divine agency are portrayed by the language, analogies, and intertextuality. Theoretically, associations between gods and celestial bodies can be traced in all literary texts. A unique *tamītu* ("oracle-query") to Šamaš and Adad, with a short hymn to a horse, will be a good, final example to show the relevance of analogical reasoning and even textual playfulness:

> *at-ta* ANŠE.KUR.RA *bi-nu-ut* KUR^{meš} KÙ^{meš} / *šar-ḫa-ta-ma i-na kal za-ap-pi* / GIM ^{d}TIR.AN.NA *ina* AN^{meš} *es-ḫe-e-ta*
>
> You (are) the horse, creature of the pure mountains, / you are magnificent among all the *Pleiades* (lit. "Bristle"), / like the rainbow you are assigned in the sky.
> (Lambert 2007: 82 ll. 15–17)

The *tamītu*, ascribed by Lambert (2007: 81) to the Middle Babylonian period (ca. 1400–1100 BC), gives divine and astral attributes to a horse that pulls Marduk's chariot. The horse is seen in its astral form in the sky as one of the stars of a *zappu*, "Bristle", i.e. the Pleiades. In no other text is a horse or a Horse-star (Gössmann 1950: 11 n. 32) associated with the Pleiades; hence, there is no astralisation of the horse. The Pleiades, referred to as *zappu*, "Bristle", are named after the mane, or the hair of Taurus, perceived as a bull in the sky (^{mul}GU$_4$.AN.NA, *alû*, "Bull of Heaven", associated with the Akkadian *is lê*, "Jaw of the Bull"; see further 6.1.1.). Only from the latter statement comes probably the substitution of the bull with the horse, and thus the *tamītu* is built on implicit or encoded analogies. As Cooley (2013: 110) stated, "the mantic power of the night sky was the core concern of those who observed it. These observers were often the same people who created narrative literature, and this literature, more often than not, featured human or divine royalty." There are, indeed, little doubts that the celestial observations heavily influenced the cultural and religious sphere, and *vice versa*.

Chapter 4

The Pleiades in Astral Compositions Dating to the Late Second or the First Half of the First Millennium BC

The Pleiades play an important role in the first so-called astral compositions of ancient Mesopotamia: MUL.APIN (see 4.1.1.), Astrolabe B and related fragments (see 4.1.2., 4.1.3.4.), and the Great Star List (see 4.1.3.–4.1.3.3.). The first part of this chapter (see 4.1.–4.1.4.) aims to discuss the role of the Pleiades by detecting specific features attributed to them in these compositions. These features represent the symbolism of the Pleiades according to the Mesopotamian conception of the cosmos: an asterism framed into a specific section of the sky and whose phenomena are a means to organise the flow of time, space, and to foresee events. Selected divinatory texts are also taken into account, especially celestial omens from the series *Enūma Anu Enlil* and *Šumma Sîn ina tāmartišu*. The second part of the chapter focuses on the role of the Pleiades, both practical and divinatory, as a reckoning device for the Mesopotamian calendar, i.e. for the intercalation system (see 4.2.–4.2.2.1.). The textual sources mentioned in this chapter are mainly preserved in copies dating to the Neo-Assyrian period (ca. 911–612 BC), with a few texts dating to the Late Babylonian period (ca. 484–30 BC) that partly represent earlier traditions.

When reading astral compositions, one should be aware that the sky observed by ancient Mesopotamians was not the same as today, though very similar.[214] The reason for that is the precession of the equinoxes: the shift of the terrestrial axis affects the apparent movement of the Sun with respect to the fixed stars. Assuming that, during the year, the Sun passes along an apparent path among the constellations, the precession causes many crucial terrestrial and celestial phenomena to occur "earlier" than before. For example, in ancient Mesopotamia in ca. 1800 BC, the vernal equinox happened in Taurus, near the Pleiades. Then it moved to Aries, while today it is in Pisces (Koch-Westenholz 1995: 22–23).

4.1. The Knowledge of the Sky

What modern scholars know about how Mesopotamians understood the sky in the first millennium BC mainly comes from scholarly compositions whose tradition dates back to at least the late second or the first half of the first millennium BC: MUL.APIN (see 4.1.1.), Astrolabe B (see 4.1.2.) and the Great Star List (see 4.1.3.). These astral compositions were compendia of data, or works of reference for Mesopotamian scholars because they included

214 "Yet the sky itself has changed only slightly over the millennia separating ourselves from the MUL.APIN astronomers, changes defined primarily by the dates of stellar phenomena, and small changes in the shape and configuration of constellations. The differences we find between MUL.APIN and modern astronomy must therefore be attributed primarily to cultural variation, rather than to variation in the physical phenomena being described." (Watson-Horowitz 2011: 11).

divinatory, religious, and "technical" information, i.e. astronomy-related information, or concerning basic astronomical concepts.[215] In these compositions, there is always a connection between stars and gods, as stars seem to be a self-standing representation of the gods. Such a cultural framework reflects the second perspective on celestial bodies in Mesopotamia (see 2.2.3.1. § 2), and the tendency to consider astralisation and anthropomorphisation as two coexistent and interchangeable *raisons d'être* (Ossendrijver 2016: 144–146). Therefore, the illusory motion, the rising and setting dates of the celestial bodies were also believed to represent the movements and the wills of the gods themselves. From the recurrence of these celestial phenomena, Mesopotamian scholars abstracted rules that helped to foresee future events.

Looking at the more technical information of the astral compositions, the ancient scholars used a terminology based on a specific conception of cosmic geography. According to this conception, the sky is divided into three horizontal bands from east to west, which are called the "paths" of Enlil, Anu, and Ea, from north to south. The sources locate a certain number of fixed stars in these three paths of the night sky, which were used to track the position of the planets, the Moon, the Sun, and even the seasons and the cardinal points in relation to such stars (Horowitz 1998: 252–258).[216] In addition, there also was a path of the Moon, that included 17 fixed stars "touched" by the Moon during its monthly motion (e.g. in MUL.APIN I iv 33–37, see Hunger-Steele 2019: 70; 173–175 Table 2).

On the whole, both the cultural and the technical background of astral compositions testify the so-called "knowledge of the sky"[217] Mesopotamians had at that time, a tie-up between schematisation and divination. Therefore, like for any celestial body, all information about the Pleiades in astral compositions derives from both divination and the schematisation of the repetitiveness of their illusory motion in the sky (i.e. their rising and setting), their position relative to other celestial body, or their visibility throughout the year (see 2.1.).

4.1.1. The Features of the Pleiades in MUL.APIN

The composition MUL.APIN is preserved in copies dating to the seventh century BC onwards. The series was first edited by Hunger and Pingree (1989) and re-edited by Hunger

215 For the meaning of "astronomy" in the context of the astral compositions until the Neo-Assyrian period (ca. 911–612 BC), see 1.4.3.

216 In Astrolabe B, the three paths of the sky are different from those of MUL.APIN. The latter frames the three paths as parallel to the celestial equator; hence, the lists of dates for the rising and setting of stars are roughly astronomically correct. In the former, the listed stars rise and set on different paths, probably because they are arranged according to religious and/or mythological reasons. (Horowitz 2014: 11–15; Hunger-Steele 2019: 170–178).

217 "The goal of the Babylonian scholars can best be called knowledge of the sky without any qualification whether it is a science or not – this question would require a different lecture" (Hunger 2011: 62). For a discussion on the meaning of the modern term science in relation to Mesopotamian astral compositions, see 1.4.3.

and Steele (2019).[218] MUL.APIN is the most complete preserved source about the knowledge of the sky dating to the late second or early first millennium BC, and it consist of a 14-sections eclectic collection of star lists, numerical schemes, and celestial omens (Hunger-Steele 2019: 1–2, 16–19).

Being one of the brightest clusters of stars in the northern hemisphere, the Pleiades are frequently mentioned in MUL.APIN. The first tablet of MUL.APIN has three lists of stars in the paths of Enlil, Anu, and Ea in the sky (see 4.1.1.): the Pleiades are listed in the path of Anu (MUL.APIN I ii 18: 23 MUL^{meš} *šu-ut* ^{d}*a-nim*, "23 stars in the path of Anu"), and they are associated with the "divine Seven, the great gods" (^{d}IMIN.BI DINGIR^{meš} GAL^{meš}, see 2.3.2., p. 30), probably the seven great gods appointed to the fates (see 3.2.5.1.), and not the demonic and warlike Sebettu who are called "great gods" only in Neo-Assyrian royal inscriptions (see 3.2.7.). In addition, the Pleiades are said to be one of the 17 stars in the path of the Moon (MUL.APIN I iv 33–37): during its monthly route across the sky, the Moon passes through all the celestial paths, and also through the region of the Pleiades (Hunger-Steele 2019: 195–196). The list of the 17 stars in the path of the Moon is arranged according to the forward westward motion of the Moon and the planets throughout the year;[219] thus, the list begins with the Pleiades, because the Moon is assumed to be located in or near this asterism at the beginning of the Mesopotamian year, in Nisannu (i.e. Month I).

Other sections of MUL.APIN (I ii 36–iii 48) provide dates and calculations for the heliacal rising, or first visibility, and setting, or last visibility,[220] of circa 35 stars. These lists are drawn from the other lists of stars in the three paths of Enlil, Anu, and Ea (MUL.APIN I i 1–ii 35), and they are more schematic than astronomically reliable from a modern perspective (Hunger-Steele 2019: 179–182). The dates for the rising and the setting of stars within certain months were based on an ideal calendar of 360 days, which did not represent the real everyday luni-solar calendar (see 4.2.1.). In MUL.APIN, it is said that the heliacal rising or first visibility of the Pleiades happens during the month Ajaru (i.e. Month II), while the Scorpion is setting:[221]

DIŠ *ina* ^{iti}GU$_4$ UD.1.KAM* MUL.MUL IGI.LÁ

¶ On the 1st day of Ajaru (i.e. Month II) the Pleiades become visible.
(MUL.APIN I ii 38, see Hunger-Steele 2019: 47)

---

218 Within this study, the quotation from MUL.APIN (i.e. number of tablets and lines) comes from its latest edition (Hunger-Steele 2019).

219 This corresponds, in astronomical terms, to an increasing celestial longitude. The celestial longitude is "the distance along the ecliptic, measured in degrees, from the vernal equinox towards the east" (Koch-Westenholz 1995: 30).

220 The heliacal rising of a celestial body is its first visibility on the eastern horizon before sunrise. On the other hand, the heliacal setting is the last visibility of a star or a planet after sunset (Koch-Westenholz 1995: 30). In MUL.APIN (Hunger-Steele 2019: 190), the heliacal rising or first visibility is indicated by the verb KUR, *napāḫu*, "to rise" (CAD N1 265–266), and the setting by ŠÚ, *rabû*, "to set" (CAD R 51–52).

221 An Astrolabe fragment from the Ashurbanipal's library (K 7931 rev. 16, see Horowitz 2014: 186–187) gives the same order.

DIŠ MUL.MUL KUR-*ma* ^{mul}GÍR.TAB ŠÚ-*bi* / DIŠ ^{mul}GÍR.TAB KUR-*ma* MUL.MUL ŠÚ-*bi*

¶ The Pleiades rise and the Scorpion sets. / ¶ The Scorpion rises and the Pleiades set.
(MUL.APIN I iii 13–14, see Hunger-Steele 2019: 52–53)

DIŠ *ina* ^{iti}GU$_4$ UD.1.KAM* GABA *šá* ^{mul}UD.KA.DUH.A *ina* MÚRU AN-*e* / IGI-*et* GABA-*ka* GUB-*ma* MUL.MUL KUR-*ḫa*

¶ On the 1st day of Ajaru (i.e. Month II), the chest of the Demon with the Gaping Mouth in(to) the middle of the sky / opposite your chest it stands still, and the Pleiades rise.
(MUL.APIN I iv 15–16, see Hunger-Steele 2019: 65)

The last of the quotations given above places the Pleiades among the so-called *ziqpu*, "apex", stars, i.e. the culmination of a star, or when it reaches its highest point in the sky, culminating simultaneously in the heliacal rising of another star. In the section about the *ziqpu* stars (MUL.APIN I iv 10–30), it is told to the reader, in second-person singular, to observe the stars by standing before sunrise with the cardinal point west to the right, and the east to the left. Therefore, "opposite your chest" means a star that culminates crossing the meridian (Hunger-Steele 2019: 190).

The rising and setting dates of stars were also used as a guideline to define the compass points in the night sky (Horowitz 1998: 198–200). Indeed, in another entry of MUL.APIN, the position of the Pleiades and the Old Man (^{mul}ŠU.GI, *šību*, i.e. modern Perseus) defines the eastern compass point as the point from where the east wind rises (see 3.1.1.1., pp. 46–47). This statement is explainable by looking at the apparent motion of the Pleiades in the sky from an ancient Mesopotamian observer's perspective. At the beginning of the Mesopotamian year, the Pleiades became visible in the eastern sky. In the second part of the year, they were visible in the western sky; later, they disappeared for around 40 days, to be visible again in the east. Thus, the position of the Pleiades at the beginning of the Mesopotamian year defined the cardinal point east. On the contrary, a wind rising and blowing from the opposite side, the west wind, defined the cardinal west.

In a schematic way, MUL.APIN frames the Pleiades (MUL.MUL) in a specific section of the sky in a specific period (i.e. path of Anu, path of the Moon, heliacal rising in Ajaru, i.e. Month II, compass point east). At the same time, it says that they are (associated with) the "divine Seven, the great gods" (^{d}IMIN.BI DINGIR^{meš} GAL^{meš}). As will be further discussed, such information – or features – of the Pleiades are found in other astral compositions as well (see 4.1.2., 4.1.3.).

4.1.2. Three-stars-each: the Features of the Pleiades in the Astrolabe B

The compositions called Three-stars-each (MUL^{meš} 3.TA.ÀM) and Astrolabe B [222] (abbreviated Alb B) are star calendars, probably composed during the second millennium BC since the tablet on which the text of Astrolabe B is found (VAT 9416) was written in the twelfth century BC (Horowitz 2014: 4–6). More specifically, the Three-stars-each is a list that names three stars – assigned to each of the three paths of the sky – for each month of the year, resulting in a list of 36 stars. On the other hand, the Astrolabe B has a bilingual menology[223] and star catalogues, based on the same list of 36 stars given in the Three-stars-each texts. The most complete edition of the sources related to Three-stars-each and the Astrolabe group has been presented by Horowitz (2014).[224]

The way in which the menology of Astrolabe B provides information about the knowledge of the sky is slightly different from MUL.APIN. For instance, in both MUL.APIN and Astrolabe B the first visibility of the Pleiades takes place in the month Ajaru (i.e. Month II); yet the text of Astrolabe B dwells on the activities performed during that month:

19. ^{iti}GU$_4$ *za-ap-pu* ^{d}IMIN.BI DINGIR^{meš} GAL^{meš}
20. *pe-tu-ú er-ṣe-ti*
21. GU$_4$^{meš} *ul-te-eš-še-rù*
22. *ru-ṭu-ub-tu up-ta-ta*
23. ^{giš}APIN^{meš} *ir-ra-aḫ-ḫa-ṣu*
24. ITI ^{d}*nin-gìr-su qar-ra-di*
25. *iš-šá-ak-ki* GAL-*i ša* ^{d}*en-líl*

19. Ajaru (i.e. Month II), the Bristle, the divine Seven, the great gods.
20. The openers of the earth,[225]
21. the bulls are set straight,
22. the wet land is opened up (i.e. is made arable),
23. the ploughs are washed.
24. (This is) the month of Ningirsu, the hero,
25. the great farmer of Enlil.

(Alb B I i 19–25, see Horowitz 2014: 34)[226]

---

222 The name "astrolabe" is used in Assyriology to define a group of cuneiform tablets. They have nothing to do with the antique circular instrument of the same name. The name "astrolabe" was first used by George Smith (1875: 407–408) to describe the fragment CT 33, 11, a tablet with the drawing of a circular diagram (Horowitz 2014: 3, 9).

223 In Assyriology, the term "menology" refers to a monthly-arranged composition, in which each month is associated with stars, gods, festivals, mythology, or various aspects of human and social life (Livingstone 1997: 59).

224 Within this study, the quotations from Astrolabe B (i.e. number of columns and lines) come from this latest edition. See previously Casaburi (2003) and Kolev (2013).

225 Or, as understood by Horowitz (2015: 34 l. 20), "the opening of the earth".

226 The Sumerian version of the same lines (Alb I i 12–18) is not quoted here since its content is only

Like in MUL.APIN, in Astrolabe B the Pleiades (*zappu*) are associated with the "divine Seven, the great gods" (^{d}IMIN.BI DINGIR^{meš} GAL^{meš}) and the month Ajaru (i.e. Month II); nevertheless, Astrolabe B goes on to describe the mundane activities associated with the month Ajaru, and proves to be a text more focused on religious and cultural details than MUL.APIN. In the lines quoted above (Alb B I I 20–25), the ploughing oxen in the watered fields symbolise the agricultural practice of ploughing flooded fields in the early spring, which was likely done by performing the following activities (Horowitz 2014: 53, 57–61):

- opening the soil
- harnessing the oxen
- ploughing the flooded fields
- washing the ploughs

These activities were performed in order to leach out the accumulated salinity (*idrānātu*) and soften up the ground for the ploughing. From a religious perspective, the heliacal rising of the Pleiades in Ajaru (i.e. Month II) reflects the cultic activities related to the god Ningirsu as the god of agriculture and war. Ningirsu/Ninurta is considered the divine author of the "Farmer's instructions",[227] a Sumerian expository composition dating to the second millennium BC. This composition is a collection of advice about farming and cultivation throughout a year; according to it, the month Ajaru is the period when the bulls are driven to the fields (Civil 1994: 79 l. 38; Verderame 2016: 113), and this is mirrored in the information given by the Astrolabe B (Alb B I I 20).

The compositions MUL.APIN and Astrolabe B share the association of the Pleiades with the "divine Seven, the great gods" and the heliacal rising in Ajaru (i.e. Month II); nonetheless, they come from different scholarly backgrounds and have a different purpose: MUL.APIN collects more technical information, i.e. astronomy-related information, like ideal dates and calculations for stars' rising and setting dates, whereas Astrolabe B includes a menology, so it is more focused on cultural information. Additionally, MUL.APIN was composed later than Astrolabe B, and this might be confirmed not only by philology but also by the precession of the equinoxes (see 4.) because solstices and equinoxes are placed one month later than the ones of Astrolabe B (Horowitz 2014: 16–18; Hunger-Steele 2019: 17–19). However, the precession of the equinoxes is relatively "slow", i.e. 1° every 72 years. It is improbable that a shift could have been potentially noted in such a "short" period: for the precession of equinoxes to be perceived, one would have to assume a gap of circa 2000 years between the information preserved in MUL.APIN and Astrolabe B. It is perhaps more likely that MUL.APIN and Astrolabe B simply attest to two different coexisting traditions.

slightly different from the Akkadian version. For a discussion on the philological differences between the two versions, see Horowitz (2014: 57–58).

227 For the first edition, see Civil (1994).

4.1.3. The Features of the Pleiades Compared to Divinatory Texts, the Great Star List, and Other Astrolabe Fragments

The following subchapters (see 4.1.3.1.–4.1.3.4.) attempt to collect the features of the Pleiades from other astrolabe fragments, the Great Star List, and divinatory texts like the series *Enūma Anu Enlil* in order to highlight similarities and discrepancies between the sources. These texts provide – in respect of MUL.APIN and the Astrolabe B (see 4.1.1.–4.1.2.) – additional and sometimes different information concerning the characteristics of the Pleiades. Sparse data from all compositions do not necessarily parallel each other, as different scholarly traditions coexisted, not only over time, but even at the same time, in different places. For instance, the Great Star List, dating to the first millennium BC, is a list of stars equated to divine entities or other synonyms, and it contains a huge variety of mythological and astronomical data on planets and stars.[228] It is still unclear to what extent there is a formal relationship between celestial divination, the Great Star List, MUL.APIN, and Astrolabe B: the rules for the interpretation of signs in celestial divination are given through written omens (see 5.1.); the Great Star List is likely a type of commentary in the form of a list (Fincke 2016: 127); MUL.APIN and Astrolabe B are compositions in their own right. Nevertheless, all these textual sources share the same plateau of information, so intertextuality can be traced among them.

4.1.3.1. Ajaru (i.e. Month II) and the Agricultural Season

The association of the Pleiades and the divine Seven with the month Ajaru (i.e. Month II) and the agricultural season is paralleled in both MUL.APIN (see 4.1.1.) and Astrolabe B (see 4.1.2.). The same is preserved in the menology of the assumed *Enūma Anu Enlil* tablet 51 (see 5.1.8.4. § EAE 50–51):

> DIŠ MUL.MUL *ina* ^{iti}GU$_4$ IGI-*mar* BAD-*ma* [MUL BI NIM-*ma* IGI] BURU$_{14}$ KUR SI.SÁ KUR SIG$_5$ IGI
>
> ¶ The Pleiades appear in Ajaru (i.e. Month II). If [their star(s) are visible early], the harvest of the country will prosper, the country will see good things.
> (Reiner-Pingree 1981: 60 X 2)

In this entry, the first visibility of the Pleiades after the beginning of Ajaru is a favourable ominous sign for the fieldwork, because it matches the beginning of the agricultural season in spring. The prediction given in the entry is quite straightforward, and it cannot be

228 The first edition was published by Weidner (1959–1960), and a more recent edition which includes new sources was published by Koch-Westenholz (1995: 187–205). Now, a new edition is being prepared by Fincke and Horowitz. An exact dating for the Great Star List is difficult, yet it might have been composed in the first half of the first millennium BC, and it was copied later, until the Late Babylonian period (ca. 484–30 BC) (Fincke 2016: 127).

excluded that the redactors of *Enūma Anu Enlil* tablet 51 worked with Astrolabe B as the main reference (Horowitz 2014: 181–182).

Two sources, the omen series *Iqqur īpuš* and a *Lipšur* litany known from a Late Assyrian copy from Nimrud, preserve something different than the sources mentioned so far. In the entries given below, it is said that the divine Seven are the gods of the Adaru (i.e. Month XII), and not Ajaru (i.e. Month II) as customary:

Iqqur īpuš

DIŠ ^{iti}ŠE *šá* ^{d}IMIN.BI DINGIR^{meš} GAL^{meš}

¶ Adaru (i.e. Month XII) (is the month) of the divine Seven, the great gods.
(Labat 1965: 196 § 105, 12)

Lipšur litany

^{iti}ŠE.GUR$_{10}$.KU$_{5}$ *lip-šur* ^{d}IMIN.BI DINGIR^{meš} GAL^{meš}

May Adaru (i.e. Month XII), (the month) of the divine Seven, the great gods, absolve.
(Wiseman 1969: 178, 59')

Considering the discrepancy between Ajaru and Adaru, and the precession of equinoxes, one could suggest that *Iqqur īpuš* and the *Lipšur* litany were composed far later than the other sources. Nevertheless, as already explained in 4.1.2., for such an hypothesis to work one would have to assume a difference of more than two millennia between the information given in MUL.APIN, Astrolabe B, and the series *Enūma Anu Enlil* on the one hand, and *Iqqur īpuš* and the *Lipšur* litany on the other hand, which is unlikely. Therefore, it is safer to assume that *Iqqur īpuš* and the *Lipšur* litany attest to a different religious tradition concerning the association between months and gods compared to other sources.

4.1.3.2. The Paths in Which the Pleiades Move

In MUL.APIN, the Pleiades are listed among the stars of the path of Anu (see 4.1.1.), whereas in Astrolabe B the Pleiades are placed in the path of Ea:

MUL *ša* EGIR.BI GUB-*zu* / MUL.MUL ^{d}IMIN.BI DINGIR^{meš} GAL^{meš}
(...)
[12] MUL^{meš} *šu-ut* ^{d}*é-a*

The star which stands behind it (i.e. the Field)[229] / (is) the Pleiades, the divine Seven, the great gods.

---

229 For the location of the Pleiades in the sky next to the Field (^{mul}AŠ.GÁN, *ikû*), see also SAA 8, 412 obv. 7–rev. 3: [DIŠ 30] TÙR NÍGIN-*ma* ^{mul}AŠ.GÁN *šá* EGIR-*šú* ⸢MUL⸣.[MUL] / *ina* ŠÀ-*šú* GUB ÁB.GU4$^{ḫi.a}$ *šá* KUR SI.SÁ / ^{mul}*ṣal-bat-a-nu* MUL KUR MAR^{ki} / ^{mul}AŠ.GÁN *šá* EGIR-*šú* MUL.MUL ^{mul}LÚ.ḪUN.GÁ, "[If the Moon] is surrounded by a halo and the Field, which is behind the Plei[ades],

(...)
[The 12] stars of Ea.
(Alb B II i 5–6, rev. 10, see Horowitz 2014: 37)

Another text for the Pleiades in the path of Anu is in CT 33, 9, a text dating to the Neo-Assyrian period (ca. 911–612 BC), which has a repertoire of 36 stars in a format similar to Astrolabe B, but the paths assigned to the stars agree with MUL.APIN (Hunger-Steele 2019: 172, 176–177):

[^{mul}*a-n*]*u-ni-tum* MUL.MUL (...) [1]2 MUL^{meš} *šu-ut* ^{d}*a-nim*

[An]unitu, the Pleiades. (...) [The 1]2 stars of Anu.
(Horowitz 2014: 210 rev. 9')

Taking into account that Astrolabe B and MUL.APIN attest to two different traditions, CT 33, 9 corroborates a relationship with both or, less likely, information which was more coherent with the contemporary situation at that time, considering the precession of the equinoxes (Koch 2015: 188–189).

4.1.3.3. The Pleiades Connected with the East and with Elam

According to the Mesopotamian conception of cosmic geography, several stars are related to the four quarters of the world or compass points because an ideal subdivision into four areas is applied to the sky as well as to the earth (Horowitz 1998: 259). It has already been mentioned how, in MUL.APIN, the Pleiades are considered markers of the eastern compass point (see 4.1.1.). Seen from a Babylonian observer's point of view, Elam is located in the east, therefore it is associated with its respective celestial compass point (Horowitz 1998: 87, 175–177, 199–200, 324–325).

These kinds of associations (e.g. Pleiades, east, and Elam) are the basis for the list of 36 stars of Elam, Akkad, and Amurru that is found in the composition Great Star List. In the latter, the list of the 12 stars of Elam was reconstructed as follows:

[^{mul}*dil-bat*	MUL.MUL
[^{mul}UR.GU.LA	^{mul}MAŠ.TAB.BA
^{mul}PAN	^{mul}UGA^{mušen}]
^{mul} [E]N.[TE].NA.[BAR.ḪUM]	^{mul}GÍR.TAB
[^{mul}UD.KA.DUḪ.A]	^{mul}GU.LA
^{mul}N[U.MUŠ.DA	^{mul}KU$_{6}$]
12 MUL^{meš}	KUR ELAM.MA^{ki}

/ stands inside it, the livestock of the land will prosper. / Mars is the star of Amurru; / the Field, which is behind the Pleiades, is Aries."

[Venus	Pleiades
Lion	Twins
Bow	Raven
Mo[use]	Scorpion
[Demon with the Gaping Mouth]	Great One
Nu[mušda	Fish]
The 12 stars	of Elam

(Koch-Westenholz 1995: 196, 197 ll. 201–207; Horowitz 2014: 208–209)

The list is mostly broken, especially in the first section; nevertheless, it is reconstructable though the comparisons with the other preserved stars associated with the lands of Akkad and Amurru (Great Star List ll. 208–221), and thanks to a fragment naming Elam, Akkad, and Amurru stars in a different order than the Great Star List. This fragment (81-7-27, 81) is a commentary of the series *Šumma Sîn ina tāmartišu*, according to its subscript.[230] As remarked by Horowitz (2014: 208), who edited this fragment, the sequence of stars in it probably follows their heliacal rising times as preserved in the Astrolabe B, and not their association with a land as in the Great Star List.

As is shown below, the commentary fragment preserves a mention of the Pleiades as "stars" of Elam:

[...] MUL.MUL MUL KUR ELAM.MA^{ki} ^{mul}ŠU.G[I...]

[...] The Pleiades (are) the star(s) of Elam. The Old M[an...].
(Horowitz 2014: 207 obv. 4')

Moreover, based on the Great Star List, and on the rest of this fragment from *Šumma Sîn ina tāmartišu* mentioned above, Horowitz (2014: 207) reconstructed the partial list of the stars of Elam, composed by six names only, according to the latter fragment as follows:

dilbat	Venus
MUL.MUL	Pleiades
^{mul}MAŠ.TAB.BA	Twins
^{mul}PAN	Bow
^{mul}UGA^{mušen}	Raven
^{mul}EN.TE.NA.BAR.ḪUM	Mouse

230 For an overview of the series *Šumma Sîn ina tāmartišu* and the commentaries of *Enūma Anu Enlil*, see 5.1.1.–5.1.3.4.

4.1.3.4. Other Features of the Pleiades in Fragments of the Astrolabe Tradition

This subchapter presents attestations for features of the Pleiades that suffer from remaining unparalleled, or whose origin is unclear. This is due, again, to the fact that different sources represent different traditions, according to their text genre and chronology. The astral compositions usually do not "explain" but only collect information, and this makes it even more difficult to get to the core of each feature. Only an intertextual approach to the sources can eventually be helpful in solving the issue, or formulating hypotheses.

The astrological text LBAT 1499 (Horowitz 2014: 124–139), dating to the Late Babylonian period (ca. 484–30 BC), gives information taken from the Astrolabe B but also from the micro-zodiac tradition (see 6.2.1.), as an anthology. This fragment also lists omens about the Akkadian *mešḫu*, "meteor", or an unusual glow.[231] The omens give numerical values corresponding to the length of night hours for the rising time of a specific constellation.[232] The entry about the Pleiades has the following:

> DIŠ *ina* ^{iti}GU$_4$ *ina še-rim* MÚL.MÚL 3,40 *meš-ḫu im-šuḫ* GU$_7$-*ti* DINGIR MU.3.KAM 3 ITI <*ina*> KUR GÁL
>
> If in Ajaru (i.e. Month II) during the morning, the Pleiades, 3,40,[233] produce a glow, there will be consumption by a god for three years and three months <in> the country. (Horowitz 2014: 127 obv. ii 14)

LBAT 1499 is an eclectic collection of information, not all of which is easy to understand. Yet, the relationship between the "consumption by a god" (GU$_7$-*ti*, *ukulti*) in the apodosis of the omen, and the Pleiades in the protasis, is not unique. When compared to other sources, such juxtaposition is found in many celestial omens – like the one quoted below – which will be presented and further discussed in chapter 5:[234]

> DIŠ ^{mul}LU.LIM *ana* MUL.MUL KUR-*ud* GU$_7$-*ti* ^{d}IMIN.BI
>
> If the Stag reaches the Pleiades, (there will be) consumption by the Sebettu. (K 5713+ obv. 13', see SIT 6: 10 source A and parallels)

---

231 The omens likely refer to an unusual luminosity of the stars (Horowitz 2014: 135–136). For a discussion on *mešḫu*, see 5.1.4. commentary to the text of entries 37–43, p. 166.

232 The same system of measurement by means of a water clock appears in the circular Astrolabes (CT 33, 11 and 12) and in MUL.APIN (II ii 21–40). For an overview of the meaning of these values, see Horowitz (2014: 18–19). For the water clock, see Brown-Fermor-Walker (1999) and Brack-Bernsen (2005).

233 The value 3,40 refers to a fixed value assigned by LBAT 1499 to the Pleiades as a star in the path of Ea first visible in Ajaru (i.e. Month II), and that follows an increasing zigzag function starting from the winter solstice (i.e. Month IX, fixed value 2) and culminating in the summer solstice (i.e. Month III, fixed value 4).

234 For an overview, see directly 5.3.–5.3.2.

Like LBAT 1499, an earlier Astrolabe fragment, BM 82923 (Horowitz 2014: 139–152), dating to the Neo-Babylonian period (ca. 626–484 BC), provides material in the form of a list over three columns. In this fragment, celestial bodies are associated with gods, divine weapons, or divine epithets:[235]

[DIŠ MUL.MU]L 3,40 GAG.ŠAR ^{d}*šár-ur*$_{4}$

[¶ The Pleiade]s, 3,40, the peg of Šarur.
(Horowitz 2014: 140, 4)

In the case of the Pleiades from the line quoted above, the "peg" (GAG.ŠAR) is probably to be understood as some kind of a pointed tool (GAG, *sikkatu*, "nail", and *edēdu*, "to be pointed", see MSL 9: 130 XXXVII 315, 319). It may refer to the pointed shape of the stars of the Pleiades in iconography (e.g. see 2.3.3. Figure 4), but the connection with Šarur, one of the divine weapons of Ninurta (Krebernik 2011), is difficult to understand. The sign SAR can also be understood as *mūšaru*, "garden plot" (Klein-Sefati 2020: 95 l. v 27, 127). Consequently, the translation of GAG.SAR ^{d}*šár-ur*$_{4}$, "the garden tool of Šarur", could also fit, as suggested by Horowitz (2014: 146). In MUL.APIN, Šarur is described as a star related to a section (i.e. the sting) of the Scorpion, the constellation which sets when the Pleiades rises heliacally and *vice versa* (see 4.1.1.):

DIŠ 2 MUL^{meš} *šá ina zi-qit* ^{mul}GÍR.TAB GUB^{meš}*-zu* / ^{d}*šár-ur*$_{4}$ *u* ^{d}*šár-gaz*

¶ The two stars which stand in the sting of the Scorpion / (are) Šarur and Šargaz.
(MUL.APIN I ii 31–32, see Hunger-Steele 2019: 45–46)

It is difficult to trace back an origin for the perception of the Pleiades as a pointed tool, and the sources never explicitly display such meaning. One can only speculate: the fragment BM 82923 might allude to an unknown-to-us mythological tradition relating to the rising of the Pleiades, and to Šarur in the Scorpion (Horowitz 2014: 146).

Two fragments, HS 1897 dating to the Middle Babylonian period (ca. 1400–1100 BC), and BM 55502 dating to the Late Babylonian period (ca. 484–30 BC) (Horowitz 2014: 101–120), first edited and discussed by Oelsner and Horowitz (1997–1998: 17), attest to a tradition of a 30-stars list combined with astrological commentaries.[236] This tradition shares several affinities with Astrolabe B, despite the fact that the Three-stars-each tradition is

235 According to Walker, Bromhead and Hunger (1977) this is a Three-stars-each text dating to the Late Babylonian period (ca. 484–30 BC). According to the colophon, it is a *mukallimtu*-commentary "to establish and set aright (the length) of night and day" (Horowitz 2014: 141, 37).

236 The 30-stars list preserves 10 stars for each of the three paths of the sky, while the Three-stars-each tradition establishes 12 stars for each path. Horowitz (Oelsner-Horowitz 1997–1998: 179–180) argued that the assumed *Enūma Anu Enlil* tablet 51 was the main reference for the fragment BM 55502, dating to the Late Babylonian period (ca. 484–30 BC), and that the 30-star lists "reflects a second millennium tradition of listing stars in groups of 10 which lived on into the first millennium in texts such as BM 55502" (Oelsner-Horowitz 1997–1998: 182).

based on a 36-stars list. In the 30-stars lists, the Pleiades are associated with a favourable price for the grain:

> DIŠ MUL *ša* EGIR-*šu* GUB-*zu* MUL.MUL ^{d}IMIN.BI DINGIR.GAL.GAL.E.NE *a-na* KI.LAM ŠE *da-ma-qí*
>
> ¶ The star which stands behind it (i.e. the Field) is the Pleiades, the divine Seven, the great gods. *It is favourable* for the price of barley (lit. for the price of the barley to be good).
> (Horowitz 2014: 102–103, Ea 2 B 3)

The association of increasing market price for barley with the Pleiades mirrors the relationship between their heliacal rising and the agricultural season. In the apodoses of celestial omens from the series *Enūma Anu Enlil*, the Pleiades are occasionally related to market rates, especially to variations of crop (see Commentary EAE 53: 1, 5, 6, 8; SIT 6: 1, 3, 7; see also 5.3.). The predictions for the business through celestial phenomena is also a common practice during the Late Babylonian period (ca. 484–30 BC), as testified by a few astrological texts (Ossendrijver 2019). Thus, the fragment dating to the Middle Babylonian period (HS 1897) might constitute an interesting forerunner for the other sources, and for the practice of predicting market rates through the stars, which will be further discussed in 6.2.3.

4.1.4. Summary of the Features of the Pleiades in Astral Compositions

Regardless of the unparalleled cases shown in 4.1.3.4., the majority of sources gives the following features of the Pleiades (MUL.MUL, *zappu*): they are associated with the "divine Seven, the great gods" (^{d}IMIN.BI DINGIR^{meš} GAL^{meš}), a divine heptad whose origin is to be sought in the seven gods of fates (see 3.2.5.1.); they first appear in the eastern sky in the month Ajaru (i.e. Month II), and so mark the east, the land of Elam by juxtaposition, and the beginning of the agricultural season in spring. What is gained from the sources discussed so far is the close relationship between astral compositions, divination and, partly, mythology. The juxtaposition of the cultural and technical framework in these sources is consistent with a multifaceted perception of the cosmos, and a modality of building, transmitting, and transforming knowledge – also the "knowledge of the sky" (see fn. 217) – based on analogical reasoning (see 2.2.3.3.). The astral compositions are so linked to a cultural background that the former cannot be considered compendia of technical rules only but tools complementary to the divination. At the same time, the predictive nature of astral compositions will arise from the parallelisms between the sources so far compared – regardless of their differences in format, language, or time – and the celestial omens discussed in chapter 5. Such a nature will begin to emerge in the next part of this chapter (see 4.2.–4.2.2.1.), where the most crucial characteristic of the Pleiades – from the point of view of "knowledge of the sky" – will be discussed: their involvement in the intercalation.

4.2. The Pleiades in the Intercalation Systems

On account of their long-lasting visibility, the Pleiades used to be a reckoning device for the Mesopotamian calendar, at least until the first half of the first millennium BC. The following subchapters (see 4.2.–4.2.2.1.) present various so-called rules concerning the appearance of the Pleiades together with the Moon, which allegedly marked the length of an upcoming year.

4.2.1. The Mesopotamian Calendar: a Short Overview

Probably already during the third millennium BC, an ideal or schematic calendar was developed in Mesopotamia. It was originally used for administrative purposes in the Ur III period (ca. 2110–2000 BC), to simplify calculations of rations, prices, and interests (Englund 1988). The ideal calendar was based on a year of 360 days divided in 12 months of 30 days each, and it became a tool to adapt the stars' rising and setting dates, and to detect discrepancies (Steele 2011). It also became a derivational scheme for bad or good ominous signs.[237]

From the third to the first millennium BC, the ideal calendar coexisted with luni-solar calendar (Brack-Bernsen 2007): the latter was built on both the Sun and the Moon's path, and operated in civil and cultic contexts (Steele 2011: 471–473). The Sun's apparent revolution around the earth takes circa 365 days (i.e. solar year); at the same time, the Mesopotamian administration was based on the concept of 12 synodic months (i.e. 12 new or full Moons) that last 354,4 days, resulting in a lunar year which is 10,6 days shorter than the solar year. Additionally, the Babylonian month was a lunar month: each month had either 29 or 30 days because it spans from the first visibility of the crescent Moon after the conjunction with the Sun until the next crescent can be detected.[238] Thus, the first visibility of the crescent determined the beginning of a new month. If the Moon was not seen due to poor weather conditions, the new month was declared after 30 days (Hunger-Steele 2019: 7–8). Considering the discrepancies between the lunar and the solar year, and the difficulties in determining a fixed length for the year and the month, the Mesopotamian experts had to find their ways to cope with such discrepancies, especially to keep the administrative year in line with the seasons.

4.2.2. Intercalation Systems

Given the differences between the lunar year and the solar year, in order to avoid the seasons shifting through the years, Mesopotamian experts added an extra month of 30 days

---

237 For an overview of Mesopotamian calendars, also outside their calendrical purpose, and previous bibliographical references, see Stern (2012: 71–124).

238 The first visibility of the crescent Moon is observed during the evening, when the crescent is seen for the first time after the new Moon, i.e. the conjunction of the Sun and the Moon according to astronomical terminology (Koch-Westenholz 1995: 25–26, 103).

from time to time. The process of inserting any additional time, be it a day, a week, or a month, into calendar years is called intercalation; in Mesopotamia, an intercalation of a complete month was in place. There was no strictly predetermined moment for adding this month, as seasonal events may also depend on meteorological factors. Nevertheless, it seems that the extra month was added circa every three years (Steele 2011: 475–478) already from the Ur III period (ca. 2110–2000 BC) onwards (Englund 1988: 122–126; Brack-Bernsen 2007: 88–90). The extra or intercalary month was called *atru* (DIRI), from Akkadian *atāru* "to exceed" (CAD A2 487–492). Sometimes, the need for intercalation was also indicated by the expression "to be left behind" (TAG4, *ezēbu*) (CAD E 416–426): a year "left behind" (TAG4-*et*, *ezbet*) is an intercalary year, or the year when intercalation will be necessary.

We can only assume the ways to calculate intercalation since there was no fixed rule, but many sources that date to the first millennium BC show that the intercalation sometimes involved the observation of the Pleiades together with the Moon. However, there is no written evidence regarding how and if such intercalation was applied in everyday life; only from the reign of Xerxes (485/484–465/464 BC) onwards, is there reliable evidence that intercalation froze in a cycle of seven intercalations every 19 years (Ossendrijver 2018: 138–151).

To calculate the intercalation, the position of the Moon and the Pleiades was observed, mainly during the first few days of the first month (i.e. Nisannu), in order to check if they were "balancing": in Akkadian the technical term for that is *šitqultu* (LÁL).[239] The balancing of the Pleiades with the Moon is comparable, in modern astronomical terms, to the Pleiades and the Moon having the same celestial longitude (see fn. 219) or being in conjunction. Yet, the word "balancing" (*šitqultu*), or the verb "to weigh" (*šaqālu*, CAD Š2 8b) from which "balancing" is derived, are used either to indicate a specific date for a conjunction of the Moon and the Pleiades (see § 1–2b below), or any of the times in which the Moon is expected to be seen close to the Pleiades in one year (see 2.1.; § 2c below).

Regardless of a possible, more practical function of checking the "balancing" for the intercalation, its ultimate purpose was divinatory. This statement is made in the "Diviner's Manual" (Oppenheim 1974), a composition dating to the seventh century BC, a "guide" for ancient scholars to interpret ominous signs and their relationship with each other. The main issue addressed by the "Diviner's Manual" is the establishment of the validity of an ominous sign, and to do so a proper date of the calendar should have been chosen. Thus,

239 The word *šitqultu* (from *šitqulu*, Gt-stem of *šaqālu*, "to weigh", see CAD Š2 8b) has been discussed by many scholars. Papke (1984) argued that LÁL should be strictly interpreted as "to be in conjunction with", in altitude. Koch (1997: 94–101) suggested that the term refer to the measurements by water clocks. Brown (2000: 118–119) argued that the word *šitqultu* should be understood in astronomical sources as "to be equally weighted", or "to be in conjunction", and its occurrence in MUL.APIN would not imply that the meaning is different, but just that the intercalation rule itself was not meant to be mathematically accurate (see also Ratzon 2016: 147). Hunger and Pingree (1999: 23 fn. 26), and Williams (2002: 478–479) suggested to translate *šitqulu* as "to be in balance". It is likely that the correct translation for LÁL, at least in the context of MUL.APIN, is "to be balanced" or "to be in balance", like the meaning of "having the same celestial longitude" in modern astronomy (Hunger-Steele 2019: 210–213).

before interpreting an ominous sign, the diviner should check if intercalation is needed or not:

> *šit-qul-ti šá* ⸢MUL⸣.M[UL *u*] ⸢^{d}30⸣ ÙRU-*ma li-pu-ul-ka-ma*
>
> Observe the balancing of the Pleiad[es and] the Moon, and it may give you an answer. (Oppenheim 1974: 200, 62)

According to the "Diviner's Manual", the diviner should have used both the relevant omen compendia and astral compositions like MUL.APIN, in order to perform celestial divination. This implies that the so-called rules to calculate the intercalation given in astral compositions did not simply detect or adjust deviations from the ideal calendar, but aid divination.[240]

The intercalation systems involving the Pleiades are collected, discussed below as rules (§ 1–2c), and then summed up in Table 2 (see 4.2.2.1.).

§ 1. The Intercalation and the Pleiades in MUL.APIN

The composition MUL.APIN dedicates two sections to rules for calculating the intercalation. One section (MUL.APIN II i 9–24) includes the checking of the rising dates of certain stars, the position of the Sun and the Moon, the length of day and night, and the equinoxes and solstices. Through the deviation from these rules, intercalation should be extrapolated. A second section (MUL.APIN II GAP A 8–ii 17) contains criteria for determining the intercalation by checking the "balancing" between the Moon and the Pleiades, the first visibility of the Pleiades and other stars, and fragmentary mathematical calculations. The sections about intercalation in MUL.APIN can be summed up by the following statement about an assumed triennial cycle for intercalation:

> *ina* 3 MU^{meš} ITI DIRI.GA *ta-qab-bi* / *ina* 12 ITI^{meš} 10 UD^{meš} DIRI^{meš} ŠID-*at* MU.AN.NA
>
> In three years (or the 3rd year) you declare an intercalary (lit. in excess) month. / In 12 months ten days are in excess, (this is) the amount of one year. (MUL.APIN II ii 11–12, see Hunger-Steele 2019: 89)

According to these two lines, the computation to calculate an intercalary month provides that every 12 months there must be "ten days in excess", for a total amount of 30 days in three years; that is, the amount of an intercalary month that must be added to the calendar

240 This has been noticed by Brown (2000: 113–122), who also wrote that "the 'ideal intercalation schemes' of MUL.APIN and related texts do not attest to a particular interest on the part of the Mesopotamian astrologer-astronomers in regulating the luni-solar calendar, despite many such opinions to the contrary. Texts such as The Babylonian Diviner's Manual indicate that the 'ideal intercalation schemes' served a largely divinatory, rather than calendrical purpose, and merely evoked an existing rule of thumb" (Brown 2000: 195).

after three years (Hunger-Steele 2019: 213–215). This statement is true for any of the rules in MUL.APIN (see § 1a–c) and not only there (see § 2): all rules place the "shifting" of any celestial phenomena – explicitly or not – one month later than their ideal date according to the ideal calendar (Brown 2000: 118–119). This statement is also supported by the structure of the early ideal calendar of the Sumerian administrative year (Englund 1988: 123–126; Brack-Bernsen 2007: 89–90).

§ 1a. *Plejaden-Schaltregel*

The second section about intercalation in MUL.APIN (II GAP A 8–ii 17) includes the most discussed rule concerning the Pleiades and the Moon; this is the intercalation system known as the *Plejaden-Schaltregel*, first discussed by Schaumberger (1935: 340–344):

8. [DIŠ *ina* ^{iti}BÁR$^{?}$ UD.1$^{?}$.K]AM MUL.MUL *u* ^{d}30 *šit-qu-lu* MU BI GI.NA-*ta*
9. [BAD-*ma*$^{?}$ *ina* ^{iti}BÁR$^{?}$ U]D.3.KAM MUL.MUL *u* ^{d}30 LÁL MU BI DIRI-*át*

8. [If on the 1^{s}]$^{\text{t?}}$ [day of Nisannu$^{?}$ (i.e. Month I)] the Pleiades and the Moon are in balance, this year is normal.
9. [If] on the 3rd [d]ay of [Nisannu$^{?}$ (i.e. Month I)] the Pleiades and the Moon are in balance, this year is intercalary (lit. in excess).

(MUL.APIN II GAP A 8–9, see Hunger-Steele 2019: 85)

According to the way this rule is restored, the Moon and the Pleiades should be in balance on the 1st day of Nisannu (i.e. Month I) of the ideal calendar. If the balancing between the Moon and the Pleiades happens later, likely on the 3rd day of Nisannu, then intercalation must take place. The dates for this rule (i.e. [1^{s}]$^{\text{t?}}$ [day of Nisannu$^{?}$]; 3rd [d]ay of [Nisannu$^{?}$]) are restored by Hunger and Steele (85, 211) based on a first, unproven interpretation of the lines by George Smith (1875: 405), in the second half of the nineteenth century, and on the compatibility with another rule (see § 1c) in MUL.APIN, assuming that – in simple words – the shift in the balancing should happen every three years (Hunger-Steele 2019: 209–213). Nevertheless, it cannot be excluded that, according to another rule (see § 2a), the dates could be differently restored as follows:

8. [DIŠ *ina* ^{iti}BÁR UD.3.K]AM MUL.MUL *u* ^{d}30 *šit-qu-lu* MU BI GI.NA-*ta*
9. [BAD-*ma*$^{?}$ *ina* ^{iti}GU$_{4}$ U]D.3.KAM MUL.MUL *u* ^{d}30 LÁL MU BI DIRI-*át*

8. [If on the 3^{r}]^{d} [day of Nisannu (i.e. Month I)] the Pleiades and the Moon are in balance, this year is normal.
9. [If] on the 3rd [d]ay of [Ajaru (i.e. Month II)] the Pleiades and the Moon are in balance, this year is intercalary (lit. in excess).

The dates "3rd day of Nisannu" and "3rd day of Ajaru (i.e. Month II)" (see § 1c) could fit instead of "1st day of Nisannu" and "3rd day of Nisannu". Indeed, other rules (§ 1b–1c) show how this or that celestial phenomenon related to the Pleiades must "shift" one month later compared to an ideal calendar date – the date for the "normal year" – in order to declare an "intercalary year", i.e. to add an extra month into the calendar. Such shift of a whole month should occur every three years, as it is noted by MUL.APIN itself (see above) and in other compositions (see § 2a).

§ 1b. First Visibility of the Pleiades

The second section about intercalation in MUL.APIN II GAP A 8–ii 17) includes one set of rules in which the Pleiades are mentioned. These rules (MUL.APIN II GAP A 10–II ii 8) deal with the first visibility in the sky at sunset (i.e. acronychal rising) of five constellations:[241] if these stars are visible on their ideal dates, the year is normal; if they are visible one month later, intercalation was necessary (Ratzon 2016: 146; Hunger-Steele 2019: 214–216). Based on the heliacal rising dates given in MUL.APIN (I ii 36–iii 12), the rule concerning the Pleiades was restored as follows by Hunger and Steele (2019: 150; previously in Hunger-Pingree 1989: 90):

10. [DIŠ *ina* ^iti^GU$_4$ UD.1.KAM] MUL.MUL IGI.LÁ MU BI GI.NA-*ta*
11. [BAD-*ma ina* ^iti^SIG$_4$ U]D.1.KAM MUL.MUL IGI.LÁ MU BI D[IRI-*át*]

10. [If on the 1st day of Ajaru (i.e. Month II)] the Pleiades become visible, this year is normal.
11. [If] on the 1st day [of Simanu (i.e. Month III)] the Pleiades become visible, this year is in[tercalary (lit. in excess)].

(MUL.APIN II GAP A 10–11, see Hunger-Steele 2019: 85–86)

According to the restoration of Hunger and Steele (2019: 210), the acronychal rising of the Pleiades is expected at the beginning of Ajaru (i.e. Month II), i.e. it coincides with the heliacal rising or first visibility date (see 4.1.1.). When the date shifted by one month (i.e. Simanu, Month III), intercalation was necessary. This rule is recalled in celestial omens, where the rising of the Pleiades "in time", i.e. according to the date of the ideal calendar, or "in late", i.e. diverging from the ideal calendar, produces a positive or a negative prediction (see, e.g. 5.1.8.4. § EAE 50–51, 5.2.4., 5.3.2.). For this rule to work, one shall again assume a triennial cycle for the intercalation.

§ 1c. The Pleiades, the Moon, and the Autumnal Equinox

The second section about intercalation in MUL.APIN also has rules for the intercalation based on the autumnal equinox (i.e. the 15th day of Tešritu, Month VII) in relation to the Moon and the Pleiades:

1. [DIŠ *ina* ^i^]^ti^DU$_6$ UD.15.KAM MUL.MUL *u* ^d^30 LÁL MU BI [GI.N]A-*át*

---

241 I.e. ^mul^KAK.SI.SÀ, *šukūdu*, "Arrow", ^mul^ŠU.PA, *nameru*, "Resplendent One", ^mul^KU$_6$, *nūnu*, "Fish", ^mul^ŠU.GI, *šību*, "Old Man". All dates for their acronychal rising are identical to the heliacal rising, except for the Arrow that received special attention (Hunger-Steele 2019: 181–182).

2. [BAD-*m*]*a ina* [iti]APIN UD.15.KAM MUL.MUL *u* [d]30 LÁL MU BI DIRI-*át*

1. [If on] the 15th day of Tešritu (i.e. Month VII) the Pleiades and the Moon are in balance, this year is normal.
2. [If] on the 15th day of Araḫsamnu (i.e. Month VIII) the Pleiades and the Moon are in balance, this year is intercalary (lit. in excess).

(MUL.APIN II ii 1–2, see Hunger-Steele 2019: 87)

The autumnal equinox happens when the length of day and night is the same,[242] when the Sun crosses the equator, which can be on any day approximately between Tešritu (i.e. Month VII) and Araḫsamnu (i.e. Month VIII). The Moon on the 15th day is a full Moon, so opposite the Sun. When the equinox shifted by one month (i.e. Araḫsamnu, Month VIII), intercalation was necessary. On an astronomical level, it seems that this rule and the *Plejaden-Schaltregel* (see § 1a) in MUL.APIN are more or less compatible. As explained by Hunger and Steele (2019: 211–212), assuming that on the 15th day of Tešritu the Moon is opposite the Sun in a normal year, on the 1st day of Nisannu (i.e. Month I) the Moon would be fairly in conjunction with the Pleiades; on the 3rd day of Nisannu, the Moon would have moved in a way that it would already be beyond the Pleiades. By contrast, in an intercalary year – three years later after a normal year – the Moon would be beyond the Pleiades only after the 3rd day of Nisannu. This rule would also be compatible with a different restoration of the *Plejaden-Schaltregel* than the one proposed by Smith and followed by Hunger and Steele: presuming that the dates should be restored as "3rd day of Nisannu", and "3rd day of Ajaru", both the rules would imply that the shift of one month for the fixed dates should occur every three years.

§ 2. The Intercalation and the Pleiades in the Series *Enūma Anu Enlil* and *Šumma Sîn ina tāmartišu*

Other intercalary schemes involving the Pleiades can be gained from divinatory texts, such as the series *Enūma Anu Enlil* and the serialised commentary *Šumma Sîn ina tāmartišu.*[243] As witnessed in the rules from MUL.APIN (see § 1), one shall assume a span of three years between a "normal year" and an "intercalary year", i.e. the addition to an intercalary month to the calendar.

§ 2a. The Pleiades and the Moon in Nisannu (i.e. Month I)

Three fragments dating to the Neo-Assyrian period (ca. 911–612 BC) show a combination of rules which does not exactly agree with MUL.APIN, though the rules are compatible (Hunger-Steele 2019: 212–213). The first fragment (K 3123, source A) is part of the assumed *Šumma Sîn ina tāmartišu* tablet 4. The second fragment (K 3923 + K 6140 + 81-7-

242 Due to the precession of the equinoxes, in ancient Mesopotamia the autumnal equinox was in Libra, in modern times it is in Virgo.

243 For an overview of the series *Enūma Anu Enlil* and *Šumma Sîn ina tāmartišu*, see 5.1.1.–5.1.3.4.

27, 149 + 83-1-18, 479, source B) has excerpts from the series *Enūma Anu Enlil*. The obverse of this fragment was partially edited by Hunger and Reiner (1975: 22–23), together with its duplicates, and it shows several rules for intercalation (see § 2c). The text of both fragments (K 3123 and K 3923+) is paralleled by another fragmentary tablet (K 2254, source C), belonging to the series *Šumma Sîn ina tāmartišu*.

List of Sources

Source:	A
Siglum:	K 3123
Copy:	Plate 43
Edition:	Wainer (2016: 57–58) (only rev. 9'–18'); see also App. B § 24 (rev. 1'–20', 25'–27')
Provenance:	Nineveh
Ductus:	Neo-Assyrian
Measurements:	8,3+ x 9,4+ x 1,2+ cm
CDLI n.:	P394818 (with photograph)
Description:	Fragment of the middle and lower part of the reverse of a tablet, with part of the right edge preserved. The obverse is lost.

Source:	B
Siglum:	K 3923 + K 6140 + 81-7-27, 149 + 83-1-18, 479 (see Commentary EAE 53 source G and App. B § 11)

Source:	C
Siglum:	K 2254
Copy:	ACh Sin 22; AAT 5; Plate 44
Edition:	ACh Sin 22; see also App. B § 25 (obv. 1, 6–8, 10, 13–17 and rev.)
Provenance:	Nineveh
Ductus:	Neo-Assyrian
Measurements:	7,6+ x 6,4+ x 2,5+ cm
CDLI n.:	P394296 (with photograph)
Description:	Fragment of the upper right part of a tablet, with part of the upper and right edge preserved. The scribe started to write the obverse on the convex surface of the tablet, which is usually assigned to the reverse.

Text

1.	**DIŠ MUL.MUL *pa-ni-ma* $^{(d)}$30 *ar-ki* MU BI *eš-ret***
A rev. 21'	[DIŠ MUL$_4$.M]UL$_4$ *pa-ni-ma* 30 ⸢*ar-ki* MU⸣ BI *eš-ret*
B rev. 1	DIŠ MUL.MUL *pa-nu-ma* ^{d}30 *ar-*[*ki*...] ⸢x x⸣ [...]
C obv. 2	[... M]UL.MUL *pa-ni-ma* 30 *ar-ki*

[If the P]leiades are ahead and the Moon is behind, that year is normal.

2. **(:) DIŠ MUL.MUL *ar-ki-ma*** (var.: ***ar-ku-ma***) **$^{(d)}$30 *pa-ni* MU BI TAG$_4$-*et***
A rev. 21’ : DIŠ MUL$_4$.MUL$_4$ *ar-ki-ma* 30 *pa-ni* MU BI TAG$_4$-⸢*et*⸣
B rev. 2 DIŠ MUL.MUL *ar-ku-ma* {^{d}}30 *pa-*[*ni*...] ⸢x⸣ [...]
C obv. 3 [... 30 *pa-n*]*i-*⸢*ma*⸣ MUL.MUL *ar-ki*

(– var.) If the Pleiades are behind and the Moon is ahead, that year is left behind (i.e. will be intercalary) (C obv. 3: [... the Moon is ah]ead and the Pleiades are behind).

3. **[DIŠ *ina* ^{iti}]BÁR UD.3.KÁM MUL$_4$.MUL$_4$ *u* 30 *ta-mur-šú-nu-ti-ma***
A rev. 22’ [DIŠ *ina* ^{iti}]BÁR UD.3.⸢KÁM⸣ MUL$_4$.MUL$_4$ ⸢*u* 30⸣ *ta-*⸢*mur*⸣*-šú-nu-ti-ma*

[If] you observe the Pleiades and the Moon [on] the 3rd day of Nisannu (i.e. Month I) and

4. ***šit-qul-lu* MU BI *eš-ret*** **[DIŠ MIN$^{?}$ *ip-pa*]*l-si-ḫu*[244] TAG$_4$-*et***
A rev. 22–23’ *šit-qul-lu* MU BI *eš-*⸢*ret*⸣ (23’) [DIŠ MIN$^{?}$ *ip-pa*]*l-si-ḫu* TAG$_4$-⸢*et*⸣
C obv. 4–5 [...] *eš-re-et* (5) [...] TAG$_4$-*et*

they are in balance, that year is normal. [If ditto$^{?}$ (i.e. you observe the Pleiades and the Moon on the 3rd day of Nisannu and) they are] apart (lit. fall down), it is left behind (i.e. will be intercalary).

5. **(DIŠ) MU.1(.KAM) SI.SÁ MU.2(.KAM NÍG.)SI.SÁ MU.3.[KA]M**
A rev. 23’ MU.1 SI.SÁ MU.2 NÍG.SI.SÁ MU.3
B rev. 3 DIŠ MU.1.KAM* SI.SÁ MU.2.KAM* SI.SÁ MU.⸢1⸣[+2.KA]M*

(¶) The 1st year is normal, the 2nd year is (almost) normal, the 3rd year

6. **TAG$_4$-*et ina mi-ni-i lu-mur-mi ta-qab-bi***
A rev. 23’ TAG$_4$-*et ina mi-ni-i lu-mur-mi ta-q*[*ab-bi*][245]
B rev. 3 ⸢TAG$_4$-*et ina mi*⸣*-n*[*i-i lu-mur-mi ta-q*]*ab-bi*

is left behind (i.e. will be intercalary). You will say: “in what can I see (this)?”.[246]

---

244 The Akkadian verb *napalsuḫu* (CAD N1 271–272), lit. “to fall down”, is understood in this context as “to be apart”, with a meaning similar to “having the same celestial longitude”, according to modern astronomy (Hunger-Reiner 1975: 24).

245 For *taqabbi*, “you say”, or “you declare”, in commentaries, see Gabbay (2016: 257–260) and Rochberg (2004: 272).

246 The reading and the translation of line 6 was suggested to the author by Hermann Hunger.

7. ***ina* ITI.1.KAM UD.1.KAM DIRI (*ina*) MU(.1.KAM) UD.12.KAM (var.: 12 u_4-*mi*)**

A rev. 24'. [*ina* ITI.1.KÁM] UD.1.KÁM DIRI MU 12 u_4-*mi*

B rev. 4 *ina* ITI.1.KAM* UD.1.KAM* DIRI *ina* MU.1.K[AM*] ⸢UD.12.KAM*⸣

In one month (or: 1st month) one day is in excess, (in one) year (var.: 1st year) 12 days (are in excess),

8. **(*ina*) MU.3.KAM ITI DIRI AN.GE_6**

A rev. 24' MU.3.KÁM ITI DIRI AN.[GE_6]

B rev. 4 *ina* MU. ⸢3⸣.K[AM* ITI DIRI A]N.GE_6

(in) the 3rd year a month is in excess, (then there will be an) eclipse.

The first rule (entries 1–2) does not mention any specific day or month, and it explains the rough "visual" difference of the balancing between the Moon and the Pleiades, expressed by the Pleiades being ahead (or in time) and behind (or late) in respect of the Moon.[247] The second rule (entries 3–4) explains that if the Pleiades were seen in balance on the 3rd day of Nisannu (i.e. Month I), the intercalation is not needed. This statement is different from the rule § 1a, as restored by George Smith and then Hunger and Steele. Ratzon (2016: 149–150) argued that the first rule (entries 1–2) should be instead referred to the second rule (entries 3–4): if on the 3rd day of Nisannu the Pleiades are ahead and the Moon is behind, then the year should be intercalary; if there is the opposite situation (i.e. the Pleiades behind and the Moon ahead), then the year is normal. In this respect, the rules in MUL.APIN (see § 1) and the fragments quoted above agree on one of the days in which the balancing should be checked (i.e. 3rd day of Nisannu, if the original restoration in MUL.APIN is correct). The fact that they would not agree on the exact day of the conjunction would only be consequential to the schematic nature of these rules. The fragments continue (entries 5–8) recalling the fluctuation of the intercalary month in the span of about three years: the statement of the "12 days in excess" for one year diverges from MUL.APIN (see § 1), where "ten days in excess" every three years amount to an intercalary month of 30 days.

§ 2b. A "Visual" Description of the Position of the Pleiades and the Moon

The rule § 2a is presented in a different way and explained in a few entries of a commentary to *Enūma Anu Enlil* tablet 53 and, in a shortened way, in *Šumma Sîn ina tāmartišu* tablet 6.[248] All the sources and their parallels will be entirely presented in score

247 In other words, the difference corresponds to the celestial longitude or latitude (Hunger-Steele 2019: 210).

248 The reading of the entry from *Šumma Sîn ina tāmartišu* tablet 6 was suggested to the author by Jeanette C. Fincke. The other entries have also been translated and commented on CAD (P 99a; N1 209 1b), and Hunger-Steele (2019: 212–213).

transliterations, translations, and commentaries in the chapter 5 (see 5.1.5.–5.1.6.), as they belong to the topic of celestial divination and omens.

List of Sources

A: K 3558 (see Commentary EAE 53 source A)
B: K 8744 (see Commentary EAE 53 source B)
C: LB 1321 (see SIT 6 source B)
D: Sm 1054 (see Commentary EAE 53 source H)
E: Sm 247 (see Commentary EAE 53 source I)
F: 81-2-4, 135 (see Commentary EAE 53 source J)
G: 79-7-8, 271 (see Commentary EAE 53 source L)

Text

1. **DIŠ *ina* SAG MU MUL.MUL *ka-ri-it ina šit-qul-ti* ^{d}30**
A obv. 19 DIŠ *ina* SAG MU MUL.MUL *ka-*[*ri-i*]*t ina šit-qul-*[*t*]*i* ^{d}30
B 6’ [DIŠ *ina* SAG MU MU]L.MUL *ka-ri-it* [...]
C rev. 32’ DIŠ *ina* SAG MU MUL.MUL *ka-rit*
D obv. 5’ [...]x uḫ *ina šit-qul-ti*

If at the beginning of the year the Pleiades are cut. (It means that) at the balancing with the Moon

2. ***im-ma-rak-ku-ma*** **MUL.MUL** ***ar-ku-ma*** **$^{(d)}$30** ***pa-ni-ma*** **[^{d}*za-ap*]*-pu***
A obv. 20–21 ⌜*im*⌝*-ma-rak-ku-ma* (21)MUL.MUL *ar-ku-ma* ⌜^{d}⌝[30 *pa-ni-ma*$^{?}$ ^{d}*za-ap-p*]*u*
C rev. 32’ 30 *pa-ni-ma* MU[L.MUL *ar-ki*...]
D obv. 5’ *im-ma-*[*rak-ku-ma*...]
G 5’ [... ^{d}*za-ap*]*-pu*

they lag behind, the Pleiades are behind and the Moon is ahead (C rev. 32’: the Moon is ahead and the Pl[eiades are behind...]), [the Brist]le

3. ***ana*** (var.: ***ina***) **^{d}UTU.È (^{d}30)** ***ana*** (var.: ***ina***) **^{d}UTU.ŠÚ(.A MIN) x[...]**
A obv. 20 *ana* ^{d}UTU.È ^{d}30 *ana* ^{d}UTU.ŠÚ.A
G 5’ *ina* ^{d}UTU.È <*ina* ^{d}UTU>.ŠÚ MIN x[...]

(is) to the east (i.e. towards sunrise), (the Moon is) to the west (i.e. towards sunset) (var.: ditto (?) ...[...]).

4. **DIŠ *ina* SAG MU MUL.MUL *šá-ti-iḫ ina šit-qul-ti* $^{(d)}$30 *i-pan-nu-ma***
A obv. 21 DIŠ *ina* SAG MU MUL.MUL *šá-*[*ti-iḫ ina šit-qul-t*]*i* ^{d}30 *i-pan-nu-ma*
B 7’ [DIŠ *ina* SAG MU MU]L.MUL *šá-ti-iḫ* [...]
D obv. 3’ [...] *ina šit-qul-t*[*i*...]
E obv. 8’ [DIŠ *ina* SAG MU MUL.M]UL *šá-ti-iḫ*
F rev. 4, 6 DIŠ MUL.MUL *šá-ti-iḫ* (6) *ina šit-qul-ti* 30 *i-*[*pan-nu-ma*]

G 6' [DIŠ *ina* SAG MU M]UL.MUL *šá-ti-i*[*ḫ*...]

If at the beginning of the year the Pleiades are elongated. (It means that) at the balancing with the Moon they are ahead.

These entries explain two "impossible" protases, i.e. protases whose literal meaning is astronomically impossible, or use – or are interpreted as made by – a non-literal and metaphorical language.[249] First, the scribe of the omens explains that the statives of *karātu*, "to strike" or "to cut", and *šatāḫu*, "to be elongated", correspond to the Pleiades respectively being behind and ahead of the Moon. Second, it is explained that "behind" and "ahead" mean that the Pleiades are seen either to the east or to the west of the Moon. These entries are based on a dichotomous scheme, a scheme which will be further discussed in the framework of celestial omens (see 5.1.5.1. commentary to the text of entries 19–23, pp. 192–193; A.2.1.2.). When comparing the rule § 2b with the rule § 2a, it can be concluded that if at the beginning of the Mesopotamian year, i.e. on 3rd day of Nisannu, Month I, the Pleiades are westward, the year was normal; it they are eastward, the year was intercalary.

§ 2c. The Pleiades and the Moon throughout the Year

Another collection of rules about the intercalation from commentaries of *Enūma Anu Enlil* tablet 53 and *Šumma Sîn ina tāmartišu* provides the means to check the position of the Pleiades and the Moon throughout the year. The obverse of the Neo-Babylonian commentary K 3923+ (obv. 6'-18b', see App. B § 11), listed as source B in § 2a, expands the rule of § 2b with an intercalation system based on whenever the Moon is close to the Pleiades in one year, i.e. once a month (see 2.1.). This rule is composed by twelve entries for each month individually, and it can be summed up as follows:

DIŠ *ina* ^iti^x UD.x.KAM MUL.MUL *u* ^d^30 IGI-*šú-nu-ti-ma iš-taq-lu* MU BI *eš-re-et ip-pal-si-ḫu* TAG$_4$-*et*

If on the xth day of month x you observe the Pleiades and the Moon, and they are balancing, that year is normal; (if) they are apart (lit. fall down), it is left behind (i.e. will be intercalary).

The "formula" quoted above, extrapolated from Hunger and Reiner (1975: 24), recalls the apparent movement of the Pleiades during an entire year. It sums up how ancient scholars calculated, for each month, a fixed position of the Pleiades in respect of the Moon. The system is simple and straightforward, and it likely diverges from the other rules because according to it the Pleiades are seen in balance with the Moon on the 3rd day of Adaru (i.e. Month XII), while according, for instance, to the restorations offered in § 1a and § 2a, they would be in balance on the 1st or the 3rd day of Nisannu (i.e. Month I):

249 For a discussion on the metaphorical language, see 2.2.3.2.; for a discussion of the word "impossible" in this context, see fn. 268.

DIŠ *ina* ^{iti}*ša-ba-ṭi* UD.3.KAM* ⸢MUL.MUL⸣ *u* ^{d}30 IGI-*šú-nu-t*[*i-ma*] *iš-taq-lu* / MU BI *eš-re-et i*[*p-pal-si*]-*ḫu* TAG$_4$-*et*

If on the 3rd day of Šabaṭu (i.e. Elamite name of Adaru, month XII, see p. 319) you observe the Pleiades and the Moon [and] they are balancing, / that year is normal, (if) they a[re ap]art (lit. fall down), it is left behind (i.e. will be intercalary).
(K 3923+ obv. 18'a–b, see App. B § 11)

Hunger and Steele (2019: 212–213) explained this shift with the difference between the placement of solstices and equinoxes in earlier texts and in MUL.APIN; this would then place the rules of K 3923+ much later than the ones given in MUL.APIN (see § 1a) and in *Šumma Sîn ina tāmartišu* tablet 4 (see § 2a).

The principle of this rule, that is, to check the Pleiades and the Moon every month, is recalled in three different sources in a shortened form, with an explanation. The first source is a commentary of the reconstructed commentary of *Enūma Anu Enlil* tablet 53 (K 3558, source A; already source A in § 2b), the second and the third are fragments from the series *Šumma Sîn ina tāmartišu* (K 2254, source B, already source C in § 2a; K 6686 + K 9234, source C).[250]

List of Sources

Source: A
Siglum: K 3558 (see Commentary EAE 53 source A)

Source: B
Siglum: K 2254
Copy: ACh Sin 22; AAT 5; Plate 44
Edition: ACh Sin 22; see also App. B § 25 (obv. 1, 6–8, 13–17 and rev.)
Provenance: Nineveh
Ductus: Neo-Assyrian
Measurements: 7,6+ x 6,4+ x 2,5+ cm
CDLI n.: P394296 (with photograph)
Description: Fragment of the upper right part of a tablet, with part of the upper and right edge preserved. The scribe started to write the obverse on the convex surface of the tablet, which is usually assigned to the reverse.

Source: C
Siglum: K 6686 + K 9234
Copy: Plate 45
Edition: App. B § 26 (obv and rev. 1'–4', 6'–11')

250 Both K 2254 and K 6686 + K 9234 were written by Nabû-zuqup-kēnu, an Assyrian scribe active between the eight and seventh century BC in Kalḫu (Hunger 1968: 90–96 nr. 293–313; May 2018: 110–120).

Provenance: Nineveh
Ductus: Neo-Assyrian
Measurements: 6,9+ x 5,7+ x 2,4+ cm
CDLI n.: P396730 (with photograph)
Description: Fragment of the right part of a tablet, with part of the right edge preserved. The fragment has a unique arrangement of the reverse, which is written in a horizontal format instead of a vertical format, like in the obverse.

Text

1. **DIŠ MUL.MUL MUL^{meš}-*šú* 12 *uš-te-te-še-rù*[251] 12 ITI^{meš}**
A obv. 18 ⸢DIŠ⸣ MUL.MUL MUL^{meš}-*šú* 12 *uš-te-te-še-rù* 12 ITI^{meš}
B obv. 11–12 [... MUL.MUL MU]L⸢^{meš}-*šú*⸣ 12 (12) [...]x ⸢UD$^{?}$ ITI⸣
C rev. 5’ [... M]UL.MUL MU[L^{me}]^{š}-*šú* 12

If the 12 stars of the Pleiades move straight ahead. (It means that) for 12 months (B obv. 12: (during that) day$^{?}$ (and that month)

2. **[K]I $^{\text{(d)}}$30 LAL^{meš}-*ma* (KIMIN) *ina* 12.KÁM ITI (*ina* ^{iti}ŠE)**
A obv. 18 [K]I ^{d}30 LAL^{meš}-*ma* KIMIN *ina* 12.KÁM ITI
B obv. 12, 9 [K]I 30 LAL^{meš}-*ma* (9) [...]x ⸢*ina* 12.KÁM ITI⸣ *ina* ^{iti}ŠE

they are balancing [wit]h the Moon. (Ditto, i.e. if the 12 stars of the Pleiades move straight ahead). (It means that) in the 12th month (or: on the 12th (day) of the month)[252] (var.: in Adaru)

3. **KI 30 LAL^{meš}-*ma* : 5 ^{d}[UD]U.IDIM^{meš} KI-*šú-nu* GUB^{me}-*zu-ma***
A obv. 18–18a KI 30 LAL^{meš}-*ma* **(18a)** : 5 ⸢^{d}⸣[UD]U.IDIM^{meš} ⸢KI-*šú*⸣-*nu* GUB^{me}-*zu-ma*
B obv. 9 KI 30 LAL^{meš}-*ma*

they are balancing with the Moon. – (var.) (it means that) the five [pla]nets stand still with them.

In a simple way, the rule recalls again the apparent movement of the Pleiades for one year: the Pleiades are expected to be seen with the Moon every month, 12 times per year (see 2.1.).- The explanation “(It means that) for 12 months they are balancing [wit]h the Moon” (entries 1–2) refers to the 12 times the Moon should be seen close to the Pleiades during one normal year. The explanation “(It means that) in the 12th month (var.: in Adaru) they are balancing with the Moon” (entries 2–3) equally refers to a situation observed during one normal year. The alternative interpretation “(It means that) on the 12th (day) of the

251 The reading *uš-te-te-še-rù* was suggested to the author by Hermann Hunger.
252 This alternative interpretation was suggested to the author by Jeanette C. Fincke.

month (var.: in Adaru) they are balancing with the Moon" (entries 2–3) is syntactically possible, but its meaning would diverge from all the other rules discussed so far, whether one interprets it as "12th day of the (1st) month (i.e. Nisannu)", or as "12th day in Adaru (i.e. Month XII)". In both instances, one should assume that the above-mentioned fragments belonged to a different tradition, compared to all the other rules.

In source A obv. 18a (entry 3), the ancient scribe even gave another explanation: the number 12 means that the Pleiades stand together with the five planets (i.e. Jupiter, Venus, Mars, Mercury, Saturn), i.e. seven stars of the Pleiades + five planets = 12 (Schaumberger 1935: 337). That could be an *a posteriori* explanation for the composition of the number 12, or a supposed occasion in which the planets and the Pleiades are all visible in the night sky.

To summarise, the fragments from the series *Šumma Sîn ina tāmartišu* (K 2254, K 6686 + K 9234, sources B and C), and the one from the commentary to *Enūma Anu Enlil* tablet 53 (K 3558, source A), sum up a rule for one normal year for the observation of the Moon and the Pleiades (i.e. when the intercalation is not needed); after that year, one should expect a shift in the rising time during the following years.

The statement that the Pleiades have 12 stars is not unique because 12 Pleiades are mentioned in Venus' omens from the series *Enūma Anu Enlil* (see 5.1.8.4. § EAE 59–61):

14. DIŠ ^{d}MIN *ina* ^{iti}GU$_{4}$ MIN ŠÈG^{meš} *u* ILLU KUD^{meš} MUL.MUL MUL$_{4}^{meš}$-*šú* 12
15. DIŠ ^{d}MIN *ina* ^{iti}SIG$_{4}$ MIN SU.GU$_{7}$ *ina* KUR GÁL-*ši* : *ina* KUR GÁL^{meš} MUL.MUL MUL$_{4}^{meš}$-*šú* 10

14. If the ditto-god (i.e. Venus) in Ajaru (i.e. Month II) ditto (i.e. has a beard), rain and flood will stop. (It means that) the stars of the Pleiades (are) 12.
15. If the ditto-god (i.e. Venus) in Simanu (i.e. Month III) ditto (i.e. has a beard), there will be famine in the country – (var.) there will repeatedly be in the country. (It means that) the stars of the Pleiades (are) 10.

(Reiner-Pingree 1998: 149, 14–15)

The beard of Venus in the entries given above is explained by Venus being surrounded by 12 Pleiades during the month Ajaru (i.e. Month II), and 10 Pleiades in Simanu (i.e. Month III). Schaumberger (1935: 339) was the first to notice this fragment and suggested that the numbers 12 and 10 refer to how many times the Moon and the Pleiades were close in a "normal" year (i.e. when the heliacal rising of the Pleiades happens in Ajaru) and in an intercalary year (i.e. when the heliacal rising of the Pleiades happens in Simanu). Therefore, the above-mentioned entries mean that if Venus is observed when the Moon and the Pleiades are balancing in Ajaru (i.e. Month II), the year is normal and the prediction is positive. If the same situation happens in Simanu (i.e. Month III) – that is, implicitly, three years later – intercalation is necessary, and the prediction is negative. This suggestion is also consistent with one of the rules in MUL.APIN (II GAP A 10–11, see § 1b), which is based on the acronychal or heliacal rising of the Pleiades. The shift in the number of stars of the Pleiades from 12 to 10[253] can be interpreted as the number of times in which the Pleiades and the Moon – ideally speaking under perfect meteorological conditions – are expected to be seen close to each other within a normal year and an intercalary year, in a span of three years.

4.2.2.1. Summary of the Intercalation Systems Related to the Pleiades

Like the Astrolabe B or the Great Star List, MUL.APIN was a compendium rather than an original composition, i.e. it was a work of reference for scholars, composed of heterogeneous material, also taken from the compendia of omens (Koch 2015: 185). MUL.APIN shares many concepts and themes with the series *Enūma Anu Enlil*, such as ideal schemes (e.g. see A.2.1.1.1.) and omens (Hunger-Steele 2019: 11). It gives schemes for the intercalation in the form of so-called rules, i.e. checking the first visibility dates of certain stars (including the Pleiades), the solstices and equinoxes, the length of day and night, and the position of the Moon and Sun relative to specific stars (including the Pleiades) (see § 1). We have no sources attesting to the actual use of these intercalation rules in everyday contexts (Hunger-Steele 2019: 213); however, it seems that the idea to check the so-called "balancing", i.e. conjunction or having the same celestial longitude (see 4.2.2. and fn. 219), between the Pleiades and the Moon on specific days of Nisannu (i.e. Month I) was the most widespread because it is found not only in MUL.APIN, but also in compendia of omens.

In the commentary to *Enūma Anu Enlil* tablet 53 and *Šumma Sîn ina tāmartišu* tablets 4 and 6 (see § 2a–b), the intercalation is calculated according to the position of the Pleiades relative to the Moon: if, at the beginning of the year, the Pleiades are ahead, in time, or seen westward, that is the standard; if they are behind, in late, or seen eastward, it means that the intercalation is needed. This simple system is given as an explanation of two entries of the commentary to *Enūma Anu Enlil* tablet 53 (see § 2b), arranged – as often happens in divinatory compendia – according to a dichotomous scheme (see A.2.1.2.). Assuming that the information given by these two entries of omens date before or contemporary to MUL.APIN, i.e. late second or early first millennium BC, one could argue that the rules derived from omens by abstraction. This is not unusual or unparalleled: other scholars have already noted how rules and models of MUL.APIN are traceable to celestial divination,[254] which widely employed a dichotomous scheme: anything that follows a "standard" means good prediction, and anything contrary means bad prediction.[255] The verbs used in the entries of omens would attest to the use of a metaphorical language (see 2.2.3.2.) alongside a more technical or specific language, that of the explanations and astral compositions: hence, "the Pleiades are cut" means they are behind the Moon, and "the Pleiades are

253 There is also a possible reference to the Pleiades having 11 stars (see Commentary EAE 53: 17 source B; see also 5.1.5.1. commentary to the text of entries 16–23, p. 192).

254 E.g. see Brown (2000: 113–122) and Rochberg (2016: 146–149, 250–251).

255 E.g. see the omens from the assumed *Enūma Anu Enlil* tablet 51 that are based on a dichotomous scheme for the first visibility dates of stars which is also given in the Astrolabe B (Reiner-Pingree 1981: 52–55, 56–58 IX 1–15).

elongated" means they are ahead of the Moon. Furthermore, within *Enūma Anu Enlil* and *Šumma Sîn ina tāmartišu*, different but complementary ways of calculating the intercalation seem to be attested. For example, the omens about the Pleiades that 12 times a year (see § 2c), i.e. every month, are seen in their ideal position, i.e. they are in balance with the Moon, mirror the gist of the rules in MUL.APIN (see § 1), according to which the intercalation has a cycle of three years.

Rule	Moment for checking if the Pleiades and the Moon are "balancing" (*šitqultu*)	Moment for checking the first visibility of the Pleiades	Type of text
§ 1a	[1^{s}]$^{t?}$ [day of Nisannu$^{?}$] (Month I) – 3rd [d]ay of [Nisannu$^{?}$]; or: [3^{r}]$^{d?}$ [day of Nisannu$^{?}$] – 3rd [d]ay of [Ajaru$^{?}$] (Month II)		MUL.APIN
§ 1b		[1st day of Ajaru] (Month II) – [1st day of Simanu] (Month III)	MUL.APIN
§ 1c	15th day of Tešritu (Month VII) - 15th day of Araḫsamnu (Month VIII)		MUL.APIN
§ 2a	3rd day of Nisannu (Month I)		SIT
§ 2b	*ina reš šatti*, "at the beginning of the year"		EAE; SIT
§ 2c	Every month (12 times) / Adaru (i.e. Month XII) / 12th day of the month (i.e. Nisannu, Month I, or Adaru)		EAE; SIT

Table 2. The intercalation rules based on the observation of the Pleiades from MUL.APIN, the series *Enūma Anu Enlil* (EAE) and *Šumma Sîn ina tāmartišu* (SIT).

All these so-called rules are not astronomically accurate but idealised and paralleled in omens.[256] They do not necessarily agree with each other also because they might have been adjusted over time; yet one point on which they all seem to agree is to check the position of the Pleiades and the Moon at the beginning of the Mesopotamian year, i.e. in Nisannu (Month I). From the acronychal or heliacal rising of the Pleiades (see § 1b), or the balancing of the Pleiades and the Moon on specific dates (see § 1a, 1c–2b) or throughout the year (see § 2c), ancient scholars derived a triennial cycle of intercalation, which adapted the ideal lunar year (ca. 354,4 days) to an approximated solar year (ca. 365 days). This calendar adaptation worked for celestial divination, as remarked by the "Diviner's Manual" (see 4.2.2.). Thus, the meaning that all the intercalation rules (see Table 2) have for the

256 E.g. see Hunger-Steele (2019: 207).

present context is that intercalation was not intended to be accurate in modern scientific terms but a tool to obtain accurate predictions, or prognostication, from celestial signs. That is the reason why these rules for intercalation are found in both MUL.APIN and celestial divination, and the ones in MUL.APIN were likely abstracted from omens: this statement indicates that the main purpose of such rules was to aid divination. Already Reiner (1999: 30) had this idea regarding the unpublished celestial omens of the Pleiades: "quite a number of fragments (which all may belong to EAE, Book 53) deal with the Pleiades, including the relationship between the Pleiades and the moon. No doubt it is from such omens that the Pleiades intercalation rule, which establishes whether an intercalary month is to be added to the year depending on the date of the conjunction of the moon and the Pleiades, was abstracted."

4.3. Conclusion for the Nature of the Pleiades in Astral Compositions

In this chapter, the features of the Pleiades given in MUL.APIN, Astrolabe B, the Great Star List, and selected entries from the series *Enūma Anu Enlil* and *Šumma Sîn ina tāmartišu* have been collected and compared. Table 3, given below, summarises these features, which also cater to the Pleiades' conceptualisation.

Divinity	First visibility	Direction/Country	Other Features
^{d}IMIN.BI DINGIR^{meš} GAL^{meš} "The divine Seven, the great gods"	1st day of Ajaru (i.e. Month II)	Markers of East/Elam	Markers of the agricultural season Relevance for harvesting and barley for business Device for Intercalation

Table 3. The features of the Pleiades recurrently found in astral compositions.

The first visibility or heliacal rising of the Pleiades in the east in Ajaru (i.e. Month II, MUL.APIN I ii 38, see 4.1.1.) marked the beginning of the agricultural season in spring and identified the cardinal point east (seen from Mesopotamia) and, consequently, the country of Elam. In omens and commentaries dating to the Late Babylonian period (ca. 484–30 BC) but representing older traditions, the relationship with the agricultural season also connects them with forecasts about harvests and trade. The conjunction, or simply the visual proximity of the Pleiades and the Moon at the beginning of the year (i.e. Nisannu, Month I), was a tool to establish the duration of the calendar for divinatory purposes. From a religious point of view, the Pleiades are associated with seven great gods, whose identity is unclear yet defined by the number seven (see 3.2.5.1.).

What can be gained from the totality of the sources discussed so far is that the Mesopotamian "knowledge of the sky" (see 4.1.) casts a light upon a tendency toward schematisation to obtain rules to ease the calendar and divination. For instance, the lists of stars with rising and setting times (e.g. see 4.1.1., 4.2.1.) were a tool to organise the flow of time and the division of space: this allowed ancient scholars to record the precise dates of phenomena and, consequently, to interpret them correctly. Similarly, intercalation was not meant to be astronomically accurate in modern terms, as the sources (see 4.2.2. § 1–2) and the "Diviner's Manual" testify (see 4.2.2., p. 111–112): whenever the alleged correct calendar is established (i.e. if there should be intercalation or not), it is also possible to confirm the date when a phenomenon (i.e. ominous sign) occurred, in order to find its correct "answer" or interpretation (i.e. prediction). Therefore, both astral compositions and the compendia of omens were necessary for a diviner to build meaningful predictions. Overall, what emerges from the sources so far discussed is a role of the Pleiades as a tool for calculating time and space, and predicting events.

Modern scholars primarily focused their efforts on understanding astral compositions as exact sciences: such an approach has been and still is extremely rewarding for the history of science. Yet, MUL.APIN, Astrolabe B, and all the other sources discussed in this chapter acquire consistency when put in perspective with the perception of the cosmos in Mesopotamia. These sources collect technical, i.e. astronomy-related, information about celestial bodies, which were also functional to divination. Many celestial phenomena recorded by the diviners in astral compositions – may they be regular occurrences or deviations that were never observed – were perceived, treated, and interpreted in divination as ominous signs, and subjected to the rules of divination. Under these premises, the features of the Pleiades (see Table 3), and the perception of the Pleiades as gained from literary texts in chapter 3 will be relevant when analysing and interpreting celestial omens and astrological texts (see 5., 6. and Appendix A).

One last remark is necessary about how we consider astral compositions against modern science. Schematisation is pre-eminent in astral compositions, instead of astronomical accuracy in modern terms; still, schematisation is attested in texts written or copied during the Late Babylonian period (ca. 484–30 BC) (e.g. see 4.1.3.–4.1.4.), combined with mathematical astronomy.[257] This fact should bring to a reconsideration of astral compositions: they were not "merely" tools of abstraction, as Brown (2000: 153–160) argued; nor did they represent a sort of primeval stage of astronomy from which Mesopotamians progressively shifted towards an increasing astronomical accuracy (Rochberg 2016: 149, 248). Rather, they were tools of prediction, complementary to the series *Enūma Anu Enlil*, sometimes included in astronomical and astrological texts dating to the Late Babylonian period because they helped to understand "how things worked", regardless of the accuracy of their results from a modern positivistic point of view.

257 E.g. see Rochberg (2010: 271–302). This statement anticipates a further discussion about astrological and astronomical texts dating to the Late Babylonian period in chapter 6.

Chapter 5

The Pleiades in Mesopotamian Celestial Divination

This chapter examines the role of the Pleiades in the celestial omens of ancient Mesopotamia. In the first part, a premise about the nature of celestial divination (see 5.1.), and how the celestial omen series *Enūma Anu Enlil* is organised (see 5.1.1.–5.1.1.2.) is presented. Such a premise is necessary to introduce the reconstruction and the edition of the assumed *Enūma Anu Enlil* tablet 52, the commentary to *Enūma Anu Enlil* tablet 53, and *Šumma Sîn ina tāmartišu* tablet 6, whose protagonists are the Pleiades (see 5.1.1.–5.1.7.). The editions of the reconstructed texts include a score transliteration and translation. The commentaries to the texts and the further comments on the texts of these editions examine not only individual philological issues, but also the working principles of the entries of omens, and individual omens within the tablets. In the second part of the chapter, omens from other edited sections of *Enūma Anu Enlil* (see 5.1.8.–5.1.8.4.) and the so-called astrological and astronomical reports of Assyrian and Babylonian scholars (see 5.2.–5.2.4.) are analysed and compared to the edited omens from *Enūma Anu Enlil* and *Šumma Sîn ina tāmartišu*, focusing on patterns and similarities.

Omens are the key sources which shed light on the perception of celestial bodies in Mesopotamian culture. Therefore, the ultimate goal of this chapter is to present possible interpretations of the celestial omens about the Pleiades. In this respect, all the features of this asterism discussed so far (see 3. and 4.) are relevant: in chapter 3, the concern was the Pleiades and the divine heptads associated with them; chapter 4 looked at their role in astral compositions as a tool to organise the Mesopotamian calendar, and in this chapter, all the features of the Pleiades will be found in celestial omens as well (see 5.3.–5.3.2.).

5.1. Celestial Divination in Mesopotamia: an Overview

Divination in ancient Mesopotamia is "a practical means of obtaining otherwise inaccessible information perceived by its users as coming from supernatural or superhuman sources" (Koch 2015: 3). In other words, it is a way to acquire knowledge about the world through the interpretation of ominous signs (GISKIM, *ittu*), which are believed to be produced and communicated by the gods. Indeed, interpreting ominous signs means to interpret the will of the gods, literally called "verdict", or "decision" (EŠ.BAR, *purussû*),[258] or their design (*uṣurtu*). These are straightforward and broad definitions for divination, yet different categorisations are currently in use.[259] In the present context, it is reasonable to follow Koch's approach (2015: 15–16), who distinguishes two categories of divination

258 The word EŠ.BAR, *purussû*, (CAD P 529–535) "decision", refers to legal verdicts, and this statement also led to the idea that the celestial bodies were perceived as judges (Rochberg 2010: 29–30).

259 "Divination is in fact so complex and multifaceted a phenomenon that I believe it would be overly reductionist to explain it with reference to a single theory" (Koch 2015: 3).

based on the *De divinatione* of Cicero (I 11–12):[260] natural divination (i.e. immediately intelligible, that is inspired signs, like prophecies and dreams) and artificial divination (i.e. ominous signs or messages to be observed and then decoded through procedures). Celestial divination is part of the latter group, as it deals with the phenomena of the sky. A celestial ominous sign is an "unprovoked" sign, which is different, for example, from an oracle or from ominous signs to be detected on a sheep's liver, which are induced by a question.[261]

The divinatory process in ancient Mesopotamia is carried out by the diviner (i.e. *āšipu*, "exorcist", and *bārû*, "seer"),[262] and it is achievable only based on the good faith between him and the client for whom the diviner conducts the divination. Therefore, given the acceptance of divination *a priori*, the motivation for a divinatory process is the need of the client to get advice for impending difficult decisions.

On cuneiform tablets, the omens are written in the form of conditional statements (Rochberg 2010: 373–397), introduced by the Akkadian word *šumma*, "if", [263] or a logogram which stands for it (Koch 2015: 7, 24–29):

If X (protasis, or antecedent), then Y (apodosis, or consequent)

In the case of celestial divination, the protasis of an omen refers to phenomena in the sky, while the apodosis is the related prognostication or prediction (Hunger-Pingree 1999: 5; see fn. 28), as in the following example:

If the Pleiades reach Pāšittu, in that year there will be famine.[264]

From the late second millennium through to the beginning of the first millennium BC, the omens were systematically organised in omen series (Koch 2015: 31–33): the most comprehensive collection of celestial omens was the series called *Enūma Anu Enlil*, "When Anu (and) Enlil" (see 5.1.1.).[265]

260 For an English translation of *De divinatione* I 11–12, see Wardle (2006: 49).

261 Mesopotamian divination deals with *omina oblativa* (i.e. unsolicited signs, natural signs from earth and heaven), except for extispicy and a few minor techniques that deal with *omina impetrativa* (i.e. solicited signs, obtained through a request and under certain conditions) (Fincke 2014a: 3–16).

262 In Mesopotamia, there are different types of scholars, whose expertise come from various fields. Celestial divination is a technique of the exorcist (*āšipu*). For an overview of all the divination scholars, their designations and field of expertise, see Koch (2015: 16–24).

263 The meaning of the logogram DIŠ, and all the other logograms used to indicate *šumma* at the beginning of the omens was (and still is) a topic of discussion. The word *šumma*, "if", also introduces laws in ancient Mesopotamia: the offence and the punishment, respectively (Roth 1997: 8–9), may be compared with the protasis and the apodosis of an omen. In this context, the omens were likely perceived as divine law. For a comparison of conditional statements in omens and laws, see Fincke (2006).

264 SIT 6: 3.

265 For an overview of the textual history of divination in general, see Koch (2015: 59–66).

Celestial omens were not necessarily based on the observation of spectacular or unusual phenomena, as one can find, for instance, in teratological omens.[266] Celestial divination was the result of the interaction between actual or practical observations, and applied theoretical schemes. More precisely, the core of celestial omens comes primarily from the observation of celestial phenomena, which was later expanded through general theories and schematisations (Maul 2005: 51–57). Indeed, omens can be based on several working principles that create a network of analogies difficult to disentangle for a modern reader. For instance, through the scheme of binary opposites, the celestial signs are usually arranged according to antithetic directions (right and left, above and below, in front of and behind, etc.), time (sunrise and sunset, on time and late or early, etc.), luminosity (bright and dark), and colours (white, black, red, and yellow/green) of celestial bodies (see A.2.1.1.). An example is given below for two consecutive omens from *Enūma Anu Enlil* tablet 52, arranged according to the scheme of binary opposites:

10'. [DIŠ ^{mul}Š]U.GI MUL^{meš}-*šú bi-rit-su-nu ma-gal* BAD-*at*
11'. [DIŠ ^{mul}Š]U.GI MUL^{meš}-*šú nen-mu-du*

10'. [If] the interspace between the stars of the [O]ld Man is very open.
11'. [If] the stars of the [O]ld Man are conjoined.
(K 3099 + K 18689 rev. 10'–11', see EAE 52: 44–45 source E_1 and parallels)

The logic behind the arrangement of these two omens was to substitute the verb in the first protasis (i.e. BAD-*at*, *petī'at*, "is open"), with another verb, whose meaning is antithetical, in the second protasis (i.e. *nenmudu*, "are conjoined"). The working principles are implicit in the way the omens are arranged; only occasionally, general principles were written down as rules (Brown 2000: 139–153). In addition, the stars and the planets were considered the representations of anthropomorphic gods (see 2.2.3.1. § 2). Consequently, religion and mythology also played an essential role in omens. Together with the many working principles, there were several analogies between celestial bodies and anthropomorphic gods with whom they were associated, resulting in omens whose meanings more often than not appear obscure to a modern reader (Fincke 2016: 122).

The celestial signs are different from the terrestrial ones because everyone can systematically observe the former in the night sky, while the latter can be observed only in a specific place (e.g. in the palace, in the streets or in the steppe, etc.) and by a limited number of people (Maul 2005: 49–50). Hence, the predictions of celestial omens concern the whole reign, the royal household, and the king, who embodies the society and the people of his country (Maul 2005: 54). The so-called astrological and astronomical reports from Assyria and Babylonia (Hunger 1992) to the kings testify this primary purpose of celestial divination: it was an essential tool for the royal court during the first millennium BC, in order to make decisions for the well-being of the country (see 5.2.). The diviners, who wrote these reports, worked side by side with the kings. They observed the sky and

266 Teratology is a technique of divination by the malformed newborns (Koch 2015: 262–273). For studies and editions of teratological omens, see Leichty (1970), and De Zorzi (2014).

chose on which phenomenon to focus, they then had to establish whether what they observed was valid for the divinatory process, and to find the appropriate omen and its prediction. The kings communicated with diviners all over the country, to rely on the different interpretations of several individual scholars. For these reasons, the astrological and astronomical reports are unique sources not only of the actual process of celestial divination in Mesopotamia, but also of the everyday life of the court.

The sources from the series *Enūma Anu Enlil* the ancient experts had at their disposal were many and often not unanimously organised, despite a process of "standardisation"[267] already started by the end of the second millennium BC (see 5.1.1.). Given the long tradition of celestial divination – according to the written documentation dating to the Old Babylonian period (ca. 2000–1500 BC) until the end of the first millennium BC – the ancient scholars needed at some point – likely from the seventh century BC onwards – to update and expand the meaning of certain omens or terms through commentaries (Fincke 2016: 117 fig. 3). In addition to the omens from the series *Enūma Anu Enlil*, there were series of excerpts and commentaries, like the serialised commentary on *Enūma Anu Enlil* called *Šumma Sîn ina tāmartišu* (Frahm 2011: 155–160). The commentaries provided not only summaries for the content of individual tablets within *Enūma Anu Enlil*, but also variants of apodoses, and protases of what we – from a Western scientific point of view – would consider "impossible", i.e. protases whose literal meaning is astronomically impossible and use, or are interpreted as using, a metaphorical language.[268] Indeed, the written omens in the series *Enūma Anu Enlil* probably trace back to an earlier oral lore; many omens concern fixed stars moving into or towards each other, then one shall assume the belief that stars moved freely, chasing each other in the sky. The ancient experts who looked at the sky to interpret its signs at some point updated and added comments to the omens of the series *Enūma Anu Enlil*. For instance, the following protasis is one example of many astronomically impossible protases; it is not always clear what it originally meant, but ancient scholars equated the protasis to another one, in which the phenomenon seems described in a more "literal" way, perhaps less "metaphorical", i.e. through factual explanations (Frahm 2011: 39–40; Koch 2015: 37–39):

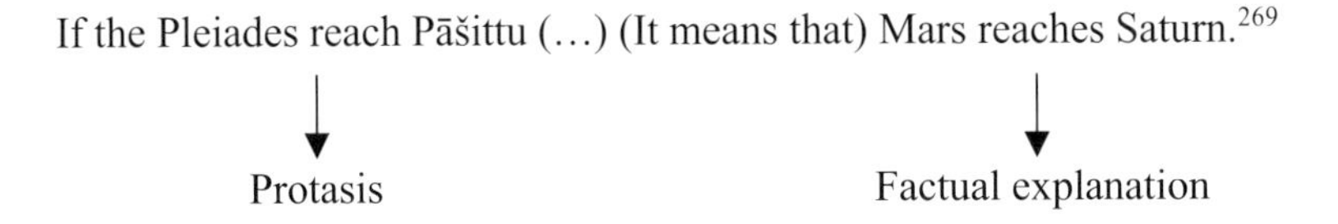
If the Pleiades reach Pāšittu (…) (It means that) Mars reaches Saturn.[269]
Protasis
Factual explanation

267 For the meaning of the word "standard" in this context, see Koch (2015: 52–54). The present author agrees with avoiding using the word "canonical", or "canonisation" (e.g. Frahm 2011: 317–322) for the process of Mesopotamian textual tradition towards a more stable form. Thus, "standard" is used to refer to the series dating to the first millennium BC used by ancient scholars as a master text (e.g. the series EAE), called *iškāru*, "assignment", in Akkadian (Fincke 2016: 119, fn. 63).

268 The word "impossible" exclusively referred to what we, modern readers, perceive as impossible from a scientific point of view. What is impossible for us was not necessarily so for the scholars of ancient Mesopotamia, and the use of terms like "possible" and "impossible" is linked to the accepted knowledge of each culture (Rochberg 2010: 387–393).

269 See SIT 6: 3.

The factual explanations of the commentaries fit the purpose of keeping the original omens – for they were considered given by the gods and could not be changed or corrected (see A.1.; Fincke 2016: 119–122) – and likely making them more consistent with the celestial mechanism by substituting the name of the stars with planets.[270]

In commentaries, the use of different hermeneutical techniques partly shows the logic of the omens. For instance, there is the use of synonyms, different readings of logograms, or allegorical approaches arising from a deeper etymological and etymographical analysis, in order to explain the omens.[271] Semantic and graphic puns, which associate the protases to the apodoses or to factual explanations, seem to be one of the criteria of the predictions: the semiotic link between the phenomena and their predictions could be a straightforward wordplay, an association of ideas and meanings, and beliefs and symbolism (see A.3.–A.3.1.6.).

Ultimately, the idea a modern reader is left with regarding the huge concern around divination, and the array of possible, probable, or improbable omens, is that the "phenomena were of interest for their potential to show change, deviation, malformation, or anomaly as against norms, or to be serviceable in re-establishing norms in the realm of the divine as in the human" (Rochberg 2016: 127).

5.1.1. The Series *Enūma Anu Enlil* (EAE)

The earliest published example of a collection of celestial omens[272] is a text dating to the Old Babylonian period (ca. 2000–1500 BC) about lunar eclipse omens (Rochberg-Halton: 1988: 9 fn. 5), but we expect that an oral tradition existed already earlier. These omens were later developed to form a standardised literary corpus that is attested from the late eighth century BC until the first century BC (Fincke 2016: 117 fig. 3). The main collection of celestial omens in ancient Mesopotamia is named *Enūma Anu Enlil* (abbreviated EAE), "When Anu (and) Enlil". This series has a bilingual mythological introduction written in Akkadian and Sumerian,[273] according to which Ea, the god of wisdom, gave the series EAE to the people, "implying that those rules for divination had divine authority" (Fincke 2016: 121). Indeed, the tablets of the series EAE were handled by the experts with the Akkadian title of *tupšar Enūma Anu Enlil*, "scribe of *Enūma Anu Enlil*"; this title refers to experts observing the sky, as what we would consider "astronomers" nowadays, and it also reflects the high status of EAE among the ancient scholars (Rochberg-Halton 2000).

270 "The impossibility of any free movement of constellations was already acknowledged by Neo-Assyrian scholars, so in their commentaries they glossed the data in the text to make it credible, in this case by taking the name of a constellations as a synonym for a planet" (Fincke 2016: 122).

271 For a discussion on hermeneutic techniques in Assyrian and Babylonian commentaries, see Frahm (2011: 59–85). There were patterns of interpretation and patterns in which signs were recorded and catalogued. Of course, patterns of interpretation allow us to have an important starting point for understanding the complex divinatory language. Greaves (2000) gives a short overview of how puns give us an insight into the criteria of predictions, whereas Noegel (2006) compares the dream exegesis in Ancient Egypt with wordplays in Mesopotamian divination.

272 For an overview of the forerunners of series EAE, see Fincke (2016: 114–115).

273 For the edition, see Koch-Westenholz (1995: 77).

5.1.1.1. The Numbering System of the Series EAE

Several scholars are associated with the study of the series EAE. It began at the end of the nineteenth century with Craig (1899), Virolleaud (1908–1912), and Weidner (1941–1944; 1954–1956; 1968–1969). Based on their studies, many scholars studied celestial divination and edited parts of the series, even though the reconstruction of individual tablets within the series remains unclear.

The tablets of EAE can be identified by their catchlines and subscripts (or rubrics), when preserved. The subscripts give the number of a tablet within the series and the title of the series (e.g. xth tablet of *Enūma Anu Enlil*). Sometimes, a catchline with the incipit of the next tablet, according to the order of the series, is added. Two fragmentary catalogues for the series EAE are known: one from Aššur dating to the eleventh century BC, and one from Uruk dating to the third century BC. The Uruk catalogue (Weidner 1941–1944: 186–189) lists only the incipits of the first 26 tablets, and the Aššur catalogue (Weidner 1936–1937; Fincke 2001) refers to tablets 39–59 of the series. Both catalogues were recently re-edited by Rochberg (2018). There is a difference of eight centuries between the two catalogues, and that makes it sufficiently clear that they do not necessarily match, as they could attest to traditions that could have changed through time.

The main issue in the reconstruction of EAE is that tablets with identical texts have different numbers in both subscripts and catalogues. Up to four different numbers for the same tablet can be traced: Weidner (1941–1944: 181) explained this situation as deriving from various "scholarly centres" (he named Babylon-Borsippa, Uruk, Aššur, Kalḫu, and Nineveh), each of which produced different tablet formats. There were three-column tablets, two-column tablets, and single-column tablets on which the omens were arranged (Fincke 2013: 583–584). Due to the different tablet formats of the different "scholarly centres", the total number of tablets of series EAE varied. The maximum number for this series is EAE 70 (from manuscripts at Nineveh), while the minimum is 63 (from manuscripts at Aššur) (Fincke 2001; Gehlken 2005: 249–268). The highest number is attested from a school that preferred oblong single column tablets, which may contain fewer omens than tablets with other formats. In addition, every scholarly centre chose the number of the omens within their sequence that they wanted to list on their tablets (Fincke 2013: 585–591). In view of the discrepancies in the numbering, Weidner (1941–1944: 185) began to provide concordances between the Aššur catalogue and the Uruk catalogue. Fincke (2001: 37) completed the task using more recent sources. More specifically, she extended the correspondences to four different possible versions of the series: a Neo-Assyrian version from Nineveh (whose numbering is the most used in editions), a Neo-Assyrian version from Aššur, a Neo-Babylonian version from Nineveh, and the "Babylonische Fassung" from Uruk. Gehlken (2005: 235–249) pointed out that it is not possible to distinguish assumed versions according to their ductus, or their formats. He argued that many texts have been copied from earlier examples whose provenance remains unknown. He stressed that there are still sources with different tablet numbers for the same text, which do not fit in any of the above-mentioned schools or versions. Ultimately, ancient scholars collected tablets from different schools and applied the numbers given in the original to their new copies without trying to form a new version with consecutive numbering (Fincke 2013: 583–584, 589–590 fn. 24).

The core issue with the reconstruction of EAE is well summed up by Rochberg (2018: 122): "the evidence suggests that tablet numbering was tied more to the local needs of the scribes than to any sense for what we would call a canonical text to be transmitted in a fixed, standardised, certainly not invariant, form across its many exemplars. Instead, there is considerable variability in the numbers assigned to tablets in catalogues, subscripts, and catchlines." In summary, a single and obligatory exact numbering system of the series EAE does not exist, but despite the different tablet formats, numbers, schools, or traditions, the incipits of the tablets always remained the same. Hence, quoting the first line of a tablet would be the only way to give an unambiguous assignment to the subject of the tablets within the series. Nevertheless, in this study, the tablets of EAE are also labelled with the numbers preserved on one of the manuscripts' traditions from Nineveh (with Neo-Assyrian ductus), the numbers used so far by the great majority of the modern scholars to quote EAE. That is only for practicality since this is the only way to understand every reference with the past editions. This means that the numbers of the tablets from the series EAE are not a statement to validate any numbering system, yet they are a necessary reference to the history of the studies.

5.1.1.2. The Internal Organisation of the Series EAE

Virolleaud (1908–1912) and Weidner (1941–1944; 1954–1956; 1968–1969) divided the subjects of the series EAE into four thematic chapters: Sin, Šamaš, Adad, and Ištar. This sequence mirrors the Babylonian pantheon as described in the creation myths (e.g. *Enūma eliš*). Sin, the Moon god, is the father of Šamaš, the Sun god; Adad is a son of Anu, and Ištar represents the younger generation of gods (Fincke 2016: 120–121). It is a modern thematic subdivision but it reflects the ancient one. For instance, the Uruk catalogue has summary lines which give a cumulative number of tablets and lines for the first "chapter" of EAE: "appearances, eclipses, (and) decision(s) of eclipse(s), and ominous portents of the Moon" (tablets 1–22) (Rochberg 2018: 134 ll. 23–24). The colophon of a commentary in Late Babylonian ductus also mentions a subchapter, the tablet of the "appearance of the Moon" (BM 45821+ rev. 16'–17', see Al-Rawi-George 1991–1992: 66). This means that the division into chapters and/or subchapters was also used in tablets of the series EAE, not only in the catalogue from Uruk.

The four chapters of EAE and the current situation of their editions are summarised as follows:

Sin (EAE tablets 1–22): Appearances of the Moon and eclipse of the Moon. Editions according to the order of tablets within the series: Verderame (2002) for EAE 1–6; see SpTU 2, 40 for EAE 7, and CTN 4, 4 and 13 for EAE 8, whose editions are available online at the website GKAB; Al-Rawi-George (1991–1992; 2006) for EAE 14; Rochberg-Halton (1988) for EAE 15–22; see also Fincke (2016a; 2017; 2019) for additions to EAE 15–22. Some tablets remain unpublished: EAE 9–13, about shape, colours, and atmospheric phenomena of the Moon. For a general overview of their content, see Verderame (2002a; 2014: 94–95).

Šamaš (EAE tablets 23–35):[274] Appearance and eclipse of the Sun. Editions according to the order of tablets within the series: Van Soldt (1995), and the additions by Fincke (2013; 2014) for EAE 23–29. The section about the eclipses of the Sun is still unedited: The solar eclipse section covers only five tablets (Weidner 1968–1969: 67–69; Fincke 2013: 591–593).

Adad (EAE tablets 36–49): Weather omens. Editions according to the order of tablets within the series: Gehlken (2005; 2008; 2012) for EAE 42–49. Six tablets are still unedited: EAE 36–41, about mist and clouds (Koch 2015: 173).[275]

Ištar (EAE tablets 50–68/70): Fixed stars and planets. Editions according to the order of tablets within the series: Reiner-Pingree (1981) for EAE 50–51; see 5.1.4.–5.1.5. in this study for EAE 52–53; Largement (1957) for EAE 56; Reiner-Pingree (1998) for EAE 59–61, with the addition of possible excerpts from EAE 62; Reiner-Pingree (1975) for EAE 63; Reiner-Pingree (2005) for EAE 64–66; see also Fincke (2013b; 2015) for additions to EAE 59–63 and EAE 56. Several tablets remain unpublished: EAE 54–55 and 57–58 about fixed stars, and 67–68/70 pertaining to *mešḫu*, "meteor" or "glow", and to celestial bodies approaching another one (i.e. TE-tablets, see Reiner 2006).

5.1.2. The Series *Šumma Sîn ina tāmartišu* (SIT)

Šumma Sîn ina tāmartišu (abbreviated SIT), literally means "If the Moon at its appearance".[276] It is labelled as a serialised *mukallimtu*-commentary on the series EAE, described in colophons as "tablet x of *Šumma Sîn ina tāmartišu*, *mukallimtu*-commentary (of *Enūma Anu Enlil*)".[277] It comprises at least six tablets whose structure and hermeneutical strategies are very similar to the ones in commentaries of EAE. However, the sequence of entries does not follow any of the EAE tablets: SIT has protases excerpted from various sections of EAE with the related explanations (Frahm 2011: 155–160). Likely, entries were chosen and grouped based on thematic principles: SIT 1 and 2 deal with lunar phenomena; the reverse of SIT 4 deals with the interpretation of phenomena like eclipses (see App. B § 24 for full bibliography and a new edition); SIT 5, according to the catchline preserved at the end of SIT 4, should be about solar phenomena (Koch-Westenholz 1995: 84–86; 1999: 150); SIT 6 focuses on the Pleiades and planetary conjunctions (see 5.1.6. further comments on the text, pp. 222–224). Furthermore, the series SIT, differently from other commentaries (see 5.1.3.4.), sometimes gives principles

274 A new edition for the Šamaš section is being prepared by Jeanette C. Fincke.

275 A reconstruction of EAE 36–37 is being prepared by Tommaso Scarpelli for his forthcoming PhD dissertation.

276 For the characterisation of this series, see Koch-Westenholz (1999: 149–152), Gehlken (2007) and Wainer, whose forthcoming PhD dissertation will provide an analysis of the preserved portions of the series SIT.

277 In the texts edited in this manuscript, the fragments K 2170+ (App. B § 18) and K 2254 (App. B § 25) have the full colophon while K 2177+ (App. B § 17) has the shortened version. For a definition of *mukallimtu*-commentaries, see 5.1.3.4.

or parameters of interpretation of phenomena in the form of schematic rules (e.g. SIT 4, K 3123, see App. B § 24; 5.1.4. commentary to the text of entries 46–60, pp. 167–168, and A.3.1.5., pp. 300–301).

The study of the series SIT is partly tied to study of EAE and all the scholars who worked on its reconstructions (see 5.1.1.1.); among the others, Weidner (1941–1944: 182–184; 308–318), who called SIT "die kommentierende Anschluss-Serie", even though SIT is not associated to any specific tablet of EAE. More recently, Koch-Westenholz (1999) edited the tablet SIT 1, and Frahm (2011: 155–160) collected all fragments known to him belonging to this series in his study about commentaries. Wainer (2016) wrote an article about SIT 4 in which he stated he is working on publishing a study and an edition of this series. An updated list of fragments belonging to SIT with related bibliography can be found online at the CCP website.[278] At present, only SIT 1 has been reconstructed and edited (Koch-Westenholz 1999), part of SIT 2 has been edited by Borger (1973: 38–39; see also 5.1.3.3.), part of SIT 4 by Koch-Westenholz (1995: 105–108), Wainer (2016: 57–58), and in this manuscript (App. B § 24), and SIT 6 also in this manuscript (see 5.1.6.).

In paragraph 5.1.1.1, it has been argued how establishing the numbering of the series EAE is inherently fraught with problems because studies have shown that different manuscript traditions coexisted, even in the same place and at the same time. Similar considerations apply to the series SIT, yet modern scholars – as well as the present author – continue to draw on numbers mentioned in colophons, not only as a matter of practicality, but also because they provide a useful tool for reconstructing the series and their contents.

5.1.3. The Reconstruction of Tablets EAE 52, Commentary EAE 53, and SIT 6

Thanks to the catalogue of celestial omens tablets and fragments compiled by Reiner (1998), and the work accomplished by Fincke for the British Museum's Ashurbanipal Library Project,[279] it was possible to create a core list of published and unpublished material for the tablets in which the Pleiades are the protagonist: the tablets EAE 52, Commentary EAE 53, and SIT 6. These are only the numbers under which these tablets are commonly known by modern scholars. Starting from the core list, the author was able to collate and identify more fragments for the reconstruction of these tablets.

278 https://ccp.yale.edu/catalogue?genre=23&page=1 accessed 25.01.2023.

279 See online the website *The Nineveh Tablet Collection* (http://www.fincke-cuneiform.com/nineveh/ accessed 25.01.2023) by Jeanette C. Fincke.

5.1.3.1. The Provenance of the Sources

The great majority of the sources used for the reconstruction of EAE 52, Commentary EAE 53, and SIT 6 come from the Ashurbanipal's library in Nineveh, and they are mainly written in Neo-Assyrian ductus. Out of 66 sources used in the score transliterations, 14 are written in Neo-Babylonian ductus, and only one (LB 1321, see Borger 1973) may represent an earlier stage of the Neo-Assyrian ductus (Frahm 2011: 25 fn. 85). Two sources certainly come from Uruk and are written in Late Babylonian ductus (W 22730/5, see SpTU 3, 101; VAT 7850 (+) AO 6486, see App. B § 27), one comes from Babylon and it is written in Late Babylonian ductus (BM 38301, see App. B § 14), and the other two fragments, whose provenance is unclear, display Late Babylonian ductus (BM 47799, see Reiner 2006; BM 44005, see App. B § 20).

5.1.3.2. The Numbers of the Reconstructed Tablets within the Series EAE

The tablets EAE 52 and 53 belong to the chapter of Ištar (i.e. fixed stars and planets) in the series EAE, and they are known with the following incipits:

EAE 52: DIŠ *ina* ^{iti}BÁR ^{mul}AŠ.GÁN *u* MUL.MUL IGI^{meš}
If in Nisannu (i.e. Month I) the Field and the Pleiades are visible

EAE 53: DIŠ MUL.MUL ^{mul}ŠUDUN KUR-*ud*
If the Pleiades reach the Yoke

These incipits are reconstructed from different sources. They are not preserved in what we have from the Uruk catalogue, and in the Aššur catalogue they are fragmentarily preserved:

8'. [DIŠ *ina* ^{iti}BÁR ^{mul}AŠ.GÁN *u* MUL.MU]L IGI^{meš} DUB.46.KÁM
9'. [DIŠ MUL.MUL ^{mul}ŠUDUN KUR-*u*]*d* DUB.47.KÁM

8'. [If in Nisannu (i.e. Month I) the Field and the Pleiad]es are visible. Tablet 46.
9'. [If the Pleiades reac]h [the Yoke]. Tablet 47.
(Fincke 2001: 24 obv. i 8'–9'; Rochberg 2018: 125)

Weidner (1925: 358; 1941–1944: 185) compared the Aššur catalogue's entries with the Uruk catalogue and concluded that these incipits corresponded to tablets 52 and 53 known from sources from Nineveh. His identification is corroborated by the following sources:

§ 1. The catchline and the subscript in one tablet from Nineveh written in Neo-Assyrian ductus, which represents the assumed tablet EAE 51, have the following:

26'. DIŠ *ina* ^{iti}BÁR ^{mul}AŠ.GÁN *u* MUL.MUL IGI^{meš} x[...]
27'. DUB.51.KAM UD AN ^{d}*en-líl*

26'. If in Nisannu (i.e. Month I) the Field and the Pleiades are visible ...[...].

27'. Tablet 51 of *Enūma Anu Enlil*
(Reiner-Pingree 1981: 59 IX J rev. 26'–27')

The number 51 refers to the text on the tablet, and the catchline gives the incipit of the next tablet, tablet EAE 52.

§ 2. A commentary from Uruk dating to the Hellenistic period (235/[?]/28 BC), edited by Weidner (1925), deals with entries from several EAE tablets. As for EAE 52, it gives its incipit, followed by a catchline with the incipit of the next tablet:

26'. [*šá* ŠÀ DIŠ *ina* ^{iti}B]ÁR ^{mul}AŠ.GÁN *u* MUL.MUL IGI^{meš} AL.TIL

27'. [DIŠ MUL.MU]L ^{mul}⸢ŠUDUN⸣ KUR-*ud*

26'. [(referring to the entries)[280] of "If in N]isannu (i.e. Month I) the Field and the Pleiades are visible" – completed.

27'. [If the Pleiade]s reach the Yoke.
(VAT 7850 (+) AO 6486 rev. 26'–27', see EAE 52 source Q, Commentary EAE 53 source T)

The tablet numbers are not given but the commentary has the same sequence of incipits known from the Aššur catalogue (i.e. of EAE 52 and 53).

§ 3. The fragmentary omen tablet from Nineveh K 2118, written in Neo-Assyrian ductus, preserves a catchline and the beginning of a subscript, broken right after the tablet number. The content of the fragment corresponds to the tablet known as EAE 52, the catchline is that of EAE 53, but the number of the tablet itself is given as 51:

9'. DIŠ MUL.MUL [^{mul}ŠUDUN KUR-*ud*...]
10'. DUB.51.K[AM* UD AN ^{d}*en-líl*...]

9'. If the Pleiades [reach the Yoke...].
10'. Tablet 51 [of *Enūma Anu Enlil*...].
(K 2118 rev. 9'–10', see EAE 52 source C, Commentary EAE 53 source U)

Therefore, this fragment follows a different numbering system in respect of § 1.

§ 4. The fragmentary commentary from Nineveh K 3558, written in Neo-Babylonian ductus, has the incipit of the tablet EAE 53:

280 Lit. "from the inside of".

[DIŠ MU]L.MUL ^{mul}ŠUDUN KUR-*ud*

[If the Ple]iades reach the Yoke.
(K 3558 obv. 1, see Commentary EAE 53 source A)

This commentary has a subscript (rev. 27'') with a severely damaged tablet number:

NÍG.PÀ[D.DA DUB.40]+⸢16.KAM*⸣ [DIŠ UD A]N ^{d}50

Mukal[*limtu*-commentary of tablet 5]6 of [*Enūma A*]*nu Enlil*.
(K 3558 rev. 27'', see Commentary EAE 53 source A)

Modern scholars restored the number of this subscript as ⸢54⸣ (Fincke 2001: 28; Frahm 2011: 133) because it would match with an assumed Neo-Babylonian version of EAE from Nineveh (Fincke 2001: 37). Nevertheless, the units preserved after the break are six, and the number should be restored as [5]6.

The identification of the incipits of the reconstructed tablets EAE 52 and 53 can be summed up as follows:

DIŠ *ina* ^{iti}BÁR ^{mul}AŠ.GÁN *u* MUL.MUL IGI^{meš}

If in Nisannu (i.e. Month I) the Field and the Pleiades are visible

Aššur	**Nineveh**	**Reconstructed number as used in this edition**
46 (see 5.1.3.2.)	51 (see § 3), (52) (see § 1)	52

DIŠ MUL.MUL ^{mul}ŠUDUN KUR-*ud*

If the Pleiades reach the Yoke

Aššur	**Nineveh**	**Reconstructed number as used in this edition**
47 (see 5.1.3.2.)	(52) (see § 3), (53) (see § 1), [5]6 (see § 4)	53

Some numbers are not attested in the sources at our disposal but inferred from the preceding numbers, so the unattested ones are given in brackets. Ultimately, as already mentioned in 5.1.1.1., the numbers chosen to label the editions in this study (i.e. EAE 52 and Commentary EAE 53) follow the number system used by scholars so far, i.e. the numbers from one of the versions of the series from Nineveh with Neo-Assyrian ductus; hence, the numbers were chosen only for a matter of consistency with the history of the studies.

5.1.3.3. The Numbers of the Reconstructed Tablets within the Series SIT

The reconstruction of the assumed tablet SIT 6 is based on the identification of two fragments, LB 1321 (Borger 1973: 38–43; SIT 6 source B) and K 2177 + K 7869 + Rm 473 (App. B § 17; SIT 6 source C). The fragmentary tablet LB 1321 is believed to come from Nineveh, and it is written in early Neo-Assyrian ductus (Frahm 2011: 156–157). The obverse has omens about lunar eclipses, while the reverse concerns stars interacting with one another, with a particular focus on the Pleiades. Borger (1973: 38–39) identified LB 1321 as the tablet SIT 2.[281] However, a careful reading of the text shows that this is only partially true: among the texts and fragments that duplicate parts of LB 1321 is K 2177+, also from Nineveh and written in Neo-Assyrian ductus.[282] Despite severe damage to the text, the remaining signs at the end of the lines make it sufficiently clear that it contains the same omens as the reverse of LB 1321, assuming that the latter is missing three omens at the beginning. Besides, K 2177+ has, after a few lines of blank space, a colophon the first line of which reads DUB.6.KAM*-*ma* DIŠ 30 *ina ta-mar-t*[*i-šu*] *mu-kal-lim-tu*$_4$, "Tablet 6 of *Šumma Sîn ina tāmartišu*, *mukallimtu*-commentary". All this suggests that LB 1321 records SIT 2 of the series on the obverse and SIT 6 on the reverse – if compared to the manuscript tradition transmitted by K 2177+ (Frahm 2011: 158).[283]

The identification of the other sources is equally complicated. For instance, the fragment K 5713 + K 7129 + Rm 2, 114 (App. B § 16; SIT 6 source A) duplicates tablet SIT 6 in the obverse, but in the reverse may present parts of another SIT tablet, or another recension of SIT 6. The same applies to K 2170 + K 3629 (App. B § 18; SIT 6 source F), which duplicates SIT 6 only in the obverse. The reverse has excerpts from one of the unidentified sections of the series EAE, or the alleged section in EAE about the planet Mars (Koch 2015: 175).

5.1.3.4. The Types of Sources and their Organisation

The editions of the tablets EAE 52, Commentary EAE 53, and SIT 6 aim to reconstruct the individual entries, rather than presenting each source individually. The score

---

281 This identification was supported by the catchline mentioned in K 4024 + K 10893 rev. 28'–30' (ACh Sin 3, 142-143), which represents the first tablet of the series SIT (Koch-Westenholz 1999: 152, 1–2).

282 Gehlken (2007), in a brief note about the series, did not mention this fragment. Instead, he argued that the tablet SIT 6 could be almost entirely reconstructed based on K 50 and its duplicates because the damaged subscript of K 50 (rev. 26') reads DUB.6.[KAM...EN].LÍL, "Tablet 6 [of... En]lil". However, contrary to K 2177+, this cannot be proven because K 50 might also belong to another series of commentaries related to series EAE. The identification of K 2177+ as SIT 6 does not exclude the possibility of the same being true for K 50, as they both could be remnants of different manuscript traditions.

283 Concerning the identification of the tablet SIT 6, see Renzi-Sepe (2021). The article was based on the publication of the fragment K 7275. Heeßel (2021: 118) came to the same conclusions as the present author regarding the identification of SIT 2 and 6, but based on VAT 9761, a commentary on lunar omens with the same layout as LB 1321.

transliterations include several types of sources, such as the so-called astrological and astronomical reports (see 5.2.), and text commentaries, as long as they are useful to reconstruct individual entries. This means that many sources do not represent the full text of EAE 52, Commentary EAE 53, and SIT 6, but only quote a few entries. All these types of texts are, indeed, different in the sequence of their entries and their contents. Only relevant and selected sources (i.e. unedited sources or those that needed to be updated, see 1.4.7.) have been presented in transliteration and translation not only in the composite texts, but also the appendix B, and copied in the plates.

One third of the sources is made of commentaries, or texts assumed to be commentaries. The text commentaries are hermeneutic texts,[284] which provide information on multiple linguistic and cultural aspects of Mesopotamian textual genres. The types, features, and purposes of the commentaries have been studied by Frahm (2011). Several commentaries relate to series EAE, and each of them deals with omens excerpted from one or more tablets of EAE. They likely attempt to unify the traditional, or standard, series EAE with the need for interpreting the ominous signs more effectively and precisely (Veldhuis 2010: 82–87). In this study, only a few types of commentaries,[285] which are listed below, have omens from tablets EAE 52 and 53 and are included in the score transliterations:

- *mukallimtu*, lit. "the one that shows/demonstrates", commentaries of the series EAE. They mainly give factual explanations, i.e. alternative interpretations of protases, sometimes more "literal" against protases that often use a metaphorical or symbolic language. Sometimes, they give remarks regarding obsolete or ambiguous words and spellings. Usually, the apodoses of the omens are not mentioned at all, or they are quoted in an abbreviated form. These commentaries can be identified by their subscript – "*mukallimtu*-commentary of tablet x of *Enūma Anu Enlil*" – which sometimes give the incipit of the corresponding EAE tablet.[286] The sequence of entries in commentaries follows the one on the EAE tablet commented upon (Frahm 2011: 42–47; 134–155).
- *ṣâtu*, lit. "distant times" (CAD Ṣ 116), commentaries. They are philological commentaries that explain difficult obsolete words or signs (Frahm 2011: 48–55). Only two commentaries edited in this manuscript belong to this typology: BM 40055 (App. B § 20) and VAT 7850 (+) AO 6486 (App. B § 27).

---

284 In Assyriology, "hermeneutic" refers to texts which pertain to the interpretation of written texts. Still, as Frahm suggested (2011: 3), "if we regard hermeneutics as a set of intellectual tools that are applied to arrive at a better understanding of the world, it would be wrong to claim that the Babylonian and Assyrian text commentaries of the first millennium BCE are the earliest testimonies of 'hermeneutical reasoning' in human history. Hermeneutics, in this most general sense, can be detected in almost every expression of the human spirit, and Mesopotamia, with its ancient lexical tradition that goes back to the birth of writing in the fourth millennium BCE, and its passion for divinatory quests, has left us large numbers of documents from much older periods to study it."

285 For a more detailed discussion on types of commentaries, see Frahm (2011: 28–58). For all the commentaries of the series EAE, see Frahm (2011: 129–166).

286 In this manuscript, the fragment K 3558 (App. B § 8) has the *mukallimtu* subscript, but it does not give the incipit of the corresponding EAE tablet.

5.1.3.5. The Organisation of the Editions

The texts belonging to the reconstructed tablets EAE 52, Commentary EAE 53, and SIT 6 (see 5.1.4.–5.1.6.) are always preceded by a list of sources. The following information is given about them: bibliography (copies and editions), provenance, ductus, measurements, and numbers assigned to the fragments on CDLI, where each source is searchable by its number, sometimes with photographs provided by the British Museum. The list also includes a description of the extrinsic features of each fragment. Parts of sources which are not edited in the reconstructed texts have been included in the appendix B; hence, in the list of sources, there are references to lines of text whose edition can be found in the appendix. For sources that belong to other parts of EAE, commentaries, or to astronomical and astrological reports, but only quote a few entries of EAE 52, Commentary EAE 53, and SIT 6 only the relevant bibliographical references to their editions are given.

The information regarding the "type of text" requires an additional explanation: whenever a source from the series EAE or SIT has a subscript with a tablet number, it is labelled "EAE or SIT number of the tablet". Whenever a source from the series EAE or SIT without a subscript can be attributed to a specific tablet because of its content, the number of the tablet is written in brackets, i.e. "EAE or SIT (number of the tablet)". Whenever a commentary is unclassified, i.e. no subscript is preserved but its content can be attributed to the series EAE or to a specific tablet, it is labelled as "EAE (number of the tablet) commentary". The fragments whose identification is not possible at the current state of this study are labelled "EAE". All additional information is written in footnotes.

After the list of sources, in the score transliteration, each reconstructed entry is written in bold and includes all variations among the sources used for the reconstruction, given in brackets as follows: (var.: ...). Then, the score transliteration and the translation follow each entry. In the reconstructed entries, half brackets are omitted to ease the reading of the transliteration. After the reconstructed text, a "commentary to the text" about philological remarks, and "further comments on the text" about general remarks on the content of the tablets are given.

5.1.4. The Reconstructed Tablet EAE 52

For the organisation of the sources, the text, and the comments, see 5.1.3.5.

List of Sources

Source:	A
Siglum:	K 3918 + K 6239
Copy:	ACh Ištar 27 (K 3918); AAT 39 (K 3918); Plate 1
Edition:	ACh Ištar 27 (K 3918)
Provenance:	Nineveh
Ductus:	Neo-Assyrian
Measurements:	8,5+ x 8,0+ x 0,4+ cm
CDLI n.:	P395307 (without photograph)

Description: Fragment of the upper left and lower right part of the obverse of a tablet, with part of the upper and left edge preserved. The reverse is lost.
Type of text: EAE (52)

Source: B
Siglum: Sm 319
Copy: ACh Suppl. 1, 46; Plate 2–3
Edition: ACh Suppl. 1, 46; App. B § 1 (reverse)
Provenance: Nineveh
Ductus: Neo-Assyrian
Measurements: 5,4+ x 4,6+ x 1,1+ cm
CDLI n.: P425351 (with photograph)
Description: Fragment of the obverse of a tablet with part of the upper and left edge preserved. The reverse is lost, except for the first signs of four lines at the lower and left edge.
Type of text: EAE (52) commentary

Source: C
Siglum: K 2118
Copy: ACh Suppl. 2, 85; Plate 4–5
Edition: ACh Suppl. 2, 85; App. B § 2 (reverse); see also Commentary EAE 53 source U
Provenance: Nineveh
Ductus: Neo-Assyrian
Measurements: 5,8+ x 3,3+ x 1,6+
CDLI n.: P394200 (with photograph)
Description: Fragment of the lower left side of a tablet with part of the left edge preserved.
Type of text: EAE 52

Source: D
Siglum: K 11632
Copy: Plate 6
Provenance: Nineveh
Ductus: Neo-Assyrian
Measurements: 6,6+ x 5,9+ x 1,4+ cm
CDLI n.: P399362 (with photograph)
Description: Fragment of the right part of the obverse of a tablet, with the lower part of the right edge preserved. The reverse is lost.
Type of text: EAE (52)

Source E_1 (+) E_2
Siglum: K 3099 + K 18689 (E_1) (+) Sm 259 (E_2)
Copy: ACh Suppl. 1, 47 (K 3099 + K 18689); ACh Suppl. 1, 57 (Sm 259); Plate 7–8

Edition: ACh Suppl. 1, 47 (K 3099 + K 18689); ACh Suppl. 1, 57 (Sm 259); App. B § 3 (E_1 obv. 1’–14’)
Provenance: Nineveh
Ductus: Neo-Babylonian
Measurements: 9,6+ x 9,2+ x 2,7+ cm (K 3099 + K 18689)
(+) 4,6+ x 3,0+ x 2,0+ cm (Sm 259)
CDLI n.: P238258 (with photograph)
Description: The two fragments do not join physically but they are part of the same tablet (Reiner 1998: 226, 275). K 3099 + K 18689 is a fragment of the upper part of the tablet, with part of the left and right edge preserved. Sm 259 is a smaller fragment of the lower left corner of the tablet, with part of the right and lower edge preserved.
Type of text: EAE (51–52)

Source: F
Siglum: Rm 100
Copy: ACh Suppl. 1, 45; Plate 9
Edition: ACh Suppl. 1, 45
Provenance: Nineveh
Ductus: Neo-Assyrian
Measurements: 8,7+ x 8,9+ x 1,6+ cm
CDLI n.: P424622 (with photograph)
Description: Fragment probably of the reverse of a tablet with part of the left edge preserved. The obverse is lost.
Type of text: EAE (52)

Source: G
Siglum: K 7214
Copy: Plate 10
Edition: App. B § 4 (ll. 1’–7’)
Provenance: Nineveh
Ductus: Neo-Assyrian
Measurements: 6,2+ x 4,4+ x 1,9+ cm
CDLI n.: P397110 (with photograph)
Description: Fragment probably of the right part of the obverse of a tablet. The reverse is lost.
Type of text: EAE (52)

Source: H
Siglum: K 11324 + K 12705
Copy: Plate 11
Edition: App. B § 5 (reverse)
Provenance: Nineveh
Ductus: Neo-Assyrian
Measurements: 2,3+ x 6,4+ x 1,6+ cm

CDLI n.: P399230 (with photograph)
Description: Fragment of the lower left part of a tablet, with part of the left and lower edge preserved.
Type of text: EAE (52)

Source: I
Siglum: K 3524
Copy: Plate 12
Provenance: Nineveh
Ductus: Neo-Assyrian
Measurements: 8,2+ x 7,1+ x 0,5+ cm
CDLI n.: P395063 (without photograph)
Description: Fragment probably of the reverse of a tablet with part of the upper and right edge preserved. The obverse is lost.
Type of text: EAE (52)

Source: J
Siglum: K 10845
Copy: Plate 10
Edition: App. B § 6 (obv. 1'–6')
Provenance: Nineveh
Ductus: Neo-Assyrian
Measurements: 3,7+ x 2,4+ x 1,3+ cm
CDLI n.: P398926 (with photograph)
Description: Fragment probably of the upper left part of the reverse of a tablet, with part of the left edge preserved. The obverse is lost.
Type of text: EAE (52) commentary

Source: K
Siglum: K 7986
Copy: Plate 13
Provenance: Nineveh
Ductus: Neo-Assyrian
Measurements: 5,8+ x 6,9+ x 1,1+ cm
CDLI n.: P397420 (with photograph)
Description: Fragment probably of the middle part of the obverse of a tablet. The reverse is lost.
Type of text: EAE (52) commentary

Source: L
Siglum: Sm 1317
Copy: Plate 14–15
Edition: App. B § 7 (obv. 4'–13'; rev. 5–9); see also Commentary EAE 53 source N, SIT 6 source N
Provenance: Nineveh

Ductus: Neo-Assyrian
Measurements: 5,6+ x 3,9+ x 1,8+ cm
CDLI n.: P425864 (with photograph)
Description: Fragment of the left part of a tablet, with part of the left and lower edge preserved.
Type of text: EAE (52–53) commentary

Source: M
Siglum: K 2246 + K 2994 + K 2324 + K 3578 + K 3605 + K 3614 + K 6152
Edition: ACh Ištar, 20; AAT 41–42 + AAT 76 (K 2324); see Largement 1957: source 2.
Provenance: Nineveh
Ductus: Neo-Assyrian
CDLI n.: P238154 (with photograph)
Type of text: EAE 56

Source: N
Siglum: 82-5-22, 46
Edition: SAA 8, 311 = RMA 234
Provenance: Nineveh
Ductus: Neo-Assyrian
CDLI n.: P336527 (with photograph)
Type of text: Report

Source: O
Siglum: K 731
Edition: SAA 8, 5 = RMA 206
Provenance: Nineveh
Ductus: Neo-Assyrian
CDLI n.: P336507 (with photograph)
Type of text: Report

Source: P
Siglum: K 3921 + DT 134 + Rm 105
Edition: Reiner-Pingree 1981: 56 IX source J
Provenance: Nineveh
Ductus: Neo-Assyrian
CDLI n.: P395310 (without photograph)
Type of text: EAE 51

Source: Q
Siglum: VAT 7850 (+) AO 6486
Copy: TCL 6, 18 (AO 6486); Plate 46 (VAT 7850)
Edition: Weidner (1925); App B § 27 (rev. 1’–2’, 5’, 10’–34’); see also Commentary EAE 53 source T

Provenance:	Uruk
Ductus:	Late Babylonian
Measurements:	11,1+ x 12,4+ x 2,8 cm (VAT 7850); 11+ x 12+ cm (AO 6486)
CDLI n.:	P363691 (AO 6486) (with photograph)
Description:	This fragment consists of a long-distance join of two fragments, one of which is housed in the Louvre Museum in Paris (AO 6486, upper part of the join), and the other one is housed in the Vorderasiatisches Museum in Berlin (VAT 7850, lower part of the join). The two fragments preserve only the reverse, the obverse is lost, and they share one line in common (rev. 21', horizontally split in two parts). VAT 7850 is a fragment of the lower part of the reverse of the tablet, with part of the lower edge preserved. The hand copy of the fragment AO 6486 was published in TCL 6, 18, and a commented edition of both AO 6486 and VAT 7850 was published by Weidner (1925).
Type of text:	EAE (52) *ṣâtu* commentary

Text

1.	**DIŠ *ina* ^{iti}BÁR ^{mul}AŠ.GÁN *u* MUL.MUL IGI^{meš} x[...]**
A obv. 1	DIŠ *ina* ^{iti}BÁR ^{mul}AŠ.[GÁN...]
B obv. 1	DIŠ *ina* ^{iti}⌜BÁR⌝ ^{mul}AŠ.GÁN *u* MUL.[MUL...]
P rev. 26' (catchline)	DIŠ *ina* ^{iti}BÁR ^{mul}AŠ.GÁN *u* MUL.MUL IGI^{meš} x[...]
Q rev. 26' (subscript)	[*šá* ŠÀ DIŠ *ina* ^{iti}B]ÁR ^{mul}AŠ.GÁN *u* MUL.MUL IGI^{meš}
	***ina* ^{iti}BÁR ^{mul}SAG.ME.GAR *u* ^{m}[^{ul}...]**
B obv. 2	*ina* ^{iti}BÁR ^{mul}SAG.ME.GAR *u* ^{m}[^{ul}...]

If in Nisannu (i.e. Month I) the Field and the Pleiades are visible, ...[...]. (It means that) in Nisannu Jupiter and the s[tar...].

2.	**DIŠ *ina* ^{iti}BÁR ^{mul}AŠ.GÁN *u* ^{mul}[... IGI^{meš}...]**
A obv. 2	DIŠ *ina* ^{iti}BÁR ^{mul}AŠ.G[ÁN...]
B obv. 3	DIŠ *ina* ^{iti}BÁR ^{mul}AŠ.GÁN *u* ^{mul}[...]
	^{mul}*dil-bat u* ^{d}GU$_4$.U[D...]
B obv. 4	^{mul}*dil-bat u* ^{d}GU$_4$.U[D...]

If in Nisannu (i.e. Month I) the Field and the star [... are visible, ...]. (It means that) Venus and Mercur[y...].

3.	**DIŠ *ina* ^{iti}BÁR ^{mul}AŠ.GÁN *u* ^{mu}[^{l}... IGI^{meš}]**
A obv. 3	DIŠ *ina* ^{iti}BÁR ^{mul}AŠ.GÁ[N...]
B obv. 5	DIŠ *ina* ^{iti}BÁR ^{mul}AŠ.GÁN *u* ^{mu}[^{l}...]
	***ḫa-rab* KUR SU.BIR$_4$^{ki} ^{m}[^{ul}...]**
B obv. 6	*ḫa-rab* KUR SU.BIR$_4$^{ki} ^{m}[^{ul}...]

If in Nisannu (i.e. Month I) the Field and the sta[r... are visible], (there will be) the devastation of Subartu. (It means that) the s[tar...].

4.	**DIŠ *ina* ^{iti}BÁR ^{mul}AŠ.GÁN *u* ^{mul}[... IGI^{meš}]**
A obv. 4	DIŠ *ina* ^{iti}BÁR ^{mul}AŠ.GÁN *u* [...]
B obv. 7	DIŠ *ina* ^{iti}BÁR ^{mul}AŠ.GÁN *u* ^{mul}[...]
	***ḫa-rab* KUR ELAM.MA^{ki} ^{m}[^{ul}...]**
B obv. 8	*ḫa-rab* KUR ELAM.MA^{ki} ^{m}[^{ul}...]

If in Nisannu (i.e. Month I) the Field and the star [... are visible], (there will be) the devastation of Elam. (It means that) the s[tar...].

5.	**DIŠ *ina* ^{iti}BÁR ^{mul}AŠ.GÁN *u* ^{mu}[^{l}... IGI^{meš}]**
A obv. 5	DIŠ *ina* ^{iti}BÁR ^{mul}AŠ.GÁN *u* [...]
B obv. 9	⸢DIŠ⸣ *ina* ^{iti}BÁR ^{mul}AŠ.GÁN *u* ^{mu}[^{l}...]
C obv. 1’	⸢DIŠ *ina*⸣ ^{i}[^{ti}BÁR...]
	[*ḫ*]*a-rab* KUR URI^{ki} ^{mul}SA[G.ME.GAR$^{?}$...]
B obv. 10	[*ḫ*]*a-rab* KUR URI^{ki} ^{mul}SA[G.ME.GAR$^{?}$...]

If in Nisannu (i.e. Month I) the Field and the sta[r... are visible, (there will be) the d]evastation of Akkad. (It means that) Ju[piter$^{?}$...].

6.	**DIŠ *ina* ^{iti}BÁR ^{mul}AŠ.GÁN *u* ^{mul}x[... IGI^{meš}]**
A obv. 6	DIŠ *ina* ^{iti}BÁR ^{mul}AŠ.GÁN *u* [...]
B obv. 11	[DIŠ *ina* ^{i}]^{ti}BÁR ^{mul}AŠ.GÁN *u* ^{mul}x[...]
C obv. 2’	DIŠ *ina* ^{iti}BÁR[...]
	[*ḫa*]*-rab* KUR DÙ.A.BI MÁŠ.ANŠE TUR [...]
B obv. 12	[*ḫ*]*a-rab* KUR DÙ.A.BI MÁŠ.ANŠE TUR [...]

If in Nisannu (i.e. Month I) the Field and the star ...[... are visible, (there will be) the de]vastation of the entire country, the herd will diminish [...].

7.	**DIŠ *ina* ^{iti}BÁR ^{mul}AŠ.GÁN *u* ^{mul}[... IGI^{meš} LUGAL^{meš} *šá*]**
A obv. 7	DIŠ *ina* ^{iti}BÁR ^{mul}AŠ.GÁN *u* ^{m}[^{ul}...]
B obv. 13	[DIŠ *ina* ^{it}]^{i}BÁR ^{mul}AŠ.GÁN *u* ^{mul}[...]
C obv. 3’	DIŠ *ina* ^{iti}BÁR[...]
	[KU]R DÙ.A.BI *in-na-da-r*[*u-ma* KÚR^{meš}]
B obv. 14	[KU]R DÙ.A.BI *in-na-da-r*[*u-ma* KÚR^{meš}]
	***ina* ^{giš}TUKUL NÍ-*šú-n*[*u*...]**
A obv. 8	*ina* ^{giš}TUKUL NÍ-*šú-n*[*u*...]
C obv. 4’	*ina* ^{gi}[^{š}TUKUL...]
D obv. 1’	[...] x[...]

If in Nisannu (i.e. Month I) the Field and the star [... are visible, the kings of the] entire [countr]y will become worri[ed and will be in enmity], under thei[r] own weapons [...].

8.	**DIŠ *ina* ^{iti}BÁR ^{mul}AŠ.GÁN *u* ^{m}[^{ul}... IGI^{meš}...]**
A obv. 9	DIŠ *ina* ^{iti}BÁR ^{mul}AŠ.GÁN *u* ^{m}[^{ul}...]
C obv. 5’	DIŠ *ina* ^{iti}B[ÁR...]
D obv. 2’	[...] x[...]

If in Nisannu (i.e. Month I) the Field and the s[tar... are visible, ...].

9.	**DIŠ *ina* ^{iti}BÁR ^{mul}AŠ.GÁN *u* ^{m}[^{ul}... IGI^{meš}...]**
A obv. 10	DIŠ *ina* ^{iti}BÁR ^{mul}AŠ.GÁN *u* ^{m}[^{ul}...]
C obv. 6’	DIŠ *ina* ^{iti}B[ÁR...]
	[... *ina* KUR] *su-a-lu* G[ÁL-*ši*]
D obv. 3’	[... *ina* KUR] *su-a-lu* G[ÁL-*ši*]

If in Nisannu (i.e. Month I) the Field and the s[tar... are visible, ... in the country] t[here will be] phlegm.

10.	**DIŠ *ina* ^{iti}BÁR ^{mul}AŠ.GÁN *u* ^{mul}[... I]GI^{meš} *ina* KUR DÙ.[A.BI *ka-ru*]-*ur-tu*$_4$**
A obv. 11	DIŠ *ina* ^{iti}BÁR ^{mul}AŠ.GÁN *u* ^{m}[^{ul}... I]GI^{meš} *ina* KUR DÙ.[A.BI...]
C obv. 7’	DIŠ *ina* ^{iti}B[ÁR...]
D obv. 4’	[... *ka-ru*]-*ur-tu*$_4$
E$_1$ obv. 15’’	DIŠ [*ina* ^{iti}]⸢BÁR ^{mul}⸣[AŠ.GÁN] *u* ⸢^{mul}⸣[...]
	GÁ[L-*ši*]
D obv. 4’	GÁ[L-*ši*]

If in Nisannu (i.e. Month I) the Field and the star [...] are [vi]sible, in the ent[ire] country there will b[e] a [voracious] hunger.

11.	**DIŠ *ina* ^{iti}BÁR ^{mul}AŠ.GÁN *u* ^{mul}*dil-bat* IGI^{meš} *ina* KUR DÙ.A.BI**
A obv. 12	DIŠ *ina* ^{iti}BÁR ^{mul}AŠ.GÁN *u* ^{mu}[^{l}*dil-ba*]*t* IGI^{meš} *ina* KUR DÙ.A.⸢BI⸣ [...]
C obv. 8’	DIŠ *ina* ^{iti}BÁ[R...]
E$_1$ obv. 16’’	DIŠ *ina* ⸢^{iti}⸣BÁR ^{mul}AŠ.GÁN *u* ^{mul}*dil-bat* ⸢IGI^{meš}⸣ [*ina*] K[UR...]
D obv. 5’	[...]
	UN^{meš} *iš-šal-l*[*a-la*]
D obv. 5’	UN^{meš} *iš-šal-l*[*a-la*]

If in Nisannu (i.e. Month I) the Field and Venus are visible, in the entire country the people will be rob[bed].

A, C, D, E$_1$ ————————————————

12.	**DIŠ *a-na* ŠÀ ^{mul}AŠ.GÁN MUL.MUL KU$_4$^{meš}-*ma* GUB^{meš} *ina* KUR DÙ.A.BI**
A obv. 13	DIŠ *a-na* ŠÀ ^{mul}AŠ.GÁN MU[L.MU]L KU$_4$^{meš}-*ma* GUB^{meš} *ina* KUR DÙ.A.BI
B obv. 15	[DIŠ *ana* ŠÀ] ⸢^{mul}AŠ.GÁN MUL.MUL⸣ K[U$_4$...]
C obv. 9’	DIŠ *a-na* Š[À...]
D obv. 6’	[... *ina* KUR D]Ù.A.BI
E$_1$ obv. 17’’	DIŠ *a-na* ŠÀ ^{mul}AŠ.GÁN MUL.MUL KU$_4$⸢^{meš} -*ma*⸣ [...]
K 1’	[DIŠ *a-na*] ⸢ŠÀ ^{mul}AŠ.GÁN MUL.MUL⸣ [...]

	***ka-ru-ur-tu*$_4$ GÁL-*ma* KUR GAL *ana* KUR TUR *ana* TIN-*ṭi* D[U-*ak*...]**
A obv. 13–14	*k*[*a-ru-ur-tu*$_4$...] (14) KUR GAL *ana* KUR TU[R] *ana* TIN-*ṭi* [...]
B	(remainder is missing)
C obv. 10'	[K]UR GA[L...]
D obv. 6'–7'	*ka-ru-ur-t*[*u*$_4$ GÁL] (7') [...]
K 2'	[*k*]*a-ru-ur-tu*$_4$ GÁL-*ma* KUR ⸢GAL *ana* KUR TUR *ana* TIN-*ṭi*⸣ D[U-*ak*...]

	(var.: ***it-t*[*al-lak*]**) **[^{d}G]U$_4$.UD : ^{d}*ṣal-bat-a-nu ina* ^{d}UTU.È**
D obv. 7'.	*it-t*[*al-lak*]
K 3'	[^{d}G]U$_4$.UD : ^{d}*ṣal-bat-a-nu ina* ^{d}UTU.È

	***ina* ŠÀ ^{mul}AŠ.GÁN GUB^{meš} [...]**
K 3'	*ina* ŠÀ ^{mul}AŠ.GÁN GUB^{meš} [...]

If the Pleiades enter into the Field and stand still, in the entire country there will be a voracious hunger, the great land [will go] to the small land for sustenance [...]. (It means that) M]ercury – (var.) Mars stands in the east (lit. sunrise) inside the Field [...].

13.	**DIŠ *a-na* ŠÀ ^{mul}AŠ.GÁN ^{mul}SIPA.ZI.AN.NA KU$_4$-*ma* GUB**
A obv. 15	DIŠ *a-na* ŠÀ ^{mul}AŠ.GÁN $^{m[ul]}$⸢SIPA⸣.ZI.AN.NA KU$_4$-*ma* GUB
C obv. 11'	[...]x [...]
D obv. 8'	[...
E$_1$ obv. 18''	DIŠ *a-na* ŠÀ ^{mul}AŠ.GÁN ^{mul}SIPA.Z[I.AN.NA...]

	AN.GE$_6$ 30 *u* 20 GAR-*ma ina* MU BI *ina* KUR DÙ.A.BI DINGIR GU$_7$
A obv. 15–16	AN.GE$_6$ 20+[10...] (16) *ina* MU BI *ina* KUR D[Ù].⸢A⸣.BI DINGIR GU$_7$
D obv. 8'	AN.G]E$_6$ 30 *u* 20 GAR-*ma ina* MU BI *ina* KUR DÙ.A.BI DINGIR [GU$_7$]

	^{d}IŠKUR MA-RAB GAR MU.3.KAM* *ana* GURUŠ KI.SIKIL
A obv. 16–17	^{d}IŠKUR MA-RAB GAR [...] (17) *ana* GURUŠ KI.SIKIL
C	(remainder is missing)
D obv. 9'	[...
E$_1$ obv. 19''	⸢^{d}⸣IŠKUR MA-RAB GAR MU.3.⸢KAM* *ana*⸣ [...]

	ŠU.GI ^{munus}ŠU.GI MÁŠ.ANŠE (*u*)	**NÍG.ZI.GÁL EDIN.NA NU *ša*[*l*$^{?}$-*ma*$^{?}$]**
A obv. 17	ŠU.G[I $^{m]unus}$ŠU.GI MÁŠ.ANŠE *u*	NÍG.ZI.GÁL EDIN.NA N[U...]
D obv. 9'	Š]U.⸢GI⸣ ^{munus}ŠU.GI MÁŠ.ANŠE	NÍG.ZI.GÁL EDIN.NA NU *ša*[*l*$^{?}$-*ma*$^{?}$]

If the True Shepherd of Anu enters into the Field and stands still, an eclipse of the Moon and the Sun will occur; in that year in the entire country the god will devour, Adad will produce MA-RAB, for 3 years for the young man, the young woman, the old man, the old woman, the herd (and) the wild animals of the steppe it will not be [good$^{?}$].

14.	**[DIŠ *a*]-*na* ŠÀ ^{mul}AŠ.GÁN $^{m[ul}$*š*]*u-ku-du* KU$_4$-*ma* GUB KÚR ZI-*ma***
A obv. 18	[DIŠ *a*]-⸢*na* ŠÀ ^{mul}AŠ.GÁN⸣ $^{m[ul}$*š*]*u-ku-du* KU$_4$-*ma* GUB KÚR ZI-*ma*
D obv. 10'	[... G]UB KÚR ZI-*ma*
E$_1$ obv. 20''	[DIŠ *a-na* Š]À ⸢^{mul}AŠ.GÁN ^{mul}⸣[...]

	MU.6.KÁM (var.: **MU.5.KÁM**) **KUR *i-da-aš* [...] KUR-*ád***
A obv. 18–19	MU.6.KÁM KUR *i-d*[*a-aš*...] (19) [...] KUR [...]
D obv. 10'–11'	MU.5.KÁM KUR *i-da-aš* (11') [...] KUR-*ád*
E	(break of unknown length)

[If] the [A]rrow enters [i]nto the Field and stands still, the enemy will rise and for 6 (var.: 5) years will trample down the country, [...] (he/she/it) will conquer.

15.	**[DIŠ *a-na* Š]À ^{mul}AŠ.GÁN ^{mul}*dil-bat* KU$_4$-*ma* GUB *ina* KUR DÙ.A.BI *siḫ-pu***
A obv. 20	[DIŠ *a-na* ŠÀ ^{mul}AŠ.GÁN ^{mul}*d*]*il-bat* KU$_4$-*ma* GUB *ina* KUR DÙ.A.BI ⸢*siḫ*⸣-*pu*
D obv. 12’	[ ]-*ma* GUB *ina* KUR DÙ.A.BI *siḫ-pu*
K 4’	[DIŠ *a-na* Š]À ⸢^{mul}⸣AŠ.GÁN ^{mul}*dil-bat* KU$_4$-*ma* GUB *ina* KUR DÙ.A.BI
	(var.: *šul-pu*) GAR-*ma* LUGAL^{meš} [...]-*šú-nu* KUR KUR LÚ LÚ
A obv. 20–21	[...] (21) [...]-⸢*šú*⸣-*nu* KUR KUR LÚ LÚ
D obv. 12’–13’	GAR-*ma* LUGAL^{meš} (13’) [... L]Ú
K 4’	*šul-pu* GA[R-*ma*...]
	ina **^{giš}TUKUL *ú-šam-qat-(ma)* ŠEŠ ŠEŠ KAR-*ʾa* [... m]e$^?$ KIMIN**
A obv. 21–22	*ina* ^{giš}TUKUL *ú-šam-qat-ma* ŠEŠ Š[EŠ...] (22) [... m]e$^?$ KIMIN
D obv. 13’–14’	*ina* ^{giš}<TUKUL> *ú-šam-*⸢*qat*⸣ ŠEŠ ŠEŠ KAR-⸢*ʾa*⸣ (14’) [...
	MU.3.KÁM *di-im-ta* NU GÁL (var.: ***ina*** **K[UR$^?$...])**
A obv. 22	MU.3.KÁM *di-im-ta* *ina* K[UR$^?$...]
D obv. 14’	M]U.3.KÁM *di*$^!$(KI)-*im-ta* NU ⸢GÁL⸣
	^{d}*dil-bat ina* ^{d}UTU.È GUB *ina* ŠÀ ^{mul}AB.SÍN *ina* te [...]
K 5’	⸢^{d}⸣*dil-bat ina* ^{d}UTU.È GUB *ina* ŠÀ ^{mul}AB.SÍN *ina* te [...]

[If] Venus enters [int]o of the Field and stands still, there will be a sweeping attack (var.: *destruction*) in the entire country, the kings [will...] them, (one) land the (other) land, (one) man the (other) man will overpo[wer] in weapon(s), (and one) brother will rob the (other) brother [...] ditto (?), for 3 years there will not be weeping/fortified area (var.: in the l[and$^?$...]). (It means that) Venus stands in the east (lit. sunrise), inside the Furrow in ...[...].

16.	**[DIŠ *a-na*] ŠÀ ^{mul}AŠ.GÁN ^{mul}ŠU.PA KU$_4$-*ma* GUB ERÍN *gu-ti-i***
A obv. 23	[... K]U$_4$-*ma* GUB ERÍN *gu-ti-i*
D obv. 15’	[... GU]B ERÍN ⸢*gu*⸣-*ti-i*
K 6’	[DIŠ *a-na*] ⸢ŠÀ ^{mul}⸣AŠ.GÁN ^{mul}ŠU.PA KU$_4$-*ma* GUB ERÍN *gu-ti-*⸢*i*⸣
	ana **KUR ELAM.MA^{ki} ZI-*ma* (**var.: **ZI-*a*) [... T]UR-*ár* LUGAL GAR(-*šú*)**
A obv. 23–24	*ana* KUR ELAM.MA^{ki} Z[I...] (24) [... T]UR-*ár* LUGAL GAR [...]
D obv. 15’–16’	*ana* KUR ELAM.MA ZI-⸢*ma*⸣ (16’) [... T]UR-*ár* LUGAL GAR-*šú*
K 7’	[*ana* KU]R [ELA]M.MA^{ki} ZI-*a*
	^{d}*ṣal-bat-a-nu ina* ^{d}UTU.È ^{mul}AB.[SÍN...]
K 7’	^{d}*ṣal-bat-a-*⸢*nu*⸣ *ina* ^{d}UTU.È ^{mul}⸢AB⸣.S[ÍN...]

[If] the Resplendent One enters [in]to the Field and stands still, the Gutean army will rise against Elam, [... he/she/it will re]duce, the king will put him in charge. (It means that) Mars in the east (lit. sunrise) of the Furr[ow...].

17.	**[DIŠ *a-na* ŠÀ] ^{mul}AŠ.GÁN ^{mul}NUN^{ki} KU$_4$-*ma* GUB *ina* KUR SU.BIR$_4$^{ki}**
A obv. 25	[... KU$_4$-*m*]*a* GUB *ina* KUR SU.BIR$_4$^{ki}
D obv. 17’	[... KU]R SU.BIR$_4$^{ki}
K 8’	[DIŠ *a-na* ŠÀ] ⸢^{mul}⸣AŠ.GÁN ^{mul}NUN^{ki} KU$_4$-*ma* GUB *ina* KUR

(var.: SU^{ki}) KIMIN (var: :) (*ina*) KUR URI^{ki} [...]x-*ti* GAR^{meš}

A obv. 25 — KIMIN [...]

D obv. 17’ — KIMIN KUR ⸢URI⸣[^{ki}]

K 8’–9’ — SU^{ki} : *ina* KUR UR[I^{ki}...] (9’) [...]x-*ti* GAR^{meš}

^{d}GU$_{4}$.UD *ina* ^{d}UTU.È *ana* ^{d}*dil-bat* DIM$_{4}$ [...]x x ma su x[...]

D obv. 18’ — [...]x ⸢x ma su⸣ x[...]

K 9’ — ^{d}GU$_{4}$.UD *ina* ^{d}UTU.È *ana* ^{d}*dil-bat* DIM$_{4}$ [...]

[If] the Eridu-star enters [into] the Field and stands still, in Subartu ditto (i.e. the Gutean army will rise against Elam, ... the child of the king will be put in charge) (var. –) (in) Akkad [the... of the]... will be established. (It means that) Mercury in the east (lit. sunrise) approaches Venus [...]... ...[...].

18. — **[DIŠ *a-na* ŠÀ ^{mu}]^{l}AŠ.GÁN ^{mul}EN.TE.NA.BAR.ḪUM KU$_{4}$-*ma* GUB *ina* KUR**

D — (remainder is missing)

K 10’ — [DIŠ *a-na* ŠÀ ^{mu}]^{l}AŠ.GÁN ^{mul}EN.TE.NA.BAR.ḪUM KU$_{4}$-*ma* GUB *ina* KUR

DÙ.[A.BI MU.x].KÁM *ana* KUR URI^{ki} LUGAL MAR.[TU^{ki}... *nam*]-*maš-še-e*

A obv. 26 — [... MU.x].KÁM *ana* KUR URI^{ki} LUGAL MAR.[TU^{ki}...]

K 10’–11’ — DÙ.[A.BI...] (11’) [... *nam*]-⸢*maš*⸣-*še-e*

EDIN ^{d}IŠKUR RA [... $^{d/mul}$... *ina*] ^{d}UTU.È *ana* ^{d}*dil-bat* DIM$_{4}$ [...]

K 11’–12’ — EDIN ^{d}IŠKUR RA [...] (12’) [... *ina*] ⸢^{d}⸣UTU.È *ana* ^{d}*dil-bat* DIM$_{4}$ [...]

[If] the Mouse enters [into] the Field and stands still, in the en[tire] country for [x years] the king of Amu[rru] against Akkad [...] Adad will devastate [the h]erds of the steppe. [... (It means that) the planet...] in the east (lit. sunrise) approaches Venus [...].

19. — **[DIŠ *a-na* ŠÀ ^{mul}AŠ.GÁ]N ^{mul}GÍR.TAB KU$_{4}$-*ma* IGI *ina* KUR D[Ù.A.BI...]**

K 13’ — [DIŠ *a-na* ŠÀ ^{mul}AŠ.GÁ]N ^{mul}GÍR.TAB KU$_{4}$-*ma* IGI *ina* KUR D[Ù.A.BI...]

[...] GAR LUGAL mu x x[...]x^{meš}

A obv. 27 — [...] ⸢GAR LUGAL mu x x⸣[...]

K 14’ — [...]x^{meš}

^{d}*dil-bat ina* ^{d}UTU.ŠÚ.A *ina* ^{mul}GÍR.TAB D[IM$_{4}$...]

K 14’ — ^{d}*dil-bat ina* ^{d}UTU.ŠÚ.A *ina* ^{mul}GÍR.TAB D[IM$_{4}$...]

[If] the Scorpion enters [into the Fiel]d and it is visible, in the e[ntire] country [...] (he/she/it) will be placed, the king ...[...]... (It means that) Venus in the west (lit. sunset) a[pproaches] the Scorpion [...].

20. — **[DIŠ *a-na* ŠÀ ^{mul}AŠ.GÁN ^{mul}Š]UL.PA.È.A KU$_{4}$-*ma* IGI *ina* KUR DÙ.A.BI x[...]**

A — (remainder is missing)

K 15’ — [DIŠ *a-na* ŠÀ ^{mul}AŠ.GÁN ^{mul}Š]UL.PA.È.A KU$_{4}$-*ma* IGI *ina* KUR DÙ.A.BI x[...]

[...^{d}S]AG.ME.GAR *ina* ^{mul}AB.SÍN [...]

K 16’ — [... ^{d}S]AG.ME.GAR *ina* ^{mul}AB.SÍN [...]

[If Š]ulpaea (i.e. Jupiter) enters [into the Field] and it is visible, in the entire country ...[... (It means that) J]upiter in the Furrow [...].

21. **DIŠ *a-na* ŠÀ ^{m}[^{ul}AŠ.GÁN ^{mul}UG]$_{5}^{?}$.GA KU$_4$-*ma* IGI [...]**
E$_2$ obv. 1’ ⸢DIŠ *a*⸣-*na* ŠÀ⸣ ^{m}[^{ul}AŠ.GÁN...]
K 17’ [DIŠ *a-na* ŠÀ ^{mul}AŠ.GÁN ^{mul}UG]$_{5}^{?}$.GA KU$_4$-*ma* IGI

URU^{me}-*šú ú-šal-pat* [...]
E$_2$ obv. 2’ URU^{me}-*šú ú-šal-*⸢*pat*⸣ [...]

***^{d}dil-bat ina* ^{mul}[...]**
K 17’ ^{d}*dil-bat ina* ⸢^{mul}⸣[...]

If [the Rav]en$^{?}$ enters into the [Field] and it is visible, [...] (he/she/it) will rise against his cities [...]. (It means that) Venus in the star [...].

22. **DIŠ *a-na* ŠÀ ^{mul}AŠ.GÁN ^{mul}AMAR.UTU KU$_4$-*ma* IGI**
E$_2$ obv. 3’ DIŠ *a-na* ŠÀ ^{mul}AŠ.⸢GÁN⸣ ^{mul}⸢AMAR⸣.[UTU...]
K 18’ [DIŠ *a-na* ŠÀ ^{mul}AŠ.GÁN ^{mul}]AMAR.UTU KU$_4$-*ma* IGI

***ina* KUR DÙ.A.[BI... ^{d}SAG].ME.GAR *ina* ^{mul}AB.S[ÍN...]**
K 18’-19’ *ina* KUR DÙ.⸢A⸣.[BI...] (19’) [... ^{d}SAG].⸢ME.GAR⸣ *ina* ^{mul}⸢AB⸣.S[ÍN...]

If the Marduk-star enters into the Field and it is visible, in the enti[re] country [... (It means that) Jup]iter in the Furr[ow...].

23. **DIŠ *a-na* ŠÀ ^{mul}AŠ.GÁN ^{mul}GU$_4$.AN.NA K[U$_4$-*ma* IGI...]x x[...]**
E$_2$ obv. 4’ DIŠ *a-na* ŠÀ ^{mul}AŠ.⸢GÁN⸣ ^{mul}GU$_4$.AN.⸢NA⸣ K[U$_4$...]
H obv. 1’ [...]x x[...]
K [...]x [...]
K (remainder is missing)

If the Bull of Heaven e[nters] into the Field [and it is visible, ...]... ...[...].

24. **DIŠ *a-na* ŠÀ ^{mul}AŠ.GÁN ^{mul}GÚ.ḪAL KU$_4$-*ma* IGI ^{mul}[...] ^{mul}AB.[SÍN$^{?}$...]**
E$_2$ obv. 5’ DIŠ *a-na* ŠÀ ^{mul}AŠ.GÁN ^{mul}GÚ.ḪAL KU$_4$-*ma* ⸢IGI ^{mul}⸣[...]
H obv. 2’ [...] ⸢^{mul}⸣AB.[SÍN$^{?}$...]

If the Throat enters into the Field and it is visible, the star [...] (It means that) the Fu[rrow$^{?}$...].

25. **DIŠ *a-na* ŠÀ $^{d/mul}$AŠ.GÁN ^{mul}KU$_6$ KU$_4$-*ma* IGI *ina* [KU]R DÙ.A.BI x[...]**
E$_2$ obv. 6’ DIŠ *a-na* ŠÀ ^{mul}AŠ.GÁN ^{mul}KU$_6$ KU$_4$-*ma* IGI *ina* [...]
H obv. 3’ [... KU]R DÙ.A.BI x[...]
J 7’ DIŠ *a-na* ŠÀ ^{d}AŠ.GÁN ^{mu}[^{l}...]

***šá* ^{d}*ṣal-bat-a-nu ina* ^{mul}[AB].SÍN [...]**
H obv. 4’ [...] ⸢^{mul}⸣[A]B.SÍN [...]
J 8’ *šá* ^{d}*ṣal-bat-a-nu ina* [...]

If the Fish enters into the Field and it is visible, in the entire [countr]y ...[...]. (It means) that Mars in the [Fu]rrow [...].

26. **DIŠ *a-na* ŠÀ ^{mul}AŠ.GÁN $^{d/mul}$*ṣal-bat-a-nu* KU$_4$-*ma* IGI MU.3.KAM***
E$_2$ obv. 7’ DIŠ *a-na* ŠÀ ^{mul}AŠ.GÁN ^{mul}*ṣal-bat-a-nu* KU$_4$-*ma* I[GI...]

H obv. 5’	[DIŠ *a-na* ŠÀ ^{mul}AŠ].⸢GÁN ^{d}*ṣal-bat*⸣-*a-nu* ⸢KU$_4$-*ma*⸣ IGI M[U.3.KÁM...]
J 9’	DIŠ *a-na* MIN ^{mul}*ṣal-bat-a-n*[*u*...]
M rev. 32’	DIŠ *a-na* ⸢ŠÀ ^{mul}⸣[AŠ.GÁN ^{mu}]^{l}*ṣal-bat-a-nu* KU$_4$-*ma* IGI MU.3.KAM*
	ŠÈG^{me}-*ni* ZAL-*ú* ILLU^{me} *ḫar-p*[*u*] GÁN.ZI SIG$_5$ (var.: **SI.SÁ) *ina* TIL MU**
E$_2$ obv. 8’	SI.SÁ *ina* TIL MU
J 10’	*ina* TIL MU
M rev. 32’–33’	ŠÈG^{me}-*ni* ZAL-*ú* ILLU^{me} *ḫar-p*[*u*] GÁN.ZI SIG$_5$ (33’) *ina* TIL [...
	ŠE.GÙN.NU *šá* KUR URI^{ki} ^{d}IŠKUR RA GÁN.BA TUR
E$_2$ obv. 8’	ŠE.GÙN.NU *šá* KUR URI^{ki} [...]
J 10’	ŠE.GÙN.N[U...]
M rev. 33’	...]x *šá* KUR URI^{ki} ^{d}IŠKUR RA GÁN.BA TUR
	[... $^{d/mul}$...]x *ina* ^{mul}AB.SÍN IGI-*ma* ^{d}UDU.IDIM.GU$_4$.UD ^{d}[...]
H obv. 6’	[...]x *ina* ⸢^{mul}⸣AB.SÍN IGI-*ma* ^{d}UDU.IDIM.GU$_4$.UD ^{d}[...]

If Mars enters into the Field and it is visible, the rain will continue for 3 years, (there will be) earl[y] floods, the cultivation will be good (var.: prosper), at the end of the year Adad will devastate the crop of Akkad, the market rate will diminish. [... (It means that) the planet...]... in the Furrow is visible, and Mercury, the planet [...].

27.	**[DIŠ] *a-na* ŠÀ ^{mul}AŠ.GÁN ^{mul}UDU.IDIM KU$_4$-*ma* IGI M[U.x.KAM*...]**
E$_2$ obv. 9’	[DIŠ] ⸢*a-na* ŠÀ ^{mul}⸣AŠ.GÁN ^{mul}UDU.IDIM KU$_4$-*ma* IGI M[U.x.KAM*...]
J 11’	[...] ⸢x x x⸣ [...]
	[...]x ÚŠ^{me} GÁL^{meš} ^{d}[...]
E$_2$ edge. 10’	[...]x ÚŠ^{me} GÁL^{meš} ⸢^{d}⸣[...]
J	(remainder is missing)

[If] a planet enters into the Field and it is visible, [for x] ye[ars...]... there will be a pestilence. (It means that) the planet [...].

28.	**[DIŠ *a-n*]*a* ŠÀ ^{mul}AŠ.GÁN ^{mul}UD.KA.DUḪ.A KU$_4$-*ma* [IGI...]**
E$_2$ edge. 11’	[DIŠ *a-na* ŠÀ ^{mul}AŠ.GÁN] ⸢^{mul}⸣UD.KA.DUḪ.A KU$_4$-*ma* [...]
H obv. 7’	[DIŠ *a-n*]*a* ŠÀ ^{mul}AŠ.GÁN ^{mul}UD.⸢KA⸣.DUḪ.A KU$_4$-*ma* [...]
	[...]^{meš} *ina* KA ^{d}*èr-ra* GÁL^{meš} ^{d}x[...]
H obv. 8’	[...]^{meš} *ina* KA ^{d}*èr-ra* ⸢GÁL⸣^{meš} ⸢^{d}⸣x[...]

[If] the Demon with the Gaping Mouth enters [in]to the Field and [it is visible], there will be [...] by the command of Erra. (It means that) the planet ...[...].

H ____

29.	**[DIŠ *a-na* ŠÀ ^{mul}AŠ.GÁN ^{mul}...]x KU$_4$-*ma* IGI ZI-*u*[*t*...]**
E$_2$ edge. 12’	[...]x KU$_4$-*ma* IGI ZI-*u*[*t*...]
H	(end of the obverse)

[If the star...] enters [into the Field]... and it is visible, (there will be) the revol[t of...].

E$_2$ ____

30. **DIŠ ^{mul}AŠ.GÁN *ina* MAŠ.SÌLA 30** (var.: **MAŠ.SÌLA-*šú*) GUB**
E_2 rev. 1 [DIŠ] ⸢^{mul}AŠ.GÁN⸣ *ina* MAŠ.SÌLA 30 GU[B...]
N obv. 9 DIŠ ^{mul}AŠ.GÁN *ina* MAŠ.SÌLA-*šú* GUB

***ina* MU BI BURU$_{14}$ KUR SI.SÁ**
N obv. 9–10 *ina* MU BI BURU$_{14}$ KUR (10) SI.SÁ

If the Field stands in the shoulder of the Moon (var.: its shoulder), in that year the harvest of the country will prosper.

31. **DIŠ ^{mul}AŠ.GÁN *ina* SI 15 ^{d}30 [GUB...]**
E_2 rev. 2 ⸢DIŠ⸣ ^{mul}AŠ.GÁN *ina* SI 15 ^{d}30 [...]

If the Field [stands] in the right horn of the Moon, [...].

32. **DIŠ ^{mul}AŠ.GÁN *ina* SI 2,30** (var.: **GÙB-*šú*) ^{d}3[0] GUB-*iz* *ina* MU BI**
E_2 rev. 3 DIŠ ^{mul}AŠ.GÁN *ina* SI 2,30 ⸢^{d}20⸣[+10...]
N obv. 7 DIŠ ^{mul}AŠ.GÁN *ina* SI GÙB-*šú* GUB-*iz* *ina* MU BI

***me-reš* KUR URI^{ki} SI.SÁ**
N obv. 8 *me-reš* KUR URI^{ki} SI.SÁ

If the Field stands in the left horn of the Mo[on], in that year the cultivation of Akkad will prosper.

33. **DIŠ ^{mul}AŠ.GÁN *ina* KI.TA [^{d}30 GUB...]**
E_2 rev. 4 DIŠ ^{mul}AŠ.GÁN *ina* KI.⸢TA⸣ [...]

If the Field [stands] below [the Moon, ...].

34. **DIŠ ^{mul}AŠ.GÁN *ina* K[I.GUB$^{?}$ ^{d}30 GUB...]**
E_2 rev. 5 DIŠ ^{mul}AŠ.GÁN *ina* K[I...]

If the Field [stands] in the p[osition$^{?}$ of the Moon, ...].

35. **DIŠ ^{mul}AŠ.GÁN [...]**
E_2 rev. 6 DIŠ ^{mul}AŠ.GÁN [...]

If the Field [...].

E_2 ———

36. **DIŠ ^{mul}[AŠ.GÁN ...]**
E_2 rev. 7 ⸢DIŠ ^{mul}⸣[...]
E_2 (end of E_2, break of unknown length)

If [the Field...].

37. **DIŠ *ṣal-[lum-mu-ú...]* A.RÁ x[...]**
E_1 rev. 1' ⸢DIŠ *ṣal*⸣-[*lum-mu-ú*...]
E_1 rev. 2' A.⸢RÁ⸣ x[...]

If a g[low (coming from) ...] ...[...], the course ...[...].

38.	**DIŠ *ṣal-lum-mu-ú* ^{mul}x[...]**
E$_1$ rev. 3’	DIŠ *ṣal-lum-*⸢*mu-ú* ^{mul}⸣x[...]

If a glow (coming from) the star ...[...].

39.	**DIŠ *ṣal-lum-m[u-ú* ^{m}]^{ul}GAL$^{?}$ IGI.DU$_8$ *ina* KUR DÙ.A.BI LUGAL^{me}**
E$_1$ rev. 4’	DIŠ *ṣal-lum-m[u-ú* ^{m}]^{ul}⸢GAL⸣$^{?}$ I[GI.DU$_8$ *ina* KUR DÙ].⸢A.BI⸣ L[UGAL...]
I 1	[... I]GI.DU$_8$ *ina* KUR DÙ.A.BI LUGAL^{me}
	ÚŠ^{me}-*ma* KUR<^{me}>-*su-nu* TUR^{meš}
I 1	ÚŠ^{me}-*ma* KUR<^{me}>-*su-nu* TUR^{meš}

If a glo[w (coming from)] the Great$^{?}$ [S]tar is seen, in all countries the kings will die, and their countries will be diminished.

40.	**DIŠ *ṣal-lum-[mu-ú]* ^{mul}TI$_8$^{mušen} IGI.DU$_8$** (var.: **IGI) *ina* KUR DÙ.A.BI**
E$_1$ rev. 5’	DIŠ *ṣal-lum-[mu-ú* ^{mul}T]I$_8$^{mušen} IGI.⸢DU$_8$⸣ *ina* KUR DÙ.A.BI
I 2	[... ^{mul}T]I$_8$^{mušen} IGI.DU$_8$ *ina* KUR DÙ.A.BI
Q rev. 3’	DIŠ MIN ^{mul}TI$_8$^{mušen} IGI
	^{munus}PEŠ$_4$^{meš} *šà* ŠÀ-*ši-na* NU SILIM^{meš} [... NÍG.Z]I.GÁL EDIN.NA NU TE
E$_1$ rev. 5’	^{munus}⸢PEŠ$_4$⸣^{m}[^{eš}...]
I 2-3	^{munus}PEŠ$_4$^{me} *šà* ŠÀ-*ši-na* NU SILIM^{meš} (3) [... NÍG.Z]I.GÁL EDIN.NA NU° TE
	: AN *ina* AB.SÍN NÍGIN-*m[a$^{?}$*...]
Q rev. 3’	⸢:⸣ AN *ina* AB.SÍN NÍGIN-*m[a$^{?}$*...]

If a glo[w (coming from)] the Eagle is seen, in the entire country pregnant women will not bring (the children) in their wombs to term, [...] (he/she/it) will not come close [to the herd of] wild animals of the steppe. (It means that) Mars is surrounded in the Furrow a[nd$^{?}$...].

41.	**DIŠ *ṣal-lum-m[u]-ú* ^{mul}UGA^{mušen} IGI.DU$_8$** (var.: **IGI) *ina* KUR DÙ.A.BI**
E$_1$ rev. 6’	DIŠ *ṣal-lum-m[u-ú* ^{mul}U]GA^{mušen} IGI.DU$_8$ *ina* KUR DÙ.A.BI
F 1’	[DIŠ *ṣal-lu-mu*]-⸢*ú*⸣ ^{mu}[^{l}...]
I 4	[... ^{mu}]^{l}UGA^{mušen} IGI.DU$_8$ *ina* KUR DÙ.A.BI
Q rev. 4’	DIŠ MIN ^{mul}⸢UGA⸣$^{?}$ IGI
	^{munus}PEŠ$_4$^{meš} [*šà* ŠÀ-*ši-n*]*a* SILIM^{meš} MÁŠ.ANŠE Ù.TU SI.SÁ KU$_6$
E$_1$ rev. 6’–7’	^{munus}[...] (7’) ⸢SI⸣.[SÁ KU$_6$
F 2’	⸢MÁŠ.ANŠE⸣ Ù.TU ⸢SI.SÁ KU$_6$⸣
I 4-5	^{munus}PEŠ$_4$^{meš} (5) [*šà* ŠÀ-*ši-n*]*a* ⸢SILIM⸣^{meš} MÁŠ.ANŠE Ù.TU SI.SÁ KU$_6$
	ina* ÍD *e-ru-ta*$_5$ ŠUB MUŠEN *ina* AN-*e* NUNUZ *u-šal-lam
E$_1$ rev. 7’	*ina* Í]D *e-ru-ta*$_5$ ŠUB MUŠEN *ina* AN-⸢*e*⸣ [...]
F 2’	⸢*ina*⸣ Í[D...]
I 5-6	*ina* ÍD (6) [...] MUŠEN *ina* AN-*e* NUNUZ *ú-šal-lam*
	: GU$_4$.UD *ina* ^{áb}AB.S[ÍN...]
Q rev. 4’	: GU$_4$.UD *ina* ^{áb}AB.S[ÍN...]

If a gl[o]w (coming from) the Raven is seen, in the entire country pregnant women will bring [(the children) in thei]r [wombs] to term, the herd (and) the offspring will prosper, the fish in the river will drop the spawn, the bird in the sky will lay the egg. (It means that) Mercury in the Furr[ow...].

42.	**DIŠ *ṣal-lum-mu-ú*** (var.: ***ṣal-lu-mu-ú***) **^{mul}ŠU.GI IGI.DU$_8$ *ina* KUR DÙ.A.BI**
E$_1$ rev. 8’	⸢DIŠ *ṣal-lum*⸣-*mu*-⸢*ú*⸣ ^{mul}⸣ŠU.GI IGI.DU$_8$ *ina* KUR DÙ.A.⸢BI⸣
F 3’	DIŠ *ṣal-lu-mu-ú* ⸢^{mul}⸣ŠU.GI IGI.DU$_8$ *ina* [...]
I 7	[... ^{mul}ŠU.G]I IGI.DU$_8$ *ina* KUR DÙ.A.BI
	***ina* MU BI [x x x]x-*ma* DU$_{11}$ NU GI.NA**
E$_1$ rev. 8’	⸢*ina* MU⸣ B[I x x x]x-⸢*ma* DU$_{11}$⸣ [...]
I 7–8	*ina* MU BI (8) [... -*m*]*a*$^?$ DU$_{11}$ NU GI.NA

If a glow (coming from) the Old Man is seen, in the entire country in that year [...]... and the speech will not be reliable.

43.	**DIŠ *ṣal-lu-mu-ú*** (var.: **[*ṣal-lum*]-*mu-ú***) **^{mul}MAR.GÍD.DA GIB-*ma* GUB-*iz***
E$_1$ rev. 9’	[DIŠ *ṣal-lum*]-*mu-ú* ^{mul}MAR.GÍD.DA GIB-*ma* GUB-*iz*
F 4’	DIŠ *ṣal-lu-mu-ú* ^{mul}MAR.GÍD.DA GIB-*ma*
I 9	[... ^{mul}MAR.GÍ]D.DA GIB-*ma* GUB-*iz*
	(var.: **GUB-*ma*) ZI-*ut* ERÍN-*man-da ana* KUR ELAM.MA(^{ki}) MU.6.KAM***
E$_1$ rev. 9’	⸢ZI-*ut* ERÍN-*man*⸣-[*da ana* KUR] ⸢ELAM⸣.MA^{ki} MU.6.KAM*
F 4’–5’	GUB-⸢*ma*⸣ Z[I...] (5’) *ana* KUR ELAM.MA^{ki} ⸢MU⸣.6.KÁM
I 9	ZI-*ut* ERÍN-*man-da ana* KUR ELAM.MA<ki>
	KUR ELAM.MA^{ki} *i-da-a-aš*
E$_1$ rev. 9’	⸢KUR⸣ [...]
F 5’	KUR ELAM.MA^{ki} [...]
I 10	[...] KUR ELAM.MA^{ki} *i-da-a-aš*

If a glow (coming from) the Wagon lies across and stands still, (there will be) an attack of the enemy horde against Elam, for 6 years it will trample down Elam.

E$_1$, F, I ————————————————————

44.	**DIŠ ^{mul}ŠU.GI MUL^{meš}-*šú bi-rit-su-nu* (*ma-gal*) BAD-*at*** (var.: ***pé-ta-at***)
E$_1$ rev. 10’	[DIŠ ^{mul}Š]U.GI MUL^{meš}-*šú bi-rit-su-nu ma-gal* BAD-*at*
F 6’	DIŠ ^{mul}ŠU.GI MUL^{meš}-*šú bi-rit-su-nu ma-gal* BAD-*at*
G 8’	[DIŠ ^{mul}ŠU.GI MUL^{m}]^{eš}-*šu bi-rit-su-nu* *pé-ta-at*
I 11	[DIŠ ^{mul}ŠU.GI MUL]^{me}-*šú bi-rit-su-nu ma-gal* BAD-*at*
Q rev. 6’	[DIŠ ^{mul}ŠU.GI MUL^{meš}-*šú*] *bi-rit*-⸢*su*⸣-*nu ma-gal* BAD-*at*
	BURU$_{14}$ *ina* MU BI BÚR-*tú* TU[KU$^?$]
E$_1$ rev. 10’	BURU$_{14}$ *ina* MU BI BÚR-*tú* [TUKU$^?$]
F 6’	BURU$_{14}$ *ina* MU BI B[ÚR-*tú* TUKU$^?$]
G 8’	BURU$_{14}$ [...]
I 11	BURU$_{14}$ *ina* MU BI BÚR-*tú* TU[KU$^?$]
	: GU$_4$.UD *u dil-bat* i ki x[...]
Q rev. 6’	: GU$_4$.UD *u dil-bat* i ⸢ki⸣ x[...]

If the interspace between the stars of the Old Man is (very) open, (there will be) harvest, in that year release/barley for shipment will be ob[tained$^{?}$]. (It means that) Mercury and Venus ...[...].

45.	**DIŠ ^{mul}ŠU.GI MUL^{meš}-*šú nen-mu-du* BURU$_{14}$ *ina* MU BI BÚR-*tú* GÁL**
E$_{1}$ rev. 11’	[DIŠ ^{mul}Š]U.GI MUL^{meš}-*šú nen-mu-du* BURU$_{14}$ *ina* MU BI BÚR-*tú* G[ÁL]
F 7’	DIŠ ^{mul}ŠU.GI MUL^{meš}-*šú* ⸢*nen*⸣-*mu-du* BURU$_{14}$ *ina* MU BI ⸢⸢BÚR⸣-*t*[*ú* GÁL]
G 9’	[DIŠ ^{mul}ŠU.GI MUL^{meš}]-*šu nen-mu-du* BURU$_{14}$ *ina* MU BI BÚR-*tu*$_{4}$ GÁ[L...]
I 12	[DIŠ ^{mul}ŠU.GI MUL]^{me}-*šú nen-mu-du* BURU$_{14}$ *ina* MU BI BÚR-*tú* GÁL
L obv. 3’	DIŠ ^{mul}ŠU.GI MUL^{meš}-*šú nen-mu-du* BU[RU$_{14}$...]
Q rev. 7’	[DIŠ] MIN MUL^{meš}-*šú* [*ne*]*n-mu-du*

	<:> *ṭe*$_{4}$*-ḫu-tú ina* ŠÀ GÁL [... *e*]*-mi-du* : *sa-na-qu*
Q rev. 7’–8’	<:> *ṭe*$_{4}$-*ḫu-tú ina* ŠÀ GÁL [...] (8’) [... *e*]-*mi*-⸢*du*⸣ : *sa-na-qu*

If the stars of the Old Man are conjoined, (there will be) harvest, in that year there will be release. (It means that) there is a close approach inside [... to com]e in contact (means) to approach.

46.	**DIŠ ^{mul}ŠU.GI TÙR (MUL^{meš}) NÍGIN(-*ma*) *ina* MU BI *ina* KUR DÙ.A.BI**
E$_{1}$ rev. 12’	[DIŠ ^{mul}Š]U.GI TÙR NÍGIN *ina* MU BI *ina* KUR DÙ.A.BI
F 8’	DIŠ ^{mul}ŠU.GI TÙR NÍGIN *ina* ⸢MU BI⸣ *ina* KUR DÙ.A.BI
G 10’	[DIŠ ^{mul}ŠU.GI TÙR NÍGI]N-*ma ina* MU BI *ina* KUR DÙ.A.BI
I 13	[DIŠ ^{mul}ŠU.GI TÙR] NÍGIN *ina* MU BI *ina* KUR DÙ.A.BI
L obv. 1’	[DI]Š ^{mul}ŠU.GI TÙR MUL^{meš} ⸢NIGIN⸣ [...]
Q rev. 8’	DIŠ MIN TÙR NÍGIN
O rev. 4	DIŠ MIN-*ma* (i.e. ^{mul}ŠU.GI ^{d}30 TÙR NÍGIN-*ma*)

	RI.RI.GA(-*a*) (var.: ŠUB-*tì*) NAM.LÚ.U$_{18}$.LU (var.: *a-me-lu-*[*tu*])
E$_{1}$ rev. 12’	RI.RI.GA NAM.LÚ.U$_{18}$.LU
F 8’	RI.RI.GA NAM.LÚ.U$_{18}$.LU
G 10’	RI.RI.GA-*a* [...]
I 13	RI.RI.GA NAM.LÚ.U$_{18}$.LU
O rev. 5	ŠUB-*tì* *a-me-lu-*[*tu*]

	***ana* ÁB.GU$_{4}$$^{\text{ḫi.a}}$ (*u*) USDUḪA NU TE(-*ḫi*)**
E$_{1}$ rev. 12’	*ana*° ÁB.GU$_{4}$$^{\text{ḫi.a}}$ *u* USDUḪA ⸢NU⸣ [TE]
F 8’	*ana* ÁB.GU$_{4}$$^{\text{ḫi.a}}$ *u* USD[UḪA NU TE]
G 11’	[...] ⸢TE⸣-*ḫi*
I 14	[*ana* ÁB.GU$_{4}$$^{\text{ḫi.a}}$] *u* USDUḪA NU TE
O rev. 6	*a-na* ÁB.GU$_{4}$$^{\text{ḫi.a}}$ USDUḪA NU TE

	(:) $^{\text{(d)}}$30 *ina* (ŠÀ) (^{mul})ŠU.GI TÙR NIGIN [...]
G 11’	30 *ina* ŠÀ ^{mul}ŠU.GI TÙR [...]
L obv. 2’	30 *ina* ŠÀ ^{mul}ŠU.GI TÙR NIGIN [...]
Q rev. 8’	^{d}30 *ina* ŠU.GI TÙR [...]

	(var.: ^{mul}ŠU.GI *ina* ŠÀ-*šú* GUB-*iz*)
O rev. 4	^{mul}ŠU.GI *ina* ŠÀ-*šú* ⸢GUB-*iz*⸣

If the Old Man is surrounded by a halo (of stars), in that year in the entire country (there will be) a downfall of people, (but the evil) will not come close to the livestock (and) the flock. (It means that) the Moon in the Old Man is surrounded by a halo [...] (var.: the Old Man stands inside it, i.e. the Moon).

47.	**DIŠ ^{mul}ŠU.GI TÙR NÍGIN-*ma* KÁ-*šú ana* IM.U$_{18}$.LU** (var.: ***ina* IM.1; U$_{18}$**)
E$_1$ rev. 13’	[DIŠ ^{mul}ŠU].GI TÙR NÍGIN-*ma* KÁ-*šú ana* IM.U$_{18}$.LU
F 9’	⸢DIŠ⸣ ^{mul}ŠU.GI TÙR NÍGIN-*ma* KÁ-*šú ana* IM.U$_{18}$.LU
G 12’	[DIŠ ^{mul}ŠU.GI TÙR NÍGIN-*m*]*a* KÁ-*šu* *ina* IM.1
I 15	[DIŠ ^{mul}ŠU.GI TÙR NÍGI]N-*ma* KÁ-*šú ana* IM.U$_{18}$.LU
Q rev. 9’	DIŠ MIN TÙR NÍGIN-*ma* ⸢KÁ⸣-*šú ana* U$_{18}$
	BAD *ina* MU BI *ina* KUR URI^{ki} KIMIN
E$_1$ rev. 13’	BAD *ina* MU BI *ina* KUR URI^{ki} KIMIN
F 9’	BAD *ina* MU BI <*ina*> KUR URI^{ki} [KIMIN]
G 12’	BAD *ina* MU BI *ina* KUR URI^{ki}
I 15	BAD *ina* MU BI *ina* KUR URI^{ki} KIMIN
Q rev. 9’	BAD
	(var.: **RI.RI.GA <NAM.>LÚ.[U$_{18}$.LU *ana* ÁB.GU$_4$]$^{⸢ḫi⸣.a}$ USDUḪA N[U TE])**
G 12’–13’	RI.RI.GA <NAM.>LÚ.[U$_{18}$.LU] (13’) [*ana* ÁB.GU$_4$]$^{⸢ḫi⸣.a}$ USDUḪA N[U TE]
	: ^{d}30 DAGAL-*šú* [...]
Q rev. 9’	: ^{d}30 DAGAL-*šú* [...]

If the Old Man is surrounded by a halo and its gate opens towards the south, in that year in Akkad ditto (i.e. there will be a downfall of people, but the evil will not come close to the livestock and the flock) (var.: there will be a downfall of pe[ople, but the evil] will n[ot come close to the livestoc]k and the flock). (It means that) the width of the Moon [...].

48.	**[DIŠ ^{mu}]^{l}ŠU.GI TÙR NÍGIN-*ma* KÁ-*šú ana* IM.SI.SÁ** (var.: **[I]M.2)**
E$_1$ rev. 14’	[DIŠ ^{mul}ŠU].GI TÙR NÍGIN-*ma* KÁ-*šú ana* IM.SI.SÁ
F 10’	[DIŠ ^{m}]^{ul}ŠU.GI TÙR NÍGIN-*ma* KÁ-*šú ana* IM.SI.SÁ
G 14’	[DIŠ ^{mul}ŠU.GI TÙR NÍGIN-*ma* KÁ-*šu ana* I]M.2
I 16	[DIŠ ^{mul}ŠU.GI TÙR NÍGIN]-*ma* KÁ-*šú ana* IM.SI.SÁ
	BAD *ina* MU BI *ina* KUR SU.BIR$_4$^{ki} KIMIN
E$_1$ rev. 14’	BAD *ina* MU BI *ina* KUR SU.BIR$_4$<^{ki}> KIMIN
F 10’	BAD *ina* MU BI <*ina*> KUR SU.BIR$_4$^{ki} [KIMIN]
G 14’	BAD *ina* MU BI *ina* KUR SU.B[IR$_4$^{ki} KIMIN]
I 16	BAD *ina* MU BI *ina* KUR SU.⸢BIR$_4$^{ki}⸣ [KIMIN]

[If] the Old Man is surrounded by a halo and its gate opens towards the north, in that year in Subartu ditto (i.e. there will be a downfall of people, but the evil will not come close to the livestock and the flock).

49.	**[DIŠ ^{mul}]ŠU.GI TÙR NÍGIN-*ma* KÁ-*šú ana* IM.KUR.RA** (var.: **[I]M.3)**
E$_1$ rev. 15’	[DIŠ ^{mul}ŠU].GI TÙR NÍGIN-*ma* KÁ-*šú ana* IM.KUR.RA
F 11’	[DIŠ ^{mu}]^{l}⸢ŠU⸣.GI TÙR NÍGIN-*ma* KÁ-*šú ana* IM.KUR.RA
G 15’	[DIŠ ^{mul}ŠU.GI TÙR NÍGIN-*ma* KÁ-*šu ana* I]M.3
I 17	[DIŠ ^{mul}ŠU.GI TÙR NÍGI]N-*ma* KÁ-*šú ana* IM.KUR.RA
	BAD *ina* MU BI *ina* KUR ELAM.MA^{ki} KIMIN
E$_1$ rev. 15’	BAD *ina* MU BI *ina* KUR ELAM.MA^{ki} KIMIN
F 11’	BAD *ina* MU BI <*ina*> KUR ELAM.MA^{ki} [KIMIN]
G 15’	BAD *ina* MU BI *ina* KUR ELAM.MA[^{ki} KIMIN]
I 17	BAD *ina* MU BI *ina* KUR ⸢ELAM⸣.[MA^{ki} KIMIN]

[If] the Old Man is surrounded by a halo and its gate opens towards the east, in that year in Elam ditto (i.e. there will be a downfall of people, but the evil will not come close to the livestock and the flock).

50.	**[DIŠ ^{mul}Š]U.GI TÙR NÍGIN-*ma* KÁ-*šú ana* IM.MAR.TU** (var.: **[IM].4)**
E_1 rev. 16’	[DIŠ ^{mul}ŠU].GI TÙR NÍGIN-*ma* KÁ-*šú ana* IM.MAR.TU
F 12’	[DIŠ ^{mu}]^{l}⌜ŠU⌝.GI TÙR NÍGIN-*ma* KÁ-*šú ana* IM.MAR.TU
G 16’	[DIŠ ^{mul}ŠU.GI TÙR NÍGIN-*ma* KÁ-*šu ana* IM].⌜4⌝
I 18	[DIŠ ^{mul}ŠU.GI TÙR NÍGI]N-⌜*ma* KÁ-*šú*⌝ *ana* IM.MAR.TU
	BAD *ina* MU BI *ina* KUR MAR.TU^{ki} KIMIN
E_1 rev. 16’	BAD *ina* MU BI *ina* KUR MAR.TU<^{ki}> KIMIN
F 12’	BAD *ina* MU BI <*ina*> KUR MAR.TU^{ki} [KIMIN]
G 16’	BAD *ina* MU BI *ina* KUR MAR.T[U^{ki} KIMIN]
I 18	BAD *ina* MU BI *ina* KUR MAR.[TU^{ki} KIMIN]

[If the O]ld Man is surrounded by a halo and its gate opens towards the west, in that year in Amurru ditto (i.e. there will be a downfall of people, but the evil will not come close to the livestock and the flock).

51.	**[DIŠ MIN *ina* TÙ]R$^{?}$ MUL GUB-*iz*** (var.: **[GUB-*m*]*a*) *ina* MU BI**
E_1 rev. 17’	[... TÙ]R$^{?}$ MUL GUB-*iz* *ina* MU BI
F 13’	[... TÙ]R$^{?}$ MUL GUB-*iz* {x} *ina* MU BI
G 17’	[... GUB-*i*]*z* *ina* MU BI
I 19	[... GUB-*m*]*a* *ina* MU BI
	***ina* KUR DÙ.A.BI DUMU^{meš} *šá* UBUR *ina* KA DUMU.MUNUS ^{d}*a-nim* ÚŠ^{meš}**
E_1 rev. 17’	*ina* KUR DÙ.A.BI DUMU^{me} *šá* UBUR *ina* KA DUMU.MUNUS ^{d}*a-nim* ÚŠ^{meš}
F 13’	*ina* KUR DÙ.A.BI DUMU^{me} *šá* UBUR *ina* KA DUMU.MUNUS ^{d}*a-n*[*im* ÚŠ^{meš}]
G 17’	*ina* KUR DÙ.A.BI DUMU^{meš} *šá* UBUR [...]
I 19–20	*ina* KUR DÙ.A.BI DUMU [*šá* UBUR] (20) [... ^{d}*a*]-*nim* ÚŠ[^{meš}]

[If ditto (i.e. the Old Man is surrounded by a halo)] (and) a star stands [in the hal]o$^{?}$, in that year in the entire country the suckling children (lit. children from the breast) will die by the command of Anu’s daughter.

52.	**[DIŠ MIN *ina* TÙ]R$^{?}$ MUL GUB-*ma* KÁ-*šú ana* IM.U$_{18}$.LU** (var.: **IM.1)**
E_1 rev. 18’	[... TÙ]R$^{?}$ MUL GUB-*ma* KÁ-⌜*šú ana* IM⌝.U$_{18}$.LU
F 14’	[... TÙ]R$^{?}$ MUL GUB-*ma* KÁ-*šú ana* IM.U$_{18}$.LU
G 18’	[...] ⌜IM⌝.1
I 21	[...] KÁ-*šú ana* IM.U$_{18}$.LU
	BAD *ina* MU BI *ina* KUR URI^{ki} KIMIN
E_1 rev. 18’	BAD *ina* MU BI *ina* KUR URI^{ki} KIMIN
F 14’	BAD *ina* MU.⌜BI⌝ [*ina*] ⌜KUR⌝ URI^{ki} [KIMIN]
G 18’	BAD *ina* MU BI ⌜*ina*⌝ K[UR...]
I 21	BAD *ina* MU BI *ina* KUR URI[^{ki} KIMIN]

[If ditto (i.e. the Old Man is surrounded by a halo)], a star stands [in the hal]o$^{?}$ and its gate opens towards the south, in that year in Akkad ditto (i.e. the suckling children will die by the command of Anu’s daughter).

53.	**[DIŠ MIN *ina* TÙ]R$^?$ MUL GUB-*ma* KÁ-*šú ana* IM.SI.SÁ** (var.: **[I]M.2)**
E$_1$ rev. 19’	[... TÙ]R$^?$ ⸢MUL⸣ GUB-*ma* KÁ-*šú ana* IM.SI.⸢SÁ
F 15’	[... M]UL GUB-*ma* KÁ-*šú ana* IM.SI.SÁ
G 19’	[... I]M.2
I 22	[... GUB-*m*]*a* KÁ-*šú ana* IM.SI.SÁ
	BAD *ina* MU BI *ina* KUR SU.BIR$_4^{ki}$ KIMIN
E$_1$ rev. 19’	BAD *ina* MU⸣ BI *ina* KUR SU.BIR$_4^{\langle ki \rangle}$ ⸢KIMIN⸣
F 15’	BAD *ina* MU [BI *ina* KUR S]U.BIR$_4^{ki}$ [KIMIN]
G 19’	BAD *ina* MU BI ⸢*ina*⸣ [...]
I 22	BAD *ina* MU BI *ina* KUR SU.BI[R$_4^{ki}$ KIMIN]

[If ditto (i.e. the Old Man is surrounded by a halo)], a star stands [in the hal]o$^?$ and its gate opens towards the north, in that year in Subartu ditto (i.e. the suckling children will die by the command of Anu’s daughter).

54.	**[DIŠ MIN *ina* TÙ]R$^?$ MUL GUB-*ma* KÁ-*šú ana* IM.KUR.RA**
E$_1$ rev. 20’	[... TÙ]R$^?$ MUL GUB-*ma* KÁ-*šú ana* IM.KUR.RA
F 16’	[... M]UL GUB-*ma* KÁ-*šú ana* IM.KUR.RA
G 20’	[...
I 23	[... GUB-*m*]*a* KÁ-*šú ana* IM.KUR.RA
	BAD *ina* MU BI *ina* KUR ELAM.MA^{ki} KIMIN
E$_1$ rev. 20’	BAD *ina* MU BI *ina* KUR ELAM.MA^{ki} ⸢KIMIN⸣
F 16’	BAD *ina* MU BI *ina* KUR ELAM.MA[^{ki} KIMIN]
G 20’	BA]D *ina* MU BI [...]
I 23	BAD *ina* MU BI *ina* KUR ELAM.MA[^{ki} KIMIN]

[If ditto (i.e. the Old Man is surrounded by a halo)], a star stands [in the hal]o$^?$ and its gate opens towards the east, in that year in Elam ditto (i.e. the suckling children will die by the command of Anu’s daughter).

55.	**[DIŠ MIN *ina* TÙ]R$^?$ MUL GUB-*ma* KÁ-*šú ana* IM.MAR.TU**
E$_1$ rev. 21’	[...] MUL GUB-*ma* KÁ-*šú ana* IM.MAR.TU
F 17’	[... TÚ]R$^?$ MUL GUB-*ma* KÁ-*šú ana* IM.MAR.TU
I 24	[... GUB-*m*]*a* ⸢KÁ-*šú ana* IM.MAR.TU⸣
	BAD *ina* MU BI *ina* KUR MAR.TU^{ki} KIMIN
E$_1$ rev. 21’	BAD *ina* MU BI *ina* KUR MAR.TU$^{\langle ki \rangle}$ KIMIN
F 17’	BAD *ina* MU BI *ina* KUR MAR.⸢TU⸣[^{ki} KIMIN]
G 21’	[... *ina* M]U [BI...]
I 24	BAD *ina* MU BI *ina* KUR ⸢MAR⸣.[TU^{ki} KIMIN]

[If ditto (i.e. the Old Man is surrounded by a halo)], a star stands [in the hal]o$^?$ and its gate opens towards the west, in that year in Amurru ditto (i.e. the suckling children will die by the command of Anu’s daughter).

56.	**[DIŠ x]x ^{d}TIR.AN.NA NÍGIN *ina* MU BI ZI-*ut***
E$_1$ rev. 22’	[DIŠ x x ^{d}T]IR.AN.NA NÍGIN *ina* MU BI ZI-*ut*
F 18’	[DIŠ x]x ^{d}TIR.AN.NA NÍGIN *ina* MU BI ZI-*ut*
G	(remainder is missing)
I 25	[...] ⸢x x x x⸣ [...] ⸢x x⸣ [...]

	ERÍN-*man-da ana* KUR DÙ.A.BI
E_1 rev. 22'	ERÍN-*man-da ana* KUR DÙ.A.BI
F 18'	ERÍN-*man-*⸢*da*⸣ [...]

[If...]... is surrounded by a rainbow, in that year (there will be) an attack of the enemy horde against the entire country.

57.	**[DIŠ MIN ^{d}TI]R.AN.NA NÍGIN-*ma*** (var.: **[DIŠ MI]N-*ma*) KÁ-*šú***
E_1 rev. 23'	[DIŠ MIN ^{d}TI]R.AN.NA NÍGIN-*ma* KÁ-*šú*
F 19'	[DIŠ MI]N-*ma* KÁ-*šú*
I	(remainder is missing)

	***ana* IM.U$_{18}$.LU BAD *ina* MU BI ZI-*ut* ERÍN-*man-da ana* KUR URI[^{ki}]**
E_1 rev. 23'	*ana* IM.U$_{18}$.LU BAD *ina* MU BI ZI-*ut* ERÍN-*man-da ana* KUR URI[^{ki}]
F 19'	*ana* IM.U$_{18}$.LU BAD *ina* MU BI Z[I-*ut*...]

[If ditto (?)] is surrounded [by a ra]inbow (var.: [If ditt]o, i.e. ... is surrounded by a rainbow), and its gate opens towards the south, in that year (there will be) an attack of the enemy horde against Akkad.

58.	**[DIŠ MIN ^{d}TIR.A]N.NA NÍGIN-*ma*** (var.: **[DIŠ MI]N-*ma*) KÁ-*šú***
E_1 rev. 24'	[DIŠ MIN ^{d}TIR.A]N.NA NÍGIN-*ma* KÁ-*šú*
F 20'	[DIŠ MI]N-*ma* KÁ-*šú*

	***ana* IM.SI.SÁ BAD *ina* MU BI ZI-*ut* ERÍN-*man-da ana* KUR S[U.BIR$_4$^{ki}]**
E_1 rev. 24'	*ana* IM.SI.SÁ BAD *ina* MU BI ZI-*ut* ⸢ERÍN-*man-da ana* KUR⸣ S[U.BIR$_4$^{ki}]
F 20'	*ana* ⸢IM⸣.[SI.SÁ BAD] ⸢*ina* MU⸣ [BI...]

[If ditto (?)] is surrounded [by a rain]bow (var.: [If ditt]o, i.e. ... is surrounded by a rainbow), and its gate opens towards the north, in that year (there will be) an attack of the enemy horde against S[ubartu].

59.	**[DIŠ MIN ^{d}TIR.AN].NA NÍGIN-*ma*** (var.: **[DIŠ MIN-*m*]*a*) KÁ-*šú ana***
E_1 rev. 25'	[DIŠ MIN ^{d}TIR.AN].NA NÍGIN-*ma* KÁ-*šú ana*
F 21'	[DIŠ MIN-*m*]*a* KÁ-*šú ana* [...]

	IM.KUR.RA BAD *ina* MU BI [ZI-*ut* ERÍN-*man-da ana* KUR ELAM.MA^{ki}]
E_1 rev. 25'	IM.KUR.RA BAD ⸢*ina* MU BI⸣ [...]

[If ditto (?)] is surrounded [by a rainb]ow (var.: [If ditto], i.e. ... is surrounded by a rainbow), and its gate opens towards the east, in that year (there will be) [an attack of the enemy horde against Elam].

60.	**[DIŠ MIN ^{d}TIR.AN.N]A NÍGIN-*ma*** (var.: **[DIŠ MIN-*m*]*a*) KÁ-*šú ana***
E_1 rev. 26'	[DIŠ MIN ^{d}TIR.AN.N]A NÍGIN-*ma* KÁ-⸢*šú ana*
F 22'.	[DIŠ MIN-*m*]*a* KÁ-*šú* [...]

	IM.[MAR.TU BAD *ina* MU BI ZI-*ut* ERÍN-*man-da ana* KUR MAR.TU^{ki}]
E_1 rev. 26'	IM⸣.M[AR.TU BAD...]
E_1	(remainder is missing)

[If ditto (?)] is surrounded [by a rainbo]w (var.: [If ditto], i.e. ... is surrounded by a rainbow), and its gate opens towards the [west, in that year (there will be) an attack of the enemy horde against Amurru].

61.	**[DIŠ x x**	**]-*ma ana* ⸢^{mul}⸣[...]**
F 23’	[DIŠ x x	]-*ma ana* ⸢^{mul}⸣[...]
F	(remainder is missing)	

[If...] and towards the star [...].

Commentary to the Text

7: The apodosis is restored based on one entry, quoted in CAD (A 106), from ACh Suppl. 1, 10 (K 2311+), a fragment re-edited by Verderame (2002: 80): [DIS 30] *ina* IGI.LÁ-*šú* AGA *a-pir ina* ITI BI LUGAL^{meš} *ša* KUR DÙ.A.BI *in-na-da-ru-ma* KÚR^{meš}, “[If] when [the Moon] appears it wears a crown, in that month the kings of the entire country will become worried and will be in enmity”.

9: For an apodosis similar to the one given in obv. 3’, see Freedman (2017: 142, 6): DIŠ ÍD GIM A *sa-aḫ-ḫi ina* KUR *su-a-lu*$_4$ GÁL, “If the river is like a swamp water, in the country there will be phlegm”.

11: The apodosis is partly paralleled in another omen of Venus: [DIŠ ^{mul}*dil*]-*bat ana* ^{mul}KU$_6$ TE ŠI.ŠI KUR GAR-*an ka-mar* UN^{meš} *ka-la-ma* UN^{meš} KUR *iš-šal-la-la*, “[If Ven]us comes close to the Fish, there will be defeat of the land, (there will be) a catastrophe for all people, the people of the country will be robbed” (in K 2226+ obv. ii 25’, see Reiner-Pingree 1998: 92, 21).

12–29: These entries refer to celestial bodies entering (KU$_4$, *erēbu*) into (*ana* ŠÀ, *ana libbi*) the Field (^{mul}AŠ.GÁN, *ikû*). The main verb is followed by GUB, *i/uzuzzu*, “to stand (still)”, in entries 12–18, and by IGI, *amāru*, “to be visible”, in entries 19–29. As shown in Table 4, the order of the celestial bodies mentioned in these entries roughly represents the order of the dates for their first visibility (i.e. heliacal rising, see fn. 220) given in MUL.APIN (Hunger-Steele 2019: 180 Table 4).

Entry	Subject of the protasis	Month of first visibility
12	MUL.MUL, *zappu*, "Bristle"	Ajaru (Month II)
13	^{mul}SIPA.ZI.AN.NA, *šitaddaru*, "True Shepherd of Anu"	Simanu (Month III)
14	^{mul}KAK.SI.SÁ, *šukūdu*, "Arrow"	Du'uzu (Month IV)
15	^{mul}*dilbat*, Venus	-
16	^{mul}ŠU.PA, *nameru*, "Resplendent One"	Ululu (Month VI)
17	^{mul}NUN^{ki}, "Eridu-star"	Ululu (Month VI)
18	^{mul}EN.TE.NA.BAR.ḪUM, *ḫabaṣīrānu*, "Mouse"	Tešritu (Month VII)
19	^{mul}GÍR.TAB, *zuqaqīpu* ,"Scorpion"	Araḫsamnu (Month VIII)
20	^{mul}ŠUL.PA.È, Jupiter	-
21	^{mul}UG$_5$.GA, *āribu*, "Raven"	-
22	^{mul}AMAR.UTU, "Marduk-star"	-
23	^{mul}GU$_4$.AN.NA, *alû*, "Bull of Heaven"	Ajaru (Month II)
24	^{mul}GÚ.ḪAL, *ur'udu*, "Throat"	-
25	^{mul}KU$_6$, *nūnu*, "Fish"	Adaru (Month XII)
26	^{mul}*ṣalbatānu*, Mars	-
27	^{mul}UDU.IDIM, *bibbu*, "planet"	-
28	^{mul}UD.KA.DUḪ.A, *ūmu nā'iru*, "Demon with the Gaping Mouth"[287]	Kislimu (Month IX)

Table 4. Order of the stars in omens from EAE 52: 12–28, and months of their first visibility or heliacal rising according to MUL.APIN.

13: The apodosis ^{d}IŠKUR MA-RAB GAR, "Adad will produce MA-RAB", is found in other celestial omens as well (e.g. 81-7-27, 65 rev. 7, see Rochberg-Halton 1988: 151 § XI.1 source C; DT 104 obv. 2, see Verderame 2002: 127). The meaning of MA-RAB, likely a weather phenomenon, remains unknown (Labat 1965: 146–147 fn. 4).

15: The meaning of *šulpu* in source K 4' (see App B § 11) is uncertain. It is either attested with the meaning of "stalk" (CAD Š3 256–257 *šulpu* A), or in negative predictions (CAD Š3 258 *šulpu* C). Considering the other sources, its meaning is likely similar to *siḫpu*, "sweeping attack" (CAD S 238–239), or *šulputtu*, "destruction" (CAD Š1 261–262; Verderame 2002: 109 l. 8'). A similar apodosis is found in VAT 9817 rev. 10 (EAE 17): *šul-pu ina* KUR GÁL, "there will be a *šulpu* in the country", followed in the next line by the apodosis ZI-*ut* UGA$^{\text{ḫi.a}}$ *ina* KUR GÁL, "there will be an attack of ravens" (Weidner 1954–1956: 76 fn. 21).
In the apodosis of entry 15, the substantive *di-im-ta* can be translated as either "weeping" (*dīmtu*, see CAD D 147–148), or "fortified area" (*dimtu*, see CAD D 144–147).

16: The second part of the apodosis in this entry ([... T]UR-*ár* LUGAL GAR-*šú*, "[... he/she/it will re]duce, the king will put him in charge") is broken and mainly logographic, therefore it is difficult to assess its correct context.

287 Though ^{mul}UD.KA.DUḪ.A is sometimes understood as *nimru*, "Panther", the latter reading is not actually attested (Beaulieu-Frahm-Horowitz-Steele 2018: 76).

24: The star ^{mul}GÚ.ḪAL, "Throat" (Gössmann 1950: 26 n. 80) is attested only in this entry. According to CAD (U/W 267–270), GÚ.ḪAL is the logogram for the Akkadian *ur'udu*, "windpipe", "throat", or "pathway".

26: GÁN.BA, *maḫīru*, is usually translated in omens as the generic "business", though it should be more precisely understood as "market rate", i.e. the reciprocal of a market price (e.g. see Ossendrijver 2019: 54).

28: A restoration of the apodosis could be suggested on the basis of the assumed tablet EAE 51: DIŠ *ina* ^{iti}BÁR ^{mul}UD.KA.DUḪ.A IGI MU.5.KAM* *ina* KUR URI^{ki} *ina* KA ^{d}*èr-ra* ÚŠ^{meš}GAL^{meš} *ana* MÁS.ANŠE NU T[E], "If in Nisannu (i.e. Month I) the Demon with the Gaping Mouth is visible, for 5 years in Akkad there will be a pestilence by the command of Erra, but (the penstilence) will not appr[oach] the herd" (K 4510 l. 6' // K 9126+ rev. 7'–9', see Reiner-Pingree 1981: 67 XIII 5, 68 XIV 4). Whereas the protasis of entry 28 only mentions the Demon with the Gaping Mouth entering into the Field ([DIŠ *a-n*]*a* ŠÀ ^{mul}AŠ.GÁN ^{mul}UD.KA.DUḪ.A KU_4-*ma*, "[If] the Demon with the Gaping Mouth enters [in]to the Field"), the entry of EAE 51 mentions the month when the Demon with the Gaping Mouth rises (DIŠ *ina* ^{iti}BÁR ^{mul}UD.KA.DUḪ.A IGI), and an additional apodosis (*ana* MÁS.ANŠE NU T[E]) because EAE 51 is a menology with excerpts concerning the first visibility of stars (see 5.1.8.4. § EAE 50–51).

30–34: These entries concern the Field (^{mul}AŠ.GÁN, *ikû*) and the Moon. In celestial omens, celestial bodies are often arranged according to their position in relation to the Moon: the interior (*libbu*), the shoulder (*naglabu*), the horn (*qarnu*), or the head (*rēšu*) were "body parts" of the Moon perceived as an anthropomorphic and theriomorphic being (see A.2.1.1.3.), and intended as directions such as left, right, above, or below. For instance, the term MAŠ.SÌLA, *naglabu*, is used in medical texts and omens with the meaning of "shoulder blade" or "scapula" of people or animals. The logogram SI, *qarnu,* "horn", is a reference to the bovine and royal features of the Moon god as a cowherd (Verderame 2014: 93; Hätinen 2021: 444–445). The personification, or theriomorphism, of the lunar sections is consistent with the use of metaphorical language in celestial omens (see 2.2.3.1. § 3).
According to Reiner and Pingree (1998: 3), when a celestial body is placed in the KI.GUB, *manzāzu*, "position" (CAD M1 237–238), of the Moon, the KI.GUB could mean the point on the horizon above which the Moon rises or sets. That is different from the KI.GUB of other celestial bodies, which could be the position of a celestial body when it is first visible the day of the observation (see 5.1.5.1. commentary to the text 41, 42, 48, pp. 193–194).

37: The logogram A.RÁ, in Akkadian *alaktu*, lit. "course" (CAD A1 297–300), could refer to an (oracular) decision, in both Akkadian and Sumerian language.[288] Since this entry is too fragmentary, the interpretation here remains uncertain.

288 See Abusch (1987) and Böck (1995).

37–43: The subject of these entries is a *ṣallummû*, coming from different celestial bodies: the Eagle (^{mul}TI$_{8}^{\text{mušen}}$, *a/erû*, see Gössmann 1950: 1–2 n. 2), the Raven (^{mul}UGA^{mušen}, *āribu*, see Gössmann 1950: 47–50 n. 132–133), the Old Man (^{mul}ŠU.GI, *šību*, see Gössmann 1950: 208–210 n. 378), the Wagon (^{mul}MAR.GÍD.DA, *ereqqu*, see Gössmann 1950: 95–97 n. 258) and the Great One (^{mul}GAL, *rabû*) (Gössmann 1950: 18 n. 62). The Akkadian *ṣallummû* is understood as a glow, a luminous phenomenon in the sky similar to a *mešḫu*, "glow" or "meteor".[289] The Great Star List explains *ṣallummû* with *mišiḫ* MUL^{meš}, "glow of stars", *ṣarār* MUL^{meš}, "brightness of stars", and *zīm* MUL, "glow of a star" (Koch-Westenholz 1995: 194, 196 ll. 178–180). The word *ṣallummû* also refers to a comet (Bjorkman 1973: 106–107; Stephenson-Walker 1985: 12–40), whereas the fragmentary *ṣâtu* commentary of tablet EAE 52 (VAT 7850 (+) AO 6486 rev. 2', see App. B § 27; EAE 52 source Q) explains *ṣallummû* (i.e. the subject of the previous broken lines) with *šarūru našû*, "to have a brillian sheen", lit. "carrying brilliance" (CAD Š2 141). Hence, the entries 37–43 probably deal with a particular brightness of the individual celestial bodies, i.e. a glow, rather than with a comet.

39: The star ^{mul}GAL, *rabû*, "Great One", or MUL GAL, lit. "a great star", is treated as a meteor or a luminous phenomenon in two astrological reports (81-2-4, 105 obv. 2–4, see SAA 8, 334; 83-1-18, 174 rev. 1–4, see SAA 8, 335). If the Great One stands for a flash, and not for a star, then this entry would deal with two luminous phenomena at the same time, the *ṣallummû* as "glow", and the Great One as "flash", which is unlikely. Because the entries 39–43 follow the same structure referring to a *ṣallummû*, "glow", "from star x", ^{mul}GAL must refer to a star, or a group of stars. Sometimes, ^{mul}GAL is also a name for Jupiter (Gössmann 1950: 18 n. 62).

40–41: The Raven (^{mul}UG$_{5}$.GA, *āribu*) and the Eagle (^{mul}TI$_{8}$ ^{mušen}, *a/erû*) are associated with pregnancy and giving birth: both entries are analysed in detail in A.4. In the factual explanation of entry 40, in source Q (VAT 7850 (+) AO 6486 rev. 3', see App. B § 27) the logogram AN is used as a name for Mars, as is usually found in texts dating to the Late Babylonian period (ca. 484–30 BC) (Brown 2000: 56).

44–45: These entries are arranged according to the scheme of binary opposites ("open" versus "conjoined"), a topic further discussed in A.2.1.2. In the apodoses, the logogram BURU$_{14}$ (*ebūru,* "harvest") is connected on a graphic and semantic level with the logogram BÚR in the second part of the apodoses. In the context of omens, BÚR is understood as *pašāru,* "to loosen" (CAD P 237 1b), or *pišertu*, "release", usually found in extispicy omens or agricultural contexts (CAD P 428–429); yet it can also stand for *napšartu,* "barley for shipment" (CAD N1 316). In the Sumerian composition "Farmer's instructions", one of the agricultural tasks to be performed after the winnowing of the grain is to "release the grain at midday". [290] The exact procedure supposedly indicated in the "Farmer's instructions" is unclear, but it possibly refers to the storage of (surplus) barley for the

289 For a discussion on the term *mešḫu,* see Horowitz (2014: 135–137).

290 u$_{4}$ sa$_{9}$-a-gin$_{7}$ še búr-ra-ab (Civil 1994: 32 l. 107); see also Civil (1994: 98).

shipment. Indeed, Rochberg-Halton (1988: 128 § II.9 D ii 29) understood the logogram BÚR as "granary" in the context of the apodoses of lunar eclipse omens.

44–50; **50–61**: The omens of entries 44–50 deal with the Old Man (^{mul}ŠU.GI, *šību*). The Old Man is associated with the god Enmešarra (Gössmann 1950: 208–210 n. 378) and located in the path of Enlil at night according to MUL.APIN (I i 3, see Hunger-Steele 2019: 30). It is not clear whether the entries 51–60 also deal with the Old Man: the beginning of the protases is not preserved, but the sign MIN, "ditto", is probably to be restored there. In this respect, see the commentary to the text of entries 46–60, 51–55 and 56–60 given below.

46: In the sentence ana ÁB.GU4$^{\text{ḫi.a}}$ (*u*) USDUḪA NU TE(*-ḫi*), lit. "(he/she/it) will not come close to the livestock and the flock", the subject of the verb is omitted, and it likely is RI.RI.GA or ŠUB(*-ti*), *miqittu*, "downfall", or "epidemic" when referred to animals only (CAD M2 101b), meant as the evil consequences of such an event.

46–60: The omens of entries 46–55 deal with TÙR, *tarbaṣu*, "halo", literally "(cattle)-pen".[291] Though TÙR rarely indicates a pen for sheep and goats, the presence of the halo in the protases relates to the herd in the apodoses of entries 46–50 by analogy.[292] The "halo" also has a "gate" (KÁ, *bābu*) in entries 47–50 and 52–55. In the protases of entries 46–55, the halo is referred to the Old Man, even though the halo is usually seen with the Sun or the Moon.[293] The term *tarbaṣu* is indeed understood as a "small ring" around the Moon or the Sun,[294] sometimes, as previously mentioned, "interrupted" by "gates" (KÁ, *bābu*) (Gehlken 2012: 83 fn. 5). Only through factual explanations, is it easier to understand the meaning of TÙR in the present context: in entry 46, the sources G, L, and Q explain that the halo of the Old Man means that the Moon, standing in the region of the Old Man, is surrounded by a halo. Consequently, the presence of a "gate" (i.e. "interruption") refers to the halo of the Moon close to the Old Man.
According to entries 47–50 (but also 52–55 and 57–60, see below), the gate of the halo is opened (BAD, *petû*, "to open")[295] towards one of the four cardinal directions. In Mesopotamian divination, the celestial directions have a direct counterpart in the terrestrial directions. Hence, the compass points introduce a correlation system between areas of the sky and earth. Several ways of individuating the compass points were in place: the rising and the setting of the Sun, stars, and winds (Horowitz 1998: 195–207). Additionally, the names of the winds can be based both on their point of origin (MUL.APIN II i 68–71, see

291 For a discussion about *tarbaṣu*, see 2.2.3.2.

292 As a learned association between the cattle-pen and the halo of the Moon, the Moon god Sin was considered a cowherd already in the earliest traditions (Krebernik 1997; Hätinen 2021: 240–247).

293 See tablet EAE 8 (CTN 4, 4 rev. i 3'–23'), Verderame (2014) and van Soldt (1995: 147 "*tarbaṣu*").

294 A halo is a ring of light which surrounds the Moon and the Sun, caused by the refraction and reflection of light passing through ice crystals in the atmosphere (Verderame 2014: 91–92).

295 In the commentary VAT 7850 (+) AO 6486 (source Q; see App. B § 27) rev. 16', the reading of BAD is given as *nesû*, "to be far" (CAD N2 184–189). Such reading is also attested in lexical lists (e.g. Ass 523 obv. i 79a, see MSL 14: 251 l. 86 source A).

Hunger-Steele 2019: 83–84) and on the destinations of the winds (Horowitz 1998: 196–198). The correlation between areas of the sky and the earth aims to create as many variables as possible in omens, even though it produces omens whose protases are not observable in reality. Therefore, the arrangement of the entries 47–50, 52–55, 57–60 follows a pattern according to which the same phenomenon is seen facing four different directions (see A.2.1.1.4., A.3.1.5.). The position of the gate follows the order of the four compass points, which reflects the priority of the four winds with which they are associated, and where the observer originally stood. The scheme abstracted from the omens in entries 47–50, 52–55, 57–60 is shown in Table 5.

Orientations of the celestial body (protasis)	**Affected area of the earth (apodosis)**
ana IM.U$_{18}$.LU (South)	KUR URI^{ki} (Akkad)
ana IM.SI.SÁ (North)	KUR SU.BIR$_{4}$^{ki} (Subartu)
ana IM.KUR.RA (East)	KUR ELAM.MA^{ki} (Elam)
ana IM.MAR.TU (West)	KUR MAR.TU^{ki} (Amurru)

Table 5. Scheme of celestial and terrestrial orientation derived from EAE 52: 47–50, 52–55, 57–60.

The four winds can also be written with numbers: IM.1 = IM.U$_{18}$.LU (*šūtu*), IM.2 = IM.SI.SÁ (*iltānu*), IM.3 = IM.KUR.RA (*šadû*), IM.4 = IM.MAR.TU (*amurru*) (Horowitz 1998: 197; Borger 2004: 389–391 n. 641). The scheme presented in Table 5 differs from another well-known "written" scheme, i.e. scheme found as a written rule.[296] The latter is found in the Great Star List (Koch-Westenholz 1995: 202 ll. 293–294), in lunar eclipse omens (Rochberg-Halton 1988: 53 fig. 4–4, 223 rev. 6–7), and in the tablet SIT 4 (K 3123 rev. 17', see App. B § 24), and it can be summed up as follows:[297]

South	=	Elam
North	=	Akkad
East	=	Subartu (+ Gutium)
West	=	Amurru

Instead, in entries 47–50, 52–55, 57–60, the scribe possibly adjusted the association scheme for the phenomenon of the halo and its gates. In effect, applying the schemes of directions and lands in various ways is not unique. This is attested, for instance, in omens about lunar eclipses: Rochberg-Halton (1988: 51–55, fig. 4–3, 4–5, 4–6) remarked that in lunar eclipse omens dating to the first millennium BC, it is possible to abstract three other different schemes apart from the one mentioned above, virtually adjusted on the four quadrants of the Moon according to the needs of the scribes.[298]

296 For a discussion on schemes of association in Mesopotamian celestial divination, see Wainer (2016).
297 See also Brown (2000: 140 A i–ii).
298 These schemes have the following associations in relation to lunar eclipses: south = Akkad, north =

51: The identity of the goddess called "Anu's daughter" (DUMU.MUNUS ^{d}*a-nim*) is given in a Babylonian commentary from Nippur as the female demon Lamaštu (AO 17661 obv. 19, see Wee 2012: 500).

51–55: The beginning of these entries is not preserved, but the broken spaces at the beginning of sources E$_{1}$ and F allow restoration between 3 and 4 (small) signs. The entries continue mentioning the KÁ, *bābu*, "gate", which, as far as known, relates to halos and, in entries 56–60, to the rainbow. The remains of the signs in E$_{1}$ rev. 17', and F 13' and 17' are likely to be read as TÙR, so the following reconstruction is possible: [DIŠ MIN *ina* TÙ]R$^{?}$ MUL GUB-*ma*, "[If ditto (i.e. the Old Man is surrounded by a halo)] and a star stands [in the hal]o$^{?}$".

56–60: The beginning of these entries has been intentionally left unsolved because the state of preservation of the signs does not allow a secure reading. Yet, there are a few options which are worth discussing here. In source F (Rm 100 ll. 19'–20'), the traces before ^{d}TIR.AN.NA (*manzât*, "rainbow") could be read as [MI]N-*ma*. Based on the size of the remaining space at the beginning of the lines (1 or 2 small signs.), it would be possible to restore the protases as follows: [DIŠ MI]N-*ma* ^{d}TIR.AN.NA NÍGIN-*ma*, "[If ditt]o (i.e. the Old Man is surrounded by a halo) and it is surrounded by a rainbow". Two weather omens have similar protases, in which a rainbow is mentioned next to ^{mul}SIPA.ZI.AN.NA, *šitaddaru*, "True Shepherd of Anu" (Rm 2, 310 obv. 9'–10', see Gehlken 2012: 93–94, 23'–24' source D and parallels).[299] Gehlken (2012: 92 fn. 25) suggested that either a faint corona or a halo is meant instead of "rainbow" in the weather omens. If this is correct, one could assume that, in entries 56–60, a faint halo called "rainbow" is mentioned with a real halo. Nevertheless, even if the Babylonians knew the position of the True Shepherd of Anu during the day, it is unlikely that a rainbow could have been observed with a halo at night. Another option would be to read the traces before ^{d}TIR.AN.NA as the sign MÚL (𒀯), resulting in the following restoration: [DIŠ MÚ]L ^{d}TIR.AN.NA NÍGIN-*ma*, "[If a sta]r is surrounded by a rainbow". In that case, one could assume that the rainbow stands for a faint corona or halo. This is, however, also unlikely, because the scribe of source F always wrote MUL (𒀭𒀭𒀭) for *kakkabu*, and never MÚL.

Further Comments on the Text

The reconstructed tablet EAE 52 deals with groups of entries concerning the Field (^{mul}AŠ.GÁN, *ikû*), a light phenomenon (i.e. *ṣallummû*) and the Old Man (^{mul}ŠU.GI, *šību*).

Subartu (+ Gutium), east = Elam, west = Amurru; south = Subartu (+ Gutium), north = Amurru, east = Elam, west = Akkad; south = Akkad, north = Akkad, east = Subartu, west = Amurru (Old Babylonian scheme).

299 E.g. Rm 2, 310 obv. 9': DIŠ ^{d}⸢IŠKUR⸣ *ina* MÚRU KIMIN GÙ-⸢*šú*⸣ ŠUB-*ma* ^{d}TIR.AN.⸢NA⸣ [G]IM *gam-lì* NÍGIN-*šú* DINGIR^{meš} *šá ana* KUR ⸢ARḪUŠ⸣ TUK⸢^{meš} *ina* SU KUR BAD⸣ [...], "If Adad thunders in(to) the middle of ditto (i.e. the True Shepherd of Anu) and it is surrounded by a rainbow [l]ike a boomerang, the gods who show mercy to the land will withdraw from the interior (lit. flesh) of the land".

The reason behind this arrangement is calendrical: according to MUL.APIN (I iii 10, 12, see Hunger-Steele 2019: 51–52), the Field and the Old Man rise in Šabaṭu (i.e. Month XI) and Adaru (i.e. Month XII), respectively, whereas in the Astrolabe B (I i 1–11, see Horowitz 2014: 33), the Field is said to rise in Nisannu (i.e. Month I), and the Old Man in Ajaru (i.e. Month II) (Alb B III 2, see Horowitz 2014: 40). The macro structure of the entries in tablet EAE 52 is summed up in Table 6.

Subject of the protasis	Date of first visibility (MUL.APIN)	Related months in Astrolabe B
^{mul}AŠ.GÁN, *ikû*, "Field"	Šabaṭu (Month XI) day 5	Nisannu (Month I)
ṣallummû	-	-
^{mul}ŠU.GI, *šību*, "Old Man"	Adaru (Month XII) day 15	Ajaru (Month II)

Table 6. Groups of entries in EAE 52 arranged by their subject in the protases, and compared to their first visibility dates or heliacal rising according to MUL.APIN and Astrolabe B.

The "shift" in months between MUL.APIN and Astrolabe B could be due to the fact that these compositions belong to different scholarly traditions and/or originate in different times because the stars' illusory movement shifts through time due to the precession of the equinoxes (see 4.1.2.). Nevertheless, what is relevant in the present context is the fact that the Field and the Old Man are subsequent in their heliacal rising: their consecutive rising time is mirrored in this consecutive arrangement of the entries.

The occurrence of commentaries among the sources used for the reconstructed text, and their corresponding factual explanations, requires comments. The omens whose factual explanations relate, with confidence, to planetary conjunctions, are 16 in total (EAE 52: 1-2, 9, 15-22, 25-26, 40-41, 44). Based on these explanations, one can infer that the Field is equated with either the Furrow (^{mul}AB.SÍN, *absinnu*) (EAE 52: 15, 16, 20, 22, 25, 26, 40, 41) – Virgo in zodiological literature (see 6.1.1.) – or with Venus ($^{d/mul}$*dilbat*) (EAE 52: 17, 18, 21, 44). All other celestial bodies mentioned refer to the other planets; more precisely, the Resplendent One (^{mul}ŠU.PA, *nameru*), the Fish (^{mul}KU$_6$, *nūnu*), and the Eagle (^{mul}TI$_8$, *a*/*erû*) are equated with Mars ($^{d/mul}$*ṣalbatānu*); the Eridu-star (^{mul}NUN^{ki}), Šulpaea ($^{d/mul}$ŠUL.PA.È), and the Marduk-star ($^{mul.d}$AMAR.UTU) are equated with Jupiter; the Raven (^{mul}UG$_5$.GA, *āribu*) is equated with Mercury; the Pleiades (MUL.MUL, *zappu*) are equated with both Mercury ($^{d/mul}$UDU.IDIM.GU$_4$.UD) and Mars ($^{d/mul}$*ṣalbatānu*).

The sources of tablet EAE 52 do not allow a complete restoration of the remaining parts at the end of the original text. Yet, two sources point to the topic of the missing omens – likely the last thirds of the text – assuming that two third of the omens have been reconstructed. First, according to the reverse of source C (K 2118 rev. 2'–8', see App. B § 2), the last omens of EAE 52 refer to the chthonian god Enmešarra (Ebeling 1938; Weidner 1938); second, the *ṣâtu* commentary from Uruk, after several entries about the Old Man (VAT 7850+ rev. 6'–13'), which refer to EAE 52: 44–55, and one entry about Enmešarra (rev. 14'), explains the connection between the latter and the Old Man as follows:

[... ^{mul}ŠU].⸢GI⸣ TA *kin-ṣi-šú* EN *a-si-di-šú* : EN.ME.ŠÁR.RA *šum-*[*šu*...]

[... the Old] Man from his knee to his heel, [his] name is Enmešarra [...].
(VAT 7850 (+) AO 6486 rev. 15', see App. B § 27; EAE 52 source Q)

This line reveals how one constellation was pictured, or conceived, in the sky: the "legs" of the Old Man, from knees to heels,[300] were named Enmešarra. It is, therefore, likely that the subject of the last entries of EAE 52 is Enmešarra, either as a name for the Old Man, or as a name for the lower part of the Old Man. Enmešarra is also associated with the Old Man in MUL.APIN (I i 3, see Hunger-Steele 2019: 30).[301]

The Old Man, or Enmešarra, and the Pleiades were seen standing very close to each other in the night sky. According to MUL.APIN, the Pleiades and the Old Man are seen rising heliacally almost one after the other in the path of the Moon:

33. [MUL.M]UL ^{mul}GU$_{4}$.AN.NA ^{mul}SIPA.ZI.AN.NA ^{mul}ŠU.GI
(...)
38. PAP *an-nu-tu*$_{4}$ DINGIR^{meš} *šá ina* KASKAL ^{d}30 GUB^{meš}-*ma*

33. The Pleiades, the Bull of Heaven, the True Shepherd of Anu, the Old Man
(...)
38. All these are the gods who stand in the path of the Moon.
(MUL.APIN I iv 33, 38, see Hunger-Pingree 2019: 70, 71)

On a symbolic level, this "closeness" of the Pleiades and the Old Man or Enmešarra could mirror the myth of the seven sons of Enmešarra, a divine heptad whose origin is difficult to trace (see 2.3.2.). They are mentioned together with Anu in a few mythological narratives related to Marduk, as part of an old divine generation. Lambert (2013: 209–217) noted that the seven sons of Enmešarra are actually "seven Enlils", seven old rebel gods defeated by Marduk and equated to seven constellations in the so-called archive of mystic heptads (KAR 142 rev. iii 3'–10', see Pongratz-Leisten 1994: 223). The traces of such a theomachy can also be seen in a cultic calendar from Aššur dating to the Neo-Assyrian period (ca. 911–612 BC):

UD.19.KAM* *ša qu-li* DU$_{11}$.GA ⸢^{d}⸣*a-num* ^{d}IMIN.BI DUMU^{meš} ^{d}EN.ME.ŠÁR.RA *ki-i* LAL-*ú*

The 19th day, which is called silence, (is when) Anu defeated the divine Seven, the sons of Enmešarra.
(SAA 3, 40 obv. 5)

300 In the so-called Uranology texts, which refer to the shape of the constellations as "drawings", the *kinṣu*, "knee", is mentioned only in connection with ^{mul}UD.KA.DUḪ.A, "Demon with the Gaping Mouth" (Beaulieu-Frahm-Horowitz-Steele 2018: 36 D i 21–22).

301 For Enmešarra as part of the Old Man, see already Schaumberger (1935: 327).

Very little is known about the mythology of Enmešarra as a chthonian, chaotic, and archaic divinity. Eventually, the textual sources available to us only point to a similarity between the Sebettu and the sons of Enmešarra, by means of their destructive and demonic features (Konstantopoulos 2015: 232–233). However, as already argued in the previous chapters (see 2.5., 3.3.), the existence of several heptads is based on the number seven as an attribute for "totality". Hence, the idea that part of the Old Man has been named Enmešarra because of an underlying mythological narrative, and his closeness to the Pleiades, must remain a hypothesis.

5.1.5. The Reconstructed Tablet EAE 53 and Its Commentary

Tablet EAE 53 has been reconstructed based on commentaries with and without subscripts. Almost all the sources preserve explanations of the protases, and only a few fragments eventually represent an original tablet from the series EAE. More specifically, the sources belong to two different types of commentaries. The first commentary of tablet EAE 53 (Commentary EAE 53) is a *mukallimtu*-commentary, and it has the assumed original content and sequence of the omens in EAE 53. It gives a collection of omens about the Pleiades, the Jaw of the Bull (^{mul}*is lê*), and the True Shepherd of Anu (^{mul}SIPA.ZI.AN.NA, *šitaddaru*). The second commentary (SIT 6) is the assumed tablet 6 of *Šumma Sîn ina tāmartišu*: its content only partly parallels EAE 53 since it duplicates the first six omens of Commentary EAE 53. SIT 6 continues with a list of stars interacting with each other, explained as planetary conjunctions in which the stars in the protasis represent a planet. On the whole, the omens of SIT 6 were likely excerpted from various chapters of the series EAE, and most of them deal with the Pleiades.

It is not possible to know precisely how much of the original tablet EAE 53 is included in both commentaries. Only one source could represent the beginning of the original tablet EAE 53: K 11001 + K 15541 (Commentary EAE 53 source F; SIT 6 source BB), a small fragment of 8 lines from the top right part of the obverse of a tablet from Nineveh, dating to the Neo-Assyrian period (ca. 911–612 BC). In this fragment, each entry consists of a protasis and an apodosis, with no explanation as would be customary in commentaries. The reverse is lost, with the exception of a few scattered damaged signs. If K 11001+ reflects the original sequence of entries from the tablet EAE 53, its omens are duplicated in both Commentary EAE 53: 1–4, 6 and SIT 6: 1, 3–7. In short, it is possible to restore only the first seven entries of the original sequence of EAE 53 according to K 11001+, as shown in the entries given below:

1. [DIŠ MUL.MUL ^{mul}ŠUDUN] KUR-*ud ina* MU BI KI.LAM TUR-*ir*
2. [DIŠ MUL.MUL ^{mul}AMAR.UTU] KUR-*ud ina* MU BI *um-mu u um-šu*$_{14}$ GÁL-*ši*
3. [DIŠ MUL.MUL ^{mul}ÉLLAG] KUR-*ud ina* MU BI *um-mu* GÁL-*ši*

4. [DIŠ MUL.MUL ^{mul}KA.MUŠ.Ì.KÚ.E] KUR-*ud ina* MU BI Š[E K]ÁR-RA[302] GÁL-*ši*
5. [DIŠ MUL.MUL ^{mul}AŠ.GÁN] KUR-*ud ina* MU BI KI.LAM TUR-*ir*
6. [DIŠ KIMIN$^?$]x ki *a-ki-*⸢*lu*$_4$⸣ GÁL ^{d}IŠKUR RA-*iṣ*
7. [DIŠ KIMIN$^?$] GUN ^{g}[^{iš}GIŠIMMA]R : NAM LAL
8. [...] ⸢x x⸣ [...]x *ina* KUR GÁL
9. [...] ⸢x x⸣
(remainder is missing)

1. [If the Pleiades] reach [the Yoke], in that year market rate will diminish.
2. [If the Pleiades] reach [the Marduk-star], in that year there will be heat and blaze.
3. [If the Pleiades] reach [the Kidney], in that year there will be heat.
4. [If the Pleiades] reach [Pāšittu], in that year there will be bar[ley ...]...
5. [If the Pleiades] reach [the Field], in that year the market rate will diminish.
6. [If ditto$^?$ (i.e. [the Pleiades] reach [the Field])]..., there will be a pest (lit. devourer), Adad will devastate.
7. [If ditto$^?$ (i.e. [the Pleiades] reach [the Field])] the yield [of the date pal]m – (var.) of the district will diminish.
8. [...] ... [...]... (it) will be in the country.
9. [...] ...
(remainder is missing)
(K 11001 + K 15541 obv., see Commentary EAE 53 source F; SIT 6 source BB)

5.1.5.1. Commentary EAE 53

The reconstruction of Commentary EAE 53 is mainly based on the sequence of entries in the fragment K 3558 (App. B § 8; Commentary EAE 53 source A). According to Frahm (2011: 133), this fragment is part of a series of commentaries in Neo-Babylonian ductus, whose subscript he labelled as "*mukallimtu* 1a": NÍG.PÀD.DA DUB.x.KAM DIŠ UD AN ^{d}50, "*mukallimtu*-commentary of the tablet x of *Enūma Anu Enlil*".

For the organisation of the sources, the text, and the comments, see 5.1.3.5.

List of Sources

Source: A
Siglum: K 3558
Copy: ACh Suppl. 2, 66; Plate 16–17
Edition: ACh Suppl. 2, 66; see also App. B § 8 (only reverse)
Provenance: Nineveh
Ductus: Neo-Babylonian
Measurements: 6,9 x 11,3+ x 2,5+ cm

302 For Š[E K]ÁR-RA, see 5.1.5.1., commentary to the text of entry 4, p. 191.

CDLI n.: P238277 (without photograph); photographs available online at CCP 3.1.52.C.b (https://ccp.yale.edu/P238277 accessed 26.01.2021)
Description: Fragment of a tablet with the upper edge and part of the left and right edge preserved.
Type of text: EAE (53) *mukallimtu*-commentary

Source: B
Siglum: K 8744
Copy: Plate 18
Edition: See App. B § 9 (ll. 15'–16')
Provenance: Nineveh
Ductus: Neo-Babylonian
Measurements: 5,7+ x 4,4+ x 1,5+ cm
CDLI n.: P238794 (with photograph)
Description: Fragment probably of the middle part of the obverse of a tablet. The reverse is lost.
Type of text: EAE (53) commentary

Source: C
Siglum: Sm 1946
Copy: Plate 18
Provenance: Nineveh
Ductus: Neo-Assyrian
Measurements: 4,5+ x 4,8+ x 1+ cm
CDLI n.: P426189 (with photograph)
Description: Fragment probably of the obverse of a tablet with a small part of the left edge preserved. The reverse is lost.
Type of text: EAE (53)

Source: D
Siglum: Sm 1349
Copy: Plate 19
Provenance: Nineveh
Ductus: Neo-Assyrian
Measurements: 3,6+ x 2,9+ x 1,4+ cm
CDLI n.: P425876 (with photograph)
Description: Small fragment probably of the left part of the obverse of a tablet. The reverse is lost.
Type of text: EAE (53)

Source: E
Siglum: Sm 197
Copy: Plate 19
Edition: See App. B § 10 (ll. 6' and 9')
Provenance: Nineveh

Ductus: Neo-Assyrian
Measurements: 2,6+ x 1,8+ x 0,5+ cm
CDLI n.: P425292 (with photograph)
Description: Small fragment probably of the left part of the obverse of a tablet. The reverse is lost.
Type of text: EAE (53)

Source: F
Siglum: K 11001 + K 15541
Copy: Plate 20
Edition: See 5.1.5., pp. 172–173; See also SIT 6 source BB
Provenance: Nineveh
Ductus: Neo-Assyrian
Measurements: 2,8 x 5,9 x 1,5 cm
CDLI n.: P399026 (with photograph)
Description: Fragment of the right part of a tablet, with part of the upper and right edge preserved. The reverse is lost, with the exception of a few scattered damaged signs.
Type of text: EAE (53)

Source: G
Siglum: K 3923 + K 6140 + 81-7-27, 149 + 83-1-18, 479
Copy: ACh Suppl. 2, 79 (K 3923); Plate 21–22
Edition: ACh Suppl. 2, 79 (K 3923); Hunger-Reiner (1975: 22–23) (obverse); see also 4.2.2. § 2a (rev. 1'–4') and App. B § 11 (obverse and rev. 5'–6')
Provenance: Nineveh
Ductus: Neo-Babylonian
Measurements: 10,4+ x 7,7+ x 2,5+ cm
CDLI n.: P237106 (without photograph)
Description: Fragment of the lower part of a tablet, with the lower edge and part of the left and right edge preserved.
Type of text: EAE (53) commentary

Source: H
Siglum: Sm 1054
Copy: Plate 23
Edition: See App. B § 12 (obv. 1'–2', 4', 8')
Provenance: Nineveh
Ductus: Neo-Babylonian
Measurements: 5,1+ x 5,0+ x 0,4+ cm
CDLI n.: P240333 (with photograph)
Description: Fragment of the obverse of a tablet, with a small part of the lower edge preserved. The reverse is lost.
Type of text: EAE (53) commentary

Source: I
Siglum: Sm 247
Copy: Plate 24
Edition: See App. B § 13 (obv. 1'–7', 9'–16')
Provenance: Nineveh
Ductus: Neo-Assyrian
Measurements: 4,7+ x 5,1+ x 2,2+ cm
CDLI n.: P425318 (with photograph)
Description: Fragment of the middle or left part of the obverse of a tablet. The reverse is largely lost, except for a few traces of signs.
Type of text: SIT (4)

Source: J
Siglum: 81-2-4, 135
Edition: SAA 8, 351 = RMA 242
Provenance: Nineveh
Ductus: Neo-Assyrian
CDLI n.: P236988 (with photograph)
Type of text: Report

Source: K
Siglum: BM 38301 (80-11-12, 183)
Copy: Plate 25
Edition: Reiner-Pingree (1981: 78–79 XVIII source V) (obverse); see also SIT 6 source HH and App. B § 14 (reverse)
Provenance: Babylon
Ductus: Late Babylonian
Measurements: 6,7+ x 5,8+ x 2,5+ cm
Description: Fragment of the lower part of a tablet, with part of the lower right edge preserved.
Type of text: EAE (53–55)

Source: L
Siglum: 79-7-8, 271
Copy: Plate 26
Edition: See SIT 6 source II and App. B § 15 (ll. 1'–4', 7', 9'–11')
Provenance: Nineveh
Ductus: Neo-Babylonian
Measurements: 3,7+ x 3,9+ x 0,4+ cm
CDLI n.: P236893 (with photograph)
Description: Small fragment probably of the reverse of a tablet. The obverse is lost.
Type of text: EAE (53) commentary

Source: M
Siglum: 79-7-8, 210 + K 6997 (l. 13); Sm 1267 (ll. 3'-4')

Edition:	Reiner-Pingree 1981: 44 IV 4a source E (79-7-8, 210 + K 6997), 28 VI 2a source G (Sm 1267)
Provenance:	Nineveh
Ductus:	Neo-Assyrian, Neo-Babylonian (K 9098)
CDLI n.:	P396954 (with photograph)
Type of text:	EAE (50)

Source:	N
Siglum:	Sm 1317 (see EAE 52 source L and SIT 6 source N)

Source:	O
Siglum:	K 8000[303]
Edition:	ACh Ištar 24; Reiner 2006: 315–318 (rev. 10’–23’)
Provenance:	Nineveh
Ductus:	Neo-Assyrian
CDLI n.:	P397428 (with photograph)
Type of text:	EAE

Source:	P
Siglum:	Sm 366 + 80-7-19, 371
Edition:	SAA 8, 491 = RMA 167
Provenance:	Nineveh
Ductus:	Neo-Assyrian
CDLI n.:	P236957 (with photograph)
Type of text:	Report

Source:	Q
Siglum:	Rm 459
Edition:	Reiner-Pingree 1981: 78 XVIII source W
Provenance:	Nineveh
Ductus:	Neo-Assyrian
CDLI n.:	P425429 (with photograph)
Type of text:	EAE (53–55)

Source:	R
Siglum:	K 12149
Copy:	Plate 35
Provenance:	Nineveh
Ductus:	Neo-Assyrian
Measurements:	1,9+ x 2,8+ x 1+ cm
CDLI n.:	P399655 (with photograph)

303 K 8000 rev. 10’–15’ is a partial duplicate of BM 47799 obv. 8’–18’ (Reiner 2006: 315–318).

Description:	Fragment probably of the middle part of the obverse of a tablet. The reverse is lost.
Type of text:	EAE (53)

Source:	S
Siglum:	W 22730/5
Edition:	SpTU 3, 101; CCP 3.1.u5[304]
Provenance:	Uruk
Ductus:	Late Babylonian
CDLI n.:	P348705 (without photograph)
Type of text:	EAE (53) *mukallimtu*-commentary

Source:	T
Siglum:	VAT 7850 (+) AO 6486 (see EAE 52 source Q)

Source:	U
Siglum:	K 2118 (see EAE 52 source C)

Text

1.	**DIŠ MUL.MUL ^{mul}ŠUDUN** (var.: **MIN$^?$)**	**KUR-*ud ina* MU BI GÁN.BA**
A obv. 1	[DIŠ MU]L.MUL ^{mul}ŠUDUN	KUR-*ud*
E 1’	[DIŠ MUL].⸢MUL⸣ [...]	
F obv. 1	[	...] KUR-*ud ina* MU BI
H 6’	[	...] MIN$^?$ KUR-*ud ina* MU BI GÁN.BA [...]
N obv. 14’	DIŠ MUL.MUL ^{mul}ŠUDUN	KUR-*ud ina* MU BI GÁN.BA
S obv. 12’	DIŠ MUL.MUL ^{mul}⸢ŠUDUN⸣	KUR-*ud ina* MU BI GÁN.BA
T rev. 27’ (catchline)	[DIŠ MUL.MU]L ^{mul}⸢ŠUDUN⸣	KUR-*ud*
U rev. 9’ (catchline)	DIŠ MUL.MUL [...]	

	(var: **KI.LAM) TUR-*ir***	**^{d}UDU.IDIM.SAG.UŠ ^{d}*ṣal-bat-a-nu* KUR-*ma***
A obv. 1		⸢^{d}⸣UDU.IDIM.SAG.UŠ ^{d}*ṣal-bat-a-nu* KUR-*ma*
F obv. 1	KI.LAM TUR-*ir*	
H 7’	[	...] ^{d}*ṣal-bat-a-nu* KUR-*ma* [...]
N obv. 14’	TUR	
S obv. 12’	TUR-*ir* ⸢^{d}⸣[...]	

If the Pleiades reach the Yoke (var.: ditto$^?$), in that year the. market rate will diminish. (It means that) Saturn reaches Mars.

2.	**DIŠ MUL.MUL ^{mul}($^{\text{.d}}$)AMAR.UTU KUR-*ud ina* MU BI *um-mu u um-šu*$_{14}$**
A obv. 2	[DIŠ M]UL.MUL $^{\text{mul.d}}$AMAR.UTU KUR-*ud*

304 The full edition is online at the website CCP (https://ccp.yale.edu/P348705 accessed 05.03.2021).

E 2’ [DIŠ MUL.M]UL ^{mu}[^{l}...]
F obv. 2 [...] KUR-*ud ina* MU BI *um-mu u um-šu*$_{14}$
S obv. 14’ DIŠ MUL.MUL ^{mul}AMAR.UTU KUR-*ud ina* MU BI *um-mu u um-*⸢*ši*⸣ [...]

(var.: ***um-ši***) **GÁL-*ši* ^{d}GU$_{4}$.UD *lu* ^{d}SAG.ME.GAR ^{mul}*za-ap-pi* KUR-*ma* BAD-*ma***
A obv. 2–3 ⸢^{d}⸣GU$_{4}$.UD *lu* ^{d}SAG.ME.GAR ^{mul}*za-ap-pi* KUR-*ma* (3) BAD-*ma*
F obv. 2 GÁL-*ši*
S obv. 14’ *um-*⸢*ši*⸣ [...]

um-ma-a-tu*$_{4}$ ^{d}UTU RA-*iṣ* BAD-*ma* EN.TE.NA ^{d}IŠKUR RA-*iṣ* ^{d}*ṣal-bat-a-nu
A obv. 3 *um-ma-a-tu*$_{4}$ ^{d}UTU RA-*iṣ* BAD-*ma* EN.TE.⸢NA⸣ ^{d}IŠKUR RA-*iṣ* ^{d}*ṣal-bat-a-nu*

GABA.RI ^{d}UDU.IDIM GABA.RI ^{d}*dil-bat* GABA.RI ^{d}UDU.IDIM.SAG.UŠ
A obv. 3–4 GABA.RI ^{d}UDU.IDIM GABA.RI (4) ^{d}*dil-bat* GABA.RI ^{d}UDU.IDIM.SAG.UŠ

^{d}*za-ap-pi* KUR-*ma* ÚŠ^{meš} GAR^{meš}
A obv. 4 ⸢^{d}⸣*za-ap-pi* KUR-*ma* ÚŠ^{meš} GAR^{meš}

If the Pleiades reach the Marduk-star, in that year there will be heat and blaze. (It means that) Mercury or Jupiter reaches the Bristle. If (in) summer, Šamaš will devastate. If (in) winter, Adad will devastate. (If) correspondingly (i.e. if in winter) Mars or a planet, or Venus, or Saturn reaches the Bristle, a pestilence will take place.

3. **DIŠ MUL.MUL ^{mul}ÉLLAG KUR-*ud ina* MU BI *um-mu* GÁL-*ši***
A obv. 5 DIŠ MUL.MUL ^{mul}ÉLLAG KUR-*ud*
E 3’ [DIŠ MUL].MUL ^{mul}ÉL[LAG...]
F obv. 3 [...] KUR-*ud ina* MU BI *um-mu* GÁL-*ši*
N rev. 4 DIŠ MUL.MUL ^{mul}ÉLLAG KUR-*ud*
S obv. 15’ DIŠ MUL.MUL ^{mul}ÉLLAG KUR-*ud ina* MU BI *um-mu* ⸢GÁL⸣ [...]

(var.: **^{d}IŠKUR [RA...]**) **^{mul}*za-ap-pi u* ^{mul}*ṣal-bat-a-nu***
A obv. 5 ⸢^{mul}⸣*za-ap-pi u* ^{mul}*ṣal-bat-a-nu*
N rev. 4 ^{d}IŠKUR [RA...]

If the Pleiades reach the Kidney, in that year there will be heat (var.: Adad [will devastate...]). (It means) the Bristle and Mars.

4. **DIŠ MUL.MUL ^{mul}KA.MUŠ.Ì.KÚ.E KUR-*ud ina* MU BI Š[E K]ÁR-RA**
A obv. 6 DIŠ MUL.MUL ^{mul}KA.MUŠ.Ì.KÚ.E KUR-*ud*
E 4’ [DIŠ MUL].⸢MUL⸣ ^{mul4}KA.M[UŠ.Ì.KÚ.E...]
F obv. 4 [...] KUR-*ud ina* MU BI Š[E K]ÁR-RA
N obv. 16’ DIŠ MUL.MUL ^{mul}KA.MUŠ.Ì.KÚ.E KUR-⸢*ud*⸣ [...]
P rev. 3 [... ^{mul}KA].MUŠ.Ì.KÚ.E KUR-*ud*
S obv. 13’ DIŠ MUL.MUL ^{mul}KA.MUŠ.Ì.KÚ.E KUR-*ud ina* MU BI [...]

GÁL-*ši* ^{d}*ṣal-bat-a-nu* ^{d}(UDU.IDIM.)SAG.UŠ KUR-*ma*
A obv. 6 ⸢^{d}⸣*ṣal-bat-a-nu* ^{d} UDU.IDIM.SAG.UŠ KUR-*ma*
F obv. 4 GÁL-*ši*
P rev. 4 [... ^{d}*ṣal*]-*bat-a-nu* ^{d}SAG.UŠ KUR-*ma*

If the Pleiades reach Pāšittu, in that year there will be bar[ley.] (It means that) Mars reaches Saturn.

5. **DIŠ MUL.MUL ^{mul}MUŠ KUR-*ud ina* MU BI GÁN.BA [TUR]**
A obv. 7 DIŠ MUL.MUL ^{mul}MUŠ KUR-*ud*
E 5' [DIŠ MU]L.MUL ^{mul}MUŠ [...]
N obv. 15' DIŠ MUL.MUL ^{mul}MUŠ KUR-*ud ina* MU BI GÁN.⸢BA⸣ [TUR]

^{d}GU$_4$.UD *lu* ^{d}SAG.UŠ ^{d}*ṣal-bat-a-nu* KUR-*ma*
A obv. 7 ^{d}⸢GU$_4$⸣.UD *lu* ^{d}SAG.UŠ ^{d}*ṣal-bat-a-nu* KUR-*ma*

If the Pleiades reach the Snake, in that year the market rate [will diminish]. (It means that) Mercury or Saturn reaches Mars.

6. **DIŠ MUL.MUL *ana* ^{mul}AŠ.GÁN TE GUN ^{giš}GIŠIMMAR**
A obv. 8 DIŠ MUL.MUL *ana* ^{mul}AŠ.GÁN TE
E 7' [DIŠ MU]L.MUL *ana* ^{mul}[...]
F obv. 7 [...] GUN ^{g}[^{iš}GIŠIMMA]R
S obv. 17' DIŠ MUL.MUL ^{mul}AŠ.GÁN KUR-*ud* GUN ^{giš}GIŠIMMAR

(var.: **: NAM) LAL ^{d}*ṣal-bat-a-nu ana* ^{d}GU$_4$.UD *lu ana* ^{mul}AB.SÍN TE-*ma***
A obv. 8 ⸢^{d}*ṣal*⸣-*bat-a-nu ana* ^{d}GU$_4$.UD *lu ana* ^{mul}AB.SÍN TE-*ma*
F obv. 7 : NAM LAL
S obv. 17' LAL ^{d}[...]

If the Pleiades come close to the Field, the yield of the date palm (– (var.) of the district) will diminish. (It means that) Mars comes close to Mercury or to the Furrow.

7. **DIŠ MUL.MUL ^{mul}SIPA.ZI.AN.NA *ina* SAG-*šú* GUB**
A obv. 9 ⸢DIŠ⸣ MUL.MUL ^{mul}SIPA.ZI.AN.⸢NA *ina* SAG⸣-*šú* GUB

^{d}UDU.IDIM.SAG.UŠ *ina* IGI MUL.MUL GUB-*ma*
A obv. 9 ^{d}⸢UDU.⸢IDIM⸣.SAG.UŠ *ina* IGI MUL.MUL GUB-*ma*

If the True Shepherd of Anu stands at the head of the Pleiades. (It means that) Saturn stands in front of the Pleiades.

8. **DIŠ MUL.MUL *u* ^{mul}MAR.GÍD.DA** (var.: **^{mul}MAR) UR.BI GUB^{me}**
A obv. 10 [DIŠ] MUL.MUL *u* ^{mul}⸢MAR.GÍD⸣.DA UR.BI GUB^{me}
E 8' [DIŠ MU]L.MUL *u* ^{mu}[^{l}...]
M MUL.MUL *u* ^{mul}MAR UR.BI GUB^{meš}

ŠÈG^{meš} *u* ILLU^{meš} DU^{meš}-*nim-ma* ŠE.GÙN.NU TUR *ina* EN.TE.NA
M ŠÈG^{meš} *u* ILLU^{meš} DU^{meš}-*nim-ma* ŠE.GÙN.NU TUR *ina* EN.TE.NA

ŠUB-*tì* [*bu-lì*] ^{d}*dil-bat* KI MUL.MUL MÚ-*ma*
A obv. 10 ^{d}*dil-bat* KI MUL.MUL MÚ-*ma*
M ŠUB-*tì* [*bu-lì*]

If the Pleiades and the Wagon stand together, rains and flood will come, the crop will be diminished, in winter (there will be) an epidemic among [the herd]. (It means that) Venus glows with the Pleiades.

9. **[DIŠ] MUL.MUL *ina* KI.GUB ^{d}UTU.ŠÚ GUB^{me} *šá ina* UD.DUG$_4$.GA-*šú-nu***
A obv. 11 [DIŠ] MUL.MUL *ina* KI.GUB ^{d}UTU.ŠÚ GUB^{me} *šá ina* UD.⸢DUG$_4$⸣.GA-*šú-nu*

la it-ba-lu
A obv. 11 *la it-ba-lu*

[If] the Pleiades stand in the position of the sunset (i.e. west). (It means) that in their specified time they do not disappear (lit. remove themselves).

10. **[DIŠ M]UL.MUL *a-dir* GIŠ.ḪUR *i-lam-mu-ma***
A obv. 12. [DIŠ M]UL.MUL *a-dir* GIŠ.ḪUR *i-lam-mu-{ma}-ma*

[If the P]leiades are dark. (It means that) a "drawing" surrounds (them).

11. **[DIŠ M]UL.MUL *meš-ḫa im-šuḫ meš-ḫu* TA ŠÀ-*šú-nu* È-*ma* SAG *ú-kul-ti***
A obv. 13 [DIŠ M]UL.MUL *meš-ḫa im-šuḫ meš-ḫu* ⸢TA⸣ ŠÀ-*šú-nu* È-*ma* SAG *ú-kul-ti*

[If the P]leiades produce a glow. (It means that) the glow comes out of them and (there will be) the beginning of the consumption.

12. **[DIŠ] MUL.MUL *iṣ-ru-ur-ma* MUL *šá* IGI-*šú* NÍGIN**
A obv. 14 [DIŠ] MUL.MUL *iṣ-ru-ur-ma* MUL *šá* IGI-*šú* NÍGIN

^{d}*ṣal-bat-a-nu* KI ^{mul}*a-ru*$^{?}$*-ú* NIGIN-*ma*
A obv. 14 ^{d}*ṣal-bat-*⸢*a*⸣*-nu* KI ^{mul}*a-*⸢*ru*$^{?}$*-ú*⸣ NIGIN-*ma*

[If] the Pleiades flare and surround a star which (is) in front of them. (It means that) Mars is surrounded with the Eagle$^{?}$.

13. **DIŠ MUL.MUL *iṣ-ru-ur-ma* ^{d}UTU NÍGIN ^{d}*ṣal-bat-[a]-nu***
A obv. 15 ⸢DIŠ⸣ MUL.MUL *iṣ-ru-ur-ma* ^{d}UTU NÍGIN ^{d}*ṣal-bat-*[*a*]*-nu*

KI ^{mul}SAG.ME.GAR NIGIN-*ma*
A obv. 15 KI ⸢^{d}SAG.ME.GAR⸣ NIGIN-*ma*

If the Pleiades flare and surround the Sun. (It means that) Ma[r]s is surrounded with Jupiter.

14. **DIŠ MUL.MUL *iṣ-ru-ur-ma ina* IGI ^{d}*dil-bat* GUB**
A obv. 16 DIŠ MUL.MUL ⸢*iṣ-ru*⸣*-ur-ma ina* IGI ^{d}*dil-bat* GUB
B 1' [DIŠ MUL.MUL *iṣ-ru-ur*]-⸢*ma ina* IGI⸣ x[...]

^{d}*ṣal-bat-a-nu il-la-kam* a-na* ^{d}*dil-bat* DU-*ma*
A obv. 16 ^{d}*ṣal-bat-*⸢*a*⸣*-nu il-*⸢*la-kam* a-na*⸣ ^{d}*dil-bat* DU-*ma*

If the Pleiades flare and stand in front of Venus. (It means that) Mars comes and moves towards Venus.

15. **DIŠ MUL.MUL [*i*]*n-na-pal ina* MU.B[I...] ^{d}*ṣal-bat-*[*a-n*]*u it-ta-na-al-lak-ma***
A obv. 17 DIŠ MUL.MUL [*i*]*n-na-pal* ^{d}*ṣal-bat-*[*a-n*]*u it-ta-na-al-lak-ma*
B 2' [DIŠ MUL.MUL *in*]*-na-pal ina* MU.B[I...]

If the Pleiades *are* [*t*]*orn down*, in th[at] year [...]. (It means that) Ma[r]s keeps going.

16. **DIŠ MUL.MUL MUL^{meš}-*šú* 12 *uš-te-te-še-rù* 12 ITI^{meš} [K]I ^{d}30 LAL^{meš}-*ma***
A obv. 18 ⸢DIŠ⸣ MUL.MUL MUL^{meš}-*šú* 12 *uš-te-te-še-rù* 12 ITI^{meš} [K]I ^{d}30 LAL^{meš}-*ma*

B 3’	[DIŠ MUL.MUL MUL^{m}]^{eš}-*šú* 12 *uš-te-te-*⸢*še-rù*⸣ [...]
	KIMIN *ina* 12.KÁM ITI KI $^{<d>}$30 LAL^{meš}-*ma*
A obv. 18	KIMIN *ina* 12.KÁM ITI KI $^{<d>}$30 LAL^{meš}-*ma*
	: 5 ^{d}[UD]U.IDIM^{meš} KI-*šú-nu* GUB^{me}-*zu-ma*
A obv. 18a	: 5 ⸢^{d}⸣[UD]U.IDIM^{meš} ⸢KI-*šú*⸣-*nu* GUB^{me}-*zu-ma*

If the 12 stars of the Pleiades move straight ahead. (It means that) for 12 months they are balancing [wit]h the Moon. Ditto (i.e. if the 12 stars of the Pleiades move straight ahead). (It means that) in the 12th month they are balancing with the Moon – (var.) (it means that) the five [pla]nets stand still with them.

17.	**[DIŠ MUL.MUL MU]L^{meš}-*šú* 11**	**[...]**
B 4’	[DIŠ MUL.MUL MU]L⸢^{meš}⸣-*šú* 11	[...]

[If the Pleiades], their [sta]rs (are) 11 [...].

18.	**[DIŠ MUL.MUL MU]L^{meš}-*šú* 10**	**[...]**
B 5’	[DIŠ MUL.MUL MU]L^{meš}-*šú* 10	[...]

[If the Pleiades], their [star]s (are) 10 [...].

19.	**DIŠ *ina* SAG MU MUL.MUL *ka-ri-it ina šit-qul-ti* ^{d}30 *im-ma-rak-ku-ma***
A obv. 19	DIŠ *ina* SAG MU MUL.MUL *ka-*[*ri-i*]*t ina šit-qul-*[*t*]*i* ^{d}30 ⸢*im*⸣*-ma-rak-ku-ma*
B 6’	[DIŠ *ina* SAG MU MU]L.MUL *ka-ri-it* [...]
H obv. 5’	[...]x uḫ *ina šit-qul-ti* *im-ma-*[*rak-ku-ma*...]
	MUL.MUL *ar-ku-ma* ^{d}[30 *pa-ni-ma*? ^{d}*za-ap*]-*pu ana*
A obv. 20	MUL.MUL *ar-ku-ma* ⸢^{d}⸣[30 ... ^{d}*za-ap-p*]*u ana*
L 5’	[... ^{d}*za-ap*]-*pu*
	(var.: ***ina***) **^{d}UTU.È (^{d}30) *ana*** (var.: **<*ina*>**) **^{d}UTU.ŠÚ(.A)** (var.: **MIN x[...]**)
A obv. 20	^{d}UTU.È ^{d}30 *ana* ^{d}UTU.ŠÚ.A
L 5’	*ina* ^{d}UTU.È <*ina* ^{d}UTU>.ŠÚ MIN x[...]

If at the beginning of the year the Pleiades are cut. (It means that) at the balancing with the Moon they lag behind, the Pleiades are behind and the [Moon is ahead?, the Brist]le (is) to the east (i.e. towards sunrise), (the Moon is) to the west (i.e. towards sunset) (var.: ditto (?) ...[...]).

20.	**DIŠ *ina* SAG MU MUL.MUL *šá-ti-iḫ ina šit-qul-ti* $^{(d)}$30 *i-pan-nu-ma* A.ŠÀ**
A obv. 21–22	DIŠ *ina* SAG MU MUL.MUL *šá-*[*ti-iḫ ina šit-qul-t*]*i* ^{d}30 *i-pan-nu-ma* (22)[A].ŠÀ
B 7’	[DIŠ *ina* SAG MU MU]L.MUL *šá-ti-iḫ* [...]
H obv. 3’	[...] *ina šit-qul-t*[*i*...]
I obv. 8’	[DIŠ *ina* SAG MU MUL.M]UL *šá-ti-iḫ* A.ŠÀ
J rev. 4–6	DIŠ MUL.MUL *šá-ti-iḫ* (6) *ina šit-qul-ti* 30 *i-*[*pan-nu-ma*] (5) A.ŠÀ
L 6’	[DIŠ *ina* SAG MU M]UL.MUL *šá-ti-i*[*ḫ*...]
	A.GÀR (var.: **K[UR]**) **(1) GUN ÍL [ÍL *na-š*]*u-ú* ÍL *šá-qu-ú* GUN ÍL [...]**
A obv. 22	A.GÀR K[UR ÍL *na-š*]*u-ú* ÍL *šá-qu-ú* GUN Í[L...]
I obv. 8’	A.GÀR GUN x[...]
J rev. 5	A.GÀR 1 GUN ÍL

If at the beginning of the year the Pleiades are elongated. (It means that) at the balancing with the Moon they are ahead, the field (and) the meadow (var.: of the c[ountry]) will produce a yield; [ÍL (means) to produ]ce, ÍL (also means) to be high, GUN ÍL (means) [...].

21.	**[DIŠ *ina* SA]G MU MUL.MUL *pur-ru-ur* [*ina šit-qul*]*-ti* ^{d}30 *im-ma-rak-ku-ma***
A obv. 23	[DIŠ *ina* SA]G MU MUL.MUL x[... *ina šit-qul*]*-ti* ^{d}30 *im-ma-rak-ku-ma*
B 8'	[DIŠ *ina* SAG MU M]UL.MUL *pur-ru-ur* [...]

[If at the beginnin]g of the year the Pleiades disperse. [(It means that) at the balancin]g with the Moon they lag behind.

22.	**DIŠ *ina* SAG MU MUL.MUL GE$_6$ IM^{meš} NU DÙG.GA^{meš}**
A obv. 24	DIŠ *ina* ⸢SAG MU⸣ MUL.MUL [...
B 9'	[DIŠ *ina* SAG MU M]UL.MUL GE$_6$ IM^{meš} NU DÙG.GA^{meš}
	Z[I^{meš}... ŠU.BI.A]Š.ÀM
A obv. 24	ŠU.BI.A]Š.ÀM
B 9'	Z[I^{meš}...]

If at the beginning of the year the Pleiades become dark, bad winds will ri[se... (It means that) di]tto (i.e. at the balancing with the Moon they lag behind).

23.	**DIŠ *ina* SAG MU MUL.MUL *um-mul* [... ŠU.BI.A]Š.ÀM**
A obv. 25	DIŠ *ina* ⸢SAG⸣ MU ⸢MUL⸣.MUL [... ŠU.BI.A]Š.ÀM
B 10'	[DIŠ *ina* SAG MU MU]L.MUL *um-mul* [...]

If at the beginning of the year the Pleiades scintillate [... (It means that) di]tto (i.e. at the balancing with the Moon they lag behind).

24a.	**DIŠ *ina* SAG MU MUL.MUL *ina* KI.GUB-*šú* ^{mul}AŠ.GÁN GUB KÚR**
A obv. 26	DIŠ *ina* SAG MU MUL.⸢MUL⸣
B 11'	[DIŠ *ina* SAG MU MU]L.MUL *ina* KI.GUB-*šú* ^{mul}AŠ.GÁN GUB KÚR [...]
	[... ^{d}*ṣal-ba*]*t-a-nu lu* ^{d}GU$_4$.UD KI-*šú-nu* G[UB...]
A obv. 26	[... ^{d}*ṣal-ba*]*t-a-nu lu* ^{d}GU$_4$.UD KI-*šú-nu* G[UB...]

If at the beginning of the year the Field stands in the position of the Pleiades, the enemy [... (It means that) Ma]rs or Mercury s[tand] with them [...].

24b.	**[DIŠ *ina* SAG MU MUL.M]UL *ina* KI.GUB-*šú* ^{mul}UD.AL.TAR GUB-*az* x[...]**
B 12'	[DIŠ *ina* SAG MU MUL.M]UL *ina* KI.GUB-*šú* ^{mul}UD.AL.TAR GUB-*az* x[...]
	^{d}UD.AL.TAR ^{d}[...] *pa-ni* [...]
A obv. 27	^{d}UD.AL.TAR ^{d}[...] *pa-ni* [...]

[If at the beginning of the year] the Heroic One stands in the position of the [Pleiad]es ...[...]. (It means that) the Heroic One (i.e. Jupiter) [...] (in) front [...].

24c.	**[DIŠ *ina* SAG MU MUL.MUL *ina*] KI.GUB-*šú* ^{mul}EN.TE.NA.BAR.ḪUM**
B 13'	[DIŠ *ina* SAG MU MUL.MUL *ina*] ⸢KI.GUB-*šú* ^{mul}⸣EN.TE.NA.BAR.ḪUM

GU[B...] ^{mul}EN.TE.NA.BAR.ḪUM ^{d}UDU.IDIM.GU$_4$.UD ^{mul}[...]x
A obv. 28 ^{mul}EN.TE.NA.BAR.ḪUM ^{d}UDU.⸢IDIM.GU$_4$.UD ^{mul}⸣[...]x
B 13’ GU[B...]

^{mul}*si-mu-[ut...]*
A obv. 28 ⸢^{mul}*si-mu*⸣-[*ut*...]

[If at the beginning of the year] the Mouse stan[ds in] the position of the Pleiades] [...]. (It means that) the Mouse (is) Mercury, the star [...]... Sim[ut (i.e. Mars)...].

24d. **[DIŠ *ina* SAG MU MUL.MUL *ina* KI.GUB-*šú* ^{m}]^{ul}MAN-*ma* GUB [...]**
B 14’ [DIŠ *ina* SAG MU MUL.MUL *ina* KI.GUB-*šú* ^{m}]^{ul}MAN-*ma* GUB [...]

^{mul}MAN-*ma* ^{d}*ṣal-bat-anu* ^{mul}[...]
A obv. 29 ^{mul}MAN-*ma* ^{d}*ṣal-bat-anu* ^{mul}[...]

[If at the beginning of the year] the Other One stands [in the position of the Pleiades]. (It means that) the Other One (is) Mars, the star [...].

25. **DIŠ MUL.MUL *ana* ŠÀ ^{d}30 KU$_4$^{meš}-*ma ana* IM.3 È x[...]**
A obv. 30 DIŠ MUL.MUL *ana* ŠÀ ^{d}30 KU$_4$^{meš}-*ma ana* IM.3 È x[...]

^{mul}NUN^{ki} ^{d}*ṣal-bat-a-nu* ^{mul}[...] *na-pal-su-ḫu a-šá-[bu...]*
A obv. 31–32 ^{mul}NUN^{ki} ^{d}*ṣal-bat-a-nu* ^{mul}[...] (32) *na-pal-su-ḫu a-šá-*[*bu*...]

AN.GE$_6$ *at-t[a-lu-ú...]*
A obv. 33 AN.GE$_6$ *at-t*[*a-lu-ú*...]

If the Pleiades enter into the Moon and come out to the east ...[...]. (It means that) the Eridu-star (is) Mars, the star [...]; to fall down (means) to sit id[ly ...], AN.GE$_6$ (means) ecl[ipse...].

26. **DIŠ ^{mul}*ís-le-e* [*ina*] ŠÀ ^{d}[30 GUB...] *ta-lit-ti* [...]**
G rev. 7 DIŠ ^{mul}*ìs-le-*⸢*e*⸣ [*ina*] ⸢ŠÀ ^{d}⸣[...] ⸢*ta-lit-ti*⸣ [...]

If the Jaw of the Bull [stands in]side [the Moon...], the offspring [...].

27. **DIŠ ^{mul}*ís-le-e ina* MAŠ.SÌLA <d>30 GUB ^{mul4}[...] lú x [...]**
A obv. 34 DIŠ ^{mul}*ís-le-e ina* MAŠ.SÌLA <d>30 GUB ^{mul4}[...]
G rev. 8 DIŠ ^{mul}*ìs-*⸢*le-e*⸣ *ina* MAŠ.SÌLA <d>10[+20...] ⸢lú x⸣ [...]

If the Jaw of the Bull stands in the shoulder of the Moon. (It means that) the star [...]... [...].

28. **DIŠ ^{mul}*ís-l*[*e*]*-e ina* SI 15 <d>10+[20 GUB...]**
G rev. 9 DIŠ ^{mul}*ìs-l*[*e*]-*e ina* SI 15 <d>10[+20...]

If the Jaw of the Bu[ll stands] in the right horn of the M[oon...].

29. **DIŠ ^{mul}*ís-le-e ina* SI 2,30 <d>10+[20 GUB...]**
G rev. 10 DIŠ ^{mul}⸢*ís-le*⸣-*e ina* SI 2,30 <d>10[+20...]

If the Jaw of the Bull [stands] in the left horn of the Mo[on...].

30. **DIŠ *ina* ŠÀ ^{mul}*ís-le-e* ^{d}30 GUB *ana* [...]**
A obv. 35 DIŠ *ina* ŠÀ ^{mul}*ís-le-e* ^{d}30 GUB *ana* [...]
G rev. 11 ⸢DIŠ *ina* ŠÀ⸣ [^{mul}*í*]*s-le-e* ^{d}[30...]

If the Moon stands inside the Jaw of the Bull. (It means that) towards [...].

31. **[DIŠ... ^{mul}*ís-l*]*e-e* [...]**
G rev. 12 [... ^{mul}*ís-l*]*e-e* [...]

If [... the Jaw of] the Bull [...].

32. **[DIŠ ^{mul}*ís-l*]*e-e meš-ḫa i*[*m-šuḫ*...]**
G rev. 13 [DIŠ ^{mul}*ís-l*]*e-e meš-ḫa i*[*m-šuḫ*...]

[If the Jaw of t]he Bull p[roduces] a glow [...].

G ————————————————————————

33. **DIŠ ^{mul}SIPA.ZI.AN.NA MUL BI GE$_6$ [...]**
A obv. 36 DIŠ ^{mul}SIPA.ZI.AN.NA MUL BI ⸢GE$_6$⸣ [...]
G rev. 14 [DIŠ ^{mul}S]IPA.ZI.AN.⸢NA⸣ [...]

If the True Shepherd of Anu – its star is black [...].

34. **DIŠ ^{mul}SIPA.ZI.AN.NA MUL BI BABBAR x[...]**
A obv. 37 DIŠ ^{mul}SIPA.ZI.AN.NA MUL BI BABBAR x[...]
G rev. 15 [DIŠ] ⸢^{mul}⸣SIPA.ZI.AN.N[A...]

If the True Shepherd of Anu – its star is white [...].

35. **DIŠ ^{mul}SIPA.ZI.AN.NA MUL BI [...]**
A obv. 38 ⸢DIŠ⸣ ^{mul}SIPA.ZI.AN.NA MUL ⸢BI⸣ [...]
G rev. 16 DIŠ ^{mul}SIPA.ZI.AN.N[A...]

If the True Shepherd of Anu – its star [...].

36. **DIŠ ^{mul}SIPA.ZI.AN.NA *ina* x[...]**
A obv. 39 ⸢DIŠ⸣ ^{mul}SIPA.ZI.AN.NA *ina* x[...]
G rev. 17 DIŠ ^{mul}SIPA.ZI.AN.N[A...]

If the True Shepherd of Anu in ...[...].

37. **DIŠ ^{mul}SIPA.ZI.AN.NA x x[...]**
A obv. 40 ⸢DIŠ⸣ ^{mul}SIPA.ZI.AN.⸢NA x⸣ x[...]
G rev. 18 DIŠ ^{mul}SIPA.ZI.AN.N[A...]

If the True Shepherd of Anu ... ...[...].

38. **DIŠ ^{mul}SIPA.Z[I.AN.NA ...]**
A obv. 41 [DIŠ] ^{mul}⸢SIPA.ZI⸣.[AN.NA...]
G rev. 19 DIŠ ^{mul}SIPA.ZI.AN.[NA...]

[If] the True Shep[herd of Anu...].

39. **DIŠ ^{mul}SIPA.[ZI.AN.NA...]**
A obv. 42 [DIŠ ^{mu}]^{l}[SIPA.ZI.AN.NA...]
G rev. 20 DIŠ ^{mul}SIPA.Z[I.AN.NA...]

[If] the True She[pherd of Anu...].

40. **[DIŠ ^{mul}]SIPA.ZI.[AN.NA...]x [...]**
A (remainder is missing)
K obv. 1' [DIŠ ^{mul}]⌜SIPA.ZI⌝.[AN.NA...]x [...]
G rev. 21 ⌜DIŠ ^{mul}⌝S[IPA.ZI.AN.NA...]

[If] the True Shep[herd of Anu...]... [...].

41. **[DIŠ ^{mu}]^{l}SIPA.ZI.AN.NA [*ina* KI.GUB$^{?}$-*š*]*ú* IM$^{!}$(MUŠ) *da-am***
K obv. 2' [DIŠ ^{mu}]^{l}SIPA.ZI.⌜AN⌝.N[A *ina* KI.GUB$^{?}$-*š*]*ú* ⌜IM$^{!}$(MUŠ)⌝ *da-am*
G (remainder is missing)
Q 1' [DIŠ ^{mul}]⌜SIPA.ZI⌝.AN.NA⌝ [...]

KUR *ana* B[AD$_4$$^{?}$ NIGIN]
K obv. 2' KUR *ana* B[AD$_4$$^{?}$ NIGIN]

[If] the True Shepherd of Anu (is) [in it]s [position$^{?}$] (and) the wind *roams*$^{?}$, the country [will gather] in a f[ortress$^{?}$].

42. **[DIŠ ^{mu}]^{l}SIPA.ZI.AN.NA *ina* È-*šú ša-qu ina* SAG I[TI-*šú* IGI MU...]**
K obv. 3' [DIŠ ^{mu}]^{l}SIPA.ZI.AN.⌜NA⌝ *ina* È-*šú ša-qu ina* SAG I[TI...]
Q 2' [DIŠ ^{m}]^{ul}SIPA.ZI.AN.NA *ina* [...]

[If] the True Shepherd of Anu is high when it comes forth. (It means that) at the beginning [of its] m[onth it is visible, the year...].

43. **[DIŠ ^{m}]^{ul}SIPA.ZI.AN.NA LI.DU[R]-*su it-ta-na-an-b*[*iṭ*...]**
K obv. 4' [DIŠ ^{m}]^{ul}SIPA.ZI.AN.⌜NA⌝ LI.DU[R]-⌜*su*⌝ *it-ta-na-an-b*[*iṭ*...]
Q 3' [DIŠ] ^{mul}SIPA.ZI.AN.NA L[I.DUR...]

[If] the nav[el] of the True Shepherd of Anu constantly shines brig[htly...].

44. **[DIŠ ^{m}]^{ul}SIPA.ZI.AN.NA LI.DU[R-*s*]*u* GE$_6$ AN.[GE$_6$ GAR-*an*...]**
K obv. 5' [DIŠ ^{m}]^{ul}SIPA.ZI.AN.⌜NA⌝ LI.DU[R-*s*]*u* ⌜GE$_6$⌝ AN.[GE$_6$...]
Q 4' [DIŠ] ^{mul}SIPA.ZI.AN.NA L[I.DUR...]

[If] the nave[l o]f the True Shepherd of Anu becomes dark, an ecli[pse will occur...].

45. **[DIŠ ^{m}]^{ul}SIPA.ZI.AN.NA KI.GUB DIM$_4$-*m*[*a*$^{?}$] KAN$_5$ BÁRA *ina* KUR [...]**
K obv. 6' [DIŠ ^{m}]^{ul}SIPA.ZI.AN.⌜NA⌝ KI.GUB DIM$_4$-*m*[*a*$^{?}$] KAN$_5$ BÁRA *ina* ⌜KUR⌝ [...]
Q 5' [DIŠ] ^{mul}SIPA.ZI.AN.NA K[I.GUB...]

[If] the True Shepherd of Anu approaches its position a[nd$^{?}$] it is dark, the sanctuary in the country [...].

46.	**[DIŠ] ^{mul}SIPA.ZI.AN.NA *ana* 15 MUL.MUL *i[q]-rib* ^{d}+*en-líl* KUR *ú-ni*$^{?}$-[...]**
K obv. 7’	[DIŠ ^{mu}]^{l}SIPA.ZI.AN.NA ⸢*ana*⸣ 15 MUL.MUL *iq-*⸢*rib*⸣ ^{d}+*en-líl* KUR *ú-ni*$^{?}$-[...]
Q 6’	[DIŠ] ^{mul}SIPA.ZI.AN.NA [...]

[If] the True Shepherd of Anu com[es] near the right side of the Pleiades, Enlil will [...] the country.

47.	**[DIŠ] ^{mul}SIPA.ZI.AN.NA *ana* 2,30 MUL.MUL *iq-rib* ^{munus}KÚR [...]**
K obv. 8’	[DIŠ ^{mul}]SIPA.ZI.AN.NA *ana* ⸢2⸣,30 MUL.MUL ⸢*iq*⸣*-rib* ^{munus}KÚR [...]
Q 7’	[DIŠ] ^{mul}SIPA.ZI.AN.NA [...]

[If] the True Shepherd of Anu comes near the left side of the Pleiades, hostilities [...].

48.	**[DIŠ] ^{mul}SIPA.ZI.AN.NA *a-dir* AN.GE$_{6}$ ^{d}30 *u* ^{d}UTU *ina* KUR DÙ.[A.BI...]**
K obv. 9’	[DIŠ ^{mul}]SIPA.ZI.AN.NA *a-*⸢*dir*⸣ AN.GE$_{6}$ ^{d}30 *u* ^{d}UTU *ina* KUR DÙ.[A.BI...]
Q 8’	[DIŠ] ^{mul}SIPA.ZI.AN.NA *a-[dir...]*
	MÁŠ.ANŠE [...]
K obv. 10’	[...] MÁŠ.ANŠE [...]

[If] the True Shepherd of Anu is dark, (there will be) an eclipse of the Moon and the Sun, in the en[tire] country [...], the herd [...].

49.	**[DIŠ ^{mul}]SIPA.ZI.AN.NA *meš-ḫu im-šu-uḫ* LUGAL EN BALA**
K obv. 11’	[DIŠ ^{mul}]SIPA.ZI.AN.NA *meš-*⸢*ḫu*⸣ *im-šu-uḫ* LUGAL EN BALA
Q 9’	[DIŠ ^{mul}SIPA.ZI.AN].⸢NA⸣ *m[eš-ḫu...]*
R 1’.	[...]x x[...]
	ina šèr-t[i-šú...] ***bu-bu-*ʾ*u-tú*** **DIRI-*ma* BA.Ú[Š]**
K obv. 11’–12’	*ina šèr-t[i-šú...]* (12’) [...] *bu-bu-*ʾ*u-tú* DIRI-*ma* BA.Ú[Š]

[If] the True Shepherd of Anu produces a glow, the king, lord of the dynasty, through [his] misdeed[s...] will become full of boil and will di[e].

K, R ____

50.	**[DIŠ] UD *ina* GUB(.BA)-*šú iš*$_{8}$*-tár* GUB**
K obv. 13’	[DIŠ] UD *ina* GUB.BA-*šú* *iš*$_{8}$*-tár* GUB
R 2’	[DIŠ UD *ina*] GUB-*šú* ⸢*iš*$_{8}$*-tár*⸣ [...]
	LUGAL KUR-*su* BAL-*su*
K obv. 13’	LUGAL KUR-*su* BAL-*su*

[¶] If Ištar (i.e. Venus) stands in its (i.e. the Moon’s?) position, the king’s country will overthrow him.

51.	**[DIŠ] UD *ina* GUB(.BA)-*šú* MUL^{meš} *ma-lu-ú* NAM.ÚŠ^{meš} GÁL^{meš}**
K obv. 14’	[DIŠ] UD *ina* GUB.BA-*šú* MUL^{meš} *ma-lu-*⸢*ú*⸣ NAM.ÚŠ^{meš} GÁL^{meš}
R 3’	[DIŠ UD *ina*] GUB-*šú* MUL^{me}[^{š}...]

[¶] If in its (i.e. the Moon’s?) position it becomes full of stars, there will be deaths.

52.	**DIŠ UD *ina* GUB(.BA)-*šú* ^{mul}*na-ka-ru* GUB BALA NAM.KÚR^{meš}**
A rev. 1’	[...]x ⸢$^{d?}$⸣ x[...]

C 1’	[DIŠ UD *ina* GUB.BA-*šú*] ^{m}[^{ul}*na-ka-ru*...]
K obv. 15’	DIŠ UD *ina* GUB.BA-*šú* ^{mul}*na-ka-ru* ⸢GUB⸣ BALA NAM.KÚR^{meš}
R 4’	[DIŠ UD *ina* GUB]-⸢*šu*?⸣ ^{mul}*na-k*[*a-ru*...]

¶ If the Hostile One (i.e. Mars) stands in its position, (there will be) a reign of hostilities.

53.	**DIŠ UD *ina* IGI MU(.KAM)** (var.: **UD.<1>.KAM**)		**$^{d/mul}$*si-mu-ut***
A rev. 2’	[DIŠ UD *ina* IGI MU.KAM*		^{mu}]^{l}*si-mu-*[*ut*...]
C 2’	[DIŠ UD *ina* IGI MU.KAM		^{d}]⸢*si*⸣-*mu-ut*
K obv. 16’	[DIŠ] ⸢UD⸣ *ina* IGI MU.KAM		
O rev. 3’	DIŠ		^{d}*si-mu-ut*
R 5’	[DIŠ UD *ina* I]GI MU	UD.<1>.KAM	

	(var.: ***ši-mu-ut*, *si-m*[*ut*...]**)	**MUL^{meš}-*šú* GE$_6$^{meš}**
C 2’		MU[L...]
K obv. 16’	^{d}*ši-mu-ut*	MUL⸢^{meš}⸣-*šú* GE$_6$^{meš}
R 5’	^{d}*si-m*[*ut*...]	

	(var.: **MUL BI GE$_6$ *l*[*u*...]**)	**ÚŠ^{meš} GÁL^{meš}**
K obv. 16’		ÚŠ^{meš} GÁL^{meš}
O rev. 3’	MUL BI GE$_6$ *l*[*u*...]	

If at the beginning of the year (i.e. in spring) (var.: the <first> day) the stars of Simut become dark (var.: its star becomes dark o[r...]), there will be a pestilence.

54.	**DIŠ [UD] *ina* IGI MU.KAM** (var.: **UD.<1>.KAM; [... MU.KA]M***)		
A rev. 3’	[DIŠ UD *ina* IGI		MU.KA]M*°
C 3’	[DIŠ UD *ina* IGI] ⸢MU.KAM⸣		
K obv. 17’	[DIŠ UD] ⸢*ina*⸣ IGI MU.KAM		
O rev. 4’	DIŠ		
R 6’	[DIŠ UD *ina* I]GI MU	UD.<1>.KAM	

	$^{d/mul}$*si-mu-ut* (var.: ***ši-mu-ut*, *si-*[*mut*...]**)		**MUL^{meš}-*šú***	(var.: **MUL BI**)
A rev. 3’–4’	^{mul}*s*[*i-mu-ut*...] (4’) [...			
C 3’	^{d}*si-mu-ut*		MUL^{m}[^{eš}...]	
K obv. 17’		^{d}*ši-mu-ut*	MUL^{meš}-*šú*	
O rev. 4’	^{d}*si-mu-ut*			MUL BI
R 6’		^{d}⸢*si*⸣-*m*[*ut*...]		

	BABBAR (var.: **ZABA[R...]**)		***na-ag-lu bar-tu$_4$* GÁL-*ši***
A rev. 4’	BABBA]R		*na-ag-lu* [...]
K obv. 17’	⸢BABBAR⸣		*na-ág-*⸢*lu*⸣ *bar-tu$_4$* GÁL-*ši*
O rev. 4’		ZABA[R...]	

¶ [If] at the beginning of the year (i.e. in spring) (var.: the <first> day) the stars of Simut (var.: that star) *glisten* with white (var.: are bronz[e...], i.e. reddish), there will be a revolt.

55.	**DIŠ UD *ina* IGI MU.KAM** (var.: **MU.1.KAM**)	**^{d}*si-mu-ut***
C 4’	[DIŠ UD] *ina* IGI MU.KAM	^{d}*si-mu-ut*
K obv. 18’	[DIŠ UD *ina* IG]I MU.KAM	
O rev. 5’	DIŠ	^{d}*si-mu-ut*
R 7’	[DIŠ UD *ina* IG]I MU.1.KAM	

(var.: *ši-mu-ut, s*[*i-mut*]) (2) MUL^{m}[^{eš}-*šú* S]IG$_{7}$^{meš} (var.: **MUL BI SIG$_{7}$-*qí***)

C 4’	2 MUL [...]
K obv. 18’	^{d}*ši-mu-ut* MUL^{m}[^{eš}-*šú* S]IG$_{7}$^{meš}
O rev. 5’	MUL BI SIG$_{7}$-*qí*
R 7’	^{d}*s*[*i-mut*...]

: (var.: *lu-u* K[I...]) MUL EGIR-*ú* SIG$_{7}$ ŠUB-*di* [...] ŠÀ x[...] KUR Ì.GÁL

A rev. 5’	[...]x EGIR-*ú* SIG$_{7}$ Š[UB$^{?}$...]
K obv. 18’–19’	: MUL EGIR-*ú* SIG$_{7}$ ŠUB-⸢*di*⸣ (19’)[...] ⸢ŠÀ⸣ x[...] KUR Ì.GÁL
K	(end of the obverse)
O rev. 5’	*lu-u* K[I...]

¶ If at the beginning of the year (i.e. in spring) (var.: of the first year) (two) stars [of] Simut are [y]ellow/green (var.: that star is yellow/green) – (var.) (var.: or ...) the star behind is sprinkled with yellow/green [...] the inside (lit. heart) ...[...] will be in the country.

56.	**DIŠ $^{d/mul}$*si-mu-ut* (var.: [*s*]*i-mut*) MUL BI** (var.: **MUL-*šu***) **SA$_{5}$ [...]**
A rev. 7’	[DIŠ ^{mul}*si-mu-ut* M]UL BI SA$_{5}$ [...]
C 5’	[DIŠ ^{mul}]*si-mu-ut* MUL-*šu* SA$_{5}$ [...]
O rev. 6’	DIŠ ^{mul}*si-mu-ut* MUL BI SA$_{5}$ [...]
R 8’	[DIŠ ^{d}*s*]*i-mut* MUL B[I...]

If Simut – its star is red [...].

57.	**[DIŠ] UD *ba-lu* GIŠGAL-*su*** (var.: **-*šú***) **DIM$_{4}$ UB 2 MU^{me} GAR [...]**
A rev. 8’	[DIŠ UD *ba-lu* GIŠ]GAL-*su* DIM$_{4}$ [...]
C 6’	[DIŠ] UD *ba-lu* GIŠGAL-*su* DIM$_{4}$ UB 2 MU^{me} GAR [...]
D 1’	[DIŠ UD *ba-lu* GIŠGAL]-⸢*šú*⸣ [...]
R 9’	[DIŠ UD *b*]*a*-⸢*lu*⸣ GIŠGA[L...]

[¶] If it (i.e. Simut?) approaches (but) apart from its position, there will be ruin(s) for two years [...].

58.	**DIŠ UD *ina* SI ZAG-*šú*** (var.: **15-*šu***) **MUL GUB [...]**
A rev. 10’	DIŠ UD *ina* SI ZAG-*šú* MUL GUB [...]
C 7’	DIŠ UD *ina* SI 15-*šu* MUL GUB [...]
D 2’	[DIŠ UD *ina* SI 1+]⸢4⸣-*šú* MUL [...]
R 10’	[x x x]x [...]

¶ If a star (of Simut?) stands in its (i.e. Moon’s) right horn, [...].

59.	**DIŠ UD *ina* SI GÙB-*šú*** (var.: **2,30-*šu***) **MUL GUB x[...]**
A rev. 11’	DIŠ UD *ina* SI GÙB-*šú* MUL GUB [...]
C 8’	DIŠ UD *ina* SI 2,30-*šu* MUL GUB x[...]
D 3’	[DIŠ UD *ina* S]I ⸢2⸣,30-*šú* MUL [...]
R	(remainder is missing)

¶ If a star (of Simut?) stands in its (i.e. Moon’s) left horn, ...[...].

60.	**DIŠ UD *ina* IGI-*šú*** (var.: **-*šu***) **MUL GAL *i-bé-eš-ma* GUB ku$^{?}$[...]**
A rev. 12’	DIŠ UD *ina* IGI-*šú* MUL GAL *i-bé-eš-ma* GUB x[...]
C 9’	DIŠ UD *ina* IGI-*šu* MUL GAL *i-bé-eš-ma* GUB ⸢ku$^{?}$⸣ [...]
D 4’	[DIŠ UD *ina* IG]I-*šú* MUL GAL *i*-[*bé-eš-ma*...]

¶ If the great star (of Simut?) (or: meteor; the Great One) moves away from its (i.e. Moon's) front and stands still, ...[...].

61. **DIŠ UD *ina* IGI-*šú*** (var.: ***-šu***) ***is-niq-ma*** **: ŠUR-*ma* GUB im[...]**
A rev. 13' DIŠ UD *ina* IGI-*šú* *is-niq-ma* : ŠUR-*ma* GUB i[m...]
C 10' DIŠ UD *ina* IGI-*šu* *is-niq-ma* : ŠUR-*ma* GUB im[...]
D 5' [DIŠ UD *ina*] IGI-*šú* *is-niq-ma* [...]

¶ If it (the great star of Simut?) approaches its (i.e. Moon's) front – (var.) flares and stands still, ...[...].

62. **DIŠ UD *ina* EGIR-*šú*** (var.: ***-šu***) ***ir-di-šú*** (var.: **UŠ-*šú***) **sal[...]** (var.: ***ir-qí*** **x[...]** or: ***ir-qí-i*[*q*$^?$...]**)
A rev. 14' DIŠ UD *ina* EGIR-*šú* *ir-qí* x[...] / *ir-qí-i*[*q*$^?$...]
C 11' ⸢DIŠ UD *ina*⸣ EGIR-*šu* UŠ-*šu* sal[...]
D 6' [DIŠ UD] *ina* EGIR-*šú* *ir-di-šú* sa[l...]

¶ If it (the great star of Simut?) follows behind it (i.e. Moon) ...[...] (var.: it hides ...[...]) or: it becomes [thin$^?$...]).

63. **DIŠ UD MUL *ina* SAG-*šú*** (var.: ***-šu***) **GUB-*iz* KUR [...]**
A rev. 15' DIŠ UD <MUL> *ina* ⸢SAG⸣-*šú* [...]
C 12' [DIŠ UD <MUL> *ina* S]AG-*šu* GUB-*iz* KUR [...]
D 7' [DIŠ UD] MUL *ina* SAG-*šú* [...]

¶ If a star (of Simut?) stands at its (i.e. Moon's) head, the country [...].

64. **[DIŠ UD M]UL^{meš}-*šú* [*in*]*a* SAG-*šú* GUB(-*ma*) *a-dir* B[ÚR$^?$...]**
A rev. 16' [... *in*]*a* ⸢SAG⸣-*šú* GUB [...]
C 13' [...] ⸢SAG⸣-*šú* GUB-⸢*ma a*⸣-*dir* B[ÚR$^?$...]
D 8' [DIŠ UD M]UL^{meš}-*šú* x[...]

[¶ If] its (i.e. Simut's?) [s]tars stand [a]t its (i.e. Moon's) head and it is dark, (there will be) r[elease$^?$...].

D ———————————————

65. **[DIŠ ^{mul}]KAK.SI.SÁ *ana* x[...] IGI KUR [...]**
C 14' [DIŠ ...] ⸢IGI⸣ KUR [...]
C (remainder is missing)
D 9' [DIŠ ^{mul}]⸢KAK⸣.SI.SÁ *ana* x[...]
D (remainder is missing)

[If the Arrow] appears towards ...[...] the country [...].

C ———————————————

Commentary to the Text

1–8: The first section of entries deals with the Pleiades addressed as MUL.MUL or *zappu* and referred to with verbs in singular. The Pleiades are seen together with other celestial bodies: they reach (KUR, *kašādu*), come close (TE, *ṭeḫû*) or stand (GUB, *i/uzuzzu*) with them. The protases of the omens are explained as phenomena of the Pleiades or as planetary movements of Mars (^{mul}*ṣalbatānu*) (see 5.1.7.).

1, 5: For KI.LAM, *maḫīru*, "market rate", or the more generic "business", see 5.1.4. commentary to the text of entry 26 (see p. 165).

2: The protasis of this entry is included in MUL.APIN but with a different apodosis (II GAP B 2): DIŠ $^{\text{mul.d}}$AMAR.UTU MUL.MUL KUR-*ud ina* MU BI ^{d}IŠKUR RA, "If the Marduk-star reaches the Pleiades, in that year Adad will devastate" (Hunger-Steele 2019: 107). In the present context, the omen has several additions. The Marduk-star ($^{\text{mul.d}}$AMAR.UTU) is a name used to indicate both Mercury and Jupiter (Brown 2000: 57), which the first explanation referred to. The ominous sign portended by Mercury or Jupiter together with the Pleiades predicts two different weather conditions according to the winter and summer seasons, meaning that the phenomenon can happen twice a year. More specifically, the first alternative apodosis, introduced by BAD-*ma*, *šumma*, has a prediction for hot weather in summer, and the second has cold weather in winter. In the second part, the entry gives another explanation followed by another apodosis: here the logogram GABA.RI, *meḫru*, "equal", "equivalent", or "correspondingly", is used to express a variant or an equation, similar to the usage of *Glossenkeile* in commentaries (Gabbay 2016: 85), or of KIMIN and MIN (Borger 2004: 414 n. 742).[305]

4: The source F obv. 4 (K 11001 + K 15541, see 5.1.5.) preserves the first half of a sign which is likely Š[E], and the second part of either [GÁ]N, [KÁ]R, or [G]Á, followed by RA. These signs could be understood as logograms, with ŠE and GÁN, KÁR, or GÁ, and RA in *status constructus*, or as two different substantives in accusative (i.e. with ŠE as *šeʾa*, "barley", and *-ra* as a phonetic complement). In both cases, the reading of the apodosis is unclear.

7, 12, 14, 62: A celestial body that is *ina* IGI, *ina pāni*, "in front of", another body, usually means that it is "below" it, and EGIR-*šú*, *arkišu*, "behind it" means "above it", if referred to evening phenomena (Reiner-Pingree 1998: 3).

9: CAD (A1 97–101) defines UD.DUG$_4$.GA, *adannu*, "a moment in time at the end of a specified period". In this entry, it means that the Pleiades are visible in a period in which they should be visible. Thus, *adannu* is the period which goes from the heliacal rising to the setting of this asterism (see fn. 220).

305 For an overview of scribal abbreviations like GABA.RI (and its usage as "correspondingly"), see Leichty (1970: 26–27), and Fincke (2022: 55–56).

10: GIŠ.ḪUR, *uṣurtu*, "drawing" (CAD U/W 293 g) around the Moon, the Sun, or a celestial body, is a disc of light and colour, called "corona" in modern astronomy (Gehlken 2012: 83 fn. 5). For a further discussion about *uṣurtu* in relation to stars, see the comments to App. B § 18 rev. 34'–36', p. 331.

11, 32, 49: The *mešḫu* is a phenomenon of light understood as "glow", or "meteor", previously discussed in 5.1.4. commentary to the text of entries 37–43, p. 166.

12–13: In these entries, the logogram NÍGIN is *saḫāru*, "to turn around" (CAD S 37–54), rather than *lamû*, "to surround" (CAD L 69–77) as it would be customary (e.g. see EAE 52: 56–60). Indeed, *lamû* is never attested with the conjunction KI, *itti*, "with", whereas *saḫāru* is at least attested in one dream omen with *itti*, even though its meaning is unclear: DIŠ NA KI GU_4 NÍGIN, "If a man has to do with an ox" (see CAD S 44a; MDP 14, 50 obv. i 10, see also Oppenheim 1956: 258). The verb *saḫāru* is often attested in divinatory and astronomical context (e.g. MUL.APIN II iv 4; Reiner-Pingree 1998: 56, 22; 60, 9; CAD S 45), but never with *itti*.

15: The verb in the protasis is *innapal*, the N-stem of *napālu*, "to dig out", or "to tear down" (CAD N1 272–275). Both AHw ("etwa abgleiten(?)", see AHw 734a N) and CAD (N1 275b 5) leave its translation open.

16–23: Entry 16 deals with one of the intercalation schemes discussed in 4.2.2. (§ 2c).[306] The 12 stars of the Pleiades represent the apparent motion of the Pleiades within a year, and the times when the Pleiades are close to the Moon within a normal year, i.e. every month (see 2.1.). The entries 17–18 are probably an extension of the same rule, with one unit less each time (i.e. the stars of the Pleiades rise 12 times, then 11 and 10 times). If entry 16 defines a normal year, the entries 17–18 define the balancing of the Pleiades and the Moon during the two following years. After the third year, intercalation was necessary. The idea behind these rules is the basic scheme for the cycle of intercalation every three years.
The omens in 19–23 preserve another intercalation rule (see 4.2.2. § 2b): intercalation is here based on the Pleiades being ahead (i.e. towards west) or behind (i.e. towards east) the Moon, similarly to the astronomical concept of having the same celestial longitude (see fn. 219).

19–23: The scheme of binary opposites is implicit in the arrangement of the statives in these entries. The verb *karātu*, "to strike" or "to cut" (CAD K 215), is in antithesis with *šatāḫu*, "to be elongated" (CAD Š2 184–185). The same is true for the logogram GE_6, *ṣalāmu*, "to become dark" (CAD Ṣ 70–71), and *wamālu* "to scintillate" (CAD U/W 401). Nevertheless, the factual explanations of these entries do not necessarily reflect a dichotomous scheme. For instance, in entries 19–20, the Pleiades "cut" and "elongated" are explained through an antithesis as the Pleiades being behind (*namarkû*, "to lag behind",

306 The interpretation of EAE 52: 16 was suggested to the author by Jeanette C. Fincke, whereas the reading of the verb *uš-te-te-še-rù* was suggested to the author by Hermann Hunger.

CAD N1 208–209) and ahead (*panû*, “to be ahead” CAD P 98–100) of the Moon. In contrast, in entries 21–23, the Pleiades “dispersed” (*purruru*, “to break up”, or “to disperse” CAD P 162–163), “dark”, or “scintillating”, are all explained as the Pleiades being behind the Moon. The reason that these specific statives are explained as the position of the Pleiades relative to the Moon is unclear. The verbs *šatāḫu*, “to be elongated”, and then *karātu*, “to cut”, and *purruru*, “to disperse”, could eventually testify the use of metaphorical language (see 2.2.3.2.): in simple words, something elongated visually recalls something ahead of something else, whereas something cut, or broken, or dispersed is something that has been left behind. A further discussion on the working principles of these omens can be found in appendix A (see A.2.1.2.).
In source H obv. 5’, the signs after the break at the beginning of the line do not allow a restoration with the verb *karātu*, “to cut”, as it would be according to the parallels in entry 19. The source H obv. 5’ preserves, at the beginning of the broken line, traces of an unidentified sign and the sign UḪ, which could also be read as *-iḫ*, as the last sign of *šá-ti-iḫ*; nevertheless, the unidentified sign before UḪ is certainly not *ti*, and the restoration [... *šá-t*]*i-iḫ* would in any case be problematic for the general meaning of the omen, if compared to entry 20 that has MUL.MUL *šá-ti-iḫ*, “the Pleiades are elongated”, and to all the parallel sources in entry 19 that have MUL.MUL *ka-ri-it*, “the Pleiades are cut”. Therefore, the reading of the beginning of source H obv. 5’ remains unclear

24a–d: These reconstructed entries mention stars explained as planets close to the Pleiades. The order of these celestial bodies is: the Field (^{mul}AŠ.GÁN, *ikû*) probably standing for Venus (Brown 2000: 59), the Heroic One (^{d}U$_4$.AL.TAR, *dāpinu*) standing for Jupiter, the Mouse (^{mul}EN.TE.NA.BAR.ḪUM, *ḫabaṣīrānu*) standing for Mercury, and the Other One (^{mul}MAN-*ma, šanûmma*) standing for Mars. The sources A obv. 26–29, and B ll. 11’–14’ provide this information differently: source A has one, main omen (i.e. protasis + apodosis) and several alternatives for the protases with the corresponding explanations; source B has four different entries, each of them built by protases and explanations.

25: The verb *ašābu*, “to sit idly” or “to sit and wait” (CAD A2 389–390), was expected here instead of *napalsuḫu*, “to fall down” (CAD N1 271–272). Given the context, as suggested by Hunger and Reiner (1975: 24), the verb *napalsuḫu* should be translated “to be apart”, because it visually recalls the Pleiades and the Moon distant from each other in the sky, i.e. when intercalation is needed (see 4.2.2. § 2a, fn. 244). The meaning of *ašābu*, “to sit and wait”, temporally recalls the Pleiades when being “late” in respect of the usual situation, i.e. when they are distant from the Moon.

26: An alternative reading of this entry could be as follows: DIŠ ^{mul}*ís-le-e* [*ana*] ŠÀ ^{d}[30 KU$_4$...] *ta-lit-ti* [...], “If the Jaw of the Bull [enters in]to [the Moon...], the offspring [...]”.

41, 42, 48: These entries are restored after Reiner and Pingree (1981: 78 XVIII 2, 3, 9). In entry 41, Reiner and Pingree read ⸢im?⸣ *da-am*, with *da-am* as from *daʾāmu*, “to darken” or “to become dark” (CAD D 1; SAD 2 1a). Nevertheless, the sign IM looks like the sign MUŠ in Neo-Babylonian ductus. The sign MUŠ could be interpreted as a logogram for *ṣēru*, “snake”, with *da-am* as a stative (*dām*), but if the latter is stative from *daʾāmu*, then

da-ʾi-im is expected, and not *da-am* (see, e.g., AHw 146a, "*daʾāmu* I"). The only other possibility left is to interpret *da-am* as a weak variant of *dâmu*, "to roam" (CAD D 80–81; AHw 146a, "*daʾāmu* II"), though parallels are unattested. Since the meaning of the logogram MUŠ hardly fits the context, one should assume that the sign is a mistake for IM, *šāru*, "wind"; then, the resulting translation, "the wind *roams*$^{?}$", even if highly uncertain, is more agreeable than "a snake *roams*$^{?}$".
Always according to Reiner and Pingree (1981: 17, 2.2.1.2.), È, *aṣû*, "to come forth" (CAD A2 356–383), in entries 41 and 42, means the first time that Venus or Mercury are seen in the evening after a period of invisibility, whereas KI.GUB, *manzāzu*, "position" (CAD M1 237–238), refers to the position of a celestial body when it is first visible the day of the observation, regardless of its visibility before. In other texts, KI.GUB refers to a point on the horizon above which a planet rises or sets (e.g. MUL.APIN I ii 13–15, see Hunger-Steele 2019: 41–42; Reiner-Pingree 1998: 3, 18–19; Hunger-Pingree 1999: 41; see also 5.1.4. commentary to the text 30–34, p. 165). The KI.GUB is also mentioned in entries 9, 24a–d, and 45.

43–44: The two entries about the "navel" (LI.DUR, *abunnatu*) of the True Shepherd of Anu corroborate the perception of this constellation as the image of a human figure in the sky, that is Papsukkal, the vizier of Anu (Beaulieu-Frahm-Horowitz-Steele 2018: 49). For further references to the navel and body parts of constellations, see A.2.1.1.3.

49: A parallel apodosis is in the unpublished fragment K 2349 obv. i 20': [...] LUGAL EN BALA *ina šèr-ti-šú bu-bu-ʾu-tú* DIRI-*ma* ÚŠ. See also the Astrolabe B-related fragment LBAT 1499 obv. ii 15 (Horowitz 2014: 127): DIŠ *ina* ^{iti}SIG$_4$ *ina še-rim* ^{múl}SIPA.ZI.<AN>.NA 4 *meš-ḫu im-šuḫ* LUGAL EN BALA-*šú* <*ina*> *šèr-tú bu-bu-ut-tu*$_4$ DIRI-*ma* ÚŠ, "If in Simanu (i.e. Month III) during the morning, the True Shepherd of Anu, 4, produces a glow, the king, lord of his dynasty, through his misdeeds will become full of boils and will die" (see 4.1.3.4. for the meaning of numerical values).

50–55, 57–64: These entries are quite different from all the others in the way that they are written down, so their translation has been intentionally left as literal as possible due to difficulty in interpreting them. Reiner and Pingree (1981: 71 XVIII) noted that the logogram UD could stand for *šumma*, "if", while the previous DIŠ would mark the beginning of the entry (¶). This shift in the written arrangement of the omens points towards a composite nature of the text, i.e. the omens were likely collected and excerpted from different sources that followed a different tradition in writing down omens.[307] Reiner and Pingree also proposed that the omens' omitted logic subject would be the Moon. That suggestion would be confirmed by the fact that, in entries 58–59, a star is said to stand in "its right horn" and "its left horn" (SI ZAG-*šú* and SI GÙB-*šú*). The third-person singular possessive suffix -*šu* must be referred to as the Moon, the only celestial body conceived as

307 For the use of UD as *šumma*, "if", in omens dating earlier than the first millennium BC, see Fincke (2006: 134).

having "horns" (see 5.1.4. commentary to the text of entries 30–34, p. 165).[308] Another option would be to interpret UD as UTU, "Šamaš", or "Sun", without any determinative. Still, this writing is never attested in solar omens, where "Sun" is either written numerically or with the determinative DINGIR (20, ᵈ20 or ᵈUTU) (e.g. see van Soldt 1995: 89, 6–10). As a last option, UD might stand as a logogram for a celestial body other than the Sun, which we do not know. In entries 53-55, Simut, usually associated with the planet Mars,[309] is said to have two stars (*simut* MULᵐᵉˢ̌-*šú*), indicating that the celestial body is perceived as a constellation or an asterism in the present context, and not as a planet. Moreover, in entry 64, a reference to "his/its stars" (MULᵐᵉˢ̌-*šu*) suggests that the sign UD, and consequently the whole entry, might refer to the stars of Simut as an asterism. One late Middle-Assyrian fragment (VAT 9765 + VAT 9913), a collection of excerpts on fixed stars, gives the same sequence as in entries 58–64: Heeßel (2021: 92, II 1'–10') interpreted the logogram UD as a sign for "ditto", like KIMIN, assuming that it refers to an unidentified star. Following Heeßel's interpretation, the implicit subject of entries 50–55, 57–64 could then be the True Shepherd of Anu (ᵐᵘˡSIPA.ZI.AN.NA) from entry 49, but it is unlikely considering that, for instance, in entries 53–55 the logogram UD cannot logically stand for the True Shepherd of Anu, but the subject is rather the star(s) of Simut. In the present context, the only safe hypothesis would be to interpret UD as *šumma*, "if", assuming that each entry refers to the previous one in an abbreviated form, i.e. celestial bodies interacting with the Moon. Consequently, the entries 57–64 would deal with the unnamed stars of Simut, which probably interact with the Moon. As a final remark, it is noteworthy that the fragments belonging to EAE 59–60 have the same writing for Venus (*iš*₈-*tár*) as in entry 50, and the sign UD as "*šumma*" (Reiner-Pingree 1998: 169–185).

50–52: The logogram GUB.BA seems to be a shortened writing for KI.GUB.BA, *manzāzu*, "position" (CAD M1 237–238, 5), though it can also be read *i/uzuzzu*, "to stand still" (CAD U/W 373–392). According to the latter reading, the protases of entries 50–52 should be understood as follows: DIŠ *ina* GUB.BA-*šú,* "If when it stands still", with the Moon as underlying subject (see 50–55, 57–63 below).
52: For the "Hostile One" (ᵐᵘˡ*na-ka-ru*) as a name for Mars, see the Great Star List (K 2067 ll. 16'–18', see Koch-Westenholz 1995: 198, 200 ll. 237–240).

53–55: The expression *ina* IGI MU(.KAM), lit. "in front", or "at the beginning of the year" means "in spring" because the Mesopotamian year began in that season (CAD P 87a 2'; Reiner-Pingree 1981: 91 *šattu*).

54: In this entry, the meaning of *na-ag-lu* remains uncertain. It comes from *nagālu*, attested in G-, D-, and N-stem, and it seems to be used mainly for celestial omens and colours (CAD N1 107; AHw 709; CDA 230). According to the attestations, *nagālu* in N-stem is a variant of *nabāṭu* in N-stem, "(be a) glow". According to Reiner and Pingree (1981: 78, 79 XVIII 15–16) its meaning could be similar to ŠUB-*di*, *innadi*, in the apodosis of entry 55

308 Nevertheless, see van Soldt (1995: 128 fn. 1).
309 See K 4386+ rev. ii 57 (MSL 17: 229 l. 309): ᵈsi-mu-ut = ᵈ*ṣal-bat-a-nu*, "Simut = Mars".

(MUL EGIR-*ú* SIG$_7$ ŠUB-*di*, "the star behind is sprinkled with green"), i.e. "to be flecked" or "to be streaked" with a colour, in this case white (BABBAR, *peṣu*). Both CDA and AHw provide a translation much closer to *nabāṭu*, probably the most appropriate, which is "to glisten".

62: In source A rev. 14', the second verb in the line can be reconstructed in two different but unclear ways. First, *ir-qí* could be the G preterite of *raqû*, "to hide" (CAD R 174-175), a verb used in celestial omens but only attested so far in the Gtn stem. Alternatively, the verb could be reconstructed as *ir-qí-i*[*q*$^?$], from *raqāqu*, "to become thin" (CAD R 167-168), a verb attested in extispicy. However, very little of the beginning of the sign iq is preserved, and it is too distant from the other two previous signs, so it could also be the beginning of another broken word.

Further Comments on the Text

Even though it is not possible to reconstruct all the omens of Commentary EAE 53, one can attempt to reconstruct the sequence of entries, and sections of entries within the tablet, assuming that the Commentary EAE 53 preserves the whole content and sequence of EAE 53:

- Omens about the Pleiades (MUL.MUL)
- Omens about the Jaw of the Bull (^{mul}*is lê*)[310]
- Omens about the True Shepherd of Anu (^{mul}SIPA.ZI.AN.NA, *šitaddaru*)
- Omens about Venus and the Hostile One (^{mul}*nakaru*, i.e. Mars)
- Omens about Simut (as an asterism or constellation)[311]
- One last omen abot the Arrow (^{mul}KAK.SI.SÁ, *šukūdu*)

As shown in Table 7, the sequence of the groups of entries mirrors the first visibility or heliacal rising of the relative asterisms in a period from Ajaru to Du'uzu (i.e. Month II and IV), according to MUL.APIN (Hunger-Steele 2019: 180 Table 4), and from Ajaru to Abu (i.e. Month II to V) according to the Astrolabe B (I i 12–ii 15, see Horowitz 2014: 43 Table 3).[312]

---

310 The Jaw of the Bull is attested as an alternative name for the Bull of Heaven (^{mul}GU$_4$.AN.NA), Taurus (Beaulieu-Frahm-Horowitz-Steele 2018: 64–65).

311 See Commentary EAE 53: 55: [DIŠ] UD *ina* IGI MU.KAM ^{d}*si-mu-ut* 2 MUL^{m}[^{eš}-*šú* S]IG$_7$^{meš}, "[¶] If at the beginning of the year two stars [of] Simut are [y]ellow/green". This proves that, in the context of the reconstructed commentary to EAE 53, Simut would not necessarily indicate the planet Mars, as it would be customary (see fn. 309), but perhaps an asterism or a constellation.

312 For the shift in the months between MUL.APIN and Astrolabe B, see 4.1.2. and 5.1.4. further comments on the text, pp. 170–172.

Subject of the protasis	Date of first visibility (MUL.APIN)	Related months in Astrolabe B
MUL.MUL, *zappu*, "Bristle"	Ajaru (Month II) day 1	Ajaru (Month II)
^{mul}*is lê*, "Jaw of the Bull"	Ajaru (Month II) day 20	Simanu (Month III)
^{mul}SIPA.ZI.AN.NA, *šitaddaru*, "True Shepherd of Anu"	Simanu (Month III) day 10	Du'uzu (Month IV)
^{mul}KAK.SI.SÁ, *šukūdu*, "Arrow"	Du'uzu (Month IV) day 15	Abu (Month V)

Table 7. Groups of entries in Commentary EAE 53 arranged by their subject in the protases, and compared to their first visibility dates or heliacal rising according to MUL.APIN and Astrolabe B.

Similar to the tablet EAE 52, the arrangement of Commentary EAE 53 points to grouping omens about celestial bodies that rise one after each other during the same time of the year. Hence, the logic behind the arrangement of the entries is the calendar.[313]

Taking into account only the omens with factual explanations, the latter refers to phenomena of the Pleiades (Commentary EAE 53: 9–11, 16–23), or to planetary conjunctions (Commentary EAE 53: 1–8, 12–15, 24a–d) in which the Pleiades (MUL.MUL) are equated to either the planet Mars (Commentary EAE 53: 1, 4–6, 12–15), or to *zappu*, "Bristle" (Commentary EAE 53; 2, 3, 7–8).

5.1.6. The Reconstructed Tablet SIT 6

SIT 6 is the reconstruction of the assumed 6th tablet of the series SIT, a collection of excerpts from various EAE tablets that duplicates the Commentary EAE 53 for six entries. Each entry of SIT 6 has a protasis, an apodosis, and an explanation, and in one instance these three components are separated on the cuneiform tablet by vertical rulings: the text of LB 1321 (SIT 6 source B) is arranged over three columns (protasis | apodosis | explanation). Where the scribe added more than one apodosis or explanation, he used a triple *Glossenkeil* to indicate the beginning of each variant, which is a unique feature in commentaries organised this way (Frahm 2011: 35). Sources where the text is written in consecutive lines, variants in apodoses and explanations are usually marked with the use of *Glossenkeile*.

For the organisation of the sources, the text, and the comments, see 5.1.3.5.

Table of Sources

Source: A
Siglum: K 5713 + K 7129 + Rm 2, 114
Copy: ACh Suppl. 1, 50 (Rm 2, 114); Plate 27–28

313 Hunger and Pingree (1999: 67–68) have noted that in the 18th book of Homer's *Iliad* (ll. 483–489) the shield of Achilles "metaphorically" shows the Pleiades, the Hyades (i.e. Jaw of the Bull), and Orion (i.e. True Shepherd of Anu), in the same order found in EAE, MUL.APIN, and Astrolabe B.

Edition:	ACh Suppl. 1, 50 (Rm 2, 114); see App. B § 16 (obv. 21'–22' and reverse)
Provenance:	Nineveh
Ductus:	Neo-Assyrian
Measurements:	12,6+ x 10+ x 2,4+ cm
CDLI n.:	P396110 (with photograph)
Description:	Fragment of the lower part of a tablet with the lower edge, and part of the left and right edge preserved. The scribe started to write the obverse on the convex surface of the tablet, which is usually assigned to the reverse.
Type of text:	SIT (6)

Source:	B
Siglum:	LB 1321
Edition:	Borger 1973: 38–43
Provenance:	Nineveh
Ductus:	Neo-Assyrian
CDLI n.:	P368698 (with photograph of the obv.); other photographs available online at CCP 3.2.2.A (https://ccp.yale.edu/P368698 accessed 04.10.2022)
Type of text:	SIT (2) obv. + SIT (6) rev.

Source:	C
Siglum:	K 2177 + K 7869 + Rm 473
Copy:	ACh Ištar 35 (K 2177); AAT 87 (K 2177); Plate 29–30; see also App. B § 17 (reverse)
Edition:	ACh Ištar 35
Provenance:	Nineveh
Ductus:	Neo-Assyrian
Measurements:	13,8+ x 8,5 x 2,2 cm
CDLI n.:	P394237 (with photograph)
Description:	Fragment of a tablet with almost all the right edge and part of the upper and left edge preserved. The reverse is largely lost and only the first half is inscribed. Traces of signs are visible at the end of eight lines, and the two last lines are better preserved and readable at the bottom.
Type of text:	SIT 6 *mukallimtu*-commentary

Source:	D
Siglum:	K 12425
Copy:	Plate 26
Provenance:	Nineveh
Ductus:	Neo-Assyrian
Measurements:	3,4+ x 2,5+ x 0,5+ cm
CDLI n.:	P399830 (with photograph)
Description:	Small fragment of the right part of a tablet.
Type of text:	SIT (6)

Source: E
Siglum: K 6484
Copy: Plate 26
Provenance: Nineveh
Ductus: Neo-Assyrian
Measurements: 4,6+ x 4,4+ x 1,0+ cm
CDLI n.: P396568 (with photograph)
Description: Small fragment of the middle part of a tablet.
Type of text: SIT (6)

Source: F
Siglum: K 2170 + K 3629
Copy: Plate 31–32
Edition: App. B § 18 (obv. 12, 14–15 and reverse)
Provenance: Nineveh
Ductus: Neo-Assyrian
Measurements: 14,8+ x 7,2+ x 2,5+ cm
CDLI n.: P394232 (with photograph)
Description: Fragment of the right part of a tablet, with part of the upper and right edge preserved. The scribe started to write the obverse on the convex surface of the tablet, which is usually assigned to the reverse. The scribe separated each entry with a horizontal line on the reverse.
Type of text: SIT *mukallimtu*-commentary (EAE Mars? obv. -SIT 6 rev.)

Source: G
Siglum: K 5277
Copy: ACh Suppl. 1, 51 (obverse); Plate 33 (obverse)
Edition: ACh Suppl. 1, 51; May (2018: 141–142 n. 33) (colophon); see also App. B § 19 (reverse)
Provenance: Nineveh
Ductus: Neo-Assyrian
Measurements: 5,0+ x 3,5+ x 1,9+ cm
CDLI n.: P395970 (with photograph)
Description: Fragment of the right corner of a tablet with part of the upper and right edge preserved. The reverse carries four broken lines of a colophon.
Type of text: SIT (6)

Source: H
Siglum: K 2138
Copy: Plate 33
Provenance: Nineveh
Ductus: Neo-Assyrian
Measurements: 5,8+ x 3,7+ x 0,9+ cm
CDLI n.: P394208 (with photograph)

Description: Fragment probably from the left part of the obverse of a tablet, with part of the left edge preserved. The reverse is lost.
Type of text: SIT (6)

Source: I
Siglum: K 7275
Edition: Renzi-Sepe 2021: 114–116
Provenance: Nineveh
Ductus: Neo-Assyrian
CDLI n.: P397154 (with photograph)
Type of text: SIT (6)

Source: J
Siglum: K 2301
Copy: ACh Suppl. 2, 88; Plate 34
Provenance: Nineveh
Ductus: Neo-Assyrian
Measurements: 12,2+ x 7,3+ x 2,2+ cm
CDLI n.: P394334 (with photograph)
Description: Fragment probably from the left part of the obverse of a tablet, with part of the left edge preserved. The reverse is lost.
Type of text: SIT (6)

Source: K
Siglum: Rm 477
Copy: Plate 35
Provenance: Nineveh
Ductus: Neo-Assyrian
Measurements: 5,6+ x 3,4+ x 0,7+ cm
CDLI n.: P424792 (with photograph)
Description: Small fragment of the middle and/or left part of a tablet.
Type of text: SIT (6)

Source: L
Siglum: BM 44005
Copy: Plate 36
Edition: App. B § 20 (obv. 6–7 and reverse)
Provenance: unclear
Ductus: Late Babylonian
Measurements: 4,3+ x 2,7+ x 1,5+ cm
CDLI n.: No
Description: Small fragment of the left part of a tablet, with part of the upper edge preserved.
Type of text: SIT (6) *mukallimtu*-commentary

Source: M
Siglum: W 22730/5 (see Commentary EAE 53 source S)

Source: N
Siglum: Sm 1317 (see EAE 52 source L and Commentary EAE 53 source N)

Source: O
Siglum: Rm 192
Copy: ACh Suppl. 1, 49; Plate 37–38
Edition: ACh Suppl. 1, 49; see also App. B § 21 (obv. 1–11, 15–16 and reverse)
Provenance: Nineveh
Ductus: Neo-Assyrian
Measurements: 10,8 x 6,4 x 2,2 cm
CDLI n.: P424644 (with photograph)
Description: Tablet with almost all the edges preserved, except for the lower right corner. The format of the tablet is horizontal or landscape, i.e. the writing is parallel to the longer side, and the ratio between the horizontal and the vertical axis is approximately 2:1, a feature typical of oracle queries or reports dating to the Neo-Assyrian period (ca. 911–612 BC) (Radner 1995: 72–74). Whereas the scribes usually started to write from the obverse continuing to lower edge, reverse and upper edge (e.g. see App. B § 20), in this tablet the scribe started to write on the upper edge and finished halfway down the reverse.[314]
Type of text: EAE (53–56) excerpts

Source: P_1 (+) P_2 (+) P_3
Siglum: K 1494a (P_1) (+) K 1494b (P_2) (+) K 1522 + K 3594 (P_3)
Copy: ACh Suppl. 2, 75 (K 1494a); Plate 39–40
Edition: ACh Suppl. 2, 75 (K 1494a); see also App. B § 22 (P_1 obv.–P_3 obv. 9', P_3 obv. 11'–17', P_3 obv. 19'- P_1 rev. 3', P_1 rev. 6'–9')
Provenance: Nineveh
Ductus: Neo-Assyrian
Measurements: 7,3 x 5,7+, x 2,4+ cm (K 1494a) (+) 2,4+ x 3,5+ x 2,2+ (K 1494b) (+) 7,3 x 5,5+ x 2,1+ cm (K 1522 + K 3594)
CDLI n.: P393902 (K 1494a); P393903 (K 1494b); P393907 (K 1522 + K 3594) (all with photographs)
Description: These fragments are part of the same tablet but do not join physically (Reiner 1998: 217). K 1494a is a fragment of the upper part of the tablet, with part of the left and right edge preserved. K 1494b is a small fragment of the right side of the tablet, with part of the right edge preserved. K

314 For a discussion on formats and contents of texts dating to the Neo-Assyrian period (ca. 911–612 BC), see Radner (1995).

	1522+ is a fragment of the lower part of the tablet, with part of the left, lower, and right edge preserved.
Type of text:	EAE (Mars?)[315]

Source:	Q
Siglum:	K 6185 + K 8901 + K 12567
Edition:	ACh Suppl. 2, 69 (K 8901)
Provenance:	Nineveh
Ductus:	Neo-Babylonian
CDLI n.:	P238629 (with photograph)
Type of text:	EAE

Source:	R
Siglum:	Sm 366 + 80-7-19, 371
Edition:	SAA 8, 491 = RMA 167
Provenance:	Nineveh
Ductus:	Neo-Assyrian
CDLI n.:	P236957 (with photograph)
Type of text:	Report

Source:	S
Siglum:	K 1343
Edition:	SAA 8, 536 = RMA 205a
Provenance:	Nineveh
Ductus:	Neo-Assyrian
CDLI n.:	P2338047 (with photograph)
Type of text:	Report

Source:	T
Siglum:	K 2330
Edition:	ACh Ištar 23; AAT 43
Provenance:	Nineveh
Ductus:	Neo-Assyrian
CDLI n.:	P394348 (with photograph)
Type of text:	EAE 57

Source:	U
Siglum:	K 3780
Copy:	ACh Suppl. 2, 78; Plate 41–42
Edition:	ACh Suppl. 2, 78; Reiner-Pingree (2005: 181) (rev. i); see App. B § 23 (obv i 1'–ii 8', ii 9'–11', ii 14'–17' and reverse)

315 For a discussion about the identification of "EAE (Mars?) tablets", see the comments of K 1494a (SIT 6 source P_1) in App. B § 22.

Provenance: Nineveh
Ductus: Neo-Assyrian
Measurements: 15,4 x 7,5+ x 2,9 cm
CDLI n.: P395231 (with photograph)
Description: Fragment of the lower part of a tablet with part of the left, lower, and right edge preserved.
Type of text: EAE (53–56) excerpts

Source: V
Siglum: BM 47799
Edition: Reiner 2006: 315–318
Provenance: unclear
Ductus: Late Babylonian
CDLI n.: No
Type of text: TE-tablet[316]

Source: X
Siglum: K 2071
Edition: ACh Ištar 32
Provenance: Nineveh
Ductus: Neo-Assyrian
CDLI n.: P394173 (with photograph)
Type of text: EAE (Mars$^{?}$)

Source: Y
Siglum: K 8000 (see Commentary EAE 53 source O)

Source: Z
Siglum: K 6415 + K 6478 + Rm 313
Edition: ACh Suppl. 2, 72 (K 6478)
Provenance: Nineveh
Ductus: Neo-Assyrian
CDLI n.: P396515 (with photograph)
Type of text: EAE[317]

Source: AA
Siglum: 79-7-8, 117 + 79-7-8, 223 + Rm 308

---

316 The "TE-tablets" have omens in which "the behavior of a star or a constellation is expressed by the verb *ṭehû*, "to come close", usually written with its logogram TE" (Reiner 2006: 313). These omens likely belong to the last sections of the series EAE (Koch 2015: 178).

317 Rm 313+ is a collection of excerpts. The incipit (obv. 1) reads: DIŠ ^{mul}*e*$_4$-*ru*$_6$ *a-na* MUL.MUL KUR-*ud*, "If the Frond comes close to the Pleiades". The identification as a collection of excerpted omens is based on the subscript of the fragment (rev. 32'): 3 *nis*° {DIŠ}-*ḫu* TIL-*a-a u* GABA.RI ^{giš}ZU *šá liq-ti šà-ṭir* [...], "3rd final section, copied from a wooden tablet with a collection of various material [...]".

Edition: ACh Suppl. 2, 68; ACh Suppl. 1, 55
Provenance: Nineveh
Ductus: Neo-Babylonian
CDLI n.: P236904 (with photograph)
Type of text: EAE[318]

Source: BB
Siglum: K 11001 + K 15541 (see Commentary EAE 53 source F)
Source: CC
Siglum: K 2894 + K 12290
Edition: ACh Ištar 28
Provenance: Nineveh
Ductus: Neo-Assyrian
CDLI n.: P394729 (with photograph)
Type of text: EAE[319]

Source: DD
Siglum: BM 98594; K 12761 + Sm 1504
Edition: Reiner-Pingree 1981: 38 II 12d–e sources B, C
Provenance: Nineveh
Ductus: Neo-Assyrian; Neo-Babylonian (BM 98594)
CDLI n.: P237589; P400040 (both with photograph)
Type of text: EAE (51)

Source: EE
Siglum: Rm 104
Edition: Gehlken 2012: 80 obv. 1, see Rm 104 catchline
Provenance: Nineveh
Ductus: Neo-Assyrian
CDLI n.: P424624 (with photograph)
Type of text: EAE 46

Source: FF
Siglum: 83-1-18, 198
Edition: SAA 8, 114 = RMA 232
Provenance: Nineveh
Ductus: Neo-Assyrian
CDLI n.: P336526 (with photograph)
Type of text: Report

---

318 The sequence of the the celstial bodies in the protases of 79-7-8, 117+ is: ^{mul}KU$_6$, "Fish", ^{mul}NU.MUŠ.DA, "Herd", ^{mul}*e*$_4$-*ru*$_6$, "Frond", $^{mul.d}$AMAR.UTU, "Marduk-star", ^{mul}UR.BAR.RA, "Wolf", ^{mul}GÁN.ÙR, "Harrow", ^{mul}APIN, "Plow", ^{mul}AB.SÍN, "Furrow", and ^{d}*e-a*, "Ea". K 6185+ (SIT 6 source Q) is a partial duplicate of 79-7-8, 117+.

319 K 2894+ is a collection of excerpts and explanations from various sections of the series EAE.

Source:	GG
Siglum:	K 2342 + K 2990 + K 12422 + K 19019
Edition:	AAT 81–82 (K 2990); ACh Ištar 21 (K 2990, composite copy)
Provenance:	Nineveh
Ductus:	Neo-Babylonian
CDLI n.:	P238166 (with photograph)
Type of text:	EAE (55) commentary[320]

Source:	HH
Siglum:	BM 38301 (see Commentary EAE 53 source K)
Source:	II
Siglum:	79-7-8, 271 (see Commentary EAE 53 source L)

Text

1.	**[DIŠ MUL.MUL ^{mul}ŠUDUN] KUR-*ud ina* MU BI KI.LAM TUR-*ir***
C obv. 1	[...] ⸢TUR-*ir*
BB obv. 1	[...] KUR-*ud ina* MU BI KI.LAM TUR-*ir*
F obv. 1	[...]
	^{d}UDU.IDIM.SAG.UŠ ^{d}*ṣal-bat-a-nu* KUR-*ma* (var.: **TE-[*ma*]**)
A obv. 2’	[... ^{d}*ṣal*]-*bat-a-nu* TE-[*ma*]
C obv. 1	^{d}UDU.IDIM.SAG.UŠ⸣ ^{d}*ṣal-bat-a-nu* KUR-*m*[*a*]
F obv. 1	^{d}UDU.IDIM.SAG.UŠ ^{d}*ṣal-bat-a-nu* KUR-*ma*
G obv. 1	[... ^{d}*ṣ*]*al-bat-a-nu* KUR-*ma*

[If the Pleiades] reach [the Yoke], in that year the market rate will diminish. (It means that) Saturn reaches (var.: comes close to) Mars.

2.	**DIŠ MUL.MUL (*ana*) ^{mul}ŠUDUN KUR-*ud ina* MU BI GÁN.BA TUR(-*ir*)**
C obv. 2	[...] *ina* ⸢MU⸣ BI GÁN.BA TUR-*ir*
F obv. 2	[...
M obv. 12’	DIŠ MUL.MUL ^{mul}⸢ŠUDUN⸣ KUR-*ud ina* MU BI GÁN.BA TUR-*ir*
N obv. 14’	DIŠ MUL.MUL ^{mul}ŠUDUN KUR-*ud ina* MU BI GÁN.BA TUR
O obv. 12	DIŠ MUL.MUL ^{mul}ŠUDUN KUR-*ud ina* MU BI GÁN.BA TUR-*ir*
U obv. ii 12’	DIŠ ⸢MUL.MUL⸣ *ana* ^{mul}ŠUDUN KUR-*ud ina* MU BI GÁN.BA TUR
	^{d}[(UDU.IDIM.)G]U$_4$.UD ^{d}(UDU.IDIM.)SAG.UŠ ^{d}*ṣal-bat-a-nu* KUR-*ma*
A obv. 3’	[... ^{d}UDU.IDIM.SAG.U]Š ^{d}*ṣal-bat-a-nu* ⸢KUR⸣-[*ma*]
C obv. 3	[...] ^{d}*ṣal-bat-a-nu* KUR-*ma*
F obv. 2	^{d}UDU.IDIM.G]U$_4$.UD ^{d}UDU.IDIM.SAG.UŠ ^{d}*ṣal-bat-a-nu* KUR-*ma*
G obv. 2	[... ^{d}*ṣa*]*l-bat-a-nu* KUR-*ma*
M obv. 12’	⸢^{d}⸣[...]

320 K 2342+ is a commentary which represents the assumed tablet EAE 55. It preserves a sequence of entries on the following subjects: ^{mul}ŠUDUN, “Yoke”, ^{mul}ŠU.PA, “Resplendent One”, ^{d}*e-a*, “Ea”, ^{mul}NUN^{ki}, “Eridu-star”, ^{mul}EN.TE.NA.BAR.HUM, “Mouse”, ^{mul}GÍR.TAB, “Scorpion”, ^{mul}ZI.BA.NI.TUM, ^{mul}ÙZ, “She-goat”, and ^{mul}MAR.GÍD.DA, “Wagon”.

O obv. 12	⸢^{d}SAG⸣.UŠ ^{d}[GU$_4$.UD...]

If the Pleiades reach the Yoke, in that year the market rate will diminish. (It means that) [Merc]ury (or) Saturn (var.: Saturn (or) [Mercury...]) reaches Mars.

3.	**DIŠ MUL.MUL (*ana*) ^{mul}KA.MUŠ.Ì.KÚ.E KUR-*ud ina* MU BI SU.GU$_7$**
C obv. 4	[...] *ina* MU BI SU.GU$_7$
F obv. 3	[... GU]$_7$
M obv. 13’	DIŠ MUL.MUL ^{mul}KA.MUŠ.Ì.KÚ.E KUR-*ud ina* MU BI [...]
N obv. 16’	DIŠ MUL.MUL ^{mul}KA.MUŠ.Ì.KÚ.E KUR-⸢*ud*⸣ [...]
O obv. 13	DIŠ MUL.MUL ^{mul}KA.MUŠ.Ì.KÚ.E KUR-*ud ina* MU BI ⸢SU⸣.GU$_7$ [...]
R rev. 3	[DIŠ MUL.MUL ^{mul}KA].MUŠ.Ì.KÚ.E KUR-*ud*
U obv. ii. 13’	DIŠ MIN *ana* ⸢^{mul}⸣KA.MUŠ.Ì.KÚ.E KUR-*ud ina* MU BI
BB obv. 4	[...] KUR-*ud ina* MU BI

	(var.: **ŠE.PAD; Š[E K]ÁR-RA**) **GÁL(-*ši*) ^{d}*ṣal-bat-a-nu* ^{d}(UDU.IDIM.)SAG.UŠ**
A obv. 4’	[...] ⸢^{d}⸣*ṣal-bat-a-nu* ^{d}UDU.IDIM.SAG.UŠ
C obv. 4	GÁL-*ši*
C obv. 5	[...] ^{d}UDU.IDIM.SAG.UŠ
D 1’	[...] ⸢^{d}*ṣal-bat-a-nu* ^{d}⸣[...]
F obv. 3	GÁL-*ši* ^{d}*ṣal-bat-a-nu* ^{d}UDU.IDIM.SAG.UŠ
O obv. 14	^{d}*ṣal-bat-a-nu* ^{d}SAG.UŠ
R rev. 4	[... ^{d}*ṣal*]-*bat-a-nu* ^{d}SAG.UŠ
U obv. ii. 13’	ŠE.PAD GÁL
BB obv. 4	Š[E K]ÁR-RA GÁL-*ši*

	KUR-*ma* (var.: **[$^{d/mul}$... ^{d}*ṣa*]*l-bat-a-nu* KUR-*ma*)**
A obv. 4’	KUR-[*ma*]
C obv. 5	KUR-*ma*
F obv. 3	KUR-*ma*
G obv. 3	[... ^{d}*ṣ*]*al-bat-a-nu* KUR-*ma*
O obv. 14	KUR-[*ma*]
R rev. 4	KUR-*ma*

If the Pleiades reach Pāšittu, in that year there will be famine (var.: barley). (It means that) Mars reaches Saturn (var.: [the planet...] reaches [M]ars).

4.	**DIŠ MUL.MUL $^{mul(.d)}$AMAR.UTU KUR-*ud ina* MU BI *um-mu u um-šu*$_{14}$**
A obv. 5’	[DIŠ MUL.MUL ^{mul}].⸢^{d}AMAR.UTU KUR-*ud ina* MU BI *um-mu u um*⸣-[*šu*$_{14}$
C obv. 6	[...] *ina* MU BI *um-mu u um-šu*$_{14}$
M obv. 14’	DIŠ MUL.MUL ^{mul}AMAR.UTU KUR-*ud ina* MU BI *um-mu u*
N rev. 1	DIŠ MUL.MUL ^{mul}AMAR.UTU KUR-*ud ina* [...]
BB obv. 2	[...] KUR-*ud ina* MU BI *um-mu u um-šu*$_{14}$

	(var.: ***um-ši***) **GÁL(-*ši*) BAD-*ma* ^{d}UDU.IDIM.GU$_4$.UD BAD-*ma***
A obv. 5’	GÁL-*ši*] ⸢BAD-*ma*⸣ ^{d}UDU.IDIM.GU$_4$.UD BAD-*ma*
C obv. 6–7	GÁL (7) [...] BAD-*ma*
F obv. 4	[... GÁL]-*ši* BAD-*ma* ^{d}UDU.IDIM.GU$_4$.UD BAD-*ma*
G obv. 4	[... ^{d}UDU.IDIM.GU$_4$].UD
M obv. 14’	*um*-⸢*ši*⸣ [...]
N rev. 2	BAD-*ma*
BB obv. 2	GÁL-*ši*

	$^{d/mul}$SAG.ME.GAR $^{d/mul}$*za-ap-pa* KUR-*ma* BAD-*ma* *um-ma-tu*$_4$
A obv. 5’–6’	^{mul}SAG.ME.[GAR] (6’) ⸢^{mul}*za-ap*⸣-*pa* KUR-*ma* BAD-*ma* *um-ma-tu*$_4$
C obv. 7	^{d}SAG.ME.GAR ^{d}*za-ap-pa* KUR-*ma*
F obv. 4	^{d}SAG.ME.GAR
G obv. 4	^{mul}*za-ap-pa* KUR-*ma*
N rev. 2	^{d}SAG.ME.GAR ^{d}*za-ap-*[*pa*...]

	^{d}UTU RA-*iṣ* BAD-*ma* EN.TE.NA ^{d}IŠKUR RA(-*iṣ*) ^{d}*ṣal-bat-a-nu* GABA.RI
A obv. 6’	^{d}UTU RA-*iṣ* BAD-*ma* EN.TE.NA ^{d}IŠKUR RA-*iṣ* ^{d}*ṣal-bat-a-nu* GABA.RI
B rev. 2’	[... EN.TE].NA ^{d}⸢IŠKUR⸣ R[A...]
C obv. 8–9	[...] BAD-*ma* EN.TE.NA ^{d}IŠKUR RA (9) [...] GABA.RI
D 1’	[...] ⸢^{d}*ṣal-bat-a-nu*⸣ G[ABA.RI]
F obv. 5–6	[... BAD-*m*]*a* EN.TE.NA ^{d}IŠKUR RA-*iṣ* (6) [...
G obv. 5	[...] ^{d}IŠKUR RA-*iṣ*
N rev. 3	^{d}IŠKUR RA-*iṣ* ^{d}*ṣal-bat-a-nu* GABA.[RI...]

	^{d}UDU.IDIM GABA.R[I] (DIŠ) ^{d}UDU.IDIM.SAG.UŠ
A obv. 6’–7’	^{d}UDU.IDIM GABA.R[I] (7’) DIŠ ^{d}UDU.IDIM.SAG.UŠ
C obv. 9	^{d}UDU.IDIM.SAG.UŠ

	^{d}*za-ap-pa* KUR-*ma* ÚŠ^{meš} GAR^{meš}
A obv. 7’	^{d}*za-ap-pa* KUR-*ma* ÚŠ^{meš} GAR^{meš}
C obv. 10	[...] ÚŠ^{meš} GAR^{meš}
D 2’	[...] ÚŠ^{meš} [GAR^{meš}]
F obv. 6	^{d}]*za-ap-pa* KUR-*ma* ÚŠ^{meš} GAR^{meš}
G obv. 6	[...] ⸢ÚŠ⸣^{meš} GAR^{meš}

If the Pleiades reach the Marduk-star, in that year there will be heat and blaze. (It means that) if Mercury (or) if Jupiter reaches the Bristle (var.: [...Mercu]ry reaches the Bristle). If (in) summer, Šamaš will devastate. If (in) winter, Adad will devastate. (If) correspondingly (i.e. if in winter) Mars, or a planet, or (if) Saturn reaches the Bristle, a pestilence will take place.

5.	**DIŠ MUL.MUL ^{mul}ÉLLAG KUR-*ud* ^{d}IŠKUR RA**
A obv. 8’	DIŠ ⸢MUL.MUL ^{mul}ÉLLAG KUR-*ud* ^{d}IŠKUR⸣ RA
B rev. 3’	[...] KUR-*ud* | ^{d}IŠKUR [...]
C obv. 11	⸢DIŠ MUL⸣.MUL ^{mul}ÉLLAG KUR-*u*[*d*] ⸢^{d}⸣[IŠKUR.R]A
M obv. 15’	DIŠ MUL.MUL ^{mul}ÉLLAG KUR-*ud*
N rev. 4	DIŠ MUL.MUL ^{mul}ÉLLAG KUR-*ud* ^{d}IŠKUR [...]
BB obv. 3	[...] KUR-*ud*

	(var.: *ina* MU BI *um-mu* GÁL-*ši*) $^{d/mul}$*za-ap-pa* (*u*) ^{d}*ṣal-bat-a-nu*
A obv. 8’	^{mul}*za-ap-pa* ^{d}*ṣal-bat-a-nu*
C obv. 11	^{d}*za-ap-pu u* ^{d}*ṣal-bat-a-nu*
D 3’	[... ^{mul}*za-ap-p*]*a u* ^{d}*ṣal-*[*bat-a-nu*]
F obv. 7	[... ^{d}]*za-ap-pa u* ^{d}*ṣal-bat-a-nu*
M obv. 15’	*ina* MU BI *um-mu* ⸢GÁL⸣ [...]
BB obv. 3	*ina* MU BI *um-mu* GÁL-*ši*

	(var.: [... ^{d}*ṣal-bat-a-nu*$^{?}$ ^{mul}*za-ap-p*]*a* KUR-*ma*)
G obv. 7	[... ^{mul}*za-ap-p*]*a* KUR-*ma*

If the Pleiades reach the Kidney, Adad will devastate (var.: in that year there will be heat). (It means) the Bristle (and) Mars (var.: [... Mars$^{?}$] reaches [the Bristl]e).

6.	**DIŠ MUL.MUL ^{mul}AŠ.GÁN KUR-*ud a-ki-lu* GÁL(-*ši*) ^{d}IŠKUR RA-*iṣ***
A obv. 9’	[DIŠ MUL.MU]L ⸢^{mul}AŠ.GÁN KUR-*ud*⸣ [*a*]-⸢*ki-lu*$_4$⸣ GÁL ^{d}IŠKUR RA-*iṣ*
B rev. 4’	[...] KUR-*ud* | *a-ki-lu* GÁL ^{d}[...]
C obv. 12	⸢DIŠ MUL⸣.MUL ^{mul}AŠ.GÁN KUR-*ud a-ki-lu*$_4$ GÁL ^{d}IŠKUR RA-*iṣ*
F obv. 8	[...]
M obv. 16’	DIŠ MUL.MUL ^{mul}AŠ.GÁN KUR-*ud a-ki-lu* GÁL-*ši* ^{d}IŠKUR [RA-*iṣ*...]
BB obv. 6	[...]x ki *a-ki-*⸢*lu*$_4$⸣ GÁL ^{d}IŠKUR RA-*iṣ*

	$^{\text{d/mul}}$UDU.IDIM.GU$_4$.UD ^{d}*za-ap-pa* KUR-*ma*
A obv. 9’	^{mul}UDU.IDIM.GU$_4$.UD ^{d}*za-ap-pa* KUR-*ma*
C obv. 12	^{d}UDU.IDIM.GU$_4$.UD ^{d}*za-ap-pa* KUR-*ma*
F obv. 8	⸢^{d}⸣UDU.IDIM.GU$_4$.UD ^{d}*za-ap-pa* KUR-*ma*
G obv. 8	[... ^{d}]⸢*za*⸣-*ap-pa* KUR-*ma*

If the Pleiades reach the Field, there will be a pest (lit. devourer), Adad will devastate. (It means that) Mercury reaches the Bristle.

7.	**DIŠ MUL.MUL (*ana*) ^{mul}AŠ.GÁN TE GUN ^{giš}GIŠIMMAR** (var.:: **NAM) LAL**
A obv. 10’	[DIŠ MUL.MUL *ana*] ⸢^{mul}AŠ.GÁN TE GUN⸣ ^{giš}GIŠIMMAR LAL
C obv. 13	⸢DIŠ MUL⸣.MUL *ana* ⸢^{mul}AŠ⸣.GÁN TE GUN ^{giš}GIŠIMMAR LAL
F obv. 9	[...
M obv. 17’	DIŠ MUL.MUL ^{mul}AŠ.GÁN TE GUN ^{giš}GIŠIMMAR LAL
BB obv. 7	[...] GUN ^{g}[^{iš}GIŠIMMA]R : NAM LAL

	$^{\text{d/mul}}$*ṣal-bat-a-nu* (*ana*) ^{mul}(UDU.IDIM.)GU$_4$.UD KUR-*ma*
A obv. 10’	^{mul}*ṣal-bat-a-nu* ^{d}GU$_4$.UD KUR-*ma*
C obv. 14	⸢^{d}*ṣal-bat-a-nu*
F obv. 9	^{d}*ṣal-bat*]-⸢*a*⸣-*nu ana* ^{mul}UDU.IDIM.GU$_4$.UD
M obv. 17’	^{d}[...]

	(var.: ***ana* [^{d}UD]U.IDIM TE-*ma*) : ^{d}*ṣal-bat-a-nu*** (var.:[**^{mul}UDU.IDIM.SAG.U]Š)**
C obv. 14	*ana*⸣ [^{d}UD]U.IDIM TE-*ma* : ⸢^{d}*ṣal*⸣-*bat-a-nu*
D 4’	[... ^{mul}UDU.IDIM.SAG.U]Š

	(:) *ana* ^{mul}AB.SÍN TE-*ma* (var.: **KUR-*ma*)**
A obv. 10’	: *ana* ^{mul}AB.SÍN TE-*ma*
C obv. 14	*ana* ^{mul}AB.SÍN TE-*ma*
D 4’	*ana* ^{mul}AB.SÍN ⸢KUR⸣-*m*[*a*]
F obv. 10	[...] TE-*ma*
G obv. 9	[... ^{mul}AB.SÍ]N KUR-*ma*

If the Pleiades come close to the Field, the yield of date palm (– (var.) of the district) will diminish. (It means that) Mars reaches Mercury (var.: comes close to [a pl]anet) – (var.) Mars (var.: [Satur]n) – (var.) comes close to (var.: reaches) the Furrow.

A ___

8.	**DIŠ $^{mul}e_{4}$-ru_{6}** (var.: **KIMIN [^{m}]$^{ul.d}$AMAR.UTU) (*ana*) MUL.MUL KUR-*ud***
A obv. 11’	⸢DIŠ ^{mul}⸣e_{4}-ru_{6} *ana* MUL.MUL KUR-*ud*
B rev. 5’	[... MUL].MUL KUR-*ud* |
C obv. 15	[DIŠ $^{mul}e_{4}$]-*r*[u_{6} *ana* MUL].MUL KUR-*ud*
L obv. 3	[DIŠ ^{m}]$^{ul}e_{4}$-ru_{6} *ana* MUL.MUL KUR-⸢*ud*⸣
M obv 18’	DIŠ $^{mul}e_{4}$-ru_{6} *ana* MUL.MUL KUR-*ud*
Q obv. 5’	DIŠ $^{mul}e_{4}$-ru_{6} KIMIN [^{m}]$^{ul.d}$AMAR.UTU *ana* MUL.MUL KUR-*ud*
S obv. 3	DIŠ $^{mul}e_{4}$-ru_{6} MUL.MUL KUR-*ud*
Z obv. 1	DIŠ $^{mul}e_{4}$-ru_{6} *a-na* ⸢MUL.MUL⸣ KUR-*ud*
AA obv. 17’	DIŠ $^{mul}e_{4}$-ru_{6} *ana* ⸢MUL⸣.[MU]L KUR-*ud*

	(*ina* MU BI) ^{d}IŠKUR RA(-*iṣ*) ^{mul}*dil-bat* $^{d/mul}$*za-ap-pu* (var.: ***za-ap-pa***) **KUR-*ma***
A obv. 11’	^{d}IŠKUR RA ^{mul}*dil-bat* ^{mul}*za*-⸢*ap-pu*⸣ KUR-*ma*
B rev. 5’	^{d}IŠKUR R[A...]
C obv. 15–16	^{d}IŠKUR RA-*iṣ* (16) [...] ^{d}*za-ap-pu* KUR-*ma*
D 5’	[... ^{mu}]^{l}*za-ap-pa* KUR-*ma*
G obv. 10	[...] KUR-*ma*
L obv. 3	[...]
M obv 18’	^{d}IŠKUR RA [...]
Q obv. 5’	*ina* MU [BI...]
S obv. 3	^{d}[IŠKUR RA]
Z obv. 1	^{m}[^{ul}...]
AA obv. 17’	*ina* MU BI ^{d}[IŠKUR RA]

	(var.: **^{mul}*dil-bat ina* ŠÀ MUL.MUL [x x]**)
D 5’	x[x x$^{?}$]
S obv. 4	^{mul}*dil-bat ina* ŠÀ MUL.MUL [x x]

If the Frond (var.: or the Marduk-star) reaches the Pleiades, (in that year) Adad will devastate. (It means that) Venus reaches the Bristle (var.: Venus inside the Pleiades [...]).

9.	**DIŠ $^{mul}e_{4}$-ru_{6}** (var.: **KIMIN $^{mul.d}$AMAR.UTU) *ana* ^{mul}UG$_{5}$.GA**
A obv. 12’	DIŠ $^{mul}e_{4}$-ru_{6} *ana* ^{mul}UG$_{5}$.GA
B rev. 20’	DIŠ $^{mul4}e_{4}$-ru_{6} *ana* ^{mul4}UG$_{5}$.GA
C obv. 17	[DIŠ $^{mul}e_{4}$-ru_{6} *ana* ^{mul}U]G$_{5}$.G[A
E 1’	[...
G obv. 11	[...]x nu[n]
L obv. 4	[DIŠ ^{mul}]e_{4}-ru_{6} *ana*
M obv 19’	DIŠ $^{mul}e_{4}$-ru_{6} *ana* ^{mul}UG$_{5}$.GA
Q obv. 6’	DIŠ $^{mul}e_{4}$-ru_{6} KIMIN $^{mul.d}$AMAR.UTU *ana*
AA obv. 18’	⸢DIŠ $^{mul}e_{4}$-ru_{6} *ana*⸣

	(var.: **^{mul}UGA^{mušen}**) **KUR-*ud* ^{še}GIŠ.Ì (NIM) SIG$_{5}$(-*iq*)**
A obv. 12’	KUR-*ud* ^{še}GIŠ.Ì NIM SIG$_{5}$
B rev. 20’	KUR-*ud* | ^{še}GIŠ.Ì SIG$_{5}$
C obv. 17	KUR-*ud*] ^{še}⸢GIŠ⸣.Ì NIM SIG$_{5}$
D 6’	[...
E 1’	...]⸢^{še}GIŠ.Ì NIM SIG$_{5}$⸣-*iq*
M obv 19’	^{še}GIŠ.Ì NIM SIG$_{5}^{!}$
Q obv. 6’	^{mul}UGA^{mušen} KUR-*u*[*d*...]
AA obv. 18’	[^{mul}UGA]⸢$^{mušen?}$⸣ KUR-*ud* ^{še}GIŠ.Ì NIM [...]

	(:) $^{d/mul}$UDU.IDIM.SAG.UŠ (*ana*) ^{mul}AB.SÍN (var.: **^{d}*dil-bat***)
A obv. 12’	^{d}UDU.IDIM.SAG.˹UŠ˺ *ana* ^{mul}AB.SÍN
B rev. 20’	⁝ ^{mul4}UDU.IDIM.SAG.UŠ *ana*
C obv. 18	[^{d}UDU.IDIM.SAG.UŠ *ana* ...] ^{d}*dil-*˹*bat*˺
D 6’	^{mul}UDU.IDIM.SAG.U]Š ^{mul}AB.SÍN
E 1’	^{d}UDU.IDIM.˹SAG.UŠ˺ ^{m}[^{ul}...]
F obv. 11	[... ^{m}]^{ul}AB.SÍN
L obv. 5	[^{mul}U]DU.IDIM.SAG.UŠ

	(var.: **^{mul4}*za-ap-pa*; ^{mul}˹*za-<ap>-pi*) KUR-*ma*** (var.: **-*ud***) (var.: **: $^{d/mul}$*dil-bat***)
A obv. 12’	KUR-*ma* : ^{d}*dil-bat*
B rev. 20’	^{mul4}*za-ap-pa* KUR-˹*ud*˺ [...]
C obv. 18	KUR-*ma*
D 6’	KUR-*m*[*a* x x$^{?}$]
F obv. 11	KUR-*ma* : ^{mul}*dil-bat*
L obv. 5	^{mul}˹*za-<ap>-pi* KUR˺-*m*[*a*...]

If the Frond (var.: or the Marduk-star) reaches the Raven, the (early) sesame will be good. (It means that) Saturn reaches the Furrow (var.: Venus; the Bristle; – (var.) Venus) [...].

10.	**DIŠ ^{mul}LU.LIM (*ana*) MUL.MUL KUR-*ud* GU$_7$-*ti*** (var.: **[*ú*]-*kul-ti***)
A obv. 13’	DIŠ ^{mul}LU.LIM *ana* MUL.MUL KUR-*ud* GU$_7$-*ti*
B rev. 6’	[DIŠ ^{mul}LU].˹LIM˺ MUL.MUL KUR-*ud* ∣ GU$_7$-*ti*
C obv. 19	[... MUL.M]UL [KUR-*ud*] GU$_7$-*ti*
E 2’	[... KUR-*u*]*d* ˹GU$_7$˺-*ti*
G	(remainder is missing)
M obv 20’	DIŠ ^{mul}LU.LIM MUL.MUL KUR-*ud* GU$_7$-*ti* [...]
P$_1$ rev. 4’	DIŠ ^{mul}LU.LIM MUL.MUL [KUR-*u*]*d* GU$_7$-*ti*
Q rev. 6	DIŠ ˹^{mul}˺LU.LIM MUL.MUL KUR-*u*[*d* *ú*]-*kul-ti*

	^{d}IMIN.BI ^{d}UDU.IDIM.SAG.UŠ $^{d/mul}$*za-ap-pa* (var.: ***zap-pa***) **KUR-*ma***
A obv. 13’	^{d}IMIN.BI ^{d}UDU.IDIM.SAG.UŠ ^{d}*za-ap-pa* KUR-*ma*
B rev. 6’	^{d}IMIN.[BI...]
C obv. 19–20	˹^{d}˺[IMIN.BI] (20) [...] ^{d}*za-ap-pa* KUR-[*ma*]
D 7’	[... ^{m}]^{ul}*za-ap-pa* [KUR-*ma*]
E 2’	^{d}IMIN.BI ^{d}UDU.IDIM.SAG.UŠ [...]
F obv. 13	[... ^{d}*za-a*]*p-pa* KUR-*ma*
P$_1$ rev. 4’	^{d}IMIN.BI KIMIN (i.e. ^{d}UDU.IDIM.SAG.UŠ ^{d}*zap-pa* KUR-*ma*)
Q rev. 6	˹^{d}˺[IMIN.BI...]

If the Stag reaches the Pleiades, (there will be) consumption by the Sebettu. (It means that) Saturn reaches the Bristle.

11.	**DIŠ ^{mul}LU.LIM KI** (var: ***ana***) **MUL.MUL *it-ten-tu id-ra-na-a-tu***
A obv. 14’	DIŠ ^{mul}LU.LIM *ana* MUL.MUL *it-ten-tu id-ra-na-a-tu*
B rev. 7’	[DIŠ] ˹^{mul}˺LU.LIM KI MUL.MUL *it-ten-tu* ∣ *id-ra-na-a-tu*$_4$
C obv. 21	[... MUL.M]UL *it-ten-tu id-ra-na-a-tu*$_4$
D 8’	[...] ˹x x x˺ [...]
E 3’	[... *it-ten*]-*tu id-ra-na-a-t*[*u*$_4$...]
P$_1$ rev. 5’	DIŠ ^{mul}LU.LIM KI MUL.MU[L] *it-ten-tu*
Q rev. 7	DIŠ ˹^{mul}˺LU.LIM MUL.MUL *it*-[*ten-t*]*u* x[...]

	(var: **KA.SES^{meš}**) ***ina* KUR GÁL^{meš} (⁝) ^{mu}[^{l}...]**
A obv. 14’	*ina* KUR ⸢GÁL^{meš}⸣
B rev. 7’	*ina* KUR GÁL^{meš} ⁝ ^{mu}[^{l}...]
C obv. 21	*ina* KUR GÁL[^{meš}]
P_1 rev. 5’	KA.SES^{meš} *ina* KUR GÁL^{meš}

	^{mul}*ṣal-bat-a-nu a-na* ^{mul}UDU.IDIM.GU$_4$.UD [TE-*ma*]
B rev. 8’	| ^{mul}*ṣal-bat-a-nu a-na* ^{mul}UDU.IDIM.GU$_4$.UD [TE-*ma*]

If the Stag together with (var.: towards) the Pleiades they go parallel to each other, there will be salinity (var.: bitter mouths) in the country. (It means that) the sta[r...] (or) Mars [comes close] to Mercury.

12.	**DIŠ ^{mul}TI$_8^{\text{mušen}}$ *ana* MUL.MUL TE ^{d}IŠKUR RA(-*iṣ*)**
A obv. 15’	DIŠ ^{mul}TI$_8^{\text{mušen}}$ *ana* MUL.MUL TE ^{d}IŠKUR RA
B rev. 9’	DIŠ ^{mul}TI$_8^{\text{mušen}}$ *ana* MUL.MUL TE | ^{d}IŠKUR RA-*i*[*ṣ*...]
C obv. 22	[... MUL.MU]L TE ^{d}IŠKUR RA-*i*[*ṣ*]
D	(remainder is missing)
E 4’	[... ^{d}IŠKUR R]A
L obv. 1	[DIŠ ^{m}]^{ul} TI$_8^{\text{mušen}}$ *ana* MUL.MU[L...]
M obv 21’	DIŠ ^{mul}TI$_8^{\text{mušen}}$ *ana* MUL.MUL TE ^{d}IŠKUR [...]
T obv. 16’	DIŠ ^{mul}TI$_8^{\text{mušen}}$ *ana* MUL.MUL TE ^{d}IŠKUR RA-*iṣ*

	^{mul}*a-ḫu-ú* (var.: **[^{mu}]^{l}*a-ru-ú*) ^{d}*ṣal-bat-a-nu***
A obv. 15’	^{mul}*a-ḫu-ú* ^{d}*ṣal-bat-*⸢*a*⸣*-nu*
C obv. 23	[...] ^{d}*ṣal-bat-a-nu*
E 4’	^{mul}*a-ḫu-*⸢*ú*⸣ [...]
L obv. 2	[^{mu}]^{l}*a-ru-ú* [...]

If the Eagle comes close to the Pleiades, Adad will devastate. (It means that) the Strange One (var.: the Eagle) (is) Mars.

13.	**DIŠ ^{mul}MAN-*ma ana* MUL.MUL TE ZÁḪ KUR BIR(-*aḫ*)**
A obv. 16’	DIŠ ^{mul}MAN-*ma ana* MUL.MUL TE ZÁḪ KUR BIR-*aḫ*
B rev. 10’	DIŠ ^{mul}MAN-*ma ana* MUL.MUL TE | ZÁḪ KUR BIR-*aḫ*
C obv. 24	[... ZÁ]Ḫ KUR BIR
E 5’	[...
V obv. 17’	DIŠ ^{mul}MAN-*ma ana* MUL.MUL TE ZÁḪ KUR [...]
Y rev. 15’	DIŠ ^{mul}MAN-*ma ana* MUL.MUL TE ZÁḪ KUR BIR [...]

	UN^{meš} ^{mul}MAN-*ma* ^{d}*ṣal-bat-a-nu* x[...]
A obv. 16’	UN^{meš} ^{mul}MAN-*ma* ^{d}*ṣal-bat-*[*a*]*-nu*
B rev. 10’	UN^{meš} | ^{m}[^{ul}...]
C obv. 24	UN^{meš} ^{mul}MAN-*ma* ^{d}*ṣal-bat-a-nu*
E 5’	UN^{me}]^{š} ^{mul}MAN-*ma* ^{d}*ṣal-bat-a-nu* x[...]

If the Other One comes close to the Pleiades, (there will be) loss of the country (and) the scattering of people. (It means that) the Other One (is) Mars ...[...].

14.	**DIŠ ^{mul}MAN-*ma ana* $^{\text{d/mul}}$AŠ.GÁN TE(-[*m*]*a*$^{?}$) A.AB.BA ḪÁD-*ma***
A obv. 17’	DIŠ ^{mul}MAN-*ma ana* ^{mul}AŠ.GÁN TE A.AB.BA ḪÁD-*ma*
B rev. 11’	DIŠ ^{mul}MAN-*ma ana* ^{mul}AŠ.GÁN TE | A.AB.BA

C obv. 25	[... T]E-[*m*]*a*? ⸢A⸣.AB.BA ḪÁD-*ma*
P$_3$ obv. 18’	⸢DIŠ ^{mul}MAN-*ma*⸣ [... A.AB.B]A ḪÁD-*ma*
V obv. 8’	DIŠ ^{mul}MAN-*ma ana* ^{mul}AŠ.GÁN TE A.AB.BA
Y rev. 11’	DIŠ ^{mul}MAN-*ma ana* ⸢^{mul}AŠ⸣.GÁN TE A.AB.BA ḪÁD-*ma* [...]
Z rev. 17’	DIŠ ^{mul}MAN-*ma ana* ^{d}AŠ.GÁN TE A.AB.B[A...]

	(var.: ***ib-bal-ma***) **MA.DAM-*šá*** (var.: ***ḫi-ṣib-šá***) **ZÁḪ (:) $^{d/mul}$*ṣal-bat-a-nu* (*lu*)**
A obv. 17’	MA.DAM-*šá* ZÁḪ ^{d}*ṣal-bat-a-*⸢*nu lu*⸣
B rev. 11’	*ḫi-ṣib-šá* ZÁḪ : ^{mul}*ṣal-bat-a-nu lu*
C obv. 25–26	*ḫi-ṣib-šá* ZÁḪ (26) [...
E 6’	[... *i*]*b-bal-ma* [...]
P$_3$ obv. 18’	*ḫi-ṣib-šá* ZÁḪ ^{d}⸢*ṣal-bat-a-nu*
V obv. 8’	*ib-b*[*al-ma*...]

	ana* ^{mul}SIM.MAḪ *lu ana* ^{mul}*a-nu-ni-tum* TE-*ma
A obv. 17’	*ana* ^{mul}⸢SIM⸣.MAḪ *lu ana* ⸢^{mul}⸣[*a-nu-ni-t*]*um* TE-*ma*
B rev. 11’	^{mul}SIM.M[AḪ...]
C obv. 26	...^{mul}] SIM.MAḪ *lu ana* ^{mul}*a-nu-ni-tum* TE-*ma*
E 7’	[...] ^{mul}*a-n*[*u-ni-tum*...]
P$_3$ obv. 18’	*ana* ^{mul}⸣[SIM.MAḪ TE-*ma*]

If the Other One comes close to the Field, the sea (will dry), its yield will perish. (It means that) Mars comes close (either) to the Swallow or to Anunitu.

15.	**DIŠ ^{mul}PAN *ana* ^{mul}TI$_8$^{mušen} KUR-*ud* ^{še}GIŠ.Ì NIM SIG$_5$ (:) ^{mul}*ṣal-bat-a-nu***
A obv. 18’	DIŠ ^{mul}PAN *ana* ^{mul}TI$_8$^{mušen} KUR-*ud* ^{še}GIŠ.Ì NIM SIG$_5$ ^{mul}*ṣal-bat-a-nu*
B rev. 12’	DIŠ ^{mul}PAN *ana* ^{mul}TI$_8$^{mušen} KUR-*ud* | ^{še}GIŠ.Ì NIM SIG$_5$: ^{mul}*ṣal-bat-a-nu*
C obv. 27	[... KUR]-*ud* ^{še}GIŠ.Ì NIM SIG$_5$
J 1’	⸢DIŠ ^{mul}⸣PA[N...]
K 1’–2’	[DIŠ ^{mul}PAN *ana* ^{mu}]^{l}⸢TI$_8$⸣^{m}[^{ušen}...] (2’) [^{mul}*ṣal*]-⸢*bat-a-nu*⸣
M obv 22’	DIŠ ^{mul}PAN *ana* ^{mul}TI$_8$^{mušen} KUR-*ud* ^{še}GIŠ.Ì NIM SIG$_5$ [...]
X ii 9’	DIŠ ^{mul}PAN *ana* ^{mul}TI$_8$^{mušen} KUR-*ud* ^{še}GIŠ.[Ì...]

	ana* ^{mul}AB.SÍN TE-*ma (var.: **^{mul}*za-ap-pa* KUR-*ma***)
A obv. 18’	*ana* ⸢^{mul}⸣A[B.S]ÍN TE-*ma*
B rev. 12’	^{mul}*za-ap-pa* KUR-⸢*ma*⸣ [...]
C obv. 28	[... KUR]-⸢*ma*⸣
E 8’	[... ^{m}]^{ul}AB.S[ÍN...]
K 2’	*ana* ^{mul}AB.⸢SÍN TE-*ma*⸣

	: *ina* ŠÀ ^{mul}AB.SÍN GUB-*ma*
A obv. 18’	: *ina* ŠÀ ^{mul}AB.SÍN [GUB]-⸢*ma*⸣
C obv. 28	: *ina* ŠÀ ^{mul}AB.SÍN GUB-*ma*
J 2’	: *ina* ŠÀ [...]
K 2’	⸢:⸣ [...]

If the Bow reaches the Eagle, the early sesame will be good. (It means that) Mars comes close to the Furrow (var.: reaches the Bristle) – (var.) stands inside the Furrow.

16.	**DIŠ ^{mul}PAN *ana* $^{d/mul}$ŠUL.PA.È DIM$_4$** (var.: **KUR-*ud***) **(KUR) ELAM.MA^{ki}**
A obv. 19’	DIŠ ^{mul}PAN *ana* ^{d}ŠUL.PA.È DIM$_4$ KUR ELAM.MA^{ki}
B rev. 14’	DIŠ ^{mul}PAN *ana* ^{mul}ŠUL.PA.È DIM$_4$ | KUR ELAM.MA^{ki}

C obv. 31 [... DI]M$_4$ KUR ELAM.[M]A^{ki}
J 3’ DIŠ ^{mul}PA[N...]
K 3’ [DIŠ ^{mul}PA]N *ana* ^{mul}ŠUL.PA.⸢È⸣ DI[M$_4$...]
M obv 23’ DIŠ ^{mul}PAN *ana* ^{mul}ŠUL.PA.È KUR-*ud* ELAM.MA^{ki}

NINDA (DÙG.GA) GU$_7$ $^{d/mul}$SAG.ME.GAR *ina* ŠÀ ^{mul}AB.SÍN GUB-*ma*
A obv. 19’ NINDA DÙG.GA ⸢GU$_7$ ^{mul}⸣SAG.ME.GAR *ina* ŠÀ ^{mul}AB.S[ÍN GU]B-*ma*
B rev. 14’ NINDA DÙG.GA GU$_7$ | ^{mul}SAG.⸢ME.GAR *ina* ŠÀ ^{mul}x x⸣ [x x]
C obv. 31–32 NINDA DÙG.GA GU$_7$ (32) [...] *ina* [ŠÀ] ^{mul}AB.SÍN GUB-*ma*
E 9’ [...] ⸢^{d}SAG⸣.[ME.GAR...]
J 4’ ^{d}SA[G.ME.GAR...]
K 4’ [NINDA DÙG.G]A GU$_7$ ^{mul}SAG.⸢ME.GAR *ina* ŠÀ ^{mul}⸣A[B.SÍN...]
M obv 23’ NINDA ⸢GU$_7$⸣ [...]

If the Bow approaches (var.: reaches) Šulpaea (i.e. Jupiter), Elam will eat (good) bread. (It means that) Saturn stands inside the Furrow.

17. DIŠ ^{mul}PAN *ana* ^{mul}KAK.SI.SÁ KUR-*ud* BURU$_{14}$ SI.SÁ KI.LAM GI.NA
A obv. 20’ DIŠ ^{mul}PAN *ana* ^{mul}KAK.SI.SÁ KUR-*ud* BURU$_{14}$ SI.SÁ KI.LAM GI.NA
B rev. 13’ DIŠ ^{mul}PAN *ana* ^{mul}KAK.SI.SÁ KUR-*ud* | BURU$_{14}$ SI.SÁ KI.LAM GI.NA
C obv. 29 [...] BURU$_{14}$ SI.SÁ KI.LAM GI.NA
E (remainder is missing)
H 1’ [DIŠ $^{m]ul}$PAN *ana* $^{mu[l}$KAK.SI.SÁ...]
J 5’ DIŠ ^{mul}PAN [...]
K 5’ [DIŠ $^{mu]l}$PAN *ana* ^{mul}KAK.SI.SÁ KUR-*u*[*d*] ⸢BURU$_{14}$ SI.SÁ⸣ K[I.LAM...]
M obv 24’ DIŠ ^{mul}PAN *ana* ^{mul}KAK.SI.SÁ KUR-*ud* BURU$_{14}$ SI.SÁ [...]
X ii 10’ DIŠ ^{mul}PAN *ana* ^{mul}KAK.SI.SÁ KUR-*ud*

(var.: **ŠE [...]**) **$^{d/mul}$UDU.IDIM.GU$_4$.UD *ina* ŠÀ(-*bi*) ^{mul}AB.SÍN GUB-*ma***
A obv. 20’ ⸢^{d}UDU⸣.IDIM.GU$_4$.UD *ina* ŠÀ ^{mul}[AB.SÍN GU]B-*ma*
B rev. 13’ | ^{mul}UDU.IDIM.[GU$_4$.UD...]
C obv. 30 [...] *ina* ŠÀ ⸢^{mul}AB⸣.SÍN GUB-*ma*
J 6’ ^{d}UDU.IDIM.[GU$_4$.UD...]
K 6’ [$^{m]ul}$UDU.IDIM.GU$_4$.UD ⸢*ina* ŠÀ-*bi*⸣ [$^{m]ul}$[A]B.SÍ[N...]
X ii 10’ ŠE [...]

If the Bow reaches the Arrow, harvest will prosper, the market rate will last (var.: barley [...]). (It means that) Mercury stands inside the Furrow.

18. DIŠ ^{mul}ÙZ (*ana*) ^{mul}AŠ.GÁN KUR-*ma* GUB *ina* MU BI ÁZAG
B rev. 15’ DIŠ ^{mul}ÙZ ^{mul}AŠ.GÁN KUR-*ma* | GUB *ina* MU BI ÁZAG
C obv. 33 [... *ina* M]U BI ÁZAG
H 2’ [DIŠ] ^{mul}ÙZ *ana* $^{mu[l}$AŠ.GÁN...]
J 7’ DIŠ ^{mul}Ù[Z...]
K 7’–8’ [DIŠ $^{m]ul}$⸢ÙZ⸣ ^{mul}AŠ.⸢GÁN KUR-*ma* GUB⸣ *in*[*a* MU BI] (8’) [ÁZAG
M obv 25’ DIŠ ^{mul}ÙZ ^{mul}AŠ.GÁN KUR-*ma* GUB *ina* MU BI ⸢Á[ZAG...]
GG rev. 12’ DIŠ ^{mul}ÙZ ^{mul}⸢AŠ.GÁN KUR⸣ [...]

GÁL(-*ši*) (DINGIR GU$_7$) ^{mul}*dil-bat* $^{m[ul}$x] x[...]
B rev. 15’ GÁL | ^{mul4}*dil-bat*
C obv. 33 GÁL-*ši* DINGIR GU$_7$
K 8’ GÁ]L ⸢^{mul}⸣*dil-bat* $^{m[ul}$x] x[...]

If the She-Goat reaches the Field and stands still, in that year there will be Asakku (and the god will devour). (It means that) Venus the s[tar...] ...[...].

19.	**DIŠ ^{mul}ÙZ ^{mul}UR.BAR.RA KUR-*ud*** (var.: **T[E...])** ***ina*** **MU BI**
B rev. 16’	DIŠ ^{mul}ÙZ ^{mul}UR.BAR.RA KUR-*ud* | *ina* MU BI
C obv. 34	[...] *ina* MU BI
H 3’	DIŠ ^{mul}ÙZ ⸢^{mul}⸣[UR.BAR.RA...]
J 8’	DIŠ ^{mul}Ù[Z...]
K 9’	[DIŠ ^{mul}ÙZ ^{mul}UR].⸢BAR⸣.RA KUR-*ud*
M obv 26’	DIŠ ^{mul}ÙZ ^{mul}UR.BAR.RA KUR-*ud* *ina* MU BI [...]
DD 12d	DIŠ ^{mul}ÙZ *ana* ^{mul}UR.BAR.RA T[E...]
GG rev. 13’	DIŠ ^{mul}ÙZ ^{mul}UR.⸢BAR.RA⸣ [...]

	(var.: ***ina*** **KUR.KUR) ŠUB-*tì bu-lì* $^{d/mul}$*dil-bat* $^{d/mul}$*ṣal-bat-a-nu* KUR-*ma***
B rev. 16’	ŠUB-*tì bu-lì* | ^{mul}*dil-bat* ^{mul}*ṣal-bat-a-nu* KUR-*ma*
C obv. 34–35	ŠUB-*tì bu-lì* (35) [...] ^{d}*ṣal-bat-a-nu* KUR-*ma*
J 9’	^{d}*dil-ba*[*t*...]
K 9’	⸢*ina* KUR.KUR⸣ [...]

If the She-Goat reaches the Wolf (var.: comes [close...]), in that year (var.: i[n] the countries) (there will be) an epidemic among the herd. (It means that) Venus reaches Mars.

20.	**DIŠ ^{mul}ÙZ $^{mul(.giš)}$GÁN.ÙR KUR-*ud ina* MU BI ŠUB-*tì bu-lì*** (var.: **ÁB.GU$_4$$^{hi.a}$)**
B rev. 17’	DIŠ ^{mul}ÙZ $^{mul.giš}$⸢GÁN.ÙR⸣ KUR-*u*[*d*]
C obv. 36	[... *i*]*na* MU BI ŠUB-*tì bu-lì*
II 4’	DIŠ ^{mul}ÙZ ^{mul}GÁN.[ÙR...]
J 10’	DIŠ ^{mul}⸢ÙZ⸣ [...]
K 10’	[...]x x[...]
K	(remainder is missing)
M obv 27’	DIŠ ^{mul}ÙZ $^{mul.giš}$GÁN.ÙR$^{!}$ KUR-*ud ina* MU BI [...]
DD 12e	DIŠ ^{mul}ÙZ ^{mul}GÁN.ÙR KUR-*ud* *ina* MU BI ŠUB-*tì* ÁB.G[U$_4$$^{hi.a}$]
GG rev. 14’	DIŠ ^{mul}ÙZ ^{mul}GÁN.Ù[R] KUR-*ud* [...]

	GABA.RI ^{d}MIN ^{d}MIN (var.: **^{d}*dil-ba*[*t*...])**
B rev. 17’	| GABA.RI ^{d}MIN ^{d}MIN | vacat
J 11’	^{d}*dil-ba*[*t*...]

If the She-Goat reaches the Harrow, in that year (there will be) an epidemic among the herd (var.: equivalent, i.e. in that year there will be an epidemic among the herd). (It means that) correspondingly, the ditto-god (and) the ditto-god (i.e. Venus reaches Mars) (var.: Venu[s...]).

21.	**DIŠ ^{mul}ÙZ (MUL<^{meš}>) KASKAL 20** (var.: **^{d}UTU) KUR-*ud*** (var.: **-*da*)**
B rev. 18’	DIŠ ^{mul}ÙZ KASKAL 20 KUR-*ud*
C obv. 37	[...]
H 5’	DIŠ ^{mul}ÙZ MUL KASKAL 20 KUR-*u*[*d*...]
J 12’	DIŠ ^{mul}ÙZ [...]
M obv 28’	DIŠ ^{mul}ÙZ ^{d}UTU KUR-*ud*
GG rev. 17’	DIŠ ^{mul}ÙZ MUL<^{meš}> KASKAL ^{d}UTU KUR-*da*

	SU.GU$_7$ *bu-lì* Ú.GUG GÁL-*ši* $^{d/mul}$*dil-bat* $^{d/mul}$ŠUL.PA.È KUR-*ma*
B rev. 18’	| SU.GU$_7$ *bu-lì* Ú.GUG GÁL-*ši* | ^{mul}*dil-bat* ^{mul}ŠUL.PA.È ⸢KUR-*ma*⸣
C obv. 37	SU.GU$_7$ *bu-lì* Ú.GUG GÁL-*ši* (38) [...] ⸢^{d}⸣ŠUL.PA.È KUR-*ma*

J 13’ ^{d}ŠUL.PA.[È...]
M obv 28’ SU.GU$_7$ *bu-lì* Ú.GUG GÁL-*ši* ^{d}*dil-bat* ^{d}ŠUL.PA.⸢È⸣ [...]
GG rev. 17’ S[U.GU$_7$...]

If the She-Goat reaches (the star(s) of) the path of the Sun, there will be famine of the herd (and) starvation. (It means that) Venus reaches Šulpaea (i.e. Jupiter).

22. **DIŠ ^{mul}UR.BAR.RA UGU-*nu* ^{mul}ÙZ GUB-*ma*** (var.: **GUB-*iz* :**)
B rev. 19’ DIŠ ^{mul}UR.BAR.RA UGU-*nu* ^{mul}ÙZ GUB-*ma*
C obv. 39 [...
H 15’ DIŠ ^{mul}UR.BAR.RA UGU-*nu* ^{mul}Ù[Z...]
Q rev. 13 DIŠ ^{mul}UR.BAR.RA [x] ka ^{mul}ÙZ GUB-*iz* :

^{mul}ÙZ KUR-*ud* KI.LAM TUR(-*ir*) BURU$_{14}$ NU SI.SÁ
B rev. 19’ ^{mul}ÙZ KUR-*ud* | KI.LAM TUR BURU$_{14}$ ⸢x x⸣ [x x]
C obv. 39 $^{mu]l}$ÙZ KUR-*ud* KI.LAM TUR-*ir* BURU$_{14}$ NU SI.SÁ
Q rev. 13 ^{mul}Ù[Z KUR-*ud*...]

If the Wolf stands above the She-Goat and (var.: or) reaches the She-Goat, the market rate will diminish, the harvest will not prosper.

23. **DIŠ ^{mul}MAR.GÍD.DA Á IM.U$_{18}$.LU** (var.: **IM.1**) **^{mul}GE$_6$ (*i-na na-ma-ri*)**
B rev. 21’ DIŠ ^{mul4}MAR.GÍD.DA Á IM.U$_{18}$.LU ^{mul}GE$_6$
C obv. 40 [...
H 6’ DIŠ ^{mul}MAR.GÍD.DA Á IM.1 ⸢^{mul}⸣G[E$_6$...]
I 2’ [DIŠ $^{m]ul}$MAR.GÍD.DA Á IM.U$_{18}$.L[U...]
J 14’ DIŠ ^{mul}MAR.GÍD.[DA...]
HH rev. 4 [DIŠ KIMI]N Á IM.U$_{18}$.LU ^{mul}GE$_6$ *i-n*[*a n*]*a-ma-ri*

KUR-*ud* ^{d}IŠKUR RA-*iṣ* (:) AN.GE$_6$ GAR-*an* (var.: ***ina en-ši* UN x[x]**) **(⁝)**
B rev. 21’ KUR-*ud* | ^{d}IŠKUR RA-*iṣ* AN.GE$_6$ GAR-*an* ⁝
C obv. 40 KUR]-*ud* ^{d}IŠKUR RA-*iṣ*
C obv. 42 [...] AN.GE$_6$
I 3’ [...]
J 15’ : AN.GE$_6$ GAR-*an* [...]
HH rev. 4 *ina en-ši* UN x[x]

$^{d/mul}$*dil-bat* $^{d/mul}$UDU.IDIM.SAG.UŠ (var.: **: ^{d}UDU.I[DIM.GU$_4$.UD]**)
B rev. 21’ ^{mul4}*dil-bat* ^{mul4}UDU.IDIM.SAG.UŠ
I 3’ ^{d}*dil-bat* ^{d}UDU.IDIM.SAG.UŠ : ^{d}UDU.I[DIM.GU$_4$.UD]

KUR-*ud lu ina* ^{d}UTU.È *lu ina* ^{d}UTU.ŠÚ.A (K[I$^{?}$...]) $^{d/mul}$*dil-bat ina* 15 2,30
B rev. 21’–22’ ⸢KUR⸣-*u*[*d lu ina* ^{d}UTU.È] | (22’) *lu ina* ^{d}UTU.ŠÚ.A ^{mul4}*dil-bat ina* 15 2,30
C obv. 41 [...] *lu ina* ^{d}UTU.È *lu ina* ^{d}UTU.ŠÚ.A
H 7’ *lu ina* ^{d}UTU.È *lu ina* ^{d}UTU.ŠÚ.A K[I$^{?}$...]
I 4’ [...
J 16’ *lu ina* ^{d}UTU.ŠÚ.A K[I$^{?}$...]

(var: **ZAG *u* GÙB**) **IGI *u* EGIR GUB^{meš} x[...]**
B rev. 22’ IGI *u* EGIR GUB^{m}[^{eš}...]
I 4’ Z]AG *u* GÙB IGI *u* EGIR GUB^{meš} x[...]

If the Wagon (var.: ditto, i.e. the Wagon) reaches the south side of the Black One (at dawn), Adad will devastate (– var.) an eclipse will occur (var.: for the weak of the people... [...]). (It means that) Venus reaches Saturn (– (var.) Mer[cury]), they stand in the east (lit. sunrise) or in the west (lit. sunset) (wi[th$^?$...]) Venus in the right (and) left, front and back ...[...].

24.	**DIŠ ^{mul}UG$_5$.GA** (var.: **^{mul}UGA^{mušen}**)	**KASKAL ^{d}UTU** (var.: **20**)	**KUR(-*ud*)**
A obv. 21’	DIŠ ^{mul}UG$_5$.GA	KASKAL ^{d}UTU	KUR-*ud*
B rev. 23’	DIŠ ^{mul4}UG$_5$.GA	KASKAL	20 KUR
H 8’	DIŠ ^{mul}UG$_5$.GA	KASKAL	20 KUR°-⸢*ud*⸣ [...]
I 5’	[DIŠ ^{m}]^{ul}UG$_5$.GA	KASKAL	20 KUR-*ud* [...]
J 17’	DIŠ ^{mul} UG$_5$.GA	KASKA[L...]	
M obv 29’	DIŠ ^{mul}UG$_5$.GA	⸢KASKAL⸣ ^{d}UTU	KUR-*ud*
U obv. ii. 16	DIŠ ^{mul}U[G]A$^{⸢mušen⸣}$	KASKAL	KUR-*ud*

	KI.LAM TUR (:) $^{d/mul}$UDU.IDIM.SAG.UŠ
A obv. 21’	KI.LAM TUR ^{d}UDU.IDIM.SAG.UŠ
B rev. 23’	| KI.LAM TUR : ^{mul4}UDU.IDIM.SAG.UŠ
I 6’	[... ^{mul}UDU.IDIM.SAG.]⸢UŠ
J 18’	^{d}UDU.IDIM.SAG.[UŠ...]
M obv 29’	⸢KI.LAM TUR⸣ [...]
U obv. ii. 16	KI.LAM TUR

	(:) $^{d/mul}$UDU.IDIM.GU$_4$.UD ^{d}ŠUL.PA.È KUR-*ma*
A obv. 21’	: ^{d}UDU.IDIM.GU$_4$.UD ^{d}ŠUL.PA.⸢È⸣ KUR-*ma*
B rev. 23’	^{mul4}UDU.IDIM.G[U$_4$.UD...]
C obv. 44	[...] ⸢^{d}⸣ŠUL.PA.È KUR-*ma*
I 6’	^{mul}UDU.IDIM.GU$_4$⸣.UD ^{d}[...]

If the Raven reaches the path of the Sun, the market rate will diminish. (It means that) Saturn – (var.) Mercury reaches Šulpaea (i.e. Jupiter).

24a.	**DIŠ ^{mul}UG$_5$.GA *ana* ^{d}ŠUL.PA.È [...]**
H 9’	DIŠ ^{mul}UG$_5$.GA *ana* ^{d}ŠUL.PA.⸢È⸣ [...]
J 19’	DIŠ ^{mul}UG$_5$.GA *ana* [...]

If the Raven [...] towards Šulpaea (i.e. Jupiter) [...].

25.	**DIŠ $^{mul.d}$AMAR.UTU *ana***	**^{mul}UG$_5$.GA** (var.: **^{mul}UGA[^{mušen}]; ^{mul}AL.LUL**)
B rev. 24’	DIŠ $^{mul4.d}$AMAR.UTU *ana*	^{mul4}UG$_5$.GA
H 11’	DIŠ $^{mul.d}$AMAR.UTU *ana*	^{mul}UGA[^{mušen}...]
J 22’	DIŠ $^{mul.d}$AMAR.UTU *ana*	^{mul}UG$_5$.[GA...]
M obv 31’	DIŠ $^{mul.d}$AMAR.UTU *ana*	^{mul}AL.⸢LUL⸣

	KUR-*ud* ^{še}GIŠ.Ì SIG$_5$ ^{d}UDU.IDIM.GU$_4$.UD (^{d}UDU.IDIM.)SAG.UŠ [...]
B rev. 24’	KUR-*ud* | ^{še}GIŠ.Ì SIG$_5$ | [...]
C obv. 45	[...] ⸢^{d}UDU.IDIM⸣.SAG.UŠ
I 7’	[... ^{š}]^e⸢GIŠ⸣.[Ì...]
I	(remainder is missing)
J 23’	^{d}UDU.IDIM.GU$_4$.UD ^{d}SAG.⸢UŠ⸣ [...]
M obv 31’	KUR-*ud* ⸢^{še}GIŠ⸣.[Ì...]

If the Marduk-star reaches the Raven (var.: the Crab), the sesame will be good. (It means that) Mercury [...] Saturn.

26.	**DIŠ ^{mul}ÉLLAG *ana* ^{mul}UG$_5$.GA** (var.: **^{mul}UGA^{mušen}**) ***i-mid***
B rev. 25’	DIŠ ^{mul}ÉLLAG *ana* ^{mul4}UG$_5$.GA *i-mid*
C	(remainder is missing)
H 12’	DIŠ ^{mul}ÉLLAG *ana* KIMIN *i-mid*
J 24’	DIŠ ⸢^{mul}ÉLLAG *ana* ^{mul}UG$_5$.GA *i*⸣-[*mid*...]
M obv 33’	DIŠ ^{mul}ÉLLAG *ana* ^{mul}UGA^{mušen} *i*-⸢*mid*⸣ [...]
	^{še}GIŠ.Ì SIG$_5$ (:) AN.GE$_6$ [GAR-*an*...]
B rev. 25’		^{še}GIŠ.Ì SIG$_5$ AN.GE$_6$ [...]
H 12’	^{še}GIŠ.Ì SIG$_5$: A[N.GE$_6$...]
J 25’	[...]x x[...]
J	(remainder is missing)

If the Kidney comes in contact with the Raven (var.: with ditto, i.e. the Raven), the sesame will be good – (var.) an eclipse [will occur...].

27.	**DIŠ ^{mul}UG$_5$.GA** (var.: **^{mul}UGA^{mušen}**) **SUKKAL** (var.: **KI**) **^{mul}SA$_5$**
B rev. 26’	DIŠ ^{mul4}UG$_5$.GA SUKKAL ^{mul4}SA$_5$
H 10’	DIŠ ^{mul}UG$_5$.GA KI ^{mul}SA$_5$
J 20’	DIŠ ^{mul}UG$_5$.GA SUKKAL ^{mu}[^{l}SA$_5$...]
U obv. ii 9’.	DIŠ ^{mul}UGA⸢^{mušen} SUKKAL⸣ ^{mul}SA$_5$
	***ana* ^{mul4}NUN.KI TE([-*ma*]) GÁN.Z[I *ina* KUR] DÙ.A.BI SI$^?$.[SÀ$^?$...]**
B rev. 26’	*ana* ^{mul4}NUN.KI TE GÁN.Z[I ...] ⸢DÙ.A.BI⸣ SI$^?$.[SÀ$^?$...]
H 10’	*ana* ^{m}[^{ul}...]
U obv. ii 9’.	*ana* ^{mul}NUN.KI T[E-*ma*]
	^{d}UDU.IDIM.GU$_4$.UD [...]
J 21’	^{d}UDU.IDIM.GU$_4$.UD [...]

If the Raven, the vizier of (var.: with) the Red One (i.e. Mars), comes close to the Eridu-star, the cultivatio[n] in the entire [country] will pr[osper$^?$...]. (It means that) Mercury [...].

28.	**DIŠ ^{mul}KU$_6$ (MUL ^{d}IŠKUR ^{mul}*nu-nu*)** (var.: **^{mul4}PAN**) **(*ana*) ^{mul4}UG$_5$.GA *i-mid***
B rev. 27’	DIŠ ^{mul4}PAN <*ana*> ^{mul4}UG$_5$.GA *i-mid*
H 13’	DIŠ ^{mul}KU$_6$ *ana* KIMIN *i-mid*
M obv 34’	[...] ^{mul}UG$_5$.GA *i-mid*
Z rev. 26’	DIŠ ^{mul}KU$_6$ MUL $^{d°}$IŠKUR ^{mul}*nu-nu* ^{m}[^{ul}...]
	KU$_6$^{meš} *u* MUŠEN^{meš} *ud-deš-šú-u* A^{meš} x[$^{d/mul}$... *ina* ^{mul}]SIM.MAḪ
B rev. 27’	KU$_6$^{meš} *u* MUŠEN$^!$(SIG)^{meš} *u*[*d-deš-šú-u*...]
H 13’	KU$_6$^{meš} *u* MUŠEN^{meš} *ud*-[*deš-šú-u*...]
M obv 34’–35’	KU$_6$^{me} ⸢*u*⸣ [MUŠEN^{me}...] (35’) [... *ina* ^{mul}]⸢SIM.MAḪ⸣
Z rev. 27’	KU$_6$^{meš} *u* MUŠEN^{meš} *ud-deš-šú-u* A^{meš} x[...]
	***lu* (*ina* ŠÀ) $^{d/mul}$*a-nu-ni-tum* GUB-*ma* : x[...]**
H 14’	*lu* *ina* ŠÀ ^{d}*a-nu-ni-tum* GUB-*ma* : x[...]
M obv 35’	*lu* ^{mul}*a-nu-ni*-⸢*tum*⸣ [...]

If the Fish (the star of Adad, the Fish) (var.: the Bow) comes in contact with the Raven (var.: ditto, i.e. the Raven), the fishes and the birds will become abundant, the water ...[...] (It means that) [the planet... in] the Swallow or (inside) Anunitu stands – (var.) ...[...].

29.	**DIŠ ^{mul}UR.BAR.RA KASKAL 20 (**var.: **^{d}UTU) KUR-*ud***
B rev. 28’	DIŠ ^{mul4}UR.BAR.RA KASKAL 20 KUR [...]
H 16’	DIŠ ^{mul}UR.BAR.RA KASKAL 20 KUR-*ud*
M obv. 37’	[...] ^{d}UTU KUR-*ud*
Q rev. 10	DIŠ ^{mul}UR.BAR.RA KA[SK]AL ^{d}UTU KUR-*ud*

	SU.GU$_7$ (var.: **: UR x[...]) *maš-ru-ú* [...]**
H 16’	⌜: UR⌝ x[...]
M obv. 37’	*m[aš-ru-ú...*]
Q rev. 10	*maš-ru-ú* [...]

If the Wolf reaches the path of the Sun, (there will be) famine (– (var.) ... ...[...]), prosperity [...].

30.	**DIŠ ^{mul}UR.BAR.RA (*ana*) ^{mul}UR.MAḪ KUR-*ud* UD^{meš} SUD^{meš}**
B rev. 29’	DIŠ ^{mul4}UR.BAR.RA ^{mul4}UR.MAḪ KUR-*ud* UD^{me} SUD^{me}
H 17’	⌜DIŠ ^{mul}⌝[...]x [...]
M obv 36’	[DIŠ ^{mul}UR.BAR.]⌜RA⌝ ^{mul}UR.MAḪ KUR-*ud* x[...]
P$_3$ obv. 10’	DIŠ ^{mul}UR.BAR.RA *ana* ^{mul}UR.M[AḪ KUR-*ud* UD^{meš} S]UD^{meš}
Q rev. 11	[DIŠ] ^{mul}UR.BAR.R[A] ^{mul}UR.MAḪ KUR-*ud* UD^{meš} SUD^{meš}

	KI.TUŠ (var.: ***šub-*[*tu*]) *ne-eḫ-tú ana* KUR x x[...]**
B rev. 29’	KI.TUŠ *n[e-eḫ-tú...*]
P$_3$ obv. 10’	KI.TUŠ *ne-eḫ-tú ana* ⌜KUR x⌝ x[...]
Q rev. 11	*šub-*[*tu...*]

If the Wolf reaches the Lion, (there will be) long-lasting days of a peaceful abode, peace for the country ... ...[...].

31.	**DIŠ $^{mul.d}$AMAR.UTU *ina* SAG MU IGI MU BI AB.SÍN-*šá* SI.[SÁ...]**
B rev. 30’	DIŠ ⌜^{mul4}⌝$^{.d}$AMAR.UTU *ina* SAG MU IGI MU BI AB.SÍN-*šá* ⌜SI⌝.[SÁ...]
H	(remainder is missing)

If the Marduk-star is visible at the beginning of the year, the furrow of that year will pros[per...].

32.	**DIŠ ^{mul}KA.MUŠ.Ì.KÚ.E ^{mul}EN.TE.NA.BAR.ḪUM K[UR$^{?}$...]**
B rev. 31’	DIŠ ^{mul4}KA.MUŠ.Ì.KÚ.E ^{mul4}EN.TE.NA.BAR.ḪUM K[UR$^{?}$...] | [...] | [...]

If Pāšittu r[eaches$^{?}$] the Mouse, [...].

33.	**DIŠ *ina* SAG MU MUL.MUL *ka-rit* 30 *pa-ni-ma* MU[L.MUL *ar-ki...*]**
B rev. 32’	DIŠ ⌜*ina*° SAG$^{!}$° MU°⌝ MUL.MUL *ka-rit* 30 *pa-ni-ma* MU[L.MUL...]
II 8’.	[DIŠ *ina* SAG MU MU]L.MUL *ka-r*[*it*$^{?}$...]

If at the beginning of the year the Pleiades are cut. (It means that) the Moon is ahead and the Ple[iades are behind...].

34.	**DIŠ ^{d}IŠKUR (*ina* MÚRU ^{d}IMIN.BI) GÙ-*šú* ŠUB-*ma* MUL^{meš} *šu-nu-tu***
B rev. 33’	DIŠ ^{d}IŠKUR GÙ-*šú* ŠUB-*ma* MUL^{meš} *šú-nu-tu*

EE rev. 12' (catchline) DIŠ ^{d}IŠKUR *ina* MÚRU ^{d}IMIN.BI ⌜ GÙ-*šú* ŠUB-*ma*

se-bet-[ti]-šu-nu (var.: **MUL^{meš} IMIN-*šu-nu*) *ana* KI *[i]t-tab-ku-ni* [...]**
B rev. 33' *se-be[t-ti-šu-nu...]*
EE rev. 12' (catchline) MUL^{meš} IMIN-*šu-nu ana* KI⌝ *[i]t-tab-ku-ni*

If Adad thunders (into the middle of the divine Seven) and those stars, their seve[n] are poured on earth [...].

35. **DIŠ ^{d}IŠKUR *ina* IM.KUR.RA GÙ-*šú* ŠUB-*di* TA IM.KUR.RA D[U...]**
B rev. 34' DIŠ ^{d}IŠKUR *ina* IM.KUR.RA GÙ-*šú* ŠUB-*di* TA IM.KUR.RA D[U...]

If Adad thunders in the west, from the west (it) go[es...].

36. **DIŠ $^{<d>}$UTU *ma-gal* SA$_5$ *nu-ḫuš* UN^{meš} MU** (var.: ***aš-šú***) ***u$_4$-ma*** (var.: **UD**)
B rev. 35' DIŠ $^{<d>}$UTU *ma*$^!$(GAL)-*gal* SA$_5$ *nu-ḫuš* UN^{meš} MU *u$_4$-ma*
CC rev. 9 [...] SA$_5$ *nu-*⌜*ḫuš* UN⌝^{meš} *aš-šú* UD

***la ti-du-ú* x[...]**
B rev. 35' *la ti-d[u-ú...]*
CC rev. 9 *la ti-du-ú* x[...]

If the Sun is very red, (there will be) plenty of the people; because you do not know the day ...[...].

37. **DIŠ ^{mul}ŠU.PA *a-rim* IDIM** (var.: ***kab-tu$_4$***) ***ina* KUR GÁL(-*ši*) MU**
B rev. 36' DIŠ ^{mul}ŠU.PA *a-rim* IDIM *ina* KUR GÁL MU
CC rev. 10 [...] *kab-tu$_4$ ina* KUR GÁL-*ši*

(var.: ***aš-šú***) **IDIM** (var.: **DUGUD**) ***la ti-du-ú* [...]**
B rev. 36' IDIM *la ti-d[u-ú...]*
CC rev. 10 *aš-šú* DUGUD *la ti-du-ú* [...]

If the Resplendent One is covered, there will be an important person in the country; because you do not know the important person [...].

38. **DIŠ ^{mul}SA$_5$ ^{mul}*ṣal-bat-a-nu* : ^{mul}*lum-nu* ^{mul}*ṣal-[bat-a-nu...]***
B rev. 37' DIŠ ^{mul}SA$_5$ ^{mul}*ṣal-bat-a-nu* : ^{mul}*lum-nu* ^{mul}*ṣal-[bat-a-nu...]*

The Red One (is) Mars – (var.) the Evil One (is) M[ars...]

39. **DIŠ ^{mul}*ṣal-bat-a-nu*** (var.: **^{mul}UDU.IDIM) (MU) 6 ITI^{meš} GUB-*ma* GIM id$^?$-[...]**
B rev. 38' DIŠ ^{mul}*ṣal-bat-a-nu* MU 6 ITI^{meš} GUB-*ma* GIM id$^?$-[...]
M rev. 3' DIŠ ^{mul}UDU.IDIM 6 ITI$^{<meš>}$

(var.: ***ina* AN-*e uš*$^!$(É)-*tab-ri-ma* [*la ir-bi*...]**)
M rev. 3' *ina* AN-*e uš*$^!$(É)-*tab-ri-ma* [...]

If Mars (that year) stands still for 6 months like ... [...] (var.: in the sky it remains visible and [it does not set...]).

40.	**DIŠ ^{mul}*ṣal-bat-a-nu ú-tan-na-at-ma* SIG$_5$ GUR$_4$-*ma*** (var.: ***ib-il-ma***)
B rev. 39’	DIŠ ^{mul}*ṣal-bat-a-nu ú-tan-na-at-ma* SIG$_5$ [...]
CC rev. 15	DIŠ ^{mul}*ṣal-*⌜*bat-a*⌝*-nu ú-tan-na-at-ma* SIG$_5$ GUR$_4$-*ma* x[...]
FF rev. 3	DIŠ ^{mul}*ṣal-bat-a-nu ú-tan-na-at-ma* SIG$_5$ *ib-il-ma*
	***a-ḫi-tú* [...] 7 *u*$_4$*-mi* 14 *u*$_4$*-mi* 21 *u*$_4$*-mi* ^{d}*bi-ib-bu* [...] 7 *u*$_4$*-mi* 14$^{!}$(13) *u*$_4$*-mi* 21**
B	(end of the reverse)
CC rev. 16–17	7 *u*$_4$*-mi* 14 *u*$_4$*-mi* 21 *u*$_4$*-mi* ^{d}*bi-ib-bu* [...] (17) 7 *u*$_4$*-mi* 14$^{!}$(13) *u*$_4$*-mi* 21
FF rev. 3	*a-ḫi-tú*
	***u*$_4$*-mi* GUR-*ma* ILLU NU GAR [...]**
CC rev. 17	*u*$_4$*-mi* GUR-*ma* ILLU NU GAR [...]

If Mars is faint, it is good; (if) it becomes bright, (there will be) misfortune [...]. (On) the 7th day, the 14th day, the 21st day, the planet (i.e. Mercury?) [...]. (On) the 7th day, the 14th day, the 21st day (Mars) recedes and the flood will not take place [...].

Commentary to the Text

1, 17, 22, 24: see 5.1.5.1. commentary to the text of entries 1, 5, p. 191.

1–2: The protasis of the omen in entry 1 is restored based on entry 2 and on Commentary EAE 53: 1 because they are two variants of the same protasis. In the apodoses, the logogram KI.LAM (see MSL 15: 26 l. 2:03, 43 l. 324, 82 l. 168) and GÁN.BA (see MSL 5: 60 l. 115; MSL 15: 162 l. 297) are two different writings for the same Akkadian word (*maḫīru,* “market rate” or “business”).

3: Pāšittu (^{mul}KA.MUŠ.Ì.KÚ.E, *pāšittu*), lit. “obliterator”, is the name of a female demon (Wiggermann 2005). According to mythology, she is associated with an illness related to the bile and to the harassment of children. The illness could have been gastroenteritis, thought to be caused by contaminated food (SAA 10, 217 obv. 11, see Parpola 1983: 140). For Š[E K]ÁR-RA, see 5.1.5.1. commentary to the text of entry 4, p. 191.

4: In source G obv. 4, the order of ^{d}SAG.ME.GAR and ^{d}UDU.IDIM.GU$_4$.UD is probably reversed, compared to the other sources.

6: For an in-depth discussion of *ākilu*, “pest”, lit. “devourer”, see Fincke (2016b).

8–9: In source Q obv. 5’–6’, the sign KIMIN, “ditto”, is used with the meaning “variant”, “or” (Borger 2004: 414 n. 742; see also 5.1.5.1. commentary to the text of entry 2, p. 191). Entries 8–9 are paralleled in MUL.APIN (GAP B 2–3, see Hunger-Steele 2019: 107).

9, 15, 25, 26: The Eagle (^{mul}TI$_8$ ^{mušen}, *a/erû*) and the Raven (^{mul}UG$_5$.GA, *āribu*) in the protases are associated with the market rate of sesame (^{še}GIŠ.Ì, *šamaššammū*) in the apodoses. This is also mentioned in a fragment dating to the Late Babylonian period (ca. 484–30 BC), where the Eagle is one of the constellations associated with the sesame: *ana* ^{še}GIŠ.Ì *ina* ŠÀ ^{mul}GÍR.TAB *u* [^{mu}]^{l}⌜TI$_8$⌝$^{\text{mušen?}}$, “For sesame: inside the Scorpion and the

Eagle?" (BM 47494 obv. 26, see Hunger 2004: 18). In BM 47494, the criteria of association derive from the way the constellations were conceived: for instance, the Hired Man, viewed as a ram, relates to the wool by analogy (Ossendrijver 2019: 66 l. 27). In the case of the Eagle and the Scorpion, the association with sesame seems less obvious.

11: In source P_1, the logogram KA.SES^{meš} replaces *idrānātu*,"salinity", the word that is found in the other sources, and by reason of that one could expect KA.SES^{meš} to be a logogram for *idrānātu*. However, KA.SES might just be a miswriting for SAḪAR.SES, *idrānu*, "salt", lit. "bitter dust" (see Lú = *ša,* MSL 12: 105 l. 53). Another option would be to assume a pseudo-logographic writing, with the sign KA read syllabically and followed by SES, *marru*, "bitter" (CAD M1 286–287). Thus, KA.SES^{meš} could be read *kamārū* "defeats", or "hunting traps" (CAD K 111–112). The most likely option would be to read both elements individually as a logogram: KA.SES^{meš} means "bitter mouths", as it is explained in lexical lists as KA *mar-rum* (*pû marru*) (MSL 13: 199 l. 318).

18: Asakku, or Asag, is a demon defeated by Ningirsu/Ninurta in the Sumerian poem "Lugal-e".[321] Like many demons, who are considered theriomorphic as a literary *topos* (see 3.1.3.), Asakku is a demon bringer of head fevers or malicious winds.[322]

20: For the usage of GABA.RI (*meḫru*, "equal", "equivalent", or "correspondingly"), see 5.1.5.1. commentary to the text of entry 2, p. 191.

23: The star ^{mul}GE$_6$, "Black One", could be a name for Saturn, based on a lexical list (Gössmann 1950: 28–29 n. 82) and, according to Brown (2000: 60–70), on several, implicit associations. However, if one simply sticks to the perception of the colour GE$_6$, *ṣalmu* (CAD Ṣ 77–78), it refers to discoloration or darkness, not only as a natural colour (Thavapalan 2020: 154–162), and Saturn is, indeed, a paler planet compared to the others, i.e. less bright.

28: Source Z rev. 26' additionally explains the Fish (^{mul}KU$_6$, *nūnu*) as the star of Adad (MUL ^{d}IŠKUR).

29: The source M obv. 37' provides two opposite predictions for the same ominous phenomenon, that is "famine", and "prosperity". This fact illuminates the composite nature of sources like source M, the assumed tablet EAE 50, which is probably a collection of omens that gives alternative apodoses from various sources (see 5.1.8.4. § EAE 50–51).

33: This entry – whose reading was suggested to the author by Jeanette C. Fincke – introduces another variant for one intercalation rule, based on the balancing between the

321 For the edition of this poem, see van Dijk (1983).

322 For the association of Asakku to diseases and negative events, and earlier bibliographical references, see Bácskay (2013).

Moon and the Pleiades (see 4.2.2. § 2b; 5.1.5.1. commentary to the text of entries 16–23 and 19–23, pp. 192–193).

34–35: Entry 34 is excerpted from the chapter of Adad in the series EAE (see 5.1.8.3.): it is the incipit of tablet EAE 46 (Gehlken 2012: 80 obv. 1), and it is related to the Pleiades. Entry 35 is also related to the weather, but it does not find any parallel in the chapter of Adad. Even though entry 35 is broken, the scribe likely chose these two omens because they mentioned the Pleiades, which could be the common topic behind the choice of most of the excerpts in the tablet SIT 6.

38–40: The names of Mars ^{mul}SA$_5$, "Red One", and ^{mul}*lumnu*, "Evil One", are also given in the Great Star List among the seven names of Mars (Koch-Westenholz 1995: 198, 200 ll. 237–240; see also 2.2.1.). Mars is sometimes just called ^{mul}UDU.IDIM, *bibbu*, "planet", like in source M rev. 3' (Brown 2000: 57). Entries 39–40 refer to the period of visibility (i.e. 6 months) and invisibility of the planet Mars (*unnutu*, "to be faint" versus *baʾālu*, "to become bright"). The visibility of Mars for 6 months is paralleled in MUL.APIN (II i 51-52), whereas the multiples of 7 as values for the visibility periods of a *bibbu*, "planet", there referred to Mercury (MUL.APIN II i 54–55, see Hunger-Steele 2019: 80–81), a planet that is sometimes addressed as *bibbu* (Brown 2000: 68). Yet, as remarked by Hunger and Steele (2019: 207) "the durations of visibility and invisibility themselves (...) are not meant to be astronomically precise." The sign GUR, *târu*, "to return" (CAD T 250–278), refers to the retrograde motion of the planets, i.e. the apparent change in their movements caused by the relative position of the earth, the Sun, and the planets themselves. The sign GUR is paralleled in this specific meaning only in the astronomical texts dating to the Late Babylonian period (ca. 484–30 BC) (Neugebauer 1955: 474–475 gur).
Entry 39 is restored from a parallel in EAE 56 (SpTU 1, 90 obv. 13; Largement 1957: 238, 12^b): DIŠ ^{mul}UDU.IDIM UD.1.KÁM UD.2.KÁM *ina* AN-*e uš-tab-ri-ma la ir-bi* ŠU.BI.AŠ.AM, "If a planet for one day (and/or) two days in the sky remains visible and it does not set, ditto (i.e. in the east and in the west a war will break out)."

Further Comments on the Text

The content of the assumed tablet SIT 6 is miscellaneous, and the entries are excerpted from various EAE tablets, as summed up and shown in Table 8.

Entry	Subject of the protasis	Entry excerpted from
1–7, 33	MUL.MUL, "Pleiades"	EAE 53
8–9	^{mul}*eru*, "Frond"	unclear
10–11	^{mul}LU.LIM, *lulīmu*, "Stag"	unclear
12	^{mul}TI$_{8}$, *a*/*erû*, "Eagle"	EAE 57
13–14	^{mul}MAN-*ma*, *šanûmma*, "Other One"	unclear
15–17, 28	^{mul}PAN, *qaštu*, "Bow"	unclear
18–21	^{mul}ÙZ, *enzu*, "She-Goat"	EAE 55, EAE 51
22, 29–30	^{mul}UR.BAR.RA, *barbaru*, "Wolf"	unclear
23	^{mul}MAR.GÍD.DA, *ereqqu*, "Wagon"	EAE 55
24	^{mul}UG$_{5}$.GA, *āribu*, "Raven"	EAE 57
25, 27, 31	$^{\text{mul.d}}$AMAR.UTU, "Marduk-star"	unclear
26	^{mul}ÉLLAG, *kalītu*, "Kidney"	unclear
32	^{mul}KA.MUŠ.Ì.KÚ.E, "Pāšittu"	unclear
34–(35)	^{d}IŠKUR, "Adad"	EAE 46
36	^{d}UTU, "Šamaš" (i.e. Sun)	unclear
37	^{mul}ŠU.PA, *nameru*, "Resplendent One"	unclear
38–40	^{mul}*ṣalbatānu*, "Mars"	unclear

Table 8. Entries in SIT 6 excerpted from other sections of the series EAE.

The entries about the Pleiades (1–5, 7) are partly excerpted from tablet EAE 53 (see Commentary EAE 53: 1–4, 6); the entries about the She-Goat (18–21) are from the assumed tablet EAE 51 and EAE 55; the entries about the Eagle (12) and the Raven (24) are from tablet EAE 57; the two entries about the weather (34–35) are from the chapter on Adad, maybe both from tablet EAE 46. The tablets EAE 55 (K 2342+, see SIT 6 source GG; BM 38301 rev., see App. B. § 14; SIT 6 source HH) and 57 (K 2330, see SIT 6 source T), lack a reconstruction of the text (Koch 2015: 176). The assumed tablet EAE 51 has been reconstructed, but its identification as belonging to EAE is still uncertain.[323]

It is not possible to indicate a precise criterion for the arrangement of the entries in SIT 6, as it was in EAE 52 and Commentary EAE 53 (see Table 6 and 7). Yet, SIT 6 has entries excerpted from at least the tablets EAE 46, and 53 to 57: the factual explanations in SIT 6 and the conclusive entries (SIT 6: 38–40) point towards the phenomena of the Pleiades and Mars or, more probably, to planetary conjunctions in general, as the criterion behind the choice of excerpts.

From factual explanations in the omens of SIT 6, it emerges that almost all phenomena are equated with planetary conjunctions, except for five entries: SIT 6: 26, in which there is a reference to an eclipse (AN.GE$_{6}$, *attalû*), in SIT 6: 33 and 34 that mention phenomena of the Pleiades, and in SIT 6: 39 and 40 which are about the visibility of the planet Mars. In omens exclusively related to the Pleiades (MUL.MUL), the latter are equated in factual explanations with *zappu*, "Bristle" (SIT 6: 4–6, 8, 10, 12–13, 33) or with the planet Mars (SIT 6: 1–3, 7, 11). In explanations that do not relate to the Pleiades, sometimes Mars is

323 See Reiner-Pingree (1981: 30, 70–71), Frahm (2011: 148–149) and Rochberg (2018: 122 fn. 5). A further discussion is in 5.1.8.4. § EAE 50–51.

unambiguously equated with other names, i.e. the Kidney (SIT 6: 5), the Eagle (SIT 6: 12), the Other One (SIT 6: 13), the Wolf (SIT 6: 19), the Harrow ($^{\text{mul(.giš)}}$GÁN.ÙR, *maškakātu*) (SIT 6: 20), the Red One (^{mul}SA$_5$, *sāmu*) and the Evil One (^{mul}ḪUL, *lumnu*) (SIT 6: 38).[324]

5.1.7. Summary for the Pleiades in Omens from EAE 52, Commentary EAE 53, and SIT 6

As seen in the reconstructed tablets EAE 52, Commentary EAE 53, and SIT 6 (see 5.1.4.–5.1.6.), not all the omens are exclusively related to the Pleiades. For instance, tablet EAE 52 and Commentary EAE 53 are structured according to the calendar, which means on first visibility dates of stars. In EAE 52, the subjects are the Field (^{mul}AŠ.GÁN, *ikû*) and the Old Man (^{mul}ŠU.GI, *šību*), rising heliacally in Nisannu and Ajaru (i.e. Month I and II), respectively;[325] in Commentary EAE 53, the subjects are the Pleiades (MUL.MUL, *zappu*, "Bristle"), the Jaw of the Bull (^{mul}*is lê*), and the True Shepherd of Anu (^{mul}SIPA.ZI.AN.NA, *šitaddaru*), rising heliacally in Ajaru, Simanu, and Du'uzu (i.e. Month II, III and IV), respectively.[326] The structure of SIT 6 is not so obvious because it is a more eclectic collection of excerpted omens, perhaps based on the common topic of the Pleiades, or on the need to equate the "impossible" protases (i.e. protases whose literal meaning is astronomically impossible, or use a metaphorical language, see p. 131 and fn. 268) to planetary conjunctions.

Among the reconstructed tablets, the Commentary EAE 53 is the most Pleiades-centred. The entries about the Pleiades mainly concern the cluster moving into or towards other celestial bodies,[327] their shape, luminosity, and position,[328] especially at the beginning of the year (i.e. Nisannu, Month I),[329] when the so-called balancing between the Moon and the Pleiades was observed to check if the intercalation was necessary.[330]

Commentary EAE 53 and SIT 6 are extremely useful because they explain, or rephrase in a more literal way or by using substitute names, many "impossible" protases. Taking into account only the planetary conjunctions in the explanations, SIT 6 is focused on the observation of the five planets (Mars, Venus, Saturn, Jupiter, and Mercury) and the Pleiades (*zappu*), Virgo (^{mul}AB.SÍN, *absinnu*, "Furrow"), and Pisces (*anunītu*, the eastern part of Pisces, and ^{mul}SIM.MAḪ, *šinūnūtu*, "Swallow", the western part of Pisces).[331] The

324 For the names of Mars in the Great Star List, see Koch-Westenholz 1995: 198, 200 ll. 237–240; see also 2.2.1.

325 According to Alb B III 2 (Horowitz 2014: 40).

326 According to Alb B I i 12–50 (Horowitz 2014: 33–34).

327 Commentary EAE 53: 1–8, 24–25.

328 Commentary EAE 53: 10–23.

329 Commentary EAE 53: 9, 16–23.

330 For a discussion on the meaning the omens have for the intercalation, see 4.2.2.–4.2.2.1.

331 E.g. Beaulieu-Frahm-Horowitz-Steele 2018: 35 D i 4–7: [MUL *šá ina*] *miḫ-rat* ^{mul}AŠ.GÁN GUB-⸢*zu*⸣ ^{mul}*sí-*⸢*nu*⸣*-nu-tú* / ⸢MUŠEN⸣ MUL KAP$^{\text{ḫi.a}}$ *mut-tap-ri-iš šá kap-pi* ⸢*ra*⸣*-šu-ú* / MUL *šá* EGIR^{meš} ^{mul}AŠ.GÁN GUB-*zu* ^{mul}*a-nun-ni-tum na-a-ru* / [^{mu}]^{l}*sí-nu-nu-tu*$_4$ *ù* ^{mul}*a-nun-ni-tum ina* KUN⸢^{meš}*-šú*⸣*-nu it-gu-ru-ú-ma*, "[The star which] stands opposite the Field (is) the Swallow, / a bird, a star (with) wings, which flies, which has the wing(s). The star which stands behind the Field (is) Anunitu, a

last three groups of stars can be seen rising during the winter and autumn months, and they are located one after the other in the path of the Moon according to MUL.APIN (I iv 33–37, see Hunger-Steele 2019: 70). In other words, they represent the modern zodiacal signs of Taurus, Virgo, Pisces (see 6.1.1. Table 11), all of them located in the same area of the celestial sphere. Therefore, the rationale behind the explanations may be the calendar.

In explanations with planetary conjunctions, sometimes the first celestial body in a protasis (i.e. the subject of the sentence) is equated with the second planet in the explanation (i.e. the object of the sentence), and the second celestial body with the first planet (see Commentary EAE 53: 1–3, 5, 7–8; SIT 6: 1–2, 4–6, 11, 15, 17). This is clear, for instance, from the entries involving the Pleiades, like the following one (SIT 6: 6):

> [DIŠ MUL.MU]L ⌜mulAŠ.GÁN KUR-*ud*⌝ (...) mulUDU.IDIM.GU4.UD d*za-ap-pa* KUR-*ma*
>
> [If the Pleiade]s reach the Field. (...) (It means that) Mercury reaches the Bristle. (K 5713+ obv. 9’, see SIT 6: 6 source A and parallels)

In simple words, the Pleiades and the Field correspond to the Bristle (i.e. Pleiades) and Mercury.[332] Additionally, the protases with verbs of movement such as KUR, *kašādu*, “to reach”, TE, *ṭeḫû*, “to come close”, and NIGIN or NÍGIN, *lamû*, “to surround”, are always equated to verbs of movement in factual explanations (e.g. Commentary EAE 53: 4, 6, 12); the same applies to verbs of state or implying a static nature, such as GUB, *i*/*uzuzzu*, “to stand (still)” (e.g. Commentary EAE 53: 7–8).[333]

Concerning the factual explanations of the Pleiades in both the Commentary EAE 53 and SIT 6, whenever the protases mention the Pleiades (written MUL.MUL), they are always and solely explained as phenomena involving the Bristle (*zappu*), or the planet Mars (*ṣalbatānu*).[334] When *zappu*, “Bristle”, is mentioned in a factual explanation, it means that the protasis has been adjusted for an observation of the Pleiades. That is, for instance, witnessed in omens regarding the intercalation,[335] which are explained as the Pleiades being

---

river. / The Swallow and Anunitu cross each other in their tails”. For the modern identification of the Furrow, Anunitu and the Swallow, which represent Virgo and Pisces, see Reiner-Pingree (1981: 10, 14).

332 The Field stands for Venus on other occasions, see Brown (2000: 59). For all the other associations of star and constellation names with planets, see Brown (2000: 54–63).

333 The only exception seems to be the omen in SIT 6: 17, whose protasis has an action-verb (KUR, *kašādu*) and the factual explanation has a state-verb (GUB, *i*/*uzuzzu*).

334 The meaning of the name *ṣalbatānu*, “Mars” has often been debated. Brown (2000: 70–72) discussed in detail the names of Mars, underlining the persistent association with the god Nergal and the colour red. Livingstone (1986: 52) noted that the epithet *muštabarrû mūtānu*, “who keeps pestilence constant”, from the Star List V R 46 (rev. 42, see Weidner 1915: 52; paralleled in K 6151 obv. 8, see CCP 7.2.u83, https://ccp.yale.edu/P238616 accessed 29.06.2022), derives from the *ṣalbatānu* itself (i.e. ZAL, *muštabarrû*, and BAD, *batānu* or *mutānu*, which is also read UG5, *mūtu*, “death”). See also Lambert (1990; 1996) for studies about the names of Nergal and Mars, and Reynolds (1998) for the names of Mars.

335 Commentary EAE 53: 16–23; SIT 6: 33.

either ahead or westward, or behind or eastward, in relation to Moon. When *ṣalbatānu*, "Mars", is mentioned in factual explanations, it means that the omen has been adjusted for an observation of Mars.

The association between the Pleiades and Mars is crucial for understanding the perception of the Pleiades in Mesopotamian culture. The planet Mars is associated with Nergal/Erra, the god of the netherworld, but also of war and plague.[336] In the Akkadian poem of "Erra and Išum" dating to the first millennium BC, the Sebettu (^{d}IMIN.BI) – with whom the Pleiades are associated in prayers (e.g. see 3.2.1.4.) and omens (e.g. SIT 6: 10) – are the warlike helpers of Erra (see 3.1.2.1.). Thus, divination mirrors mythology: the relationship between Nergal/Erra and the Sebettu is paralleled by analogy in celestial divination. This analogy seems to be the principle which lies behind the choice for the Pleiades (written MUL.MUL) as a substitute name for Mars (Brown 2000: 62–63).[337] And not only that: as shown in one apodosis (and discussed in further subchapters, see 5.3.), such analogy also lies behind the associations of phenomena with predictions. Indeed, in SIT 6: 10, the Pleiades are explicitly equated with the Sebettu, and the prediction of the omen is GU$_7$-*ti* ^{d}IMIN.BI, *ukulti sebetti*, "consumption by the Sebettu", because the Sebettu are devoted to destroying the herd in the poem of "Erra and Išum".

To sum up, the omens about the Pleiades are part of a bigger structure within each tablet, where groups of entries, or factual explanations, are arranged according to the calendar. The omens about the Pleiades concern mainly their alleged appearance, luminosity, and position at the beginning of the Mesopotamian year (i.e. Nisannu, Month I): that is because the most important feature of the Pleiades was as a tool to organise the Mesopotamian calendar (see 4.2.2.). The "impossible" protases, i.e. whose literal meaning is astronomically impossible, in which the Pleiades (written MUL.MUL) are said to move forward or into other celestial bodies, are explained as planetary conjunctions in which a planet is approaching them (written *zappu*, "Bristle), or they are a substitute for *ṣalbatānu*, "Mars", approaching another celestial body. The reason for associating the Pleiades in the protases with Mars in factual explanations is to be sought in an analogy with the poem of "Erra and Išum" (see 3.1.2.1.4.): Mars is perceived as the counterpart of Erra, just as the Pleiades are the counterpart of Sebettu.

5.1.8. The Pleiades in Omens from Other Sections of the Series EAE

Before going further into a summarising discussion on the nature of the Pleiades in celestial omens (see 5.3.–5.3.2.), it is necessary to scrutinise all the other omens in which the Pleiades feature, in other chapters of the series EAE (see 5.1.8.1.–5.1.8.4.), and the so-called astrological and astronomical reports of Assyrian and Babylonian scholars (see 5.2.–5.2.4.). The focus of the following subchapters will be on underlining the features of the Pleiades, their association with the divine heptad, and their role in the intercalation systems, with a look at the structure and the logic behind the arrangement of the omens.

336 For the association between Nergal/Erra and Mars, see von Weiher (1971: 76–83), Brown (2000: 56, 63, 70–74, 256), and Wiggermann (1998–2001: 222–223).

337 For a discussion of symbolic associations as working principles in celestial omens, see A.3.1.5.

5.1.8.1. The Pleiades in the Chapter of Sin, the Moon

The first six tablets of the series EAE deal with the appearance of the Moon and the position of celestial bodies in respect to the Moon. The first section of the tablet EAE 6 is entirely dedicated to the Pleiades; this section of entries mentions the Pleiades standing "within" the Moon (i.e. occultation), standing in the "horns" of the crescent, and "in front" of the Moon:[338]

1. DIŠ 30 *ina* IGI.LÁ-*šú* MUL.MUL *ina* ŠÀ-*šú* GUB LUGAL ŠÚ-*ta*$_5$ DÙ-*uš* [KUR] BI GAM-*su*
2. DIŠ MUL.MUL *ina* ŠÀ-*šú* GUB^{meš} ÚŠ^{meš} GAR^{meš}-*ma* ^{d}IMIN.BI KUR GU$_7^{meš}$
3. DIŠ MUL.MUL *ina* ŠÀ SI^{meš}-*šú* GUB^{meš} ÚŠ^{meš} GAR^{meš}-*ma* ^{d}IMIN.BI GA[M...]
4. DIŠ MUL.MUL *ina* SI ZAG-*šú* (15-*šú*) GUB^{meš} *ina* KUR MAR.TU^{ki} ÚŠ^{meš} *sad-ru*
5. DIŠ MUL.MUL *ina* SI GÙB-*šú* GUB^{meš} (2,30-*šú*) *ina* KUR URI^{ki} ÚŠ^{meš} *sad-ru*
6. DIŠ MUL.MUL *ana* IGI-*šu* GUB^{meš} KUR UR.[BI...]

1. If when the Moon appears, the Pleiades stand inside it, the king will exercise the power, that [country] will submit to him.
2. If the Pleiades stand inside it (i.e. the Moon), a pestilence will take place, the Sebettu will devour the country.
3. If the Pleiades stand inside its (i.e. the Moon's) horns, a pestilence will take place, the Sebettu will subm[it (the country)...].
4. If the Pleiades stand in its (i.e. the Moon's) right horn, a pestilence will regularly occur in Amurru.
5. If the Pleiades stand in its (i.e. the Moon's) left horn, a pestilence will regularly occur in Akkad.
6. If the Pleiades stand in its (i.e. the Moon's) front, the country alto[gether...].

(Verderame 2002: 176–177 § 1–6 sources f, g, and parallels)

In these entries, the apodoses imply an association of the Pleiades with the Sebettu (^{d}IMIN.BI) and pestilence (ÚŠ, *mūtānu*), the latter being a feature of Nergal/Erra (see 5.3.1.). One apodosis concerns the well-being of the kingship (i.e. "the king will exercise the power, that country will submit to him"): the Moon god Sin is the symbol of the royal power,[339] so the prediction of lunar omens often concerns the king himself.

---

338 For a discussion of the areas of the Moon as working principles of omens, see A.2.1.1.3.

339 The role of Sin as the Moon god in the royal cultural and religious background during the first millennium BC is testified by a huge array of references (Hätinen 2021: 150–228). To cite some of them: the *bīt rimki* (Læssøe 1955: 28–89), the Akkadian *šuʾilla* prayers against eclipses (Mayer 1976: 100–102), and the substitute king (*šar pūḫi*) ritual (Parpola 1983: xxii-xxxii). See Hall (1985: 878–899) for an overview of this role earlier than the first millennium BC.

5.1.8.2. The Pleiades in the Chapter of Šamaš, the Sun

In this chapter of the series EAE, the Pleiades, as well as other celestial bodies of the night sky, are not particularly involved because the phenomena of the Sun bring daylight. Only one broken, and unclear omen can be mentioned, about the Sun being "surrounded" by the Pleiades:

> [DIŠ MAN KUR-*ma* MUL.M]UL NÍGIN ŠUB-*tì* KUR-*šú* GÁL-*š*[*i*]
>
> [If the Sun rises and] it is surrounded by [the Pleiad]es, there will b[e] the downfall of his country.
> (K 4025 rev. 6, see van Soldt 1995: 70 I 7 source B).

5.1.8.3. The Pleiades in the Chapter of Adad, the Weather

The incipit of tablet EAE 46 (Gehlken 2012: 80–81) concerns seven divine beings, and Adad who thunders among them. This incipit, also partly quoted in the tablet SIT 6 (see SIT 6: 34), gives a long and notable apodosis:

> If Adad thunders in(to) the middle of the divine Seven and their stars, their seven, are poured over the earth, Adad the locks of the country – (var.) of the sea will mix up / he will smash the tamarisk of [...] the temples of the gods of the highest rank; a city to a city, a household to a household, a brother to his brother, a man to a man, an irrigation canal to a river, / a small branch of an irrigation canal to its irrigation canal, an irrigation ditch to its small branch of an irrigation canal will show hostilities, for 55 years a man will eat flesh of a man, a man will dress in the skin of a man.
> (see 3.1.2.2.2., pp. 63–64)

In this long entry, an analogy between the Pleiades and the divine Seven is implied. Such analogy is commented upon in K 68, a commentary of this specific tablet of EAE (i.e. EAE 46) (Frahm 2011: 147–148; Gehlken 2012: 76):

> [DIŠ... MUL^{meš}-*šú-nu se-bet-t*]*i-šú-nu* ŠUB^{meš}-*ni* ^{d}IŠKUR BÁRA ^{d}IMIN.BI RA-*iṣ* :
> DIŠ 30 *u* MUL$_4$.MUL$_4$ *šit-qu-lu* ^{d}IŠKUR ⸢GÚ-*šú* ŠUB-*ma*⸣
>
> [If... their stars], their [seve]n, fall down, Adad will devastate the sanctuary of the divine Seven. – (var.) If the Moon and the Pleiades are balancing (and) Adad thunders.
> (K 68 rev. 18, see AAT 56; ACh Adad 17)

The commentary entry quoted above shows two variants of the same protasis. The first one is an "impossible" protasis, i.e. it is astronomically impossible in its literal meaning (see p. 131 and fn. 268), whereas the second one is possible. Seven stars in the first protasis symbolise the sanctuary of the divine Seven in the apodosis. A thunder during the balancing of the Moon and the Pleiades is a variant or explanation for the seven stars "falling down" because of Adad.[340] These analogies are likely the same ones behind the incipit of tablet EAE 46: the seven stars of the divine Seven "poured on earth" while Adad thunders, means a thunder while the Pleiades and the Moon are in balance.

It cannot be excluded that the incipit of EAE 46 points further to seven demons named ᵈIMIN.BI, instead of seven gods. Its long apodosis is, indeed, very reminiscent of the consequences of a demonic attack, which subvert the regular cosmic order by threatening and destroying (Wiggermann 1992: 151–152). That is, as mentioned in chapter 3 (see 3.1.2.2.2.), similar to the consequences brought by the demonic heptads in the incantation series UDUG.ḪUL (see 3.1.2.2.). One of these heptads, seven servants of Adad, is also said to bring the occultation of the Moon in the so-called lunar eclipse myth (see 3.1.2.2.1.). The analogy between the incipit of EAE 46 and the lunar eclipse myth is speculative; still, the scribes of the series EAE were also *āšipū*s, exorcists (Koch 2015: 20–21), so they would have been familiar with an incantation series like UDUG.ḪUL. One cannot wholly exclude that many analogies unclear to us might have been obvious, if not implicit, for ancient scholars.

5.1.8.4. The Pleiades in the Chapter of Ištar, the Stars

The content of the chapter of Ištar in the series EAE is quite heterogeneous. The tablets (or subchapters) are fragmentary, and not all of them have yet been reconstructed. The first group of tablets deals with fixed stars (EAE 50–58), followed by tablets about Venus (EAE 59–63), Jupiter (EAE 64–66) and supposedly Mars (EAE 67–68/70) (see 5.1.1.2.).

§ EAE 50–51: Fixed Stars

The identification of the assumed tablets EAE 50 and 51 is hypothetical because their reconstruction is mainly based on commentaries and excerpts. The structure of EAE 50 is unique, if compared to the other tablets of the series, because it is a list of stars related to terrestrial phenomena and omens together (Reiner-Pingree 1981: 28–34; Koch 2015: 175). Besides, the tablet EAE 51 is reconstructed mainly through sources similar to Astrolabe B (Reiner-Pingree 1981: 52–55): the tablet has a full 12-month menology arranged into binary omens, concerning the rising time of stars (Reiner-Pingree 1981: 60–63). This menology preserves 12 sections of texts for each month, similar to a shortened version of the Astrolabe B. Horowitz (2014: 48–51) argued that some parts of the assumed tablet EAE 51 represent a 13-month menology written in Sumerian (Reiner-Pingree 1981: 62–63 X 24–50) and arranged in a different order than Astrolabe B and its parallels. It could, then,

340 For a possible meaning of ŠUB, *maqātu* "to fall", in this context, see Commentary EAE 53: 25, where *napalsuḫu*, "to fall down" (CAD N1 271–272), is equated to *ašābu*, "to sit idly" (CAD A2 386–408). A possibility is to understand *maqātu* as a synonym of *napalsuḫu* in its astronomical meaning related to the intercalation and the balancing between the Moon and the Pleiades, i.e. "to be apart" (see fn. 244).

derive from an earlier, different composition, or tradition, as it includes the unusual 13th intercalary month called Adaru.[341]

In tablet EAE 50, the entries associate the Pleiades with epidemics of the flocks:

2. DIŠ ^{d}IMIN.BI *a-na* GU$_{7}$-*ti bu-lì* :
2a. MUL.MUL *u* ^{mul}MAR UR.BI GUB^{meš} ŠÈG^{meš} *u* ILLU^{meš} GUB^{meš}-*nim-ma* ŠE.GÙN.NU TUR *ina* EN.TE.NA ŠUB-*tì* [*bu-lì*]
2b. ^{mul}UDU.IDIM *ana* MUL.MUL KUR-*ud* (var.: [T]E : KUR-*ud*) ^{d}IMIN.BI KUR GU$_{7}^{meš}$

2. ¶ The Sebettu (are) for the consumption of the herd:
2a. (If) the Pleiades and the Wagon stand together, rains and floods will come (and) the crop will be diminished. In winter, (there will be) an epidemic among the [herd].
2b. (If) a planet reaches (var.: [com]es close or reaches) the Pleiades, the Sebettu will devour the country.
(Reiner-Pingree 1981: 48 VI 2–2b; 44 IV 4–4b; see also 3.1.2.1.2.)

The tablet EAE 51 has entries about the Pleiades' heliacal rising:

¶ In Ajaru (i.e. Month II), the Pleiades, the divine Seve[n, the great gods. If] they rise [at] their [specified ti]me, the great gods will assemble and give good counsel to the country, [good] wind[s will blow. If] they do not rise [at] their specified time, [the great gods will assemble and] give bad counsel to the country, [evil] winds will blow (and) there will be grief among the people.
(see 3.2.5.1., p. 91)

The quoted entries from the tablets EAE 50 and 51 roughly sum up how the Pleiades are perceived in divination as a whole. The entries from tablet EAE 50 focus on planetary conjunctions involving the Pleiades, and suggest one analogy: the Pleiades (written MUL.MUL) are associated with the Sebettu (written ^{d}IMIN.BI) from the poem of "Erra and Išum", who are devoted to the consumption of the herd (GU$_{7}$-*ti bu-lì*, *ukulti būli*) (see 3.1.2.1.2.).[342] Whereas the entry from tablet EAE 51 focuses on the period of the heliacal rising of the Pleiades, the month Ajaru (i.e. Month II). It is known from astral compositions that ancient experts had fixed and specific times in which a star was supposed to rise (e.g. see 4.1.1.).[343] Thus, if the rising of the Pleiades was not at the "specified time"

341 Horowitz (2014: 169–183) re-collected all the fragments edited by Reiner and Pingree and commented on them in comparison with the other known menologies.

342 See the Akkadian epithets of the Sebettu, *šumqutu būl Šakkan*, "to destroy the herd of Šakkan" (Erra I 43, see Cagni 1969: 62 l. 43), and *mūšamqitū būli*, "who consume the herd" (Zappu 1, K 2369+ rev. 26, see Jiménez 2014: 108).

343 The heliacal rising of the Pleiades in Ajaru (i.e. Month II) is also found in the Astrolabe B (I i 1, 12, see Horowitz 2014: 33), and MUL.APIN (I ii 38, see Hunger-Steele 2019: 47).

(UD.DUG$_4$.GA, *adannu*), i.e. they rose later or earlier and diverged from the rules, a negative prediction was expected (see also 4.2.2. § 1b, 5.2.4., 5.3.2.). Additionally, the Pleiades in the entry from EAE 51 do not seem to relate to the tradition of the Sebettu: the fact that they are called "divine Seven, the great gods" (written ^{d}IMIN.BI DINGIR^{meš} GAL^{meš}) who gather in an assembly to deliver counsels (GALGA^{meš}, *milkū*) is more reminiscent of an association with the seven gods who are determiners of fates. This association has already been discussed, as is found in *šuʾilla* prayers, rituals, and various exegetical texts (see 3.2.5.1).

§ EAE 56: A Planet

The tablet EAE 56 was first reconstructed and edited by Largement (1957), and editions of new fragments are available online at the website GKAB.[344] The subject of the tablet is UDU.IDIM, *bibbu*, a "(generic) planet". Only a few entries can be mentioned in this context:

22'. DIŠ ^{mul}UDU.IDIM *ina* ŠÀ MUL.MUL GUB-*iz* ŠUB-*tì* SIG$_5$ MÁŠ.ANŠE : DIŠ ^{mul}UDU.IDIM *ina* MÚRU MUL.MUL GUB-*iz* AN.GE$_6$
23'. DIŠ ^{mul}UDU.IDIM KI MUL.MUL *in-né-tu id-ra-nu* GÁL-*ši* : DIŠ ^{mul}UDU.IDIM *u* MUL.MUL TE^{meš} NIMUR^{meš} [...]
24'. DIŠ ^{mul}UDU.IDIM MUL.MUL KUR-*ud* ^{d}IMIN.BI KUR GU$_7$^{meš}

22'. If a planet stands inside the Pleiades, (there will be) an epidemic of the best (part) of the livestock. – (var.) If a planet stands in(to) the middle of the Pleiades, (there will be) an eclipse.
23'. If a planet goes parallel to the Pleiades, there will be salinity. – (var.) If a planet and the Pleiades come close to each other, salinity [...].
24'. If a planet reaches the Pleiades, the Sebettu will devour the country.
(AO 6450 rev. 22'–24', see Largement 1957: 250, 83^{a}–85^{a})

The apodoses of these entries are rather generic,[345] but the association between the Pleiades and the Sebettu is clearly one of the working principles, as the Sebettu (^{d}IMIN.BI) are mentioned in the last apodosis (AO 6450 rev. 24'). The reference to the salinity of the ground is also found in SIT 6: 11, and in Astrolabe B (see 4.1.2.).

§ EAE 59–61: Venus

The tablets EAE 59–61, edited by Reiner and Pingree (1998), give a few omens about an assumed conjunction between the planet Venus and the Pleiades, a phenomenon that

344 W 22307/83 // W 22307/84, see SpTU 1, 91 (http://oracc.org/cams/gkab/P348512 accessed 05.03.2021) and SpTU 1, 92 (http://oracc.org/cams/gkab/P348513 accessed 05.03.2021). More sources of tablet EAE 56 are listed and edited in Fincke (2015).

345 For a discussion on generic or "stock" apodoses, see 5.3.

happens, in reality, every eight years. In several entries, Venus is said to reach, enter into, and stand in the Pleiades for one or two days:

> DIŠ ^{mul}*dil-bat ina* ŠÀ (var.: *ana*) MUL.MUL UD.2.KAM (var.: UD.1.KAM) DU-*ma* DIB(-*iq*) GALGA KUR MAN-*ni* ^{d}*dil-bat u* ^{d}*ṣal-bat-a-nu* ⌜TE ⌜DIB$^{?}$: ^{d}⌝[...]
>
> If Venus on the 2nd day (var.: 1st day) goes inside (var.: into) the Pleiades and passes by, the counsel of the country will change. (It means that) Venus and Mars come close ... – (var.) the planet [...].
> (Reiner-Pingree 1998: 46, 61; 68, 6)[346]

In the factual explanation of this omen, the Pleiades are a substitute for Mars, as widely witnessed in Commentary EAE 53 and SIT 6 (see 5.1.7.). Therefore, a conjunction between Venus and the Pleiades was actually not intended there.

One commentary (K 137 and parallels) mentions the planet Venus with a "beard" (SU$_6$, *ziqnu*), a phenomenon that has been already discussed in 3.1.1.3.1.:

> [If] Venus in Nisannu (i.e. Month I) rises (and) has a beard, the gods [abun]dance / in the country they will pour out, (there will be) prosperity of the harvest, expansion of Nisaba (i.e. grain), prolongation / of the days of the prince. (It means that) the 12 Pleiades (in) front$^{?}$, ditto (i.e. the Pleiades) stand at her side. / (It means) that in the east (lit. sunrise) she is re[d] and she shines brightly.
> (see 3.1.1.3.1., p. 51)

20. DIŠ ^{d}MIN *ina* ^{iti}BÁR SU$_6$ *zaq-na-at* UN^{meš} KUR NITA^{meš} Ù.TU^{meš}
21. *ina* ŠÀ MU BI KI.LAM TUR SU$_6$ *zaq-nu* SU$_6$ *na-ba-ṭu ba-ʾa-lat né-bat*
22. MUL.MUL *ina* ^{d}UTU.ŠÚ.A *ana* IGI-*šá* KIMIN *ina* Á^{meš}-*šá* GUB^{meš}-*ma*
23. DIŠ ^{d}MIN *ina* ^{iti}GU$_4$ MIN ŠÈG^{meš} *u* ILLU KUD^{meš} MUL.MUL MUL$_4$^{meš}-*šú* 12
24. DIŠ ^{d}MIN *ina* ^{iti}SIG$_4$ MIN SU.GU$_7$ *ina* KUR GÁL-*ši* : *ina* KUR GÁL^{meš} MUL.MUL MUL$_4$^{meš}-*šú* 10

20. If the ditto-god (i.e. Venus) in Nisannu (i.e. Month I) has a beard, the people of the country will give birth to male children,
21. inside (i.e. in the middle) of the year the market rate will diminish. SU$_6$ (means) "beard", SU$_6$ (means) "to be bright", she is bright, (and) she shines.
22. (It means that) the Pleiades stand in the west (lit. sunset) at her front, ditto (i.e. the Pleiades stand) at her sides.
23. If the ditto-god (i.e. Venus) in Ajaru (i.e. Month II) ditto (i.e. has a beard), rain and flood will stop. (It means that) the stars of the Pleiades (are) 12.

346 See the other parallels in Reiner-Pingree (1998: 174, 39 and 44; 194, 3).

24. If the ditto-god (i.e. Venus) in Simanu (i.e. Month III) ditto (i.e. has a beard), there will be famine in the country – (var.) will repeatedly be in the country. (It means that) the stars of the Pleiades (are) 10.

(Reiner-Pingree 1998: 149, 13–15 source A 20–24)

These entries are relevant on a metaphorical level and contribute to improving our understanding of how the cosmos was perceived in Mesopotamia. First, Venus, called *dilbat*, is the astral representation of Inanna/Ištar, the goddess of love and war. In literary texts and iconography, she is sometimes described with an androgynous appearance or depicted as a woman with a beard. She was considered male when she appeared in the sky as an evening star (Groneberg 1986); hence, Venus, just as her anthropomorphic counterpart Inanna/Ištar, can have a "beard". Therefore, the entries quoted above testify to personification in omens and represent the third perspective on celestial bodies as defined by Rochberg (see 2.2.3.1. § 3). Second, the protases of the entries are explained as the Pleiades having 12 stars in Ajaru and 10 stars in Simanu (obv. 23–24). That is a reference to the difference between a normal year and an intercalary year according to intercalation rules (see 4.2.2. § 2c): when the heliacal rising of the Pleiades took place in Ajaru (i.e. Month II), that was considered a normal year, whereas if it took place in Simanu (i.e. Month III), that was an intercalary year (see 4.2.2. § 1b). The numbers 12 and 10 likely refer to the number of times the Pleiades are seen close to the Moon within a normal year and an intercalary year. It is interesting how the scribes of omens and commentaries reasoned here: they applied a symbolism shared by the anthropomorphic and astral realms but also associated "impossible" protases, i.e. protases whose literal meaning is metaphorical, like a bearded Venus, with important calendrical events, i.e. intercalation, by schematising months and numbers. Ultimately, it is difficult to assess why the scribes explained that the beard of Venus is the Pleiades; we could seek a reason in the visual resemblance of the Pleiades – meant as a group of many bright stars – to a beard "made of light" in the night sky. A less likely option would be, as previously discussed in 3.1.1.3.1. that the explanation references Inanna as Venus and her seven torches, or weapons, in the Sumerian myth "Lugalbanda in the Wilderness".

5.2. The Pleiades in Omens Quoted in Reports of Assyrian and Babylonian Scholars

The letters and the so-called astrological and astronomical reports of Assyrian and Babylonian scholars are considered a priceless insight into the mentality of ancient scholars and the royal court during the first half of the first millennium BC. Scholars communicated in order to warn the king, the palace, and the country of the predictions of negative ominous signs, and to protect them with precautionary measures. Hence, letters and reports give proof of what phenomena were actually observed because they refer to real situations in the sky. They also testify how celestial observations were interpreted, quoting omens from the series EAE, MUL.APIN, and other compositions (Fincke 2016: 127–128). The great majority of the letters and reports to the Neo-Assyrian kings come from Nineveh, written during the eighth and the seventh century BC (Hunger 1992: XV).

While the letters addressed the king personally and included heterogeneous information, the reports did not explicitly address the king but only quoted the omens referring to what they observed (Hunger-Pingree 1999: 23–26). More specifically, the experts chose freely, to some extent, which observation to report. The series EAE was used as the primary source for celestial divination, and the experts had to find the appropriate omen, or omens, for their observation among all the sources at their disposal. They also needed to check whether an ominous sign had a negative or positive meaning, and to whom it referred to, whether to their own country or to another one. Since an ominous sign in Mesopotamia is considered a warning, more than an incontrovertible act, the experts sometimes proposed solutions to avert the potential danger of an ominous unpropitious sign. For instance, in the letters, the measures suggested were the *namburbi*s, the substitute king ritual, the *bīt rimki*, or the *bīt salā᾽ mê*, of which *šu᾽illa* prayers to celestial bodies were part (see 3.2.–3.2.5.). The reports usually contain omens and sometimes explanations of protases, apodoses, or individual words. They also have statements about the observation of phenomena and their predictions, sometimes with personal additions, such as special requests from the scholar to the king (e.g. SAA 8, 244 and 474).

The letters and reports to the Neo-Assyrian kings were first published by Thompson (1900). The letters were collected and studied by Parpola (1983; 1993 [SAA 10]), and the reports by Hunger (1992 [SAA 8]). Only one letter (SAA 10, 63 rev. 5) mentions the Pleiades only in passing, whereas the reports are richer in mentions of the Pleiades. The omens commented on by the reports represent the instances in which the Pleiades were particularly visible, or observed, in the night sky. The reports that attest to observations of the Pleiades are grouped in the next subchapters (see 5.2.1.–5.2.4.), according to the main subjects of the omens excerpted from the series EAE. The reports also shed light on how ancient scholars reasoned to interpret the observations and how they perceived celestial phenomena related to the Pleiades.

5.2.1. The Pleiades and the Moon

The Moon is the closest observable celestial body from earth, and when the weather conditions were good, the position of the stars was checked in relation to the Moon. Indeed, omens about the Pleiades and the Moon (see 5.1.8.1.) are often found in reports, mostly because the position of the Pleiades and the Moon was a means to establish the intercalation, and was relevant to determine the validity of any celestial sign (see 4.2.2.).

When the Pleiades were seen standing in the halo of the Moon, the following omen was once quoted:

> DIŠ 30 TÙR NÍGIN-*ma* MUL.MUL *ina* ŠÀ-*šú* GUB^{meš} *ina* MU BI MUNUS^{meš} NITA^{meš} Ù.TU^{meš} LUGAL KUR-*su* KÚR-*šú nu-šur-re-e* ŠE-*im*
>
> If the Moon is surrounded by a halo and the Pleiades stand inside it, in that year the women will give birth to male children, the king's country will defect from him, (there will be a) diminution of barley.
> (SAA 8, 5 rev. 2–3; SAA 8, 273 rev. 3–5; SAA 8, 529 obv. 1–5; SAA 8, 531 obv. 1–3)

The quoted omen is from tablet EAE 8 (ND 5495 obv. i 27–30),[347] which is part of the Sin chapter of the series EAE. The apodosis is divided into three sentences about pregnancy, kingship, and harvest, that recall analogies between the Moon and the menstrual cycle, the royal power, the cattle, and the fertility (Hall 1985: 883–884; Hätinen 2021: 229–247). Already the birth incantations dating to the Old Babylonian period (ca. 2000–1500 BC), such as the composition "A Cow of Sin",[348] pointed out the importance of the role of the Moon god for women and fertility. In the glyptic dating to the Middle and Neo-Assyrian period (ca. 1400–612 BC), the Pleiades, the Moon, and the omega symbol are often depicted together, to symbolise fertility (Steinert 2017: 214–215). For example, a seal from Tyre dating to the late second millennium BC (Steinert 2017: 215 fig. 13) portrays a woman in front of a statue of the weather god:[349] the woman has the omega symbol on top of her head, and she is surrounded by the lunar crescent, seven dots representing the Pleiades, and the eight-pointed star representing the goddess Ištar/Venus. Though it cannot be excluded that the Pleiades, the Moon, and Ištar could only represent a nocturnal setting depicted in the seal (e.g. see 2.3.3.), the presence of the omega symbol in this context is, indeed, noteworthy.[350]

The relationship between the Moon god and the royal power is the best attested (Hall 1985: 884–886; Hätinen 2021: 196–228). Several omens show such analogy, like the following one, quoted in a report, about the Pleiades standing "on top of the Moon" (*ana* UGU 30, *ana muḫḫi Sin*):

> DIŠ MUL.MUL *ana* UGU 30 DAR-*ma* GUB-*iz* / LUGAL ŠÚ-*tú* DÙ-*uš* KUR-*su* DAGAL-*iš*
>
> If the Pleiades are elongated towards the top of the Moon and stand still, / the king will exercise the power, his country will expand.
> (SAA 8, 296 obv. 1–2; SAA 8, 455 obv. 1–3; SAA 8, 548 obv. 3–4)

In the protasis of this omen, the logogram DAR, *šatāḫu*, "to be elongated", refers to a specific position of the Pleiades, i.e. ahead of the Moon or westward, as explained in Commentary EAE 53: 20. This position stands for an average year, i.e. a year when the intercalation in the calendar is not needed because the Pleiades and the Moon set where and when expected (see 4.2.2. § 2b). As for the meaning of the apodosis of the omen mentioned above, the prediction is positive for the king and his military expansion. The Pleiades might be implicitly perceived as the representation of the Sebettu, shaped in their warlike aspect by the royal Assyrian cultural heritage: the same heritage according to which the Sebettu accompanied the king for a successful war campaign in royal inscriptions (see 3.2.7.).

347 The edition of this fragment is in CTN 4, 4, and it can be found online at the website GKAB (http://oracc.org/cams/gkab/P363419 accessed 05.03.2021).

348 For the edition, see Veldhuis (1991).

349 "Although the weather god does not play a role in human birth in Mesopotamian mythology, he is, by sending rain, responsible for the fertility of fields and vegetation" (Steinert 2017: 215).

350 For the association between the omega symbol and a star called Womb ([mul]ŠÀ.TUR.RA.ŠÈ, see Reiner-Pingree 1981: 42 III 23–24b), see also Niederreiter (2008: 64).

The report SAA 8, 351 fully describes the balancing (*šitqultu*) between the Moon and the Pleiades, the most critical observation of the Moon and the Pleiades together. In this report, the Pleiades were again in the expected position, according to the rules of intercalation (see 4.2.2. § 2b), i.e. they are ahead of the Moon or westward. The ancient expert found three fitting omens for this situation:

Transliteration

Obverse
1. DIŠ MUL.MUL *ana* ŠÀ [30 KU$_4$]
2. KUR DIŠ-*niš* ZÁḪ ITI ⸢*uš-ta*⸣-[x x]
3. ELAM.MA^{ki} ^{lú}KÚR [x x]
4. DIŠ MUL.MUL *ana* ŠÀ 30 KU$_4$$^{me?}$-[*ma*]
5. *ana* IM.SI.SÁ È^{meš}-*ni* [x]
6. ŠÀ KUR URI^{ki} DÙG-*ab* LUGAL URI[^{ki}]
7. KALAG.GA-*ma* GABA.RI NU TUK-*ši*

Lower edge
8. IM.SI.SÁ DU-*ma*
9. *ul-tu* 30 *ana* ŠÀ MUL.MUL

Reverse
1. *i-ru-bu* IM.SI.SÁ DU-*ak*
2. EN.NUN.UD.ZAL.LA KUR ELAM.MA^{ki}
3. ḪUL *šá* ^{lú}KÚR *šu-ú*
4. DIŠ MUL.MUL *šá-ti-iḫ*
5. A.ŠÀ A.GÀR 1 GÚN ÍL
6. *ina šit-qul-ti* 30 *i*-[*pan-nu-ma*]
7. EN LUGAL^{meš} *lu-ú* ⸢*da*⸣-[*ri*]
8. *šá* ^{m}*a-šá-ri*-[*du*] ⸢*qa*⸣-[*at-nu*]

Translation

Obverse
1. If the Pleiades [enter] into [the Moon],
2. the country altogether will perish, a month ... [...],
3. the enemy [...] Elam.
4. If the Pleiades enter into the Moon [and]
5. come out towards the north [...],
6. Akkad will become happy, the king of Akkad will become
7. strong and will have no rival.

Lower edge
8. The north wind blows.
9. Since the Moon entered into the Pleiades,

Reverse
1. the north wind is blowing.
2. The morning watch (is for) Elam,
3. this is bad for the enemy.
4. If the Pleiades are elongated,
5. the field will produce the yield.
6. (It means that) at the balancing with the Moon t[hey go ahead].
7. May the lord of kings be eve[r lasting]!
8. From Ašare[du] the y[ounger].

(SAA 8, 351)

The first omen (obv. 1–3) is not well preserved, yet it mentions the land of Elam, the land associated with the Pleiades in the Great Star List (see 4.1.3.3.). The observation happened during the morning, and a "morning watch"[351] is also associated with Elam (rev. 2–3) in the Great Star List (Koch-Westenholz 1995: 202 l. 292), and tablet SIT 4 (K 3123 rev. 13', see App. B § 24). The expert chose the second omen (obv. 4–7, see also SAA 8, 443) because the north wind was blowing during the observation (obv. 5 and 8–rev. 1). According to the Mesopotamian conception, the north wind is identified with the land of Akkad, the land where the king dwells (see 5.1.4. commentary to the text of entries 46–60, pp. 167–168; K 3123 rev. 17', see App. B § 24). The last omen and its explanation (rev. 4–6) are quoted from Commentary EAE 53: 20. On the whole, the report SAA 8, 351 is an example of how ancient scholars reasoned to interpret the signs, and how sky and earth were entangled in a dense network of analogies, according to the Mesopotamian view of the cosmos. Since when the Pleiades are observed they are in the expected position, regularity means a positive sign for the king (i.e. Akkad) and a negative one for the enemy (i.e. Elam).

5.2.2. The Pleiades as Mars

The name MUL.MUL, "Pleiades", is not connected solely to the observation of Pleiades: it is likely to find, in reports or commentaries, explanations that imply that MUL.MUL is used as a substitute name for the planet Mars (*ṣalbatānu*). Indeed, the factual explanations of omens usually explain that fixed stars "moving towards" each other are meant as planetary conjunctions (see 5.1.7.). For instance, an explanation of the Pleiades as Mars occurs in SAA 8, 376, where an omen about the Pleiades in the halo of the Moon is given:

6. DIŠ 30 TÙR NÍGIN-*ma* MUL.MUL *ina* ŠÀ-*šú* GUB
7. *ina* ^{iti}GU$_4$ ^{munus}PEŠ$_4$^{meš} NITA^{me} *ul-la-da*
8. LUGAL ŠÚ* KUR-*su* KÚR-*šú-ma* e ⌜x⌝ [x x x]
9. MUL.MUL ^{d}*ṣal-bat-a-nu*

351 The day and night are each divided into three watches. The "watch of the night" (EN.NUN GI$_6$) means the entire night, which is again divided into three watches. The "watch of the day" (EN.NUN *u*$_4$*-mi*) is the third watch of the night, and it refers to the entire length of the day(light) (Hunger-Steele 2019: 200).

6. If the Moon is surrounded by a halo and the Pleiades stand inside it,
7. in Ajaru (i.e. Month II) pregnant women will give birth to male children,
8. the king of the world's country will defect from him and ... ...[...].
9. (It means that) the Pleiades (are) Mars.

(SAA 8, 376 obv. 6–9)

In this report, the expert chose an omen about the Pleiades – very similar to the one already discussed in 5.2.1. (p. 234) – and adapted it to the situation he observed in the sky at night, i.e. Mars being close to the Moon, by adding a factual explanation.

On several other occasions, the Pleiades are said to be substitutes for Mars in reports and letters, specifically when Mars is observed close to the Moon.[352] The reason for this choice can be sought in an association of ideas, which was probably obvious to ancient experts. The planet Mars is associated with the god Nergal/Erra:[353] Erra and the Sebettu, protagonists of the poem of "Erra and Išum" (see 3.1.2.1.), are associated with Mars and the Pleiades, respectively (see 5.1.7.); thus, the Pleiades are substitutes for Mars by analogy.

5.2.3. The Pleiades and Venus

There are only a few cases in which the Pleiades are observed and reported together with Venus, and they are all similar. For instance, in SAA 8, 282, the prediction for such a celestial phenomenon is the flood:

1. DIŠ *ina* SAG MU MUL.MUL
2. *ina* GÙB$^{?}$ ^{mul}*dil-bat* GUB
3. ^{lú}KÚR : ILLU BURU$_{14}$ *ú-tál-lal*

1. If at the beginning of the year the Pleiades
2. stand on the left$^{?}$ of Venus,
3. the enemy – (var.) a flood will disrupt the harvest.

(SAA 8, 282 rev. 1–3)

This report refers to the beginning of year (Nisannu, i.e. Month I), when the Moon sets before the Pleiades, among which Venus appears.

In SAA 8, 461, Venus is said to stand in the Bull of Heaven, whose prediction is the flood as well:

1. MUL.MUL *a-na* UD ŠEG$_{6}$.GÁ ⸢UD ŠEG$_{6}$⸣.GÁ [x x x]
2. ^{mul}*dil-bat ina* MUL.MUL ⸢GUB⸣-[*ma*]
3. [DIŠ] ^{d}*iš-tar* AGA KÙ.BABBAR *ap-rat* IL[LU] *ku-li-li* ⸢DU⸣-[*kam*]

352 SAA 8, 50: obv. 1–9; SAA 8, 72 obv. 1–6, rev. 2; SAA 8, 491 rev. 3–4; SAA 10, 63 rev. 5–8.

353 The analogy between Mars and Nergal/Erra also emerges more or less explicitly from the reports SAA 8, 114 and 502.

4. [ILLU] *ku-li-li* ILLU *gap-*[*šu*]
5. [^{mul}]*dil-bat ina* ^{mul}GU$_4$.AN.NA GUB-*ma*

1. The Pleiades (are) for a boiling day; boiling day (means) [...].
2. (It means that) Venus stands in the Pleiades.
3. [If] Ištar wears a silver crown, a flo[od] of dragonflies will c[ome].
4. [A flood] of dragonflies (means) a massi[ve] flood.
5. (It means that) Venus stands in the Bull of Heaven (i.e. Taurus).

(SAA 8, 461 obv. 1–5)

In both the above-mentioned reports, the expert chose what more closely related to what he observed at night: that is, an omen about Venus and the Pleiades, and one about Venus and the Bull of Heaven. The latter is the name of the zodiacal sign Taurus, in which the Pleiades are a cluster of stars (see 2.1. Figure 1). The predictions regard a flood which can be destructive (SAA 8, 282 rev. 3) but it is also necessary when rain becomes scarce. For instance, in SAA 8, 461 (obv. 7–rev. 5), the scribe suggests performing a ritual against drought before Adad. In order to underline the importance of the ritual, the scribe even makes a quotation from the Akkadian myth *Atra-ḫasīs* (II ii 11'–13', 16', 19', see Lambert-Millard 1968: 74). According to Lambert and Millard (1968: 28), quotes like this are examples of how ancient scholars heavily refer to the principles proposed by epics and mythology: that is, everything that has happened in the epic might be repeated in history. Cases like this one testify to the relevance of intertextuality applied to Mesopotamian literacy.

5.2.4. The First Visibility of the Pleiades

The omens about the first visibility of heliacal rising of stars are based on the straightforward assumption that everything happening during the scheduled time is positive, while any divergence from the rule is negative. In this respect, two reports quote a shorter version of a well-known omen about the heliacal rising of the Pleiades from EAE 51, already discussed in 5.1.8.4. (§ EAE 50–51):

> [DIŠ *ina*] ^{iti}GU$_4$ MUL.MUL ^{d}[IMIN.BI] / [DINGIR]^{meš} GAL^{meš} *šum-ma ina a-dan-ni-*[*šú-nu*] / [KUR^{meš}-*ni*] DINGIR^{meš} GAL^{meš} NIGIN^{meš}-[*m*]*a* / [GALGA KUR *ana*] SIG$_5$-*tì* GALGA^{meš} / [IM^{meš}] DÙG.GA^{meš} DU^{meš}
>
> [If in] Ajaru (i.e. Month II), the Pleiades, the divine [Seven], / the great [god]s, [rise] at [their] specified time, / the great gods will assemble [a]nd / give a good [counsel to the country], / good [winds] will blow.
> (SAA 8, 275 obv. 6–rev. 3; SAA 8, 507 rev. 1–4)

5.3. Conclusion for the Nature of the Pleiades in Celestial Divination

The role of the Pleiades in the celestial omens examined in this chapter suggests the following conclusions. In total, 58 omens involve the Pleiades; the protases mostly deal with astronomically impossible phenomena, or phenomena described through metaphor, e.g. "the Pleiades move into other stars", or "the Pleiades are cut". In commentaries, these phenomena are sometimes described as planetary conjunctions (see 5.1.7.); in these cases, the Pleiades, written MUL.MUL in protases, can stand either for their Akkadian name (*zappu*, "Bristle") or for the planet Mars (*ṣalbatānu*), the astral form of Nergal/Erra, as shown in the following two examples (see also A.3.1.5., A.4.1.):

[DIŠ MUL.MU]L ⸢^{mul}AŠ.GÁN KUR-*ud*⸣ (...) ^{mul}UDU.IDIM.GU$_4$.UD ^{d}*za-ap-pa* KUR-*ma*

[If the Pleiade]s reach the Field. (...) (It means that) Mercury reaches the Bristle.
(K 5713+ obv. 9', see SIT 6: 6 source A and parallels)

Pleiades = Bristle; Field = Mercury

[DIŠ MU]L.MUL ^{mul}ŠUDUN KUR-*ud* ⸢^{d}⸣UDU.IDIM.SAG.UŠ ^{d}*ṣal-bat-a-nu* KUR-*ma*

[If the Plei]ades reach the Yoke. (It means that) Saturn will reach Mars.
(K 3558 obv. 1', see Commentary EAE 53: 1 source A and parallels)

Pleiades = Mars; Yoke = Saturn

The role of the Pleiades as a tool to organise the calendar – as also witnessed in astral compositions (see 4.3.) – is the major topics of omens. To accomplish intercalation (see 4.2.2.), the balancing of the Moon and the Pleiades was checked at the beginning of the year. Therefore, the protases about the position, the shape, and the luminosity of the Pleiades at the beginning of the year often stand for a specific position of the Pleiades concerning the Moon, as shown by the example given below:

If at the beginning of the year the Pleiades are cut. (It means that) at the balancing with the Moon they lag behind, the Pleiades are behind and the [Moon is ahead$^{?}$, the Brist]le (is) to the east (i.e. towards sunrise), (the Moon is) to the west (i.e. towards sunset).
(see 4.2.2. § 2b, pp. 119–120; Commentary EAE 53: 19)

The rising of the Pleiades in Ajaru (i.e. Month II) also had relevance because it marked the beginning of the agricultural season, a critical time for predicting the outcome of the year and securing a positive response from the gods, as shown in the following omen, quoted in two reports:

[If in] Ajaru (i.e. Month II), the Pleiades, the divine [Seven], the great [god]s, [rise] at [their] specified time, the great gods will assemble [a]nd give a good [counsel to the country], good [winds] will blow.
(See 5.2.4., p. 239)

If one considers the context in which the signs of the Pleiades are mentioned, the latter were considered "ominous" during a specific time of the year, which goes from the beginning of the Mesopotamian year (i.e. Nisannu, Month I) until Ajaru, the month of the ploughing (i.e. Month II) (see 4.1.2.). The role of the Pleiades in that period emerges from the arrangement of the reconstructed tablets EAE 52, Commentary EAE 53, and SIT 6 where the sequence of the groups of entries mirrors the calendar (see A.2.1.1.1.).

If the protases are astronomically impossible, the apodoses have events which can happen in real life, like a good harvest or famine, a good or bad year of business for barley, or the well-being of the flock. These are called "stock apodoses", very recurrent, short, and straightforward predictions (Koch 2019: 232), as shown in the following omen:

DIŠ MUL.MUL ^{mul}ŠUDUN KUR-*ud ina* MU BI GÁN.BA TUR

If the Pleiades reach the Yoke, in that year the market rate will diminish.
(Commentary EAE 53: 1, SIT 6: 2)

In addition to such "stock apodoses", which are mainly about harvesting and its outcome, there is a more limited number of apodoses (seven out of 58 omens) dealing with the well-being of the royal court, the king, and the land, like in the example given below:

If when the Moon appears, the Pleiades stand inside it, the king will exercise the power, that [country] will submit to him.
(see 5.1.8.1., p. 227, 1)

The apodoses in the last two examples given above belong to two distinct realms, and roughly sum up the topic of the predictions related to the Pleiades: farming and harvesting on the one hand, and rulership on the other hand. Thus, in celestial omens, the role of the Pleiades is defined not only by their role in establishing the calendar, the agricultural season, and all that goes with it (e.g. predictions of epidemics or trade trends, etc.), but also by something related to the reign's decisional power. An interpretation of the latter type of apodosis is to be sought by analysing the semiotic link between the protases (i.e. ominous signs) and the apodoses (i.e. predictions) of omens (see A.3.). In the case of the Pleiades, the semiotic link between a protasis and an apodosis has been proved to be often an association between the Pleiades and a divine heptad (see A.3.1.5., A.4.1.). Sometimes this association is explicit, as in the following omen:

If the Pleiades stand inside it (i.e. the Moon), a pestilence will take place, the Sebettu will devour the country.
(see 5.1.8.1., p. 227, 2)

Associations involving the Pleiades have already been found in other literary genres of Mesopotamia: for instance, the Pleiades are associated with the Sebettu, or with seven great gods, i.e. the divine Seven, in myths, prayers (see 3.), and astral compositions (see 4.). When comparing the perception of the Pleiades in celestial omens with other textual genres, the evidence points towards two traditions (Verderame 2016: 115): the following subchapters (see 5.3.1.–5.3.2.) sum up the semiotic link between protases and apodoses, and the content of the omens about the Pleiades in two macro-categories mirroring their perception, i.e. the Pleiades as the Sebettu, or as the divine Seven.

5.3.1. The Pleiades Perceived as the Sebettu (^{d}IMIN.BI)

Specific seasonal activities affected in apodoses	**Generic events affecting seasonal activities**
	Famine (SU.GU$_7$, *ḫušaḫḫu*)
Harvest (BURU$_{14}$, *ebūru*)	Plague (*ākilu*)
Salinity (*idrānu*)	Consumption (GU$_7$, *ukultu*)
Herd (MÁŠ.ANSE, *būlu*)	Pestilence (ÚŠ^{meš}, *mūtānu*)
Market rate (KI.LAM, *maḫīru*)	Downfall, epidemic (ŠUB, *miqittu*)
Yield (GUN, *biltu*)	Voracious hunger (*karurtu*)
Barley (ŠE, *šeʾu*)	Heat (*ummu*) and blaze (*umšu*)
Crop (ŠE.GÙN.NU, *šegunû*)	Stormy weather
Date palm (^{giš}GIŠIMMAR, *gišimmaru*)	(^{d}IŠKUR RA, *Adad iraḫḫiṣ*, "Adad will devastate")

Table 9. The topics of the apodoses related to the Pleiades perceived as the Sebettu.

The ominous signs of the Pleiades usually relate to predictions affecting the herd, the ploughing, the harvest of the grain and the date palm as major crops, and consequently the business and prices related to them (see Table 9). The reason for that is the calendar, as already implied in the astral compositions discussed in chapter 4 (see Table 3): the Pleiades rise heliacally after the beginning of the Mesopotamian year, and this fact marks the beginning of the agricultural season. Therefore, the activities affected by the predictions refer mainly to agriculture, farming, and their outcome. The reference to salinity (*idrānu*) is also important for the agricultural activities performed before the summer, i.e. at the beginning of the Mesopotamian year. The salinity of the ground influenced the success of the ploughing, a statement also remarked in Astrolabe B and in the Sumerian composition "Farmer's instructions" (see 4.1.2.).

The outcome of harvesting, farming, and marketing is usually influenced by epidemics, famine, or pestilences considered to be brought by divine anger. The "consumption" (*ukultu*) is one of the most generic yet frequent predictions for the ominous signs of the Pleiades. Such predictions imply an association of the Pleiades with the Sebettu (^{d}IMIN.BI) (see A.3.1.5., A.4.1.). For instance, the following omen mentions a "consumption" brought about by the Sebettu (GU$_7$-*ti* ^{d}IMIN.BI, *ukulti* ^{d}*sebetti*):

> If the Stag reaches the Pleiades, (there will be) consumption by the Sebettu.
> (see 4.1.3.4., p. 107)

The analogy between the Pleiades, the Sebettu, and the consumption of the herd (*ukulti būli*) is also corroborated by one entry of the assumed tablet EAE 50 (see 5.1.8.4. § EAE 50–51):

> ¶ The Sebettu (are) for the consumption of the herd: (If) the Pleiades and the Wagon stand together, rains and floods will come (and) the crop will be diminished. In winter, (there will be) an epidemic among the [herd]. (If) a planet reaches (var.: [com]es close or reaches) the Pleiades, the Sebettu will devour the country.
> (see 5.1.8.4. § EAE 50–51, p. 230)

Frahm (2011: 148–149) considers the assumed tablet EAE 50 a commentary. Following his interpretation, the entries quoted above have one omen (i.e. "If the Pleiades and the Wagon[354] stand together"), associated with its symbolic meaning (i.e. "The Sebettu are for the consumption of herd"), and one exemplary omen to show the link between them (i.e. "If a planet reaches the Pleiades, the Sebettu will devour the country"). Consequently, the entries would associate the Sebettu (ᵈIMIN.BI) with the Pleiades (MUL.MUL), and both with the prediction for the "consumption of the herd" (*ukulti būli*), and the "epidemic among the herd" (*miqitti būli*) but also stormy weather (*zunnū u mīlū*, "rains and floods") and problems with the crop (*šegunû*). The semiotic link between the protases and the apodoses is rendered by the analogy of the Sebettu with the Pleiades. It is possible to derive that analogy through intertextuality: in the poem of "Erra and Išum" (see 3.1.2.1.) dating to the first millennium BC, the Sebettu are devoted to destroying the herd of Šakkan, the pastoral god (see 3.1.2.1.2.). Therefore, the assumed EAE 50 succeeds in "summarising" that the Pleiades can portend the destruction of the herd, as well as their divine counterparts do in "Erra and Išum".

A further example of intertextuality which attests the analogy between the Pleiades, the Sebettu, and the herd can be found in a hemerology[355] for the 18th day of Nisannu (i.e. Month I) dating to the Neo-Assyrian period (ca. 911–612 BC):

> NÍG.GIG ᵈGÌR *i-ba-ar-ma* / É-*su nam-maš-šu-u* ZÁḪ / KIMIN ŠUK-*su ana* MUL.MUL GAR-*ma ma-ḫir*
>
> If he hunts what is sacred to Šakkan, / his house and his herd will be destroyed. / Ditto (i.e. if he hunts what is sacred to Šakkan), he should place his food offering for the Pleiades, and it will be acceptable.
> (Livingstone 2013: 112 obv. ii 43–45)

---

354 In this occasion one should assume that ᵐᵘˡMAR(.GÍD.DA), *ereqqu*, "Wagon", stands for a planet (ᵐᵘˡUDU.IDIM, *bibbu*). However, the relationship between the planet and the prediction for the herd is consistent with one explanation of ᵐᵘˡUDU.IDIM as *mušmīt būli*, "killer of cattle" (K 6151 obv. 7, see CCP 7.2.u83, https://ccp.yale.edu/P238616 accessed 29.06.2022).

355 In Assyriology, the term "hemerology" refers to compositions in which favourable or unfavourable days and months of the year are listed (Labat 1972–1975: 317).

If someone's livestock was negatively affected, the owner of the herd should offer bread to the Pleiades to please them, because they are the astral counterpart of the Sebettu, who destroy the herd of Šakkan, the pastoral god, in mythology. It is interesting that here the Pleiades (written MUL.MUL) are mentioned, and not the Sebettu (written dIMIN.BI): it is possible that the offering was supposed to take place during the night. Ultimately, there is little doubt that the hemerology of Nisannu shares the same *plateau* of information with mythology and the series EAE.

The association between the Pleiades in protases, the "famine" (SU.GU$_7$, *ḫušaḫḫu*) in apodoses, and Mars in factual explanations is also relevant. The god Nergal/Erra is *bēl šibṭi u šaggašti*, "lord of plague and carnage" (CAD Š2 387–388; CAD Š1 69–70 c). A reference to the famine as a plague occurs, for instance, in a gloss in the following terrestrial omen from the series *Šumma ālu* tablet 80:

> DIŠ UR.GI$_7$ *ana* UDU ⌜*ú*⌝-[*ḫa*]-*an-ni-iṣ ú-kul-ti* d*èr-ra ina* KUR GÁ[L-*ši*]
> ⌜SU.GU$_7$⌝
>
> If a dog r[ub]s himself against a sheep, there will b[e] consumption – (that is) SU.GU$_7$ (i.e. famine) – by Erra in the country.
> (CT 39, 26–27 obv. 8–8a)

Roberts (1971) pointed out textual evidence according to which Erra should be properly considered the god of famine, other than the god of plague. Though this association is blurred, famine and consumption are extensively associated with Nergal/Erra, as well as with the Pleiades. Jeyes (1980: 108–109; 1989: 105–106, 121) also noted that the association between Nergal/Erra and the pestilence (*mūtānu*) already existed in the early Old Babylonian period (ca. 2000–1900 BC), as shown in a letter to the king Lipit-Ištar given below:

> *mu-ta-nu* / *a-nu-um-ma* / *i-na a-li-im* / *i-ba-aš-šu-ú* / *mu-ta-a-nu ú-la*! *ša* dNÈ.IRI$_{11}$.GAL
>
> Now / there is / a pestilence / in the city, / (but) the pestilence is not of Nergal.
> (CT 29, 1 obv. 5–10; Jeyes 1989: 121, 11')

In ancient commentaries and reports from Assyrian and Babylonian scholars, the association between the Pleiades, the Sebettu, Mars, and Nergal/Erra seems to be based on the same analogy. Whenever an ancient expert needed to find an omen for an observation of Mars, he would either choose omens about Mars (*ṣalbatānu*) or the Pleiades (MUL.MUL) (see 5.1.7., 5.2.2.). Whenever the Pleiades are substitutes for the planet Mars, and whenever an ominous sign of the Pleiades predicts famine, plague, or consumption, this can be a choice dictated by an analogy between divination and mythology. Ultimately, it must be stressed that, even though the Pleiades in factual explanations are often substitutes for Mars, this does not mean that the semiotic link between the protasis and the apodosis always has the same origin as the explanations. It might be argued that the predictions can be based on more and different analogies, also at the same time, and unknown to us (see A.3.1.5., A.4.1.).

5.3.2. The Pleiades Perceived as the "Divine Seven, the Great Gods" (^{d}IMIN.BI DINGIR^{meš} GAL^{meš})

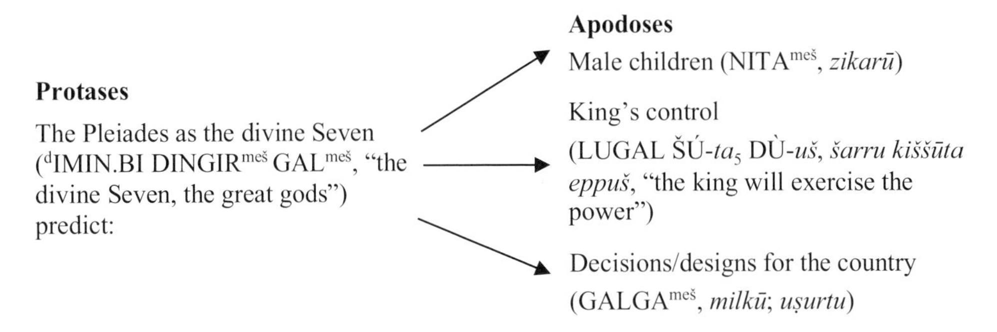

Table 10. The topics of the apodoses related to the Pleiades perceived as the divine Seven.

Whenever ominous signs of the Pleiades predict the well-being of the king and the country (see Table 10), especially from protases about the balancing of Moon and the Pleiades (see 5.1.8.1., 5.2.1.), or the heliacal rising of the Pleiades in Ajaru (i.e. Month II) (see 4.2.2. § 1b, 5.1.8.4. § EAE 50–51, 5.2.4.), the Pleiades are associated with the divine Seven, the seven great gods. This association, as opposed to the one between the Pleiades and the Sebettu, was already noted by Verderame (2016: 115): "The Pleiades are associated with the main divine gods who sit in council and establish the fates of the country. So in omens, the Pleiades forecast positive responses in relation to the harvest and the fate of the country."

One entry of the assumed tablet EAE 51 equals the Pleiades to seven gods, who are thought to determine the decisions for the country:

> ¶ In Ajaru (i.e. Month II), the Pleiades, the divine Seve[n, the great gods. If] they rise [at] their [specified ti]me, the great gods will assemble and / give good counsel to the country, [good] wind[s will blow. If] they do not rise [at] their specified time, [the great gods will assemble and] / will give bad counsel to the country, [evil] winds will blow (and) there will be grief among the people.
> (see 3.2.5.1., p. 91)

In this long entry, the prediction for the earlier or later rising of the Pleiades is good or bad winds (IM^{meš}, *šārū*), as well as good or bad decisions (GALGA^{meš}, *milkū*). Seven gods assembling and deliberating decisions in heaven, as well as human beings gather to deliberate decisions on earth, are a metaphor of the Pleiades rising heliacally in Ajaru: the semiotic link between the protases and the apodoses is based on the analogy between the Pleiades and seven gods of fates (see A.3.1.5., A.4.1.).

In one report to Esarhaddon (ca. 680–669 BC), the ancient scholar Nabû-aḫḫe-eriba explains an omen about the Pleiades and the Moon, emphasising the association between the Pleiades and the "great gods", who are not the Sebettu:

1. DIŠ MUL.MUL *ana* IGI ^{d}30 TE-*ma* GUB-*iz*
2. DINGIR^{meš} GAL^{meš} / *uṣ-rat* / *ma-a-ti*
3. *a-na* / *da-mì-iq-ti* / *uṣ-a$^{?}$-ru*
4. ^{d}30 / *ab-bu-ut-ti* / LUGAL
5. *a-na* DINGIR^{meš} *iṣ-ṣa-bat*
6. LUGAL *šá-li-im*

1. If the Pleiades come close to the front of the Moon and stand still,
2. the great gods the (divine) design of the land
3. for the well-being they will draw.
4. The Moon for the king
5. with the gods interceded.
6. The king is safe.

(SAA 8, 72 obv. 1–6)

In this report, the omen and its factual explanation are based on a specific symbolism. The Pleiades in the protasis are associated, in the apodosis, with a group of great gods who establish the destiny (*uṣurtu*, "design", see CAD U/W 290–295) of the country. The Pleiades approaching the Moon are associated, in turn, with the Moon interceding with the great gods.

The divine Seven, or the seven great gods, and their actions in the apodoses are reminiscent of the "seven gods who decide the fates" (dingir nam tar-ra imin-na-ne-ne) from the Sumerian poem "Enlil and Ninlil", or from the Akkadian composition *Enūma eliš* (see 3.2.5.1.). This group of gods symbolises the importance of the numerical hierarchy within the pantheon, regardless of the identities of the individual gods belonging to it. As shown by the many examples given in 3.2.5.1., the names of the gods within the group are not fixed. They are probably the main gods of the panthea who also had a pre-eminent astral counterpart: Anu, Enki/Ea, Enlil, Inanna/Ištar, Nanna/Sin, and Utu/Šamaš, with the inclusion of another god among Aššur, Marduk, Nergal, Ninurta, or Adad. Thus, the divine Seven (^{d}IMIN.BI DINGIR^{meš} GAL^{meš}) must be distinguished from the Sebettu (^{d}IMIN.BI): the Sebettu were never traditionally part of the great gods, they were demonic in their actions and called "great gods" only in the framework of the Neo-Assyrian military royal propaganda (see 3.2.7.). This fact may create an ambiguous picture for a modern reader; hence, only the context may clarify which heptad was intended in any text.

As a final note, the role of the Pleiades in celestial omens, based on the features already discussed in the previous chapters (see 2., 3., and 4.), can be outlined as follows:

- The Pleiades are a means of calendar organisation to foresee events.
- The Pleiades mark the beginning of the agricultural season, so they are thought to influence its outcome.
- The Pleiades are an asterism associated with the number seven, i.e. a "complete" group, and accordingly with two heptads from mythology with whom they share either a negative or a positive symbolism: the demonic Sebettu who destroy livestock, and the seven great gods of fate who make decisions on the political future of the country.

Chapter 6

The Pleiades in Texts Dating to the Late Babylonian Period

This chapter focuses on texts dating to the Neo-Babylonian (ca. 626–484 BC) and mostly Late Babylonian period (ca. 484–30 BC). These sources chronologically follow or are roughly contemporary to the ones discussed in chapters 4 and 5. They testify to new textual genres and a progressive accuracy of mathematical calculations and astronomical predictions, leading to what modern scholars would distinguish as "astronomy" and "astrology".[356] However, their content partly draws on divination and analogical reasoning.

In the previous chapters, it has been shown that the astral compositions dating to the late second and first half of the first millennium BC are heavily based on schematisations and analogies: compendia like MUL.APIN and Astrolabe B (see 4.) were meant to individuate stars and planets in the sky and to develop an ideal calendar through them. The development of schemes and rules contributed to a process of "standardisation" of celestial omens and related matters (see 5.); the need to organise knowledge, and to predict events (or to "interpret" messages from the gods) more accurately ran parallel. These facts contributed to creating a robust tradition, which persisted until the end of the first millennium BC – when the series *Enūma Anu Enlil* (EAE) and *Šumma Sîn ina tāmartišu* (SIT) were still copied by the same scholars who practised what we call astronomy.[357] Even after the Parthian period (ca. 247 BC–224 AC) and the end of cuneiform writing, the tradition caused an aftermath in other cultures.[358]

Although there is no trace or comment in the sources regarding a "shift" from celestial divination and the first astral composition towards astronomical and astrological texts, modern scholars tried to assume that some scientific development took place before the first half of the first millennium BC. For instance, Brown (2000: 126–207) strongly distinguished an "EAE paradigm" made out of ideal schemes found in astral compositions (see 4.) and celestial omens (see 5.), and a "PCP (= prediction of celestial phenomena) paradigm", developed in the Late Babylonian period along with mathematical astronomy. As he based his methodological approach on Kuhn's idea of scientific revolutions (Kuhn 1996), he assumed that the ideal schemes produced a progressive disillusionment with divination and a break with the past. Nevertheless, Brown's paradigms suffer from the comparison by default of the conceptual framework of cuneiform culture with Western

356 Nevertheless, the differentiation between astronomy and astrology "would have made no sense to a Babylonian. The goal of the Babylonian scholar can be best-called knowledge of the sky without any qualification whether it is a science or not" (Hunger 2011: 62). Consequently, the use of words such as "astronomical" and "astrological" texts is a modern classification, and its only purpose is to label the sources which were developed during and after the Neo-Assyrian period (ca. 911–612 BC) (see 1.4.3.).

357 The latest attestation of the series EAE known to us is an unpublished tablet on lunar eclipses dating to the first century BC (Fincke 2016: 116 fn. 47).

358 E.g. see the "Mandean Zodiac Book" (Rochberg 2010: 233–235).

positivist thought. Indeed, as Rochberg and Ossendrijver started proving,[359] divination, astrology, and astronomy shared several predictive modes and aims, and their traces are even found in the neighbouring Greek tradition.[360] As Rochberg (2016: 236) put it, "the distinction between astronomical and astrological thus made is in accordance with the aims of particular text genres, not a matter of demarcating between science and nonscience, or even astronomy and astrology, which cuneiform texts did not do."

With these preliminaries, the first part of chapter 6 focuses on an overview of astronomical texts and the zodiac (see 6.1.–6.1.3.) where the Pleiades stands as a name of Taurus. The second part discusses astrological texts in which the Pleiades are named, and the issues related to the context and the purpose of these texts (see 6.2.–6.2.4.).

6.1. Astronomical Texts: an Overview

The astronomical sources dating to the Late Babylonian period (ca. 484–30 BC) are usually divided by modern scholars into mathematical (MATs) [361] and non-mathematical astronomical texts (NMATs). Regardless of the level of accuracy in calculations and predictions achieved by the sixth century BC, it is already possible to gather the embryos of two models from MUL.APIN, both in use during the Late Babylonian period: the zodiac (see 6.1.1.) and the Normal Stars (see 6.1.2.), or the use of certain constellations to keep track of the path of the Moon and the Sun in the sky. At the same time, the astronomical texts dating to the Neo- and Late Babylonian period (ca. 626–30 BC) do not refrain, like celestial divination, from analogical reasoning as a modality of building the knowledge, as is witnessed in the Astronomical Diaries (see 6.1.3.), for instance.

6.1.1. The Zodiac

The zodiac[362] was not foreign to Mesopotamia, even though it is not clear when it came to be so important to involve computed data and become the main spatial framework in astral

359 "Both late astronomical and astrological traditions were inclusive of methods and models carried over from the past. Where we would perhaps view the older methods as incompatible with the newer, clearly the ancient scribes did not" (Rochberg 2016: 252). For cases of similarities between celestial divination and astrological texts, see also Rochberg (2010: 271–302; 2016: 269–273) and Ossendrijver (2012: 1–16; 2021)

360 The influence of Babylonian astronomy and astrology on Greece has not been systematically studied yet, though we are aware that a transmission happened, at least according to the sources of Ptolemy and Hipparchus (Britton-Walker 1996). A thorough study on the cross-cultural transmission of the zodiac from Babylonia to the Graeco-Roman world is conducted by Mathieu Ossendrijver's ERC project "ZODIAC-Ancient Astral Science in Transformation" in Berlin.

361 The MATs were published and studied by Neugebauer (1955), Swerdlow (1998), and Ossendrijver (2012).

362 The Sun makes an apparent movement through the stars roughly eastward within a year. Starting from the vernal equinox, the ecliptic of the Sun is theoretically divided into 12 parts of 30° each (i.e. 360°, a full circle), that is, the zodiacal signs. The Moon travels through all of them (i.e. the zodiacal belt) in 27,3 days to complete its sidereal month.

science. Scholars nowadays agree that its proper introduction into the literature (i.e. uniform zodiac) happened probably at the end of the fifth or beginning of the fourth century BC (Sachs in Neugebauer 1969: 140; Britton 2010).

We know that the Babylonian month began with the first appearance of the lunar crescent, and afterwards the Moon was thought to pass through 17 parts of the sky or through 17 stars or constellations. These stars, called "stars in the path of the Moon", are "proto-zodiacal" constellations and groups of stars, and the Pleiades are part of them according to MUL.APIN (see 4.1.1.). Later, the uniform Babylonian zodiac of 12 zodiacal signs (i.e. the zodiacal belt) would have been developed as a shortened form of the 17 stars in the path of the Moon. The Babylonian zodiac is the same one we know today, except for the names of the zodiacal signs; it was included in Greek astronomy and astrology, even though the scientific and philosophic framework of the latter is quite different from the Babylonian one (Koch-Westenholz 1995: 177–179).[363]

The Babylonian zodiacal signs, as shown in Table 11, are usually logographically written with the names of the constellations which represent them:[364]

Aries	LÚ.ḪUN.GÁ; ḪUN; LÚ; LU	The Hired Man
Taurus	GU_4.AN.NA; GU_4; MUL.MUL; MÚL.MÚL	Bull of Heaven, Pleiades
Gemini	MAŠ.TAB.BA.GAL.GAL; MAŠ.MAŠ	The Great Twins
Cancer	AL.LUL; ALLA	The Crab
Leo	UR.GU.LA; A; UR.A; UR	The Lion
Virgo	AB.SÍN; ABSIN	The Furrow
Libra	RÍN; ZI	The Scales
Scorpio	GÍR.TAB; GÍR	The Scorpion
Sagittarius	PA.BIL.SAG; PA	Pabilsag
Capricorn	SUḪUR.MÁŠ; MÁŠ	The Goat-Fish
Aquarius	GU.LA; GU	The Great
Pisces	KUN^{meš}; ZIB^{meš} ; IKU	The Tails, The Field

Table 11. The 12 Babylonian zodiacal signs (Koch 2015: 202–203 Table 17).

Taurus is either called GU_4.AN.NA (abbreviated as GU_4), "Bull of Heaven (lit. sky)", or MÚL.MÚL, lit. "Pleiades" (see Table 11). The Pleiades were perceived as belonging to the Taurus, i.e. they were a *pars pro toto* (Rochberg 1998: 29). In zodiological literature, the name of the Pleiades almost always indicates Taurus (see 6.2.2.): the reason for such

363 For the influence of the Babylonian astronomy and astrology in Greece, see Neugebauer (1975: 589–614) and Rochberg (2010: 143–165; 2020: 147–159; 2020a).

364 Zodiacal signs (i.e. 12 bands of the ecliptic of 30° each) and constellations (i.e. areas of the sky defined by specific boundaries) coincided when the system of signs was defined. Due to the precession of the equinoxes, today the sign Aries covers the area of the sign Pisces, and so on (see 4.).

ambivalence is because the Pleiades – even though they are a relatively small group of stars – were one of the "proto-zodiacal" celestial bodies in the path of the Moon (MUL.APIN I iv 33, see Hunger-Steele 2019: 70), and were important as a calendrical tool for the intercalation (see 4.2.2.) before the zodiac was introduced. This statement is corroborated by the famous graphic representation of the Pleiades, the Taurus, and the Moon in the *Gestirndarstellung* text VAT 7851 (see Figure 4 and 6.2.1.).

6.1.2. The Normal Stars

Another important astronomical concept attested in texts dating to the Late Babylonian period (ca. 484–30 BC) is the group of Normal stars (MUL ŠID^{meš}, *kakkabū minâti*, "stars of counting", or "predictable stars"). They were used as reference stars for measuring and expressing the positions of the Moon and the planets in the sky, and to predict planetary and lunar phenomena for a specific year.[365]

There are usually 25 Normal Stars ("the core 25") which appear in lists mainly dating to the Late Babylonian period (Steele 2017: 14). Their forerunners are the *ziqpu*, "culminating", stars, a list of 14 stars already given in MUL.APIN (I iv 1–9, see Hunger-Steele 2019: 61–63). The Pleiades were one of the Normal Stars: their name is usually written MÚL.MÚL, and they might have been identified by their brightest star, Alcyone or η Tauri (Sachs-Hunger 1988: 17), although there is no written evidence that a specific star among the cluster was strictly intended (Jones 2004: 483).

6.1.3. The Astronomical Diaries

The increasing importance of the knowledge of the sky in Mesopotamia, especially testified by the so-called astrological and astronomical reports dating to the Neo-Assyrian period (ca. 911–612 BC) (see 5.2.), led to types of texts in which computed data were mixed with regular observations. This is the case of the Astronomical Diaries: they are recordings of day-by-day observations of the sky, with predictions for the Moon, Sun, stars, weather, and relevant data for the economy of the country (e.g. water level of rivers, other astronomical and non-astronomical ominous events).[366] Their origin can be sought between the second half of the seventh century and the first half of the sixth century BC, yet they do have forerunners. The first regular records (*maṣāru ša ginê*, "regular watch") of celestial phenomena were those of the lunar eclipses dating to the second half of the eighth century BC onwards, which also attest to the use of an astronomically specialised terminology. Already by the first half of the seventh century BC, records of astronomical observations with accounts of historical events began to appear. Some of these annalistic texts can be considered the forerunners of the Astronomical Diaries. With the addition of the records of

365 E.g. the Goal-Year texts (235/234 BC–40/39 BC) list the passing of Jupiter, Venus, Saturn, and Mars by the group of Normal Stars. For the edition of these texts, see Hunger (2006). For overviews of the Goal-Year methods, see Brack-Bernsen (2020) and Hunger (2020).

366 For the editions of Astronomical Diaries, see Sachs and Hunger (1988; 1989; 1996), and Hunger (2001; 2006; 2014).

planetary movements, the terminology became more and more standardised, until a more or less complete standardisation in texts dating to the fifth century BC.[367] The recording of phenomena and prices in Astronomical Diaries was the basis for later textual genres, such as Almanacs (200/199–79/80 BC) and Horoscopes (ca. 405–68 BC).[368]

Even if still debated, many scholars suggest that some of the characteristics of divination survived in the Astronomical Diaries;[369] among the others, Pirngruber (2013) discussed how the Astronomical Diaries mirror some principles of divination, such as the schemes for the compass points (see A.2.1.1.4.). The Diaries surely have a different purpose and predictive assumption compared to divination, and their accurately predicted celestial phenomena seem to be much more related to historical and mundane affairs (Tuplin 2019: 84–107). However, regardless of the accuracy of the predictions, this does not substantially exclude the presence of elements belonging to earlier divinatory systems, such as ideal schemes and analogies.

6.2. Astrological Texts: an Overview

After the Neo-Assyrian period (ca. 911–612 BC), the divinatory corpora like the series EAE and SIT continued to be copied for a long time,[370] alongside the production of astronomical texts (see 6.1.–6.1.3.). The astrological texts dating to the Late Babylonian period (ca. 484–30 BC) show the coexistence and a re-elaboration of old elements (i.e. celestial divination, medicine, magic), combined with new ones, the zodiac above all.

Whereas in celestial divination the concern of omens was the reign and the king (see 5.1.), the concern of astrological texts is the life of individuals (Rochberg 2004: 62–63). This fact is explainable when astrological texs are framed in their political and historical context, as suggested by Wainer (2016: 71): "with the rise of Cyrus and the end of the Neo-Babylonian empire, Mesopotamia would be controlled by successive rulers who reigned from afar. (...) At the same time, scholars seem to have broadened the focus of celestial divination so that it would appeal to the private individual, who was mainly concerned with the health, well-being, and fortunes of himself and those around him".

Concerning their working principles, astrological texts are based on a different predictive assumption than divination, that is, to forecast events based on computed celestial phenomena, with the zodiac as a spatial framework. At the same time, they are also related to the predictive modes of divination that are based on analogical reasoning. Therefore, the aim of the following subchapters (6.2.1.–6.2.4.) is to trace similarities and differences in astrological texts with celestial omens in which the Pleiades are involved, to underline tradition or a change in their perception.

---

367 For an overview of the history of the Astronomical Diaries and their forerunners, see Steele (2019).

368 For the Almanacs, see Hunger (2014); for the Horoscopes, see Rochberg (1998).

369 E.g. see van der Spek (1993: 94–95), Pirngruber (2013), or Stevens (2019: 203–205). While according to Hunger and Pingree (1999: 139–144), the Astronomical Diaries only have an astronomical purpose, rather than divinatory.

370 The sources of these two compositions were mainly found in Uruk, Babylon (Oelsner 1986: 176–179, 212–214), and Sippar (Al-Rawi-George 2006).

6.2.1. Precursors of Greek Astrology: the Micro-Zodiac

The tablet VAT 7851, also known as the "man in the Moon" tablet,[371] preserves unique depictions of celestial bodies. The tablet was first edited and called by Weidner (1967: 12–28) *Gestirndarstellung* text, and recently re-worked by Monroe (2016: 131–132). On the obverse, VAT 7851 (see 2.3.3. Figure 4) shows a drawing of the Pleiades as seven pointed stars – with the caption MUL.MUL – and a bull, whose bristles, or tuft of hair, are emphasised on the back of its neck, similar to the shape of a comb; a "man in the Moon" is placed between the bull and the seven stars.

On both an iconographic and interpretative level, the Pleiades depicted as seven stars on VAT 7851 is the only rendition of this asterism together with its name, MUL.MUL (see 2.3.3. Figure 4). The Akkadian reading of MUL.MUL is *zappu*, "Bristle": the hair drawn on the back of the bull may symbolise the Pleiades as *pars pro toto* for the sign Taurus (^{mul}GU$_4$.AN.NA, *alû*, lit. "Bull of Heaven", see 6.1.1.). The whole drawing of the seven stars matches the glyptic dating to the first millennium BC, where a strong presence of the seven stars as dots and the crescent can be found.[372] The drawing is also considered the depiction of the forerunner concept of the *hypsoma* (from the Greek ὕψωμα), "(planetary) exaltation", "the association of a particular constellation with each of the seven planets[373] as the place where it attained special significance" (Koch-Westenholz 1995: 178). The Greek concept of *hypsoma* is similar to the Akkadian *ašar* or *qaqqar niṣirti*, "secret place", or *bīt niṣirti*, "secret house" (Weidner 1919; Rochberg 2010: 147–148 fn. 13): such secret places are positions in the sky where planets – when entering into them – were thought to have a special power, or influence on mundane events. Thus, the drawing on VAT 7851 represents the secret place of the Moon, i.e. when it is close to the Pleiades and Taurus.[374] This statement is understandable, if we consider the secret place as the balancing of the Moon and the Pleiades, when intercalation was supposed to be checked (Hunger-Pingree 1989: 147 Table X; Stol 1992: 254–255). Although we know that a 19-year cycle of intercalation (see 6.1.2.) was in use in the Late Babylonian period (ca. 484–30 BC), the balancing of the Pleiades and the Moon (see 4.2.2.) could be a reference to divination in the past or be an event still relevant to astrology.

The obverse of VAT 7851 preserves the beginning of a micro-zodiac scheme.[375] A micro-zodiac scheme represents one zodiacal sign, i.e. one month, 30°, introduced by a relevant omen: for instance, in VAT 7851, the initial omen is about a lunar eclipse in Taurus, in Ajaru (i.e. Month II). Then, Taurus is divided into twelve parts, or signs, corresponding to two and a half days a month, and each part is named after a zodiacal sign.

---

371 For the "man in the Moon" and the practice of seeing figures on the face of the Moon, see Beaulieu (1999) and Horowitz-Andre-Kritsch (2018).

372 According to the catalogue of seals by Konstantopoulos (2015: 321–367), more than half (circa 64%) of the seals depict seven dots or stars together with the crescent, and the great majority of them date to the Neo-Assyrian period (ca. 911–612 BC).

373 I.e. Mars, Jupiter, Mercury, Venus, Saturn, the Sun, and the Moon.

374 For the identification of the Babylonian *bīt niṣirti* of the planets, see Rochberg (2010: 147–155).

375 For a list of the known micro-zodiac texts, see Monroe (2016a: 127).

The micro-zodiac scheme gives a sort of zodiac within the zodiac, and it could be the forerunner concept of the *dodekatemoria* (from δώδεκα, lit. "twelve"), from Greek astrology.[376] The iconographical setting of VAT 7851 relates to the micro-zodiac text on the tablet: the drawing of the bull, the Pleiades, and the Moon stands for a situation which we assume to be observable in the night sky in Ajaru – the month of Taurus and the heliacal rising of the Pleiades (see fn. 220) – to which the micro-zodiac text is related.

The micro-zodiac scheme is also applied to other types of texts, in various arrangements, for instance in the *Kalendertexte* (i.e. iatromathematical calendar text)[377] a type of composition with a heterogeneous content. In the *Kalendertext* of Iqīšâ – an exorcist from Uruk who lived in the fourth century BC – each zodiacal sign within the micro-zodiac scheme corresponds to several ingredients for medical ointments, to prepare and use in a specific month. The ingredients chosen for the Taurus, written MÚL.MÚL (see 6.1.1. Table 11), are related to a bull, as shown in the quote which is given below:

> MIN 5 MÚL.MÚL 5 ÚŠ GU$_4$ *lu* Ì.GIŠ GU$_4$ *lu* SÍG GU$_4$ MIN
>
> Ditto (i.e. Month IV) (day) 5, Taurus 5°, ditto (i.e. you smear with) bull-blood or bull-fat or bull-hair.
> (SpTU 3, 104 obv. 5; Steele 2011a: 336; Wee 2016: 178)

The *materia medica* of Taurus has been chosen because of the imagery of that zodiacal sign, i.e. the bull; since the content of the micro-zodiac texts is exclusively based on the zodiac, the Pleiades are not intended here, apart from being the name for Taurus.

To sum up, the fragment VAT 7851 is perhaps the best-known evidence for the iconography of the Pleiades as seven stars in the Taurus. Content-wise, the micro-zodiac texts have some of the working principles of the older omens, like the analogical reasoning: the analogy between Taurus and the *materia medica* from a bull is a straightforward example of that. These texts drew on new concepts, like the zodiac and the *dodekatemoria*, but also consolidated strategies. However, they are zodiac-centred and never dwell on individual stars or asterisms like the Pleiades.[378] The latter are not relevant in these texts; instead, they resurface in the depiction in VAT 7851, which is likely a link to divination and/or represents the perception of the Pleiades at that time, according to Mesopotamian scholars.

---

376 "The *dodekatemoria* scheme models lunar motion by advancing the location of the moon by 13 degrees every day" (Monroe 2016: 268). For the numerical relationship and the differences between the *Kalendertexte* and the *dodekatemoria*, see Brack-Bernsen and Steele (2004), Wee (2016), and Schreiber (2020a: 38–41).

377 The *Kalendertexte* can be divided into two types, the hemerological calendar texts and iatromathematical calendar texts (e.g. SpTU 3, 104, see 6.2.1.). This subdivision is proposed by Marvin Schreiber in his unpublished PhD thesis "Die astrologische Medizin der spätbabylonischen Zeit" (2017, Humboldt-Universität zu Berlin).

378 Reiner (1995: 117) wrote about the *Kalendertexte*: "the punning relationship between the prescription and the corresponding sign of the zodiac can be understood only with reference to the classical zodiac."

6.2.2. Medicine and Celestial Bodies: Astral Medicine

Together with the ominous signs from the sky and the earth, diseases were a real concern in Mesopotamia. Diagnostic-prognostic divination and the healing process of patients was a matter shared between (practical) medicine (*asûtu*, performed by the *asû*, "physician") and the art of the exorcist (*āšipūtu*, performed by the *āšipu*, "exorcist"), i.e. divination, and magic (Scurlock 2014: 1–4; Koch 2015: 273–282). Disease was considered a consequence of having fallen into disgrace before the gods: this can be seen in medical texts where the expression *īpuš Ea ipšur Ea*, "Ea made it, Ea loosened it", is recurrent. In other words, it means that the gods were considered responsible for both causing and healing diseases (Reiner 1995: 81). Expressions such as *qāt* DN, "hand of DN", a disease caused by the "touch" of a god (Heeßel 2000: 53–54), are also common diagnoses that attest to the idea of corporeality, i.e. divine beings having a bodily existence and metaphorical ability to touch human beings.[379]

If celestial signs deliver the decisions of the gods, as well as diseases which are sent by the gods, then celestial bodies, like the anthropomorphic gods, can portend and heal diseases too. That is the gist of astral medicine, based on the belief that stars and planets can act as "mediators" for, communicate with, and influence the human world. This argument, addressed first by Reiner (1995: 15–24), is corroborated by the expression *ina kakkabi tušbāt*, "you have (it) spend the night under the stars", which is the usual way to refer to the preparation of medicaments (Reiner 1995: 48). On the one hand, "to spend the night under the stars" refers to the act of leaving the medicine – usually made with herbs and organic ingredients – sitting overnight to let it macerate. On the other hand, a sort of "astral irradiation" (i.e. the influence of the stars on medicaments) is likely intended, because there are occurrences in which the exposure of the medicine to this or that specific celestial body is required (Heeßel 2008: 4).[380] These premises put astral medicine into the second perspective on celestial bodies as established by Rochberg (see 2.2.3.1. § 2), according to which celestial bodies represent anthropomorphic gods.

The Pleiades are never mentioned in diagnostic-prognostic omens dating to the first millennium BC. However, the Sebettu (^{d}IMIN.BI) are said to cause diseases in some entries of the series SA.GIG, *sakikkû*, "ill strands", whose standard version comes from Ashurbanipal's library:[381]

> DIŠ *ina* ÉLLAG-*šú ša* 2,30 SÌG-*iṣ* UŠ$_{4}$-*šú* NU DIB MÚD BURU$_{8}$ ŠU ^{d}IMIN.[BI GAM]

379 For bibliography and references to corporeality, see fn. 122.

380 Wee (2014a) argued how, in the myths of Lugalbanda (see 3.1.1.3.), the exposition to stars, and the use of medicaments represent two stages of a celestial healing process, i.e. the removal of the negative influences and the restoration of health by eating plants and drinking water. His point would extend the concept of "astral irradiation" to mythology, and it would place it at least in the Old Babylonian period (ca. 2000–1500 BC).

381 The editions can be found in Labat (1951), Heeßel (2000) and Schmidtchen (2021). A reconstruction and a summary can be found in Scurlock (2014).

If he is injured on his left kidney, he is not in possession of his faculties (and) he vomits blood, (it is the) hand of the Sebet[tu, he will die].
(TDP, 31 rev. ii 22, see Scurlock 2014: 95, 85'')

DIŠ *ina* GIG-*šú* IGI^{II}-*šú* NU ÍL-*ši ina* IGI^{II}-*šú* KIR$_{4}$-*šú* KA-*šú* GESTU^{II}-*šú u* GÌŠ-*šú* MÚD TÉŠ.BI È.ME-*a* / ŠU ^{d}IMIN.BI : UD.31.KÁM ŠU ^{d}MAŠ.TAB.BA

If during his illness, he does not raise his eyes, blood comes out of his eyes, his nose, his mouth, his ears, and his penis at the same time, / (it is the) hand of the Sebettu – (var.) (If he has been sick for) thirty-one days, (it is the) hand of the Twin Gods.
(Heeßel 2000: 197, 25–26)

As shown in these entries, certain diseases are caused by *qāt Sebetti* (ŠU ^{d}IMIN.BI), "the hand of Sebettu". More specifically, the entries say that the Sebettu are related to haemorrhagic fevers (i.e. shock from blood loss) caused by infections (i.e. by viruses and fevers), or wounds (Scurlock-Andersen 2005: 76–77, 473–474). An implicit reference to the Pleiades, as the astral form of the Sebettu, cannot be excluded. For instance, the injuries described in the first entry given above (TDP, 31 rev. ii 22) are related to the "left kidney" (ÉLLAG-*šu ša* 2,30, *kalītišu ša šumēli*). In the bilingual lexical list Ura = *ḫubullu* 22, one of the Akkadian readings of MUL.MUL ("Pleiades") is *kalītum* (ÉLLAG), "kidney" (see 2.3.1.). Another hint for this analogy is the famous incantation *anāku nubattu aḫāt Marduk*, "I am the vigil, the sister of Marduk", dating to the first millennium BC. The incantation is preserved in two medical manuscripts, with the purpose of healing the *kalīti šumēli*, the "left kidney".[382] This incantation (see 3.2.4.) is a complex mixture of celestial and magic elements: *nubattu*, the vigil, is the name for the seventh night of the month, when the Moon is "kidney-shaped", and it is, at the same time, the name of the daughter of the Pleiades and Bālu (i.e. Mars or Orion). A commentary from Nippur dating to the Achaemenid period (ca. 547–331 BC) tells us that the kidney was thought to be governed by the planet Mars (i.e. Nergal):

DIŠ NA ÉLLAG-*su* KÚ-*šú* ^{d}U.GUR *šá* E-*u* / ^{mul}ÉLLAG : ^{d}*ṣal-bat-a-nu*

If a man's kidney hurts him, (the disease is from) Nergal. / The Kidney (is) Mars.
(Civil 1974: 337 ll. 20–21)

Arranged in chronological order, the sources quoted so far imply a tradition made of analogies and associations, in which the Pleiades are related to the Kidney, the Sebettu, and Nergal/Mars. Such analogies can be traced back roughly to the lexical lists dating to the second millennium BC and attested up until the start of the Greek astrological tradition. Indeed, Mars was believed to govern the kidney until the Greek-Byzantine period (Reiner 1993).[383]

382 E.g. KA.INIM.MA ÉLLAG 2,30, "Wording (of an incantation) for the left kidney" (Abusch-Schwemer 2016: 54 J rev. 8).

383 See the *De septem stellarum herbis* (Kroll 1903: 83), dating to the twelfth century.

Just a few sources dating to the Neo-Assyrian period (ca. 911–612 BC) could be considered forerunners of astral medical texts: a connection between celestial and diagnostic-prognostic omens seems to have been already in place at that time (Schreiber 2020: 119–123). These few sources could be the earliest outcome of what Heeßel (2008: 16) defined as an "astrologisation" of medicine. Only after the Neo-Assyrian period, medicine and astrology "fuse" together in a zodiac-oriented astral medicine, where celestial bodies, in a zodiac-based spatial framework, were mixed with diseases, events, and objects. Astral medicine is very likely the precursor of the Hellenistic concept of *melothesia* (μελοθεσία), i.e. the influence of zodiac on specific parts of the human body, as attested in Ptolemy's *Tetrabiblos* (Geller 2014: 70–71, 77–80).

Good examples of astral medicine are two texts from Uruk, BRM 4, 19 (MLC 1886) and BRM 4, 20 (MLC 1859), dating to the Hellenistic period (ca. 323–63 BC), drawn up on an earlier hemerology,[384] STT 300, dating to the eighth century BC.[385] While BRM 4, 19 and 20, re-edited by Geller (2014: 27–46), involve the zodiac and the *dodekatemoria* (see 6.2.1.), the earlier hemerology STT 300 – the so-called Exorcist's Almanac – was composed before the invention of the zodiac, because it assigns incantations and rites to calendar dates instead of zodiacal signs (Geller 2014: 47). Both BRM 4, 19 and BRM 4, 20 seem to be a re-arrangement with integrations of STT 300;[386] in every entry, zodiacal signs are associated with witchcraft, diseases, or misfortunes, indicating the propitious time when averting a hostile fate, or invoking harmful magic to counteract it (Geller 2014: 27–28). Likely, the magic was thought to be effective when the Sun passed into a particular zodiacal sign, that is, the texts would reflect the motion of the Sun into a micro-zodiac scheme (Wee 2016: 176); indeed, BRM 4, 19 even includes a micro-zodiac scheme (see 6.2.1.).

A few relevant entries have been chosen among the above-mentioned astral medical texts to show the heterogeneous nature of their structure:

STT 300 obv. 7, 8 DIŠ *ina* ^{iti}GU$_4$ (...) (8) UD 13.KÁM LÍL.LÁ.AN.NA KI.SIKIL.LÍL.LÁ.EN.NA ⌜ZI⌝-*ḫi* DÍM-*ma* AL.SILIM

CBS 562 obv. ii 2'–4' DIŠ *ina* ^{iti}G[U$_4$ ...] (3') : ⌜12⌝[(+1$^?$).KAM...] (4') KI.SIKIL.E K[A$^?$]

On the 13th day of Ajaru (i.e. Month II), to remove Lilû and Lilītu, make (incantations/rituals), and it will be well.
(For STT 300, see Geller 2014: 48; for CBS 562, see Rutz 2018: 103, 107)

384 For the meaning of "hemerology" in Assyriology, see fn. 355. For the structure of STT 300 as a hemerology, see also Schreiber (2020a: 35 fn. 4).

385 Partial duplicates of STT 300 are CBS 562 (Rutz 2018: 101–109), and BM 37080 (Schreiber 2019). A partial duplicate of BRM 4, 19 is BM 39279 (Fincke 2021), while BRM 4, 20 is duplicated by SpTU 5, 243 and LBAT 1626 (= BM 35537) (Geller 2014: 58–60). For further studies and bibliography on the Exorcist's Almanac, see Scurlock (2005) and Fincke (2021).

386 BRM 4, 20 is a *ṣâtu* commentary of STT 300, as stated at the end of the fragment (Schreiber 2020a: 37).

BRM 4, 20 obv. 29–30 LÍL.LÁ.EN.NA KI[387] MUL.MUL *šá-niš* KI ^{mul}GIŠ.RÍN (30) KI.SIKIL.LÍL.LÁ KI MUL.MUL *šá-niš* KI ^{mul}MAŠ.MAŠ *šal-šiš* KI ^{mul}P[A.BI]L.[S]AG

SpTU 5, 243 obv. 3–4 x KI.SIKIL.LÍL.E.NE KI ^{múl}<ŠU.GI>[388] (4) [...] KI MÚL.MÚL

(For) Lilû, region of the Pleiades, (or) secondly the region of the Scales. (For) Lilītu, (only SpTU 5, 243: region of the <Old Man>, i.e. Taurus), region of the Pleiades, (or) secondly the region of the Great Twins, (or) thirdly the region of Pabilsag.
(Geller 2014: 29 ll. 27–28; 59 obv. 2)

2 12 ⸢LÍL.LÁ.EN.NA KI.SIKIL.LÍL⸣.LÁ.⸢EN.NA⸣ ZI-*ḫi*[389] / DÍM-*ma* AL.SILIM 2$^{!}$ 12 7 6 RÍ[N *š*]*á* MÚL.MÚL ZI

(In month) 2 (day) 12, to remove Lilû and Lilītu / you make (incantations/rituals), and it will be well. 2$^{!}$ 12 7 6 Micro-Libra of Taurus is the distance.
(BRM 4, 19 rev. 1'–2', see Geller 2014: 41 rev. 27–28)

All these sources given above mention either Taurus (written MÚL.MÚL) as a zodiacal sign, the Pleiades (written MUL.MUL) as a part of the Taurus, or a calendrical reference (i.e. 13th day of Ajaru). The sources agree on the association of Ajaru (i.e. Month II) with Lilû and Lilītu (or Ardat-lilû), two demons usually mentioned together and possibly – according to their etymology – related to the wind (LÍL) or the evening (*līlâtu*, see CAD L 184–185) (Farber 1987–1990).[390]

In BRM 4, 20, the Pleiades are also associated with aphasia (KA.DAB.BÉ.DA, *kadabbedû*):

KA.DAB.BÉ.DA K[I MUL.MUL][391]

(For) Aphasia, the region of [the Pleiades]
(BRM 4, 20 obv. 43, see Geller 2014: 31 l. 36)

387 In Late Babylonian astronomical and astrological texts, KI, *qaqqaru*, lit. "territory", refers to the portion of the night sky where constellations are located, or the position of the Sun, the Moon, or the planets relative to it, a position which usually "denotes the coordinate along the ecliptic" (Ossendrijver 2016: 154).

388 The integration was suggested by Reynolds in SpTU 5, 243 § 2.

389 As part of a micro-zodiac scheme, BRM 4, 19 rev. 1' should be understood as follows: "Libra (i.e. micro-zodiacal sign) of the Taurus (i.e. main zodiacal sign)" (Wee 2016: 169 rev. 20'). For the meaning of the numerical values given in these texts, see Schreiber (2020a: 38–41).

390 An incantation dating to the Old Babylonian period (ca. 2000–1500 BC) dedicated to Ardat-lilî is the only attestation of a connection with Erra: KA.INIM.MA [*w*]*a-ar-da-at li-li-i-im* / *re-e-di-it* ⸢*i*⸣*-li-im* ⸢*èr*⸣*-ra*, "Prayer: Ardat-lilî, / attendant of the god Erra" (Farber 1989: 16 obv. 1–2).

391 Restoration by Ungnad (1944: 259, 43 and 272, 43).

This association was already found in one *šuʾilla* prayer addressed to the Pleiades dating to the first millennium BC. This prayer, entitled "Zappu 2", refers to the Pleiades specifically to counteract spells of aphasia sent to the petitioner (see 3.2.1.2.). Therefore, BRM 4, 20, attests to a link between anti-witchcraft incantations and rituals, and astral medicine dating to the Hellenistic period (ca. 323–63 BC).

Other astral medical texts relate the zodiac to a more specific medical knowledge. For instance, the fragment BM 41583 (LBAT 1597) deals with many diseases and their celestial influences.[392] The text is not explicit, and it must be assumed that the illnesses were contracted (or averted) when a planet passed into a constellation (Wee 2016: 155 fn. 46):

> DIS KI MÚL.MÚL TU.RA KÌLIB.BA *u* NAM.ÚŠ *šib-ṭu ṣi-bit* KÚM *ḫe-pí*
>
> If (it is in the) region of Taurus, (it refers to the) totality (of diseases) and death, plague, attack of fever *break*.
> (Geller 2014: 80 obv. 5')

This entry mentions death (NAM.ÚŠ, *mūtu*) and plague (*šibṭu*) in association with Taurus (written MÚL.MÚL). This association resonates in the poem of "Erra and Išum": Nergal/Erra himself is the *bēl šibṭi u šaggašti*, "lord of plague and carnage", joined in the myth by the Sebettu, with whom the Pleiades are associated (see 3.1.2.1., 5.3.1.).

From the texts discussed so far in this subchapter, one can conclude that astral medicine is strictly embedded into the zodiac. A concept of corporeality already existed in diagnostic-prognostic omens, where the touch of the Sebettu caused kidney diseases – perhaps by the analogy between the Pleiades, the Sebettu, and the kidney. However, the idea of the touch of stars directly causing disease is found almost exclusively in connection with constellations. Indeed, just like in micro-zodiac texts (see 6.2.1.), the Pleiades are only regarded as a name of the zodiacal sign Taurus. Some instances could recall features and symbolism of the Pleiades (e.g. Lilû and Lilītu in Ajaru, i.e. Month II, aphasia in Kislimu, i.e. Month IX, plague in Taurus); nevertheless, it is clear that in astrological texts the Pleiades are consistently less involved and substituted by the zodiacal belt.

6.2.3. Fortune-telling and the Stars: Astral Magic

The tradition of celestial omens – combined with more detailed predictions and the zodiac – also led to the elaboration of what Erica Reiner (1995) called astral magic texts. This type of text shows how stars were thought to influence the performative act (i.e. magic rituals) in different fields of knowledge or expertise.[393]

---

392 The text was first edited by Leibovici (1956) and partially translated by Heeßel (2008: 8), Geller (2014: 80–83), and Wee (2016: 199–200). It can be classified as a commentary because of the presence of the *Winkelhaken* and the sign MU, *aššu*, "concerning", "because" (Geller 2014: 80 fn. 4; Gabbay 2016: 144–165). The fact that there is also the gloss *ḫepi*, "break", means that the text has been copied from an older one (Gabbay 2016: 63–64).

393 E.g. the Egalkura (É.GAL.KU$_4$.RA), "Entering of the palace", is a magic ritual to have a favourable

The involvement of the Pleiades with magic is testified by a unique type of divinatory text. The tablet YBC 9863, dating to the Neo-Babylonian period (ca. 626–484 BC), and edited by Hallo (2010: 650–658), is a collection of celestial omens and related incantations. It is a close parallel to the more famous text from Ḫuzirīna, STT 73, dating to the Neo-Assyrian period (ca. 911–612 BC), and edited by Reiner (1960a).[394] Both texts show a mixture of celestial omens and magic: for this reason, Reiner (1995: 70) labelled the text from Ḫuzirīna "fortune-telling", as a type of divination accessible to everyone that probably did not require scholarly expertise.[395] These fortune-telling texts, prayers, ritual instructions, and omens aim to obtain from the gods a favourable answer (i.e. an oracle) to specific problems addressed in the prayers (i.e. healing from sickness or the success of a plan). These texts are unique also because they testify to an interest in *omina impetrativa* in Mesopotamia, next to extispicy.[396] The ominous signs they quote are supposed to be obtained through specific divinatory techniques: dream incubation, the course of shooting stars, and the sprinkling of an ox with water.

In both texts, YBC 9863 and STT 73, the stars are invoked as divine beings *tout court*. Prayers and rituals are performed at night, when stars and planets are visible, to have either a dream or the passing of a shooting star. Only in YBC 9863 there is a formulaic prayer, logographically written, to the Pleiades:

4. [MU]L.MUL SI.SÁ MUL.MUL SI.SÁ MUL.MUL GIN.NA MUL.MUL GIN.NA
5. [MUL].MUL GUB.GUB MUL.MUL GUB.GUB MUL.MUL TUKU.TUKU MUL.MUL TUKU.TUKU
6. [MUL.MU]L DU_8.DU_8 MUL.MUL DU_8.DU_8 MUL.MUL SIG_5.GA MUL.MUL SIG_5.GA
7. [MUL.MU]L GIŠ.TUKU MUL.MUL GIŠ.TUKU TE.ÉN

---

8. [KA.INIM].MA KA.AŠ.BAR BAR.RE

---

4. [Pl]eiades proceed! Pleiades proceed! Pleiades come! Pleiades come!

legal judgement on someone's case. It is mentioned in the hemerology STT 300 (see 6.2.2.) and in the astral medical text BM 47457. For the Egalkura magic and other rites that can be considered "astral magic" of the Late Babylonian period (ca. 484–30 BC), see Schreiber (2020a).

394 See also KAL 4, 53 for parallels.

395 Koch (2015: 143 fn. 384) includes STT 73 among the few "everyday oracles" attested in Mesopotamia.

396 The *omina impetrativa* are solicited signs obtained through a request and under certain conditions. In Mesopotamia, the interest was mainly in *omina oblativa* (i.e. unsolicited signs, natural signs from earth and heaven), except for the extispicy, which had huge fame (Fincke 2014a). Techniques such as the flight of the birds were usually in the repertoire of Hittite divination (Sakuma 2014; see also De Zorzi 2009). The shooting star technique is not strictly for *omina impetrativa*, though, because their appearance is accidental. Indeed, omens about shooting stars are preserved in one of the unedited sections of the *Šumma ālu* series (Babyloniaca 4, 125) and in the dream omens (Oppenheim 1956: 283–284).

5. [Ple]iades stand still! Pleiades stand still! Pleiades take possession! Pleiades take possession!
6. [Pleiade]s release (the evil)! Pleiades release (the evil)! Pleiades be kind! Pleiades be kind!
7. [Pleiade]s listen! Pleiades listen!

---

8. [Incantati]on to make a decision.

(Hallo 2010: 652 rev. 4–8)

Through this incantation, the priest who recited it asked the Pleiades to come to the place where the ritual – sweeping the roof, sprinkling water, and burning plants – should have been performed (Hallo 2010: 652 rev. 9–13). Texts like the one in YBC 9863, related to simple and everyday divinatory practices, perhaps really belonged to a more popular context (Koch 2015: 143). Still, they are linked to a cultural framework in which stars are invoked both for having an ominous sign (i.e. they are mediators), and because they directly produce ominous signs, like the shooting stars. In other words, they were considered divine agents (see 2.2.3.1. § 2). That is not unusual: a similar behaviour has been witnessed in the prayer to the gods of the night, in which celestial bodies were considered responsible for the outcome of extispicy (see 3.2.2.). In this context, the Pleiades are just one asterism among the many stars that were thought to have a direct, magical influence on mundane affairs.

6.2.4. Predictions for Weather, Market, Cities, and Mundane Events

Weather and business were two primary concerns of the Mesopotamian people: for instance, a considerable variety of omens about the weather were organised in the chapter of Adad in the series EAE (see 5.1.8.3.), and other sources;[397] predictions for market rates (KI.LÁM or GÁN.BA, *maḫīru*) were also so frequent to be considered "stock apodoses" (Koch 2019: 232). Later, the Astronomical Diaries recorded regular observations of weather and planetary phenomena, the prices of commodities, astronomical phenomena, the water level of rivers, historical events, and other ominous events (see 6.1.3.).

A few sources dating to the Late Babylonian period (ca. 484–30 BC) attest to the merging of the zodiac with predictions for weather, prices, cities, and mundane events: these astrological texts seem to carry a variety of references from divination, astronomical, and other astrological texts. These texts are few but very complex and cannot be framed in any of the known textual genres. However, what they have in common is undoubtedly their focus on predicting specific events such as trade trends or the weather. In this subchapter, three examples (see below § BM 47494, § AO 6449, and § AO 6455) will be presented

397 E.g. MUL.APIN (II iii 33–34, GAP B 7–8, see Hunger-Steele 2019: 104, 108), or the apodoses of *Iqqur īpuš* (Labat 1965: 172–195 § 88–104). See also De Zorzi (2009).

because they include references to the Pleiades from which it is possible to draw parallels to divination and omens.

§ BM 47494

The first example is a fragment with various types of predictions, BM 47494, from Babylon or Borsippa, written by the scribe Iprāya before 337 BC (Hunger 2004; Ossendrijver 2019: 65–68; 2021: 225–226). The obverse of the fragment has first a list of stars related to cities and countries, specific events, weather predictions and market activities, while the reverse has entries about the appearance of the Moon and planets in specific sections of the sky. In each entry of the obverse of BM 47494 (obv. 1–15), one star or constellation is associated with one or more cities. In that section, the Pleiades are addressed as follows:

DIŠ MUL.⸢MUL ÉN⸣.ŠÁR$^{⸢\text{ki}⸣}$

The Pleiades (are for) Keš.
(Hunger 2004: 16 obv. 1)

As a comparison, only in one Neo-Babylonian fragment, an eclipse of the Moon in the Pleiades is said to affect the city of Der or Uruk:

DIŠ *ina* KI MUL.MUL *ú-lu* ^{mul}ŠU.GI *a-dir* EŠ.BAR BÀD.AN^{ki} DUR.AN.KI : UNUG^{ki}

If (the Moon) is dark in the region of the Pleiades or of the Old Man, (it is a) prediction (lit. decision) for Der (*called*) DUR.AN.KI[398] – (var.) (for) Uruk.
(Steele 2015: 208 rev. 38)

The association between stars and cities may reflect a focus on geographical area, as it often happens in divination, where a geographical area of the sky affects the same area on earth and *vice versa* (see A.2.1.1.4.). As noted by Wainer (2016: 68–71) and Ossendrijver (2019: 66–67), several omens in the reverse of BM 47494 (rev. 7–22) are shaped on an older scheme given in the tablet SIT 4 (K 3123 rev. 13’–16’, see App. B § 24): this scheme provides months in which an eclipse of the Moon should affect a particular geographic area, whereas in BM 47494 (rev. 17–22) the months are substituted by the corresponding zodiacal signs. At any rate, the analogies between stars and cities given by BM 47494 could be references to divination in the past, unfitting for the time when the text was written. Indeed, as remarked by Steele (2015: 213), “by the Late Babylonian period some of these cities, such as Girsu and Eshnunna, had long since been abandoned, suggesting that the interest among Babylonian scholars was, at least in part, either in fitting into long-standing scholarly traditions or in glorifying the past (or perhaps both) rather than in providing a practical tool for astrological interpretation.”

398 Usually Nippur is called dur.an.ki (see, e.g., van der Meer 1939: 179 text 88 obv. i 2), so the juxtaposition with Der is unclear in this line.

In BM 47494 obv. 16–38, specific events are assigned to three constellations.[399] It must be assumed that these events were predicted when a celestial body, probably a planet, was in these constellations (Ossendrijver 2021: 234, 241). As shown in the following entries, the Pleiades are said to be bringers of pestilence, flood, and storms when a celestial body passes next to them:

19. *ana* ÚŠ^{meš} *ina* ŠÀ ⸢MUL⸣.MUL ^{mul}GU$_4$.AN.NA *u* ^{mu}[^{l}SI]PA.ZI.AN.NA
(...)
28. *ana* ŠÈG A.GU$_4$ *ina* ŠÀ ^{mul}KU$_6$ ^{mul}GU.LA *u* MUL.MUL
(...)
31. *ana* ZI *me-ḫe*[-*e* UD.D]È$^?$.RA.RA *ri-iḫ-ṣu ina* ŠÀ MUL.MUL *u* ^{mul}LÚ.ḪUN.GA

19. For the pestilence: inside the Pleiades, the Bull of Heaven and the [Tr]ue Shepherd of Anu.
(...)
28. For the rain (and) flood: inside the Fish, Gula and the Pleiades.
(...)
31. For the rising of a violent stor[m$^?$, devast]ation (and) destruction: inside the Pleiades and the Hired Man.
(Hunger 2004: 17–18 obv. 19, 28, 31)

The associations between the Pleiades and some of the events mentioned above from the entries of BM 47494 were already in apodoses of celestial omens:

If the Pleiades stand inside it (i.e. the Moon), a pestilence will take place, the Sebettu will devour the country.
(see 5.1.8.1., p. 227, 1)

(If) the Pleiades and the Wagon stand together, rains and floods will come (and) the crop will be diminished. In winter, (there will be) a pestilence among the [herd].
(see 5.1.8.4. § EAE 50–51, p. 230, 2a)

DIŠ ⸢MUL.MUL ^{mul}ÉLLAG KUR-*ud* ^{d}IŠKUR⸣ RA

If the Pleiades reach the Kidney, Adad will devastate.
(K 5713+ obv. 8’, see SIT 6: 5 source A and parallels)

The pestilence (ÚŠ^{meš}, *mūtānu*), rain (ŠÈG^{meš}, *zunnū*), and devastation (*riḫṣu* and RA, *raḫāṣu*) were predictions for the Pleiades already given in celestial omens, and based on the working principles of divination, that is, in this instance, an analogy between the Pleiades and the Sebettu (see 5.3.1., A.3.1.5., A.4.1.).

399 Brown (2018: 438) explains that these constellations have “astrological values”, irrespective of which planet stands in them.

Ossendrijver (2019) compared BM 47494 to the Astronomical Diaries and earlier celestial omens, and concluded that there seems to be no relationship between BM 47494 and the Astronomical Diaries, even though their content (i.e. weather and market predictions) is similar; instead, BM 47494 uses more elements from divination, but it develops them in a different way, i.e. by associating events to individual constellations. He further suggested that the predictive assumption of BM 47494 might perhaps be "a form of divination that could be used to construct *a posteriori* astrological explanation for market developments in the past" (Ossendrijver 2019: 73), and that texts like BM 4749 "explicitly combine period-based methods for long-term prediction of astronomical phenomena with inferential methods for predicting weather phenomena" (Ossendrijver 2021: 251).

§ AO 6449

A second example is provided by AO 6449 (TCL 6, 19),[400] which attest to a Babylonian astrometeorology[401] with two other fragments, AO 6455 (TCL 6, 11) and AO 6448 (TCL 6, 20).[402] These texts were first studied by Hunger (1976) and then by Ossendrijver (2021), who illuminated how these examples of astrometeorology combined long-term astronomical computations (i.e. conjunctions of planets, simultaneous phenomena) and the logic of the omens.

The fragment AO 6449 mentions the Pleiades and relate them to bad weather, as shown in the entry given below:

> ^{d}GU$_{4}$.UD *u* AN *ina* ŠÚ *ina* ^{múl}ŠU.GI GUB^{meš} (…) *ina zap-pi* ^{múl}MÁŠ *ina* KUR *lu ina* ŠÚ GUB^{meš}-*ma* ŠED$_{7}$ *dan-nu*
>
> If Mercury and Mars stand still at the east of the Old Man (…) stand in the Bristle (or) Capricorn in the east or west, (there will be) a strong cold.
> (Hunger 1976: 249 rev. 15, 19)

This entry from AO 6449 is very reminiscent of a long one from the reconstructed Commentary EAE 53:

> [DIŠ M]UL.MUL $^{mul.d}$AMAR.UTU KUR-*ud* ⌜^{d}⌝GU$_{4}$.UD *lu* ^{d}SAG.ME.GAR ^{mul}*za-ap-pi* KUR-*ma* / BAD-*ma um-ma-a-tu*$_{4}$ ^{d}UTU RA-*iṣ* BAD-*ma* EN.TE.⌜NA⌝ ^{d}IŠKUR RA-*iṣ* ^{d}*ṣal-bat-a-nu* GABA.RI ^{d}UDU.IDIM GABA.RI / ^{d}*dil-bat* GABA.RI ^{d}UDU.IDIM.SAG.UŠ ⌜^{d}⌝*za-ap-pi* KUR-*ma* ÚŠ^{meš} GAR^{meš}
>
> [If the P]leiades reach the Marduk-star, in that year there will be heat and blaze. (It means that) Mercury or Jupiter reaches the Bristle. / If (in) summer, Šamaš will

---

400 Duplicated by BM 36647 (Schreiber 2018: 741–748).

401 For astrometeorology in Mesopotamia, see Ossendrijver (2021). There might be traces of the influence of the Babylonian astrometeorology on the Greek one, the latter being extensively attested (Lehoux 2004: 228–233), contrary to the former.

402 The fragments AO 6449 and AO 6448 are parts of the same series (Hunger 2019: 182–183).

devastate. If (in) winter, Adad will devastate. (If) correspondingly (i.e. if in winter) Mars, or a planet, or Venus, or Saturn reaches the Bristle, a pestilence will take place. (K 3558 obv. 2–4, see Commentary EAE 53 source A and parallels)

In the entry from Commentary EAE 53, the Pleiades and the Marduk-star, a name of Mercury (see A.4.1.), are explained as Mercury and other planets approaching the Pleiades. The apodoses concern both the summer and the winter: the devastation of Adad in winter (EN.TE.NA ᵈIŠKUR RA-*iṣ*, *kuṣṣu Adad iraḫḫiṣ*) caused by Mercury or Jupiter in the Pleiades resurfaces in the "strong cold" (SED *dan-nu*, *kuṣṣu dannu*) caused by Mercury or Mars in the Pleiades in AO 6449 rev. 19.

§ AO 6455

The combination of old and new elements is also testified by a third and last example, AO 6455 (TCL 6, 11), a text which deals with weather predictions, constellations, and Goal-Year methods (see fn. 365). The text was edited by Hunger and Brack-Bernsen (2002: 7–23), who also suggested the combination of earlier elements of astronomy and divination (e.g. MUL.APIN, *Enūma Anu Enlil*) with more advanced concepts, such as the "Lunar Six".[403] First, AO 6455 quotes the incipit of EAE 53 but gives a different explanation:

EAE 53

DIŠ MUL.MUL ᵐᵘˡŠUDUN KUR-*ud* *ina* MU BI GÁN.BA TUR-*ir* ᵈUDU.IDIM.SAG.UŠ ᵈ*ṣal-bat-a-nu* KUR-*ma*

If the Pleiades reach the Yoke, in that year the market rate will diminish. (It means that) Saturn reaches Mars.
(Commentary EAE 53: 1, SIT 6: 2; see also 5.1.3.2.)

AO 6455

MÚL.MÚL ᵐᵘˡŠUDUN KUR-*ud* MÚL.MÚL *šá* UD.1.KAM ᵐᵘˡŠUDUN ᵐᵘˡGU$_4$.AN *šá* UD.1.KAM IGI.LÁ EN UD.20.KAM NU IGI.LÁ

"(If) The Pleiades reach the Yoke. (It means that) the Pleiades of the 1st day (are) the Yoke, the Bull of Heaven which becomes visible on the 1st day, it does not become visible until the 20th day"
(Hunger-Brack-Bernsen 2002: 7–8 obv. 5)

Second, AO 6455 has entries whose nature is way more schematic, as shown in the following ones:

403 The Lunar Six is the way to calculate the motion of the Moon through six intervals, see Sachs (1948).

6. *šá* GU$_4$ MÚL.MÚL *šá* SIG ^{múl}UG$_5$.GA^{meš} *šá* ŠU *šu-ku-du šá* NE ^{múl}MAR.GÍD.DA *šá* KIN ^{múl}*e*$_4$-*ru*$_6$ *šá* DU$_6$ ^{múl}UR.GI$_7$
7. *šá* APIN ^{múl}GÍR.TAB *šá* GAN ^{múl}MÁ.GUR$_8$ *šá* AB AN *šá* ZÍZ *ḫar-ri-ri šá* ŠE AŠ.GÁN *an-nu-ú* MÚL.MÚL *šá* ITI^{meš} *gab-bi* IZI.ŠUB
8. ME.A GAR-*an* BAD-*ma* MÚL.MÚL *e-la-nu* ^{d}UDU.IDIM IZI.ŠUB RA BAD-*ma* KI.TA *i-ṣa*

6. Of Ajaru (i.e. Month II) the Pleiades, of Simanu (i.e. Month III) the Raven, of Du'uzu (i.e. Month IV) the Arrow, of Abu (i.e. Month V) the Wagon, of Ululu (i.e. Month VI) the Frond, of Tešritu (i.e. Month VII) the Dog,
7. of Araḫsamnu (i.e. Month VIII) the Scorpion, of Kislimu (i.e. Month IX) the Boat, of Ṭebetu (i.e. Month X) Mars, of Šabaṭu (i.e. Month XI) the Vole, of Adaru (i.e. Month XII) the Field; those *stars* of the months (mean) any stroke of lightning.
8. You make a prediction: if the *stars* are above a planet, a stroke of lightning will devastate; if below, it will be small.

(Hunger-Brack-Bernsen 2002: 8 obv. 6–8)

The entries of AO 6455 show an interesting way to predict weather phenomena on the basis of associations with stars. The "stroke of lightning" (*miqitti išati*, lit. "strike of fire", or IZI.ŠUB, *izišubbû*) is a meteorological phenomenon mentioned in *Šumma ālu* (Freedman 2017: 91), in letters to Assyrian and Babylonian scholars (SAA 10, 42 and 69), and *namburbi*s (Maul 1994: 117–156), but nowhere it is associated with other celestial phenomena.[404] The scribe of AO 6455 followed a pattern of stars that roughly rise heliacally in each of the given months, and associated them with the predictions for lightning (Hunger-Brack-Bernsen 2002: 18, 5 and 6f.). The strength of the stroke is said to depend on the position of a planet moving towards those stars (i.e. above or below). The use of fixed schemes and patterns to predict phenomena, such as the heliacal rising of stars or the binary opposites (i.e. above versus below) is among the working principles of divination par excellence (see A.2.1.1.1., A.2.1.2.) and, in AO 6455, coexists with astronomical computations.

On a final note, the schematic nature of the predictions in AO 6455 is perhaps stressed by the choice of the scribe to use MÚL.MÚL, *zappu*, "Pleiades", or *kakkabū*, "stars", to indicate all the other stars rising in each month (AO 6455 obv. 6–8). Usually, *kakkabū*, "stars", in celestial omens is written MUL^{meš} and never MÚL.MÚL, but MUL.MUL can also be read *itanbuṭum*, "to be constantly bright" (see 2.3.1.), or be taken as the reduplication of MUL, *kakkabu*, "star", to indicate its plural. In AO 6455, probably the scribe deliberately chose first to write MÚL.MÚL for "Pleiades" in obv. 6, and then again MÚL.MÚL but for *kakkabū*, "stars", in obv. 7–8, because these two lines refer to a plurality of stars and not to one group of stars.

404 See also *Iqqur īpuš* (Labat 1965: 134–137 § 65, 66): IZI ŠUB-*ut*, "a fire falls (over something)".

6.3. Conclusion for the Nature of the Pleiades in the Late Babylonian Period

This chapter explored the astronomical and astrological texts dating to Neo- and mainly Late Babylonian period (ca. 626–30 BC) in which the Pleiades are mentioned. The discussion focused on understanding the gist of these texts: they show significant changes in how ancient scholars predicted celestial phenomena, new textual genres, and at the same time, elements from the divinatory tradition. Indeed, EAE continued to be copied until the first century BC by the same scholars acquainted with astronomy.

The increasing importance of the knowledge of the sky held under the Neo-Assyrian court developed into astronomical texts on the one hand (see 6.1.) and astrological texts on the other (see 6.2.). Computed data, regular celestial observation, and recordings of prices were the basis for the Astronomical Diaries (see 6.1.3.). The influence of the zodiac (see 6.1.1.) can be found in the micro-zodiac texts (see 6.2.1.), astral medicine (see 6.2.2.), astral magic (see 6.2.3.), and various types of astrological texts (see 6.2.4.). Within these sources, the "celestial divination changed in primary purpose from being concerned primarily with the public domain to include the life of the individual" (Koch 2015: 198). The predictive assumption of these texts is different from divination. Before the zodiac (i.e. end of the fifth or beginning of the fourth century BC), ancient scholars chose an omen for a specific celestial observation; hence, ominous signs predicted future events. In Late Babylonian astrology, computed astronomical phenomena predicted future events. Nonetheless, "relationships can be made between the epistemic and predictive goals of the early knowledge of the heavens and the later history of astronomy" (Rochberg 2016: 85). Astronomical and astrological texts show how ancient scholars still understood phenomena in relation to analogies and associations, like in celestial omens (Rochberg 2016: 231–273). This concept contrasts with the Kuhnian idea of scientific revolutions proposed by Brown (2000: 153–161), the idea that the astral compositions of the late second and early first millennium BC (see 4.) were "merely" tools of abstraction, or showed a "scientific break" with the past compared to the Late Babylonian astronomy.

Regarding the Pleiades, in sources dating to the Neo- and Late Babylonian periods (ca. 626–30 BC), it is possible to trace both old and new patterns. These texts still cater to the association of the Pleiades with witchcraft, illnesses of the kidney, pestilence, plague, rain, and flood (see 6.2.2., 6.2.4.) – all features found in previous traditions – depending on the needs on which the texts focused, i.e. disease, trade, or weather. The Pleiades are also depicted in their most important position in the sky, balancing with the Moon, a conjunction on which the Mesopotamian ideal calendar used to be based (see 4.2.2., 6.2.1.). However, astrological texts testify to a "minor" role of the Pleiades compared with the Neo-Assyrian period (ca. 911–612 BC). Of course, that could also be due to the preservation of the sources at our disposal, but it seems clear that the significance of stars, their names, and explanations became obscured by the zodiac as the primary spatial and theoretical framework. Consequently, the Pleiades are almost exclusively seen as *pars pro toto* for Taurus in zodiological literature: new analogies arise from the zodiacal sign Taurus (e.g. the *materia medica* of Taurus in 6.2.1.), and they partially exclude the tradition of celestial omens where several analogies between stars and gods (e.g. the Pleiades associated with the Sebettu or the divine Seven, see 5.3.1.–5.3.2.) arose. In this scenario, the Pleiades act as the heritage of an earlier tradition.

The cuneiform and the Classical sources testify – even though this is based on limited material dating to the Achaemenid (ca. 547–331 BC) and Hellenistic period (ca. 323–63 BC) – the influence of the Babylonian astrology on the Hellenistic one. The zodiac, the *hypsoma*, the *dodekatemoria*, and the *melothesia* (see 6.1.1., 6.2.1.–6.2.2.) could be products of re-interpreted Babylonian elements. Rochberg (2016: 273) further suggests that if the same predictive modes are attested in omens, astronomical and astrological texts, and their heritage is found in Greece, then we should include cuneiform knowledge in the history of scientific thought: "Each branch of cuneiform predictive knowledge was tied to programs of observation and a tradition for recording and dating those observations. From our point of view, observation, prediction, and explanation belong within the purview of science. Although the methods of investigation into the world of perception and experience, and ideas of what was usual, unusual, regular, irregular, normative, and anomalous were determined by the particular phenomena – many but not all ominous phenomena – that interested the Assyrian and Babylonian scholars over time, in terms of the observational, predictive, and explanatory dimensions of cuneiform knowledge, particularly in its persistent attempts to grasp an order of things and to resolve what is anomalous into a system, it is hardly possible not to see the features of its kin in the later history of science."

Chapter 7

Conclusions

The case of the Pleiades proved to be a fruitful way to study the perception of celestial bodies in Mesopotamia. This study has been set in a theoretical and methodological framework crucial for such an investigation, due to intertextuality and the concept of metaphor in cognitive science and semiotics (chapter 1). When starting from such ground premises, cuneiform knowledge immediately deserves epistemological attention: it interlaces culture, science, and religion, and it is often built by analogical relationships, on both an ontological and a philological level. First, there was the effort of reconciling two cultural units (i.e. Sumerian and Akkadian language) in one, which created a mixed syllabic-logographic writing system, and a network of associations between words and signs. Second, such a way of building knowledge was consistent with the overall perception of the cosmos in Mesopotamia, which is multifaceted. The words of Heeßel (2005: 17–18) epitomise this way of reasoning: "Das Denken in Analogie geht davon aus, daß zwischen bestimmten Teilen des Kosmos offene und verborgene Entsprechungen existieren, die der Kundige lesen und entziffern kann."[405] Therefore, the perception of the celestial bodies and the Pleiades in Mesopotamia have been treated in this study as products of a culture – Mesopotamian culture – where the modalities of building knowledge are widely based on analogical reasoning, without confronting them by defect with a positivistic conception of culture and science.

This study began with a discussion on how the Pleiades were perceived within the lexical and cultural Mesopotamian repertoire (chapter 2). According to lexical lists, MUL.MUL, lit. "stars", was intended as anything extremely or continuously bright (i.e. equated to the Akkadian verbs *napāḫu* and *nabāṭu*, "to glow" and "to shine brightly"). At least from the Old Babylonian period (ca. 2000–1500 BC) onwards, the Pleiades are identified as a group of stars called *zappu*, "Bristle", conceived as the mane of Taurus pictured in the sky as a bull. The Akkadian *zappu* has then been associated with the logogram MUL.MUL, merging two different traditions. The Pleiades were also tied up with the "mystical" meaning of the number seven (IMIN, *sebe*), as they were perceived as a group of seven stars.[406] Already in the third millennium BC, the number seven was a relevant number not only in Mesopotamia, but in the entire Ancient Near East,[407] so that it became a synonym for "totality" (*kiššātu*), or "completeness". Therefore, the Pleiades

405 "Thinking by analogies starts from the assumption that there are explicit and implicit correspondences between certain parts of the cosmos that an expert can read and decipher."

406 See the debate regarding the number of the Pleiades (see 2.1.).

407 It is not possible to say with certainty why the number seven came to be so popular in the entire Ancient Near East (see 2.4.–2.4.2., and Reinhold-Golinets 2008). This leads to the chicken and the egg issue because the recurrent use of the septenary structure is challenging to assume without an empirical phenomenon and *vice versa* (Negretti 1973: 52 fn. 44). The same issue applies to the Pleiades, who are seven to ten or 12 stars with naked eye but are called seven by a fixed attribute in many ancient cultures (see fn. 8).

became intrinsically embedded, by analogy, in the meaning of the seven, and the number of the Pleiades was a fixed attribute, regardless of how many of their stars were observable with the naked eye. In chapter 2, the Pleiades have also been included in a wider discussion about the perception of celestial bodies and gods in Mesopotamia. Following the arguments of Rochberg (2009; 2010: 317–338; 2011), the Pleiades can be understood in light of three perspectives according to different types of sources: gods described as celestial bodies in mythology (i.e. astralisation), and celestial bodies as representations of the gods and/or being personified in prayers and astral science (i.e. astral compositions, divination, astrology).

The discussion continued with the Pleiades in Mesopotamian literary texts (chapter 3). Mythology is full of scattered divine, heroic, and demonic heptads with astral features but most of them share nothing with the Pleiades but the number. For instance, compositions such as the Sumerian "Hymn of Ḫendursaĝa", and the UDUG.ḪUL incantations, attest to a literary *topos* of seven hybrid or theriomorphic entities of the Mesopotamian pandemonium, rather than an association with the Pleiades. Generally speaking, mythology does not disclose much about the Pleiades; there is only one possible mention in the Sumerian myth "Lugalbanda in the Wilderness" dating to the Old Babylonian period (ca. 2000–1500 BC), where "seven torches of the battle" (izi-ĝar mè imin-me-eš), or the "pure weapon" (šíta kù) of Inanna in the night sky are called mul-mul, "Pleiades", or "starry". From the first millennium BC onwards, the Pleiades are associated in various contexts with the seven demonic and warlike Sebettu (^{d}IMIN.BI, *sebettu*) of Akkadian mythology. They are part of the religious framework of the Neo-Assyrian (ca. 911–612 BC) military propaganda, co-protagonists of Erra in the Akkadian poem of "Erra and Išum". Through intertextuality, we learn that they were considered the successors of an "older" heptad, the seven heroes (ur-saĝ imin) of the Sumerian composition "Gilgameš and Ḫuwawa"; these heroes had the ability to guide travellers at night by the starlight – maybe because they embodied a group of stars. According to another parallel tradition, it seems that the Pleiades were associated with at least another heptad: seven great gods belonging to Mesopotamian pantheon, responsible for making decisions for the well-being of humankind. These "divine Seven, the great gods" (^{d}IMIN.BI DINGIR^{meš} GAL^{meš}), sometimes called "determiners of fates" (*mušimmū šīmāti*), appear as a group in the Sumerian cosmogony of "Enlil and Ninlil", and in the Akkadian composition *Enūma eliš*. The fact that the Pleiades could have been associated with this or that heptad is especially testified by prayers dating to the first millennium BC. For instance, the *šu'illa* prayers attest to the coexistence of astral, warlike, demonic, and divine aspects of the Pleiades, sometimes by quoting epithets or passages from mythology. The fact that in the sources discussed in chapter 3 the Pleiades were perceived either as divine agents – capable of influencing rituals or extispicy, like the gods of the night (*ilū mušīti*) – or as the representation of the gods is always consistent with the multifaceted perception of the cosmos, and with the perspectives on celestial bodies discussed in chapter 2. Therefore, a re-definition of astralisation has been provided at the end of chapter 3: astralisation should not be understood as a religious tendency *per se* but as one of the ways of conceiving and expressing nature and divine agency through language according to the Mesopotamian culture.

The first astral compositions (chapter 4), such as MUL.APIN or Astrolabe B, describe the Pleiades in their celestial context, and show that, when in close proximity to the Moon (i.e. LÁL, *šitqultu*, "balancing"), they determined the length of the Mesopotamian year, i.e. determined the intercalation. These compositions testify to procedures or so-called rules allegedly used by ancient scholars to calculate intercalation for an ideal calendar, at least from the late second or early first millennium BC onwards. The intercalation was very likely not meant to be used in practice, but it aided divination by adjusting the calendar, so that ancient diviners could have made correct predictions. The fact that references to the intercalation are found as celestial omens means that the rules in astral compositions have been likely derived by abstraction from them; thus, intercalation had a predictive purpose, and not "merely" technical. Regardless of intercalation, astral compositions show several features, or symbolism, attributed to the Pleiades: the association with the divine Seven, their heliacal rising (see fn. 220) in Ajaru (i.e. Month II) matching with the beginning of the yearly agricultural activities, and the connection with the cardinal point east and Elam. These features are recurrent in other sources like the Great Star List, celestial omens, and a few sources dating to the Late Babylonian period (ca. 484–30 BC) that represented older traditions.

The celestial divination of Mesopotamia was the key topic to analyse the perception of the Pleiades (chapter 5). Divination produced a bank of information based on the omens or the logic of conditionals. The first celestial omens date to the Old Babylonian period (ca. 2000–1500 BC), and they appear in the form of a series called *Enūma Anu Enlil* in sources dating to the first millennium BC. Considering that the Sumerian pantheon was already based on associations between gods and stars – and considering that *Enūma Anu Enlil* is a final product of a way older, longer and re-worked oral tradition – it is reasonable to believe that the omens dating to the first millennium BC were already a mixture of various cultural heritages. As for their content, celestial omens provided predictions with different shades of probability, built through associations between celestial and terrestrial events, given in the protases and the apodoses of the omens, respectively. More specifically, the semiotic links between protases (i.e. ominous signs) and apodoses (i.e. predictions) of celestial omens were built on associations of any kind, with a focus on analogical reasoning over the physics of nature. The existence of "impossible" protases, i.e. protases whose meaning is metaphorical, or whose literal meaning is astronomically impossible, testify to this statement. All omens were copied, preserved, and then matched with factual explanation, i.e. alternative protases in which phenomena are described mainly as planetary ones, until the end of the first millennium BC. First, because those omens were still consistent with the modes of building cuneiform knowledge (i.e. through analogical reasoning). Second, because they were considered by every means a form of divine and celestial writing, and therefore untouchable. Intertextuality is strongly involved because one can find references to celestial bodies in mythology and literary texts, and mythological and religious quotes in divinatory texts at the same time. At any rate, the Mesopotamian divination is the product of a strong literate tradition: the diviners (*ašīpu*, "exorcist") were certainly trained in mythology, likewise other scholars knew about the principles of divination and astral compositions.[408] Analogies between mythology and divination were probably obvious to Mesopotamian scholars, who knew how to interpret each text – that is, the meaning of intertextuality in a nutshell.

Given the heterogeneous nature of the working principles of celestial omens, a specific discussion is included in order to develop the necessary groundwork to interpret the omens (Appendix A). The rationale of omens – or the essence of the semiotic link between protases and apodoses – is surely based on analogical reasoning, but it is exercised in different ways, some of which have been also witnessed in astral compositions and astrological texts. In other words, omens are based on a network of working principles, which are examined through examples in the appendix A, and arranged according to the principles of linguistics. The structure of the series of omens themselves – a sort of thematic "proto-encyclopaedias" – proves how these working principles were epistemologically valid in Mesopotamia, and should be considered relevant for gnoseology and the history of scientific thought.

In this framework, the role of the Pleiades in celestial omens – in the reconstructed *Enūma Anu Enlil* tablet 52, the commentary to tablet 53, the assumed *Šumma Sîn ina tāmartišu* tablet 6, other edited *Enūma Anu Enlil*'s sections, and the so-called astronomical and astrological reports (chapter 5, appendix B and plates) – can be summed up as follows: the Pleiades and their phenomena (i.e. conjunctions, luminosity, risings and settings) were seen as a tool to organise the calendar, to mark the beginning of the agricultural season in spring, and thus to foresee positive or negative events related to the latter. The Pleiades were also perceived as the representation of two divine heptads, with which they shared specific features:

- The Pleiades (MUL.MUL) named in protases are associated with the Sebettu (^{d}IMIN.BI) in the apodoses. Whenever the association is not explicit, it can be implicit in the events predicted, which deal with the herd and the harvest. The most recurrent events are the consumption (GU$_{7}$, *ukultu*) or the epidemic (ŠUB, *miqittu*) of the herd (*būlu*), which are quotes of the actions of the Sebettu in the Akkadian poem of "Erra and Išum".[409] Famine (SU.GU$_{7}$, *ḫusaḫḫu*) and pestilence (ÚŠ^{meš}, *mūtānu*) as predictions are strictly related to Nergal/Erra, who is *bēl šibṭi u šaggašti*, "lord of plague and carnage". The Pleiades were sometimes intended as substitutes for Mars (*ṣalbatānu*) in ancient commentaries of omens because of the analogy between them and the Sebettu. This association is unique and built on another association between Nergal/Erra and Mars.

- The Pleiades (MUL.MUL) named in the protases are associated with seven great gods (^{d}IMIN.BI DINGIR^{meš} GAL^{meš}, lit. "divine Seven, the great gods") in the apodoses. This association, when not explicit, can be implied by the events predicted, which deal

408 Indeed, we know that the exorcists considered *Sîn-lēqe-unnini*, the famous author of the epic of Gilgameš (George 2003: 28–33), to be their ancestor, and he was an exorcist too (Lambert 1957; 1962).

409 E.g. *šumqutu būl Šakkan*, "to destroy the herd of Šakkan" (Erra I 43, see Cagni 1969: 62 l. 43).

with the well-being of the king and the country granted by the decisions (GALGA[meš], *milkū*, or *uṣurtu*, "divine designs") of the great gods, or the actions of seven gods "determiners of fates", a powerful group of divine beings within the hierarchy of Mesopotamian panthea.

The association of the Pleiades with these two different heptads is only attested from the first millennium BC onwards. We cannot assess whether these two parallel traditions are the legacy of older beliefs, or whether they were learned associations made *a posteriori* by the scholars who wrote celestial omens.

The astronomical and astrological texts dating to the Late Babylonian (and partly already Neo-Babylonian) period (ca. 626–30 BC) have been discussed in chapter 6 because they presume a different predictive framework than the earlier astral compositions, due to an increasing mathematical accuracy in predicting celestial phenomena, and an increasing focus on the life of individuals versus the well-being of the reign. At least from the end of the fifth or beginning of the fourth century BC, zodiac-based systems substituted the heterogeneous panorama of stars, with their names and divine counterparts. At the same time, Late Babylonian scholars still used the older principles of divination, together with computed data, to predict events. A *fil rouge* tied to the earlier tradition in the perception of the Pleiades is still detectable, for instance in astral medicine and astrometeorology. These texts are based on analogical reasoning and the same schemes found in celestial omens and astral compositions. Therefore, the Pleiades are still associated with the kidney (ÉLLAG, *kalītu*),[410] pestilence (ÚŠ[meš], *mūtānu*), rain (ŠÈG[meš], *zunnū*) and devastation by Adad, the weather god ([d]IŠKUR RA, *Adad iraḫḫiṣ*), all features already known in omens, that appear in astrological texts as a reference to earlier divination, and because they served the purposes of those texts, i.e. to predict disease, trade trends, or weather. However, within the zodiac belt as the main spatial framework of Babylonian astrology, the Pleiades appear only as a *pars pro toto* of Taurus (GU$_4$.AN.NA, *alû*, "Bull of Heaven"), or as its name. The sources of zodiological literature at our disposal provide a sufficiently clear impression that the Pleiades played a major role only in celestial divination.

A last remark is necessary concerning one of the research questions posed in chapter 1: what is the relationship between the Pleiades and the concept of "god"? Considering the totality of the sources analysed, the Pleiades are framed into different perspectives, as Rochberg (2009; 2010: 317–338; 2011) argued regarding any celestial body. An astralisation of seven beings as the Pleiades is only in the background (e.g., the seven torches in "Lugalbanda in the Wilderness"); most of all, they are seen either as the representation of a group of gods (e.g. in prayers, omens, and astral compositions), or they act like divine agents. The latter perspective is attested in contexts where the Pleiades are directly addressed by their proper name (MUL.MUL or *zappu*, in *šuʾilla* prayers), and are capable of influencing the outcome of an incantation or a ritual (e.g. the prayer to the gods of the night, or the incantation "I am the vigil, the sister of Marduk"). There seems to be a progressive shift towards the celestial bodies seen as directly responsible for influencing

410 See the Pleiades associated with the kidney in lexical lists (see 2.3.1.), and incantations dating to the first millennium BC (see 3.2.2., 3.2.4.).

mundane affairs in the astrological texts. However, the references closely related to the Pleiades are few due to the predominance of the zodiacal signs, and also the gods related to them, be they the Sebettu or the divine Seven, are never mentioned.

The questions posed in this study have by no means exhausted the study of the overall perception of celestial bodies in Mesopotamia, as the Pleiades were only a case-study. If anything, this study opens several possibilities for further comparative investigations. Many questions on how the role of the Pleiades in Mesopotamia should be included in wider historical, ethnoastronomical, and anthropological inquiries remain untouched. Indeed, a peculiar significance of the Pleiades is witnessed not only in Mesopotamian culture (Young 1987: 8734–8735): there is the possibility that the cult of the Pleiades and the stories related to them were one of those archaic tales whose origin precedes writing and literacy. Already in a panel in the prehistoric cave of Lascaux in France (ca. 15.300 BC), modern scholars recognised a depiction of the Pleiades with a bull, as shown in Figure 5 (Rappenglück 1997).

Figure 5. Lascaux cave panel in the "Salle des Taureaux" (Rappenglück 1997: 218 fig. 1).

Due to their high visibility throughout the globe and basically all year round, the relevance of the Pleiades as a seasonal or calendrical marker is a pattern that recurs in the first stages of several cultures, and thus has fascinated modern scholars. The perception of the Pleiades is similar in various civilisations, due to the distinctive position of the cluster near the ecliptic, and their particular brightness. Among the parallels between the Pleiades in Mesopotamia and other cultures, the conjunctions with the Moon, with Venus, the use of the Pleiades as a spatio-temporal marker and as a tool for astrometeorology, and the identification with the ploughing stand out particularly (Rappenglück 2008). Lévi-Strauss (1969: 216–246), in his famous book "The Raw and the Cooked", assumed that the mythical contrast between Orion and the Pleiades (i.e. motif of pursuit; Campion 2012: 28–

29) in the religions of South America and Greece mirrored the continuity of the Pleiades and the discontinuity of Orion in their heliacal rising. The Pleiades also marked the beginning of the agricultural season in Greece (Toomer: 1996: 68), in India[411] (Pingree 1996: 123; Brown 2018: 585–601), in Mexico among the Aztecs and Incas (Aveni 1996: 296), in Africa (Warner 1996: 306–311), and the beginning of the dry season in Northwest Amazonia (Hugh-Jones 1979). Therefore, it is hoped that this study will serve not only as an incentive for other case-studies about celestial bodies in Mesopotamia, but also as a starting point for comparative studies in the future.

411 Numerous motifs of Indian astronomy might find their forerunners in Mesopotamia (Pingree 1987 and 1996; Puhvel 1991). Perhaps, there is a motif similar to Erra and the Sebettu in Hinduism: the Pleiades, Kṛittikā in Sanskrit, conceived Kārttikeya, or Skanda, the young Hindu god of war, also identified with Mars (Mannikka 1996: 190), like Nergal/Erra.

Appendix A

Interpreting Celestial Omens

This appendix aims to describe and explain some of the principles behind the arrangement of the omens dating to the first millennium BC, in order to understand their rationale. In the first part of the appendix (see A.1–A.1.2.) one finds an overview of the ongoing debate regarding celestial divination. Is it science or not? And, should it be included in the history of science? These issues are introduced and discussed with respect to the structure and the logic behind the arrangement of the omens. In the second part of the appendix (see A.2.–A.4.1.), the working principles of celestial omens are categorised and commented on based on examples, not only from the omens about the Pleiades, but also from other sources. This has been done because the reconstructed *Enūma Anu Enlil* tablet 52, the commentary to tablet 53, and *Šumma Sîn ina tāmartišu* 6 are only fragmentarily preserved, and the necessary components for a more in-depth analysis are sometimes lacking.

Many of the discussed working principles are preliminary and still need further studies because they were never collected systematically.[412] Therefore, this appendix does not aim to provide a comprehensive description of all the principles of divination, as this would require a study in its own right; instead, it works towards introducing a fruitful way to approach celestial omens, in the light of old and new viewpoints on Mesopotamian divination.

A.1. Sense and Non-sense, Science and Religion

In the Mesopotamian world, the concern of the ancient scholars was to interpret the messages of the gods, which were thought to be expressed through natural phenomena. The phenomena are considered ominous signs (GISKIM, *ittu*) and interpreted as if nature is something intelligible or even readable, like a cuneiform tablet.[413] In this conceptual framework, the goal of divination is to interpret the will of the gods, or their "designs" (*uṣurtū*), as portrayed by ominous signs. The ominous signs and their predictions were written down in the form of omens, therefore the collections of omens, such as the series *Enūma Anu Enlil* (EAE) or *Šumma Sîn ina tāmartišu* (SIT), constitute a bank of information on celestial phenomena.

For a modern scholar, the reading of Mesopotamian celestial omens can raise doubts about their meaning: at first glance, they can appear repetitive and devoid of logical sense, written with an extensive use of logograms. For instance, the protases of celestial omens describe both astronomically possible and impossible phenomena, and the terminology

412 "The challenge, in short, is to find systematic interconnections between the protases of omens and their apodoses" (Guinan 1996: 5).

413 E.g. see the concept of *šiṭir šamê*, "writing of the sky" (see 2.2.3.2.).

used to describe them seems somewhat obscure. One of the omens of the Pleiades is a good example to explain this issue:

> If at the beginning of the year the Pleiades are elongated. (It means that) at the balancing with the Moon they are ahead.
> (see 4.2.2. § 2b, pp. 119–120, 4)

The Pleiades are a cluster of stars, and by no means can this cluster be "elongated" (*šatiḫ*). In the protasis, the logic, which we would call scientific, is missing, but the factual explanation equates the protasis to an observable phenomenon, or a phenomenon differently described, perhaps with a more literal, technical, or updated terminology. The balancing between the Moon and the Pleiades is a celestial conjunction, a phenomenon that was well known by the diviners because it was used as a reckoning device for the intercalation in the Mesopotamian calendar (see 4.2.2.–4.2.2.1.). A first question spontaneously arises: Why did the ancient scribes use omens that were written in a way so that were "impossible to happen" in reality? Generations of ancient experts were able to create the basis for astronomical computations by the second half of the first millennium BC (see 6.1.), and at the same time they continued to copy and work on the series EAE, including astronomically impossible protases, and produced commentaries in order to modify or update them. It is challenging for us to understand how that happened, yet the reason may be explained only through the examination of cultural features in a more detailed way.[414]

Evidence from Sumerian literary, religious, and iconographic tradition shows that celestial bodies were considered gods moving freely across the sky, without making any specific distinction between planets and stars (see 2.2.1.). No omens written in Sumerian dating to the third and second millennium BC are preserved though. The first celestial omens we know of date to the Old Babylonian period (ca. 2000–1500 BC), their protases were based on recurring phenomena, such as lunar, solar, and weather phenomena, and they already combined several subjects through fixed working principles (Fincke 2016: 114–115). These facts point towards a scenario in which the scribes collected – probably by the end of the second millennium BC – all the omens at their disposal, and wrote down the omens from the oral tradition in order to produce a more standardised reference text such as the series EAE.[415]

The series EAE was thought, at least during the Neo-Assyrian period (ca. 911–612 BC), to be given by Ea, the god of wisdom (Rochberg 1999: 419). According to another

---

414 "Perhaps the reason for our puzzlement is that for too long we have failed to see how a notion of the order of nature was fundamentally absent from and irrelevant to cuneiform divination" (Rochberg 2016: 107).

415 "Mesopotamian astronomy-astrology, contrary to current opinion, drew heavily on an intellectual tradition in which the nature of the universe had been explained since 'Sumerian' times in the third millennium BC in terms of 'divine powers', a cosmic 'design' in space (broadly, celestial 'writing') and time (broadly, 'ideal periods') and the production of 'signs' for the benefit of mankind" (Brown 2000: 239).

contemporary tradition, EAE was believed to be written by the mythic sage Adapa.[416] Even the practice of divination itself was thought to be revealed to the sage Enmeduranki by Šamaš and Adad (Lambert 1998). From these statements, it is clear that the knowledge transmitted by the omens had a sort of divine authority. Therefore, even when copying tablets from EAE or another omen series, the original omens were never changed, yet they were expanded, selected, and commented on by ancient scholars to obtain as many variables as possible to cover any contingency. Ancient scholars pursued the so-called "secret" knowledge (*niṣirtu* or *pirištu*), given by the gods and taught in schools (e.g. temples), or by private scribes. The "exclusivity" of Mesopotamian scholarship, remarked in rubrics and colophons, leaned on the scholars who considered themselves the descendants of the mythic sages (Frahm 2011: 47; Koch 2015: 53–54). As Fincke (2013a: 178; 2016: 119–122) stated, the legitimacy for the new additions and expansions came from the expertise of these highly educated scholars who generated them; these additions and expansions were incorporated into the main sequence of older omens, which was considered divine and even more reliable than the newer omens. This way, the whole tradition survived until the first millennium BC, and even later, until the Hellenistic period (ca. 323–63 BC), when the series EAE still continued to be copied (Fincke 2016: 119).

To briefly sum up, the first written omens dating to the Old Babylonian period (ca. 2000–1500 BC) might be seen as an initial attempt to organise a bigger and prolific oral tradition. When the standard series of omens (i.e. EAE, see 5.1.1.) was found on tablets dating to the first millennium BC, it was already a much-varied collection of omens (Fincke 2016: 114–117, fn. 39). At any rate, the overall picture for Mesopotamian omens – especially celestial omens – is a sort of *vade mecum* for understanding and cataloguing phenomena in an encyclopaedic way, in order to decipher the will of the gods.

We are aware that the transmission of written knowledge came side by side with the creation of a lineage of expert scribes, concerned with their status, the accuracy of the transmission, and hermeneutics.[417] Yet, Mesopotamian scribes rarely wrote down the principles of the omens: regardless of the relatively few explanations provided in hermeneutical texts, we shall assume that the great majority of information unclear to us was obvious to Mesopotamian scholars. "As with other ancient Near Eastern literature, a proper understanding of the texts assumes knowledge that is not given on the tablet itself but could be found elsewhere or had been transmitted orally. The reader is given the most obvious aspects of the texts to grasp, but the overall context and specific terminology is taken for granted" (Fincke 2016: 124).

With such preliminaries in mind, a second question which arises is whether divination is more akin to religion or science. The recent debate on including cuneiform knowledge into the history of scientific thought comprises Rochberg (2016: 17–37), who followed the theories of historical epistemologists, according to whom the Western positivistic approach

416 See the Verse Account of Nabonidus (Schaudig 2001: 570 V 12'), Lambert (1962: 64 I obv 1), Machinist-Tadmor (1993), and Koch-Westenholz (1995: 74–76).

417 For divine authorship in Mesopotamia in general, see Lambert (1957; 1962); for the canonicity in celestial omens, see Rochberg-Halton (1984); for the scribes involved in hermeneutics, see Frahm (2011: 12–23).

contributed to creating a misleading picture of the Ancient Near East, as only dominated by religion and superstition. It might be more fruitful to look at science of the past as the process of acquiring knowledge and explaining it through local and cultural features, as this would be in line with pluralism and localism within the history of science. If this way of reasoning is applied to cuneiform knowledge, it would be opposed to the older mythopoetic vision of Frankfort (Frankfort-Frankfort-Wilson-Jacobson 1946): according to his view, which became very popular among Assyriologists, ancient thought would be a speculative and thought.[418] The debate on how to include cuneiform knowledge in the history of science is the counter trend of that popular vision, yet it is extremely fruitful for the present context – the context of celestial omens. Before discussing the debate any further, it is necessary to explain the structure of omens (see A.1.1.), and their rationale (see A.1.2.).

A.1.1. A List for Everything

If we read the omens one by one, they tell us a little about how they worked and what they really meant. However, if we include them in a broader context, they begin to disclose interesting hints. Therefore, a first focus should be on "the practice of how knowledge is used in an epistemic context" (Graßhoff 2011: 34), more than on understanding the meaning of each omen.

Divination is a tool used to gain and collect messages from the gods, and omens are pieces of information in the form conditional statements (see already 5.1.):

If X, then Y

In logic, this type of conditional statement is expressed by the formula P→Q, which means that the preposition P, or the antecedent (i.e. If X), is sufficient for the truth of the preposition Q (i.e. then Y), or consequent. Rochberg (2010: 373–397) was the first to propose the idea of understanding Mesopotamian omens as inferences. An inference, based on the *modus ponens*, or deductive logic, means to derive conclusions from known premises: If P→Q is true, and P is true, then Q is also true. Only if Q is false, then P→Q is also false: this means that the validity of P→Q does not depend on the inherent truth or falsity of P and Q, but only on the relationship between these two.[419] Rochberg (2010: 385–386) rightly pointed out that "the written body of omens can be construed as a list of conditional statements that offer inferences from established premises". A knowledge built on inferences is not necessarily based on the physics of nature, like modern science, but on differently oriented logic connections (i.e. semiotic link) between an antecedent and a consequent. This means that the core logic of written omens is widely based on how they are written and arranged.

418 For a detailed critique of mythopoetic thought and how it influenced generations of Assyriologists, see Rochberg (2016: 38–58).

419 One practical example of *modus ponens* is the following: "(P) If it rains, (Q) the road is wet". Even if P did not happen, this does not mean Q cannot happen anyway; but if P happened and Q did not occur, then the statement "If it rains, the road is wet" is false.

On clay tablets, the omens are organised in one or more columns, like lists. The "list" format is considered a distinctive element of Mesopotamian textual culture: lists of any kind are ubiquitous, and the oldest example is that of lexical lists. In lexical lists, names were given to objects, activities, or concepts, and framed into categories using determinatives or classifiers. Visually speaking, the internal organisation of lexical lists works vertically (e.g. between individual entries) and horizontally (e.g. signs equated to words):

udu niga	fattened sheep	X = A
udu niga sag$_{10}$	good quality fattened sheep	X_1 = B
udu niga ĝír-gu-la	fattened sheep that has been plucked with a knife	X_2 = C, etc.

(MSL 8/1: 83 ll. 1–3)

KU (...) *kak-ku*	KU = weapon	X = A
KU.KU (...) *ṣa-la-a-lu*	KU.KU = to lie down	X_1 = B, etc.

(MSL 14: 184–185 ll. 158–159)

In practice, each entry of a lexical list is based on the same sign, together with many additions or qualifications applied to all the possible signs in different contexts. For instance, the first archaic lexical lists[420] – which were likely a teaching tool, or one of the tools for the "construction of the scribal identity" (Veldhuis 2014: 59) – covered the basic topics useful to administrative purposes (e.g. animals, metals, trees, officials, etc.). Yet, it has been noted that there were already combinations of signs without them having necessarily some correspondence in reality or practical use. One could assume that scholars created new signs to prevent the creation of new ones in the future, or that lexical lists were created by abstraction from the administrative use of writing (Veldhuis 1999; 2014: 43).

As discussed by Veldhuis (2014: 16–23), the arrangement in lists itself has been first understood as the urge to organise knowledge: the terms *Listenwissenschaft* and *Ordnungswille*, coined by von Soden (1965, orig. 1936), are widely adopted in Assyriological studies, and they express that urgency. The sense of both terms, as initially meant by von Soden in 1936, was a product of the racial ideology of that time, implying that the differences between Mesopotamian and Greek literary production lay in race. Later, the racial implications were rightly exposed and rejected, and the terms have been used to indicate the epistemic practices which led to lexical lists and similar works (Hilgert 2009; Veldhuis 2014: 22). The anthropologist Goody (1986) included Mesopotamian lexical lists in his theory about literacy, based on earlier writings of Oppenheim (1964): Goody's theory dwells on the idea that the writing extended from practical needs to the religious field. This "step" made possible a more significant development of thought, producing new questions and needs. Writing fixed the concepts, the ideas, and the formulas which were handed down. Writing established the so-called "stream of tradition", and the earliest fixed formula

420 For an overview of archaic lexical lists, see Veldhuis (2014: 32–48).

we have evidence of is the list format. Goody suggested that writing shaped cognitive patterns, our way of thinking, as well as probably shaped the earliest forms of Mesopotamian divination before omens were written down the way we read them today. Based on these assumptions, the idea of tracing back the origins of thought from Mesopotamian lexical lists was born. More specifically, many scholars kept suggesting that the lexical lists, as well as divination, may reflect the *Ordnungswille*, or a systematic "will of order" of the things of the world, as a catalogue of the universe.[421]

Nevertheless, there seems to be room for fallacy in that suggestion. Veldhuis (2014: 55) remarked that the idea of an "order of the world" is weak, because lexical lists do not represent the necessity to catalogue knowledge based on an assumed nature of things (i.e. ontological taxonomy) but provide a philological taxonomy. For instance, Veldhuis (2006) noted how, in lexical lists, the names of animals are organised on linguistic principles and not on biologic taxonomy.[422] A distinction between wild and domestic animals is not even relevant in the literature, even though the use of determinatives can distinguish these two categories. Cuneiform signs and words are listed together according to similarities in their composition of wedges, an ordering principle which has nothing to do with their intrinsic properties. Rochberg (2016: 274–284) argued how these similarities are dependent on the overall goal of Mesopotamian thought: that is, to create norms and detect anomalies based on epistemologically true and meaningful categories. That goal is embedded with the idea of the world according to Mesopotamian scribes (i.e. conceptual and cultural framework): the divine agency permeates the world, and what has been designed by the gods in heaven is true for the earth as well. Therefore, the world should be intelligible, just like the cuneiform signs of a tablet (see 2.2.3.3.). What we perceive as ambiguous or impossible in cuneiform culture is only so for our modern perception because the lexical lists are not organised according to the "nature of things", or to express a "will of order", but based on other needs, convenient for the scribes' purposes, for instance, for administrative purposes, and so on (Rochberg 2016: 95–102, 159).

Lexical lists and omen collections are different in their formal structure and purpose.[423] However, what they have in common is their systematic arrangement. The first important step in understanding the arrangement of the lexical lists of Mesopotamia was made by Schuster (1938) and Cavigneaux (1976): they pointed to the fact that there were many "principles" behind the vertical arrangement of lexical lists, that is, semantic and

421 For an overview of scholars who discussed the *Listenwissenschaft* and the *Ordnungswille* in Mesopotamian lexical tradition, see Veldhuis (2014: 21–23); for the *Ordnungswille* in lexical lists and divination, see Larsen (1987).

422 E.g. the logogram AM, "wild ox", is listed together with other animals whose names begin with the sign AM, and not in the section for GU_4, "ox". This example was also discussed by Rochberg (2016: 99–100).

423 The influence of lexical lists on omen series has been discussed by Goody (1977: 74–111), Leichty (1993), and Brown (2000: 76 fn. 203), as a literate phenomenon. Fincke (2016: 124 fn. 90) argued that omens formally have nothing to do with the *Listenwissenschaft* since the vertical wedge DIŠ at the beginning of each entry in omen collections is not an indication of a list but represents the beginning of a conditional clause (*šumma*, "if"); hence, it is similar to Mesopotamian laws (Fincke 2006).

graphemic principles like the shape of the signs, the recurrence of the vowels *u-a-i*, and the meaning of the signs. The principles of organisation of contiguous omens are comparable to the principles of organisation of the entries in a lexical list; thus, an attempt to individuate their working principles has already been made by several scholars. For example, Brown (2000: 130–131), Rochberg (2010: 381 fn. 13), and Winitzer (2017: 28–30) borrowed the terminology of the structuralist linguistic school (de Saussure 1916) to describe the structure of Mesopotamian omens in a vertical and a horizontal level:

– The vertical level is the paradigmatic relationship between groups of entries and contiguous entries in omens collections. For instance:

DIŠ *ina* SAG MU MUL.MUL *ka-ri-it*	If X (then Y)
DIŠ *ina* SAG MU MUL.MUL *šá-ti-iḫ*	If X_1 (then Y_1)

If at the beginning of the year the Pleiades are cut.
If at the beginning of the year the Pleiades are elongated.
(Commentary EAE 53: 19–20)

Within the vertical level, one omen is followed by another, different omen, based on working principles which will be discussed in A.2.

– The horizontal level is the syntagmatic relationship between the protasis (X, X_1, etc.) and the apodosis (Y, Y_1, etc) of a single omen. This relationship is dictated by several working principles, discussed in A.3., and according to which every protasis is followed by a specific apodosis.

On both the vertical and the horizontal level, the use of different principles creates new combinations of protases and apodoses, as well as creating new cuneiform signs and readings in lexical lists. The words of Frahm (2011: 20) sum up the relationship which ties the omens to the lexical lists very well:

"The omen compendia contain full sentences and not just individual words like the lexical lists, but share with the lists the bipartite structure of their entries. Like the lists, the compendia are gigantic collections of equations, based on correspondences of various types. Protases and apodoses are related to each other through etymology, etymography, mythological and symbolic association, indigenous conceptions of the laws of nature and culture, and many other principles that vary in the different branches of Mesopotamian divination."

Many of the individual working principles of Mesopotamian omens have been identified and studied. For instance, Brown (2000: 105–207) arranged the celestial omens into several types, distinguishing between observed and invented protases and apodoses. Each of them implies a syntagmatic relationship (horizontal) and a metaphoric relationship (vertical). In lunar omens (Rochberg-Halton 1988: 36–63; 2010a), extispicy (Larsen 1984; Winitzer 2017), teratological omens (De Zorzi 2011), physiognomic omens (Böck 2010) and terrestrial omens (Streck 2001) anaphora, paronomasia, associations of ideas and binary opposites have also been detected as working principles of omens.[424]

A.1.2. The Rationale of Omens

The rationale, or the semiotic link, between protases and the apodoses needs to be scrutinised further. Scholars like Koch-Westenholz (1995: 13–19), Brown (2000: 108–113), and Rochberg (2015; 2016) have particularly stressed that the criterion of the internal organisation of omens is simply not in line with Western positivism. The arrangement according to different working principles produces various degrees of combinations, from the possible to the highly improbable. Accordingly, the protases of omens, although they appear to be devoid of logical sense for the Western imagination, are epistemologically true and possible because they are encoded within a given format.

The working principles are based on analogical reasoning, which is also what mainly builds cuneiform knowledge. Analogy is the expression of "inductive or deductive reasoning. It does not represent an inherently irrational cognitive process" (Rochberg 2016: 144). In this respect, McNamara's (2009: 10) words about religion in neuroscience fit in describing how modern scholars had and have to reason for an in-between case like Mesopotamian divination:

> "By observing how something works, we can sometimes make reasonable inferences as to its function. Although it is clear that we can often better understand the mechanism of a thing by first understanding its function, in many instances we do not know the function of the thing. All we have before us is the thing itself or some basic knowledge about its workings or mechanisms. We can sometimes observe the workings of a thing to get clues about its functions. In these cases we can "reverse engineer" a mechanism to discover clues as to its function."

As already argued, human thought and language are metaphorical par excellence according to cognitive studies and semiotics (see 2.2.3.2.–2.2.3.3.). Analogy and metaphor influenced not only poetic, but also scientific thought (Dunbar 2001). This means that analogy has been – and still is – an inherent part of logic and science in the history of knowledge for a long time (Lloyd 1966). The history of Western philosophy has only recently begun to argue the meaning of "rationality" beyond positivism. For instance, Hautamäki (2020) in his viewpoint on relativism argued that knowledge is based on "points of view" (P), which are defined by three elements: a subject (S), an object (O) of the point of view, and features (A) representing the object to the subject. Consequently, every statement can be true

424 For instance, the hermeneutic techniques detected in Babylonian and Assyrian commentaries are synonyms and synonym chains, explanations of logograms, pronunciation of syllabic signs and logograms, phonological and morphological variants, antonyms, paraphrases, figurative interpretations, etymology, etymography, and gematria (Frahm 2011: 59–85). For a general discussion on the analogy as a working principle in Mesopotamian divination, see also Glassner (1984).

according to P, if the statement is true of O *qua* (i.e. interpreted as) A.[425] Reasonings like this one lead to a re-definition of knowledge in terms of consensus in epistemic communities, to ontological relativism and to core rationality – the universal principles of rationality – which includes different sciences and cultures. Indeed, the principles of rationality, according to Hautamäki, are consistency, deduction, induction, and evidence, all four of them being features of Mesopotamian culture. The words of Francesca Rochberg (2016: 102) explain at best how, although ancient Near Eastern thought is not in line with Western positivism, it does not mean it is out of rationality:

> "We find ourselves confronted with classifications and categories, even phenomena, that sometimes confound our own sense of the order of nature. Still, as conceived in the cuneiform world, the overriding goal of the observation and interpretation of phenomena was to determine norms and anomalies within meaningful categories, and using those categories as vehicles, to find an order of things."

On these premises, we might be left with the idea that the omens depended more on their encoding principles than on real phenomena. However, this does not mean that Mesopotamian divination lacks an empirical basis. For instance, Graßhoff (2011), in his interpretation of the "Diviner's Manual", argued that omens could be understood as *ceteris paribus* statements, or statements that can be true unless unknown conditions occur when observed. While according to Rochberg (2010: 385, fn. 21), written omens are inferences and ancient scholars followed the *modus ponens*; hence, they are the expression of deductive logic.[426] Generally speaking, it is safe to say that the strategy of reasoning of the omens is the analogy. The goal of omens was not necessarily the understanding of the effects of physical causality, as science is for us today: the (semiotic) link between an ominous sign (protasis) and its prediction (apodosis) is differently oriented, and validated by the religious and cultural tradition. The latter statements do not imply that there is a total lack of causality: there is no physical causality, but there is a causal language (Rochberg 2010: 424) because the relationship between cause and effect is portrayed through words, symbols, or metaphors. Perhaps this is especially possible in a culture where two semantic systems (i.e. languages or cultural systems, the Sumerian, and the Akkadian) coexisted side by side for a long time (see 2.2.3.2.). On the whole, the goal of divination is consistent with the idea of cosmos according to the scribes of Mesopotamia, i.e. the divine permeates the world, and what has been designed by the gods in heaven is true for earth as well. Thus, natural phenomena are thought to be ominous signs which should be readable and interpretable in order to establish norms. This is written explicitly, for instance, in the "Diviner's Manual" (Oppenheim 1974: 204, 38–40): "The signs on earth just as those in the sky give us signals. Sky and earth both produce portents though appearing separately, they are not separate (because) sky and earth are related."

---

425 That is: P = [S, O, A].

426 If an omen is an inference, the *modus ponens* results in: if $P \rightarrow Q$ is true and P is true, then Q is also true. If $P \rightarrow Q$ is false and P is true, then Q is false (see A.1.1.).

The conceptual and cultural framework of the cuneiform culture is unique, yet the purpose of divination is not so distant from the essence of our modern science, despite being compared to Western thought only by defect. The knowledge disclosed by the omens was a tool to challenge the infinite possibilities of the future: Mesopotamian scholars developed their way to plan them, creating "possible visions", even highly improbable ones, which were the starting point from where they began to exclude, include, prevent, or cure possible consequences (Maul 2018: 254–256); these possible visions further served to establish norms and anomalies from which ancient scholars abstracted rules and schemes, as testified by astral compositions (see 4.). The working principles of the omens are among what we could define "modes of establishing norms" (Rochberg 2016: 102). From this point of view, the Mesopotamian divinatory apparatus cannot be labelled and downgraded as *superstitio*, and although it does not own the characteristics of modern science, it should be at least included in the history of scientific thought.[427] That is possible if we define "science" as the purpose, and not the accuracy of the results or, more specifically, the "putting nature to the question" (Rochberg 2016: 274).

A.2. The Vertical Level

Why are omens arranged in the way they are arranged? The vertical level of organisation of omens, or paradigmatic relationship, is based on the way the protases (X) are formulated and listed. The beginning of each entry is indicated with a logogram for *šumma*, "if", followed by one or more words (i.e. elements).[428] The topic of a protasis always belongs to a specific "thematic" section. For instance, in celestial omens these sections are arranged according to a celestial body or phenomenon in the protasis (A = 1st element).

In a protasis, the word order does not reflect the usual order of an Akkadian sentence (i.e. subject-object-indirect object-predicate). The celestial body or phenomenon (A) is often placed among the first positions, without necessarily being the subject of the verbal sentence, because it denotes the principle by which the omens are arranged. The other elements which compose the protasis of a celestial omen are usually another celestial body (B = 2nd element), space and time variables (C), or actions and/or qualifications (D = verbs or adjectives in a nominal sentence):[429]

427 For discussions about whether divination is science or not, see, e.g., Cryer (1994), Koch (2015: 1–4; 2016, Delnero (2016), Scurlock (2016), and Rochberg (2016: 17–58).

428 For the use of logograms for *šumma*, "if", see Fincke (2006: 134–137).

429 Additional elements to B, C, and D could be added to better describe a phenomenon. See, for instance, the long protases in the lunar eclipse omens dating to the Old Babylonian period (ca. 2000–1500 BC) (Rochberg-Halton 1988: 19–22).

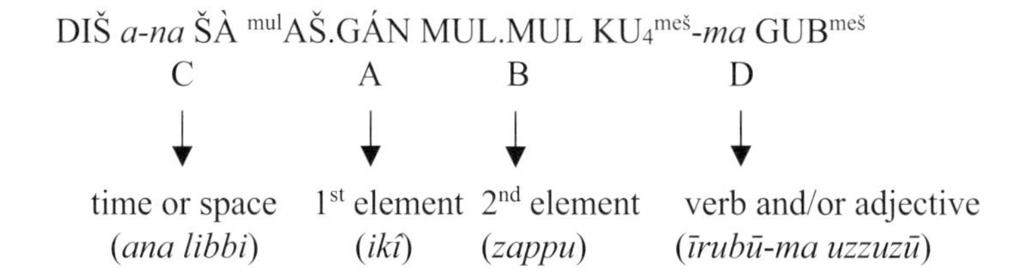

If the Pleiades enter into the Field and stand still.

It is possible to assign each part of the protasis to one of these four elements or variables. The variables have a somewhat mobile position in the sentence, except for the verb, which is always in the last position. Within the same group of contiguous omens, the first words of the protases are the same, but at least one other element always changes (e.g. A $B_xC_xD_x$). This type of arrangement helps the expert in his search for the omen that explains the phenomenon he had observed. By changing or replacing one element of the previous omen, a new set of combinations is created. In the example given below, which is excerpted from the reconstructed tablet EAE 52, contiguous entries of omens are arranged by changing the 2nd element (B) of the protasis:

DIŠ *a-na* ŠÀ ^{mul}AŠ.GÁN MU[L.MU]L KU$_4$^{meš}-*ma* GUB^{meš} — If X (then Y)
C A B D
DIŠ *a-na* ŠÀ ^{mul}AŠ.GÁN $^{m[ul]}$⸢SIPA⸣.ZI.AN.NA KU$_4$-*ma* GUB — If X_1 (then Y_1)
C A B_1 D

[DIŠ *a*]-⸢*na* ŠÀ ^{mul}AŠ.GÁN⸣ $^{m[ul}$ *š*]*u-ku-du* KU$_4$-*ma* GUB — If X_2 (then Y_2)
C A B_2 D

If the Ple[iad]es enter into the Field and stand still.
If the True Shepherd of Anu enters into the Field and stands still.
[If] the [A]rrow enters [i]nto the Field and stands still.
(K 3918 + K 6239 obv. 13, 15, 18, see EAE 52: 12–14 source A and parallels)

Groups of entries on the same tablet are also defined by changing at least one element of the protases. For instance, by changing the thematic subject (i.e. 1st element, A):

DIŠ (...) MUL.MUL, "If the Pleiades" — If A_1
(Commentary EAE 53: 1–25)

DIŠ (...) ^{mul}*ís-le-e*, "If the Jaw of the Bull" — If A_2
(Commentary EAE 53: 26–32)

DIŠ ^{mul}SIPA.ZI.AN.NA, "If the True Shepherd of Anu" — If A_3, etc.
(Commentary EAE 53: 33–49)

Sometimes, the changing of the thematic subject (i.e. A) is highlighted by the scribe himself by drawing a horizontal line of separation between the different sections.

A.2.1. Working Principles

The reason that the scribes changed the elements of a protasis is not obvious to us. First, the tablets are often damaged, and the protases are not always completely preserved. Brown (2000: 105–160; 2006) discussed many types of omens and distinguished between observed and invented protases. He defined five groups of working principles – that he called the "EAE Code" – which can be applied on both the vertical and the horizontal level:

- The four directions or countries (north, south, east, west; Akkad, Subartu, Elam, Amurru).
- The three watches (evening, middle, and morning watch, from sunset to sunrise).
- The binary code (based on opposites such as right versus left, up versus down, in front versus behind, etc.).
- Appearances of planets (colours, brightness, association with the royal family, or cities).
- Miscellaneous (stars associated with cities; phenomena behaving according to the ideal calendar or not).

These five groups can be narrowed down into two groups, i.e. fixed sequences, and binary opposites, if applied to the vertical level of omens. These working principles will be presented in the next subchapters (see A.2.1.1.–A.2.1.2.) by examples from the reconstructed tablets EAE 52, Commentary EAE 53, SIT 6, and other divinatory series whenever necessary.

A.2.1.1. Fixed Sequences

The order of contiguous entries, or groups of contiguous entries in omens collections, is arranged mostly based on fixed sequences. These sequences (see A.2.1.1.1.–A.2.1.1.5.) are drawn from different perceivable aspects of reality, encoded into fixed schemes, but the order of elements within a group is not necessarily fixed.

A.2.1.1.1. First Visibility of Stars

Recurrent celestial phenomena were observed in order to find a criterion for the Mesopotamian ideal calendar (see 4.2.1.). For instance, the beginning and the end of each month were detected through the ideal first visibility dates, or heliacal rising (see fn. 220), of individual stars arranged in lists, like in MUL.APIN (Hunger-Steele 2019: 180 Table 4). The rising time of the stars may also be the gist of several omens, and the rationale behind the arrangement of groups of entries: in EAE 52 (12–29), contiguous entries within one group are arranged by the substitution of the 2nd element (B) of the protasis, a star or a constellation which is the subject of the protasis. As previously shown in Table 4 (see

5.1.4., p. 164), the order of stars in the protases represents the progressive heliacal rising (see fn. 220) of 17 asterisms, from Ajaru (i.e. Month II) to Kislimu (i.e. Month IX), following the tradition preserved in MUL.APIN.

The principles according to which groups of entries are arranged in EAE are not always clear to us because this would require almost fully-preserved sources. Only one example can be traced in the reconstructed Commentary EAE 53 (see 5.1.5.1. Table 7, p. 197): similar to the above-mentioned example from EAE 52, the order of the groups of entries in Commentary EAE 53 mirrors the heliacal rising time of stars and constellations – the Pleiades, the Jaw of the Bull, the True Shepherd of Anu, and the Arrow, from Ajaru (i.e. Month II) to Du'uzu (i.e. Month IV), or Abu (i.e. Month V) as seen in the traditions in MUL.APIN and Astrolabe B.

The logic behind these arrangements mentioned above is always the calendar: the order of the entries echoes the heliacal rising of the stars involved. These arrangements allowed the diviners to check and easily find a correct omen for the observed phenomenon at any specific time of the year.

A.2.1.1.2. Colours

A fixed sequence for the colours is crystallised by the scribal culture in lexical lists, and in omens in general (Thavapalan-Stenger-Snow 2016: 200; Thavapalan 2020: 27–28).[430] In celestial omens, the scribes can change the qualification of a celestial body (i.e. verb or adjective D) by changing its colours, which can affect a specific meteorological or celestial phenomenon, or they can be intrinsic features of the planets (Brown 2000: 143), as in the following examples:

16'. [DIŠ] ⸢UD⸣ *ina* IGI MU.KAM ^{d}*ši-mu-ut* MUL$^{⸢meš⸣}$-*šú* GE$_{6}$^{meš}
17'. [DIŠ UD] ⸢*ina*⸣ IGI MU.KAM ^{d}*ši-mu-ut* MUL^{meš}-*šú* ⸢BABBAR⸣ *na-ág-*⸢*lu*⸣
18'. [DIŠ UD *ina* IG]I MU.KAM ^{d}*ši-mu-ut* MUL$^{m[eš}$-*šú* S]IG$_{7}$^{meš}

16'. [¶] If at the beginning of the year the stars of Simut become dark (i.e. black).
17'. [¶ If] at the beginning of the year the stars of Simut are flecked with white.
18'. [¶ If at the beginni]ng of the year the stars [of] Simut are [y]ellow/green.
(BM 38301 obv. 16'–18', see Commentary EAE 53: 53–55 source K and parallels)

10. ⸢DIŠ⸣ ^{mul}*dil-bat* AGA GE$_{6}$ *a*[*p-rat*]
11. DIŠ ^{mul}*dil-bat* AGA BABBAR *ap-*[*rat*]
12. DIŠ ^{mul}*dil-bat* AGA SIG$_{7}$ *ap-rat*
13. DIŠ ^{mul}*dil-bat* AGA SA$_{5}$ *ap-rat*

430 For fixed sequences of colours in celestial omens, see Rochberg-Halton (1988: 55–57); for terrestrial omens, see Streck (2001: 219–222) and Geller (2011: 120); for teratological omens, see De Zorzi (2014: 56–57); for *šumma immeru* omens, i.e. omens of the behaviour of sacrificial sheep or lambs, see Cohen (2007: 248); for physiognomic omens, see Fincke (2018: 210).

10. If Venus w[ears] a black crown.
11. If Venus we[ars] a white crown.
12. If Venus wears a yellow/green crown.
13. If Venus wears a red crown.
(Reiner-Pingree 1998: 58, 10–13)

30. [DIŠ ^{mul}]*dil-bat* AGA BABBAR *ap-rat*
31. [DIŠ ^{mul}]*dil-bat* AGA SA_5 *ap-rat*
32. [DIŠ ^{mul}]*dil-bat* AGA [GE_6] *ap-rat*

30. [If] Venus wears a white crown.
31. [If] Venus wears a red crown.
32. [If] Venus wears a [black] crown.
(see Reiner-Pingree 1998: 215, 30–32)

As seen in these examples, the colours either affect a celestial body (i.e. ^{d}*ši-mu-ut*, "Simut"), or a phenomenon related to it (i.e. AGA, *agû*, "crown", or halo of Venus);[431] the order within the groups of colours is not necessarily fixed (i.e. black-white-green, black-white-green-red, or white-red-black). It is improbable that stars, with the naked eye, could have been always perceived with this wide range of different shades of colours but the scribes used all the colours they could observe in the sky and applied them as working principles, in order to produce as many omens (and predictions) as possible.

A.2.1.1.3. Areas of Celestial Bodies

Another fixed sequence of elements is based on the position of a celestial body (i.e. time or space, C) toward another body. For instance, the tablet EAE 6 deals with celestial bodies that stand in specific areas of the Moon, such as the inside (ŠÀ, *libbu*, lit. heart), the horn (SI, *qarnu*), the shoulder (MAŠ.SÌLA, *naglabu*), the front (IGI, *pānu*), the back (EGIR, *arkatu*), or the side (Á, *idu*) (Verderame 2002: 169):

2. DIŠ MUL.MUL *ina* ŠÀ-*šú* GUB^{meš}
3. DIŠ MUL.MUL *ina* ŠÀ SI^{meš}-*šú* GUB^{meš}
4. DIŠ MUL.MUL *ina* SI ZAG-*šú* (15-*šú*) GUB^{meš}
5. DIŠ MUL.MUL *ina* SI GÙB-*šú* GUB^{meš} (2,30-*šú*)
6. DIŠ MUL.MUL *ana* IGI-*šu* GUB^{meš}

2. If the Pleiades stand inside it (i.e. the Moon).

431 The crown is usually associated to the Moon, and it is a description of its shape, an atmospheric phenomenon, or the earthshine, i.e. when the the grey part of the Moon is visible with the crescent (Verderame 2002: 60; Hätinen 2021: 61–68). The crown of Venus might be here interpreted as a luminous phenomenon of Venus, or as other planets standing in front of Venus (Reiner-Pingree 1998: 12).

3. If the Pleiades stand inside its (i.e. the Moon's) horns.
4. If the Pleiades stand in its (i.e. the Moon's) right horn.
5. If the Pleiades stand in its (i.e. the Moon's) left horn.
6. If the Pleiades stand in front of it (i.e. the Moon).

(Verderame 2002: 176–177 VI § 2–6 sources f, g, and parallels)

Similar fixed sequences about the areas of the Moon are found in EAE 52 and Commentary EAE 53:

1. [DIŠ] ⸢^{mul}AŠ⸣.GÁN *ina* MAŠ.SÌLA 30 GU[B]
2. ⸢DIŠ⸣ ^{mul}AŠ.GÁN *ina* SI 15 ^{d}30 [GUB]
3. DIŠ ^{mul}AŠ.GÁN *ina* SI 2,30 ⸢^{d}20⸣[+10 GUB]

1. If the Field stand[s] in the scapula of the Moon.
2. [If] the Field [stands] in the right horn of the Moon.
3. If the Field [stands] in the left horn of the Moo[n].

(Sm 259 rev. 1–3, see EAE 52: 30–32 source E_2 and parallels)

9. DIŠ ^{mul}*ís-l*[*e*]-*e ina* SI 15 10+[20 GUB...]
10. DIŠ ^{mul}⸢*ís-le*⸣-*e ina* SI 2,30 10+[20 GUB...]

9. If the Jaw of the Bu[ll stands] in the right horn of the M[oon...].
10. If the Jaw of the Bull [stands] in the left horn of the Mo[on...].

(K 3923+ rev. 9–10, see Commentary EAE 53: 28–29 source G)

The areas of the Moon are named as body parts (e.g. "heart" or "horn"), pointing towards a representation of the Moon as a bull with horns.[432] As is the case with the colours (see A.2.1.1.2.), the order and the choice of the areas within the sequence is not fixed.

The areas of the Moon can also be associated with cardinal points in order to create even more possibilities for protases, as can be seen in the omens about lunar eclipses. For instance, in tablet EAE 20, an eclipse of the Moon can be placed on the side "south above" (*šūti eliš*), or "north below" (*iltāni šapliš*) of the lunar disc (Rochberg-Halton 1988: 179, 1–2).

Regardless of the Moon, other celestial bodies are less frequently referred to as having areas or body parts. For example, the True Shepherd of Anu (^{mul}SIPA.ZI.AN.NA, *šitaddaru*) is said to have a LI.DUR, *abunnatu*, "navel" (Commentary EAE 53: 43–44), the Bow (^{mul}PAN, *qaštu*) has a front, a back, and a navel (App. B § 23 rev. i 10–12), Venus has a head (SAG.DU, *qaqqadu*) and a back (Reiner-Pingree 1998: 16–17), and Enmešarra has a knee (*kinṣu*) and a heel (*asīdu*) see 5.1.4. further comments to the text, pp. 170–171).[433]

432 For a Sumerian hymn to the Moon god Nanna/Ašimbabbar as a cowherd, see Hall (1986).

433 For a further discussion of body parts associated to celestial bodies in Uranology texts, see Beaulieau-Frahm-Horowitz-Steele 2018: 77–80. For references to the concept of corporeality in Mesopotamia, see fn. 122.

A.2.1.1.4. Compass Points

There were several ways to determine the compass points, which were usually named after the four primary winds recognised in Mesopotamia (Horowitz 1998: 193–207). The orientation or the movement towards cardinal directions and the location in one of them is one of the fixed sequences applied to protases (i.e. by changing the element time or space C):

9'. ⸢DIŠ⸣ ^{mul}ŠU.GI TÙR NÍGIN-*ma* KÁ-*šú ana* IM.U$_{18}$.LU BAD
10'. [DIŠ ^{m}]^{ul}ŠU.GI TÙR NÍGIN-*ma* KÁ-*šú ana* IM.SI.SÁ BAD
11'. [DIŠ ^{mu}]^{l}⸢ŠU⸣.GI TÙR NÍGIN-*ma* KÁ-*šú ana* IM.KUR.RA BAD
12'. [DIŠ ^{mu}]^{l}⸢ŠU⸣.GI TÙR NÍGIN-*ma* KÁ-*šú ana* IM.MAR.TU BAD

9'. If the Old Man is surrounded by a halo and its gate opens towards the south.
10'. [If] the Old Man is surrounded by a halo and its gate opens towards the north.
11'. [If] the Old Man is surrounded by a halo and its gate opens towards the east.
12'. [If] the Old Man is surrounded by a halo and its gate opens towards the west.
(Rm 100 ll. 9'–12', see EAE 52: 47–50 source F and parallels)

18'. [DIŠ MIN *ina* TÙ]R$^{?}$ MUL GUB-*ma* KÁ-⸢*šú ana* IM⸣.U$_{18}$.LU BAD
19'. [DIŠ MIN *ina* TÙ]R$^{?}$ ⸢MUL⸣ GUB-*ma* KÁ-*šú ana* IM.SI.⸢SÁ BAD⸣
20'. [DIŠ MIN *ina* TÙ]R$^{?}$ MUL GUB-*ma* KÁ-*šú ana* IM.KUR.RA BAD
21'. [DIŠ MIN *ina* TÙ]R$^{?}$ MUL GUB-*ma* KÁ-*šú ana* IM.MAR.TU BAD

18'. [If ditto (i.e. the Old Man is surrounded by a halo)], a star stands [in the hal]o$^{?}$ and its gate opens towards the south.
19'. [If ditto (i.e. the Old Man is surrounded by a halo)], a star stands [in the hal]o$^{?}$ and its gate opens towards the north.
20'. [If ditto (i.e. the Old Man is surrounded by a halo)], a star stands [in the hal]o$^{?}$ and its gate opens towards the east.
21'. [If ditto (i.e. the Old Man is surrounded by a halo)], a star stands [in the hal]o$^{?}$ and its gate opens towards the west.
(K 3099 + K 18689 rev. 18'–21', see EAE 52: 52–55 source E$_{1}$ and parallels)

22'. DIŠ ^{mul}SAG.ME.GAR *ṣir-ḫa ana* IM.U$_{18}$.LU GAR
23'. DIŠ ^{mul}SAG.ME.GAR *ṣir-ḫa ana* IM.SI.SÁ GAR
24'. DIŠ ^{mul}SAG.ME.GAR *ṣir-ḫa ana* IM.KUR.RA GAR
25'. DIŠ ^{mul}SAG.ME.GAR *ṣir-ḫa ana* IM.MAR.TU GAR

22'. If Jupiter places a flare towards the south.
23'. If Jupiter places a flare towards the north.
24'. If Jupiter places a flare towards the east.
25'. If Jupiter places a flare towards the west.
(Reiner-Pingree 2005: 136 obv. 22'–25')

Contrary to the sequences mentioned previously (see A.2.1.1.1.–A.2.1.1.4.), the order of the compass points is fixed (i.e. south-north-east-west), because their hierarchy was based on a specific priority of the winds blowing during the year in Mesopotamia (Horowitz 1998: 197):

IM.1 = IM.U_{18}.LU (*šūtu*) = south
IM.2 = IM.SI.SÁ (*iltānu*) = north
IM.3 = IM.KUR.RA (*šadû*) = east
IM.4 = IM.MAR.TU (*amurru*) = west

Thus, the compass points were sometimes written with their respective numbers and their order is fixed in other texts genres as well.[434]

A.2.1.1.5. Thematic Sequences

Already the archaic lexical repertoire was organised into thematic lexical compositions, so that the theme of each list was based on words and objects useful to early administrative purposes. In the Early Dynastic period (ca. 2900–2340 BC), standardised lists of professions, as well as lists of trees, fishes, or birds were composed (Veldhuis 2014: 72–102). In the Old Babylonian period (ca. 2000–1500 BC), thematic word lists were also included in the comprehensive collection Ura = *ḫubullu*, where additional themes were included, such as the list of stars (Veldhuis 2014: 149–157).

Similar to lexical lists, one of the strategies of listing consequential omens is by following a certain theme. The topic of the protases can vary by changing one of the elements of the protases (i.e. 1st element A, 2nd element B, or verb or adjective D) but within the same theme, i.e. semantic domain. This is the case for the two protases given below:

5’. DIŠ *ṣal-lum-*[*mu-ú* ^{mul}T]$I_8^{mušen}$ IGI.⸢DU_8⸣
6’. DIŠ *ṣal-lum-m*[*u-ú* ^{mul}U]$GA^{mušen}$ IGI.DU_8

5’. If a glo[w (coming from) the Ea]gle is visible.
6’. If a glo[w (coming from) the R]aven is visible.
(K 3099 + K 18689 rev. 5’–6’, see EAE 52: 40–41 source E_1 and parallels)

The 2nd element (B) of the first protasis is a constellation which bears the name of a bird (i.e. Eagle). Therefore, in the consecutive entry, the 2nd element (B) is replaced with a constellation with the name of another bird (i.e. Raven). Since only these two stars are named after birds, the length of such thematic sequence is limited.

A similar strategy of listing omens is found in the following two protases from the Adad section of EAE:

434 E.g. in lexical lists, see Erimḫuš II (MSL 17: 30–31 ll. 82–85); see also Borger (2004: 389 n. 641).

17. DIŠ ^{d}IŠKUR GÙ-*šu* GIM ANŠE.KUR.RA ŠUB
18. DIŠ ^{d}IŠKUR GÙ-*šu* GIM ANŠE ŠUB

17. If Adad thunders like a horse.
18. If Adad thunders like a donkey.
(Gehlken 2012: 17–18, 17–18 source A)

The group these two entries belong to refers to various shades of the sound, or entity of the sound (i.e. verb or adjective D), of a thunder (i.e. 2nd element B) produced by Adad (i.e. 1st element A) (Gehlken 2012: 15–20, 1–30). The two quadrupeds in the protases are consecutive, as well as other animals which are listed together in the subsequent sections: a scorpion and a snake (Gehlken 2012: 15, 2–3), a dog and a puppy (Gehlken 2012: 15–16, 5–6), and a lion and a wolf (Gehlken 2012: 16–17, 11–12), constitute consecutive pairs of reptiles, domestic mammals, and wild mammals.

More examples of thematic sequences can be found in terrestrial omens, where the subjects of the protases are more varied because they belong to both the human and the animal realm. In the following entries taken from the terrestrial omen series *Šumma ālu*, "If a city", the thematic sequences are based on physical characteristics of human beings and wild animals, respectively (i.e. 1st element A):

8'. DIŠ *ina* URU ⸢^{lú}⸣SA$_5$^{meš} MIN (i.e. *ma-a'-du*)
9'. DIŠ *ina* URU *bi-ir-du*^{meš} MIN
10'. DIŠ *ina* URU Ú.ḪUB^{meš}
11'. DIŠ *ina* URU IGI.NU.TUK^{meš} MIN

8'. If in the city *red* men ditto (i.e. are numerous).
9'. If in the city pockmarked men ditto (i.e. are numerous).
10'. If in the city deaf men ditto (i.e. are numerous).
11'. If in the city blind men ditto (i.e. are numerous).
(Freedman 1998: 32, 92–95)

1. DIŠ AM *ina* IGI ABUL IGI.DU$_8$
(...)
10'. [DIŠ] AM.SI *ina* IGI ABUL ⸢IGI⸣.[DU$_8$]
(...)
13'. ⸢DIŠ⸣ *ú-ma-am* EDIN *ina pa-*⸢*ni*⸣ [ABUL IGI.DU$_8$]
(...)
16'. ⸢DIŠ *ú*⸣*-ma-am* KUR.RA *ana* ŠÀ UR[U KU$_4$]

1. If a wild ox is seen in front of the city gate.
(...)
10'. [If] an elephant is s[een] in front of the city gate.
(...)
13'. If an animal of the steppe [is seen] in front of [the city gate].
(...)

16’. If a mountain animal [enters] into a cit[y].
(Freedman 2017: 34, 1, 10’, 13’, 16’)

A.2.1.2. Binary Opposites

Binary opposites are the most evident example of analogical reasoning in divination (Rochberg 2016: 159). In the following sequence of entries, the protases about a star with specific qualifications (i.e. verb or adjective D) are listed according to a dichotomous scheme:

10’. [DIŠ ^{mul}Š]U.GI MUL^{meš}-*šú bi-rit-su-nu ma-gal* BAD-*at*
11’. [DIŠ ^{mul}Š]U.GI MUL^{meš}-*šú nen-mu-du*

10’. [If] the interspace between the stars of the [O]ld Man is very open.
11’. [If] the stars of the [O]ld Man are conjoined.
(K 3099 + K 18689 rev. 10’–11’, see EAE 52: 44–45 source E_1 and parallels)

The working principle of these entries is based on the antithesis: “very open” is the opposite of “conjoined”. Likewise, in the Commentary EAE 53 (19–20, 22–23, see 5.1.5.1. commentary to the text of entries 19–23, pp. 192–193), contiguous entries of omens about the Pleiades are listed by a dichotomy of the stars’ qualifications (i.e. verb or adjective D):

karātu, “to cut”	versus	*šatāḫu*, “to be elongated”
ṣalāmu, “to become dark”	versus	*wamālu*, “to scintillate”

Binary opposites work both for the vertical arrangement of entries and the meaning of omens in general (see A.4.). For instance, in the case of extispicy, every section of the lungs and the liver is divided into left and right parts, each having a negative and a positive meaning (Meyer 1987: 81–93). In celestial omens, this rule – positive versus negative – is assumed to be applied through the juxtaposition of directions, times, and qualifications (i.e. left versus right, up versus down, in front of versus behind, on-time versus not on-time, bright versus dim, to have versus not to have, etc.) of celestial bodies and phenomena (Brown 2000: 142), by changing elements within the protases (i.e. time or space C, verb or adjective D).

A.3. The Horizontal Level (Antecedent and Consequent)

The horizontal level of the organisation of the omens, or the syntagmatic relationship between protases and apodoses, is based on several principles according to which the apodoses (Y) follow the protases (X). The apodoses of celestial omens usually describe events which happened in real life, as witnessed in the apodoses of the omens about the Pleiades (see 5.3.); or they quote mythology, referring to specific features of the gods with whom the stars are associated. In the latter case, the semiotic link between the protasis and the apodosis may arise from intertextuality, as witnessed in omens about the Pleiades and the Sebettu (see 5.3.1.); concerning the former case, the great majority of the apodoses can be defined "stock apodoses", or "apodoses which are found throughout an omen genre or even throughout all kinds of divination" (Koch 2019: 232). Such apodoses (e.g. KI.LAM TUR, "the market rate will diminish", or $BURU_{14}$ SI.SÁ, "the harvest will prosper") are very common for a great number of protases because they can be associated with such a variety of phenomena that the semiotic link appears at first meaningless and are relevant only for their positive or negative connotation (Koch 2019: 232–234).[435]

Generally speaking, the semiotic link between a protasis and an apodosis lies in how scribes wrote the omens. Analogical relationships of any kind often reside between two or more elements of an omen, based on cultural ideology, and therefore difficult to grasp for us. This fact also emerges from hermeneutics in Mesopotamia: we are aware that the scribes applied hermeneutical techniques at a certain level of writing expertise, such as the use of synonyms, different Akkadian renderings of logograms and allegorical approaches arising from a deeper etymological and etymographical analysis (see fn. 79; Frahm 2011: 20–23). In many instances, even subconscious associations seem to be in place, "provoked by certain words, whose specific connotations imparted to them a favourable or an unfavourable character, which in turn determined the general nature of the prediction" (Oppenheim 1964: 211).

A.3.1. Working Principles

Several scholars have discussed the working principles of the horizontal level of organisation of omens in the context of various divinatory series. Koch (2019: 223), in applying the terminology of semiotics, described an omen as made up of a signifier (i.e. protasis) and a signified element (i.e. apodosis). She then considered the working principles between a signifier and a signified element of various nature, for instance symbolic (i.e. cultural, linguistic, or social convention), iconic (i.e. the signifier resembles the signified), and indexical (i.e. the signifier is assumed to be caused by the signified). Whereas Brown (2000: 75–81) has discussed textual playfulness and learned associations in celestial omens: he identified etymological, ideographic, semantic, and graphic connections on a horizontal level, assuming that names of stars and constellations are all substitute names for planets (see A.4.1.).

The above-mentioned categorisations have been summed up into six working principles, according to which one can recognise the logic behind a protasis and its prediction: graphic, phonetic, morphologic, semantic, symbolic, and factual principles. These working principles will be discussed in the next subchapters (see A.3.1.1.–A.3.1.6.) using examples taken from the reconstructed tablets EAE 52, Commentary EAE 53, SIT 6, and other divinatory series whenever necessary.

435 More specifically, Koch suggested that the combination of protases with stock apodoses in celestial omens may derive from the urge to understand only if the ominous sign was good or bad and for whom, and not from the semiotic coherence of the prediction. For earlier bibliography on this topic, see also Brown (2000: 132, fn. 334).

A.3.1.1. Graphic Principle

The omens dating to the first millennium BC – or "standard omens" – are mainly written in an abbreviated way using logograms. The scribes of the omens knew the logograms, the lexical lists, and the arrangement of the cuneiform signs within the lists, to the point that the shape of a cuneiform sign or the composition of a grapheme in a protasis could have had influenced the choice of another sign in an apodosis. Several puns in omens were built precisely on the "endless" combination of cuneiform signs, especially in logograms. Many interesting examples of such a graphic principle behind the semiotic link between the protases and the apodoses of omens have been discussed by Noegel (2006: 97–100), together with many other examples of exegetical punning. The following two examples – from a dream omen and a liver omen – have been individuated by Noegel (2006: 97–98):

> DIŠ *ana* ID-*ra-an*^{ki} DU *a-ra-an-šú* D[U$_8$]
>
> If (a man dreams that) he goes to Idran/Arranu, his crime will be r[emoved].
> (Oppenheim 1956: 313, x+21)
>
> DIŠ KÁ.É.GAL *ir-ta-pi-iš ir-bu-um a-na* É.GAL *i-ir-ru-ub*
>
> If the Palace Gate has become wide, an income will enter into the palace.
> (YOS 10, 23 obv. 3, see Winitzer 2017: 230)

In the first omen, the scribe linked the protasis to the apodosis using a pun: the writing ID-*ra-an*, if it is to be read *á-ra-an*, is phonetically related to *aranšu*, "his crime". The city of Arranu is elsewhere written *ár-ra-nu* (SAA 15, 162 obv. 11, 14); thus, the writing ID-*ra-an* can be either a contrived writing for Arranu, or the name of another city (Idran or Aran). In the second omen, the part of the liver called KÁ.É.GAL, *bāb ekallim*, "Palace Gate",[436] in the protasis, resurfaces in the apodosis as a reference to the palace (É.GAL, *ekallu*).

Graphic puns have not been found in the reconstructed tablets EAE 52, Commentary EAE 53, and SIT 6; yet they can be found in other celestial omens, for instance in the following one:

> DIŠ ^{d}IŠKUR *ina* MÚRU 30 GÙ-*šú* ŠUB-*ma* ^{d}TIR.AN.NA *šá* MÚŠ-*šá ma-diš* SA$_5$ 3[0 GIM *gam-lì* NIGIN] / ŠUB-*tì* ^{giš}MÁ^{meš}
>
> If Adad thunders in(to) the middle of the Moon and a rainbow whose appearance is foremost red surrounds the Moon like a boomerang, / (there will be) downfall of ships.
> (Gehlken 2012: 106, 53')

This entry, from the tablet EAE 46 in the chapter of Adad, is part of a group of entries arranged by different colours of the rainbow (see A.2.1.1.2.). In the protasis, the logogram

436 The Palace Gate "is the umbilical fissure that divides the *lobus sinister* from the *lobus quadratus*" of a liver (Koch-Westenholz 2000: 46).

SA$_5$, 𒋛𒀀, (si + a), for, *sāmu*, "red", is similar to the logogram MÁ, 𒈣, *eleppu*, "ship", in the apodosis: the sign SA$_5$ has one more vertical wedge in addition to MÁ.

A.3.1.2. Phonetic Principle

Many scholars have noticed various figures of speech in omens and hermeneutics.[437] Simple assonances (e.g. *a-i*) can constitute the semiotic link between the protasis and the apodosis, like in the following omen commented by Oppenheim:

> DIŠ UGA^{mušen} GU$_7$ *ir-*⸢*bu* KU$_4$⸣-[*ub*]
>
> If (a man dreams that) he eats a raven, an income will ent[er].
> (Oppenheim 1956: 316, y+10)

In the omen quoted above, the assonance is made between the reading of UGA^{mušen}, *arbu*, "raven", and *ir-bu* KU$_4$-*ub*, *irbu īrub*, "an income will enter".

One can also find more complex phonetic similarities that establish associations between protases and apodoses. Among the figures of speech attested in omens, there is the metathesis – the transposition of phonemes within one or more words – as seen in the following report dating to the Neo-Assyrian period (ca. 911–612 BC):

> [DIŠ] ^{d}*iš-tar* AGA KÙ.BABBAR *ap-rat* IL[LU] *ku-li-li* ⸢DU⸣-[*kam*] / [ILLU] *ku-li-li* ILLU *gap-*[*šu*]
>
> [If] Ištar wears a silver crown, a flo[od] of dragonflies will c[ome]. / [A flood] of dragonflies (means) a massi[ve] flood.
> (SAA 8, 461 obv. 3–4)

In this report, though the Akkadian reading of AGA, *agû*, "crown", is the most widespread, another reading might have been intended: *kilīlu*, "headband" (CAD K 358). If this were true, a wordplay was made by the scribe between the reading *kilīlu*, "headband", and *kulīlu*, "dragonflies": the dragonflies, common insects in Mesopotamia, usually living on the surface of the water.[438]

In tablet EAE 52, only one instance of a phonetic association (i.e. *a-e*) between the protasis and the apodosis was found:

---

437 For puns in Mesopotamian divination, see Greaves (2000) and Noegel (2006: 97–100), while for puns in hermeneutics, see Frahm (2011: 70–76).

438 There is a group of contiguous entries of omens about dragonflies carried by the river in *Šumma ālu* tablet 61 (Freedman 2017: 147–148, 111–120). A passage from *Atra-ḫasīs* (III iv 6) even mentions a metaphor of dragonflies as a multitude of corpses: *ki-ma ku-li-li* / *im-la-a-nim na-ra-am*, "They (i.e. the dead bodies perished in flood) filled the river / like dragonflies" (Lambert-Millard 1968: 96 iv 6–7).

DIŠ *ṣal-lum-[mu-ú* ^{mul}T]I$_8$^{mušen} IGI.DU$_8$ *ina* KUR DÙ.A.BI ^{munus}PEŠ$_4$^{meš} *šà* ŠÀ-*ši-na* NU SILIM^{meš}

If a glo[w (coming from) the Ea]gle is visible, in the country altogether pregnant women will not bring (the children) in their wombs to term.
(K 3099 + K 18689 rev. 5' // K 3524 l. 2, see EAE 52: 40 sources E$_1$, I and parallels)

In this entry quoted above, the scribe played with the words TI$_8$, *a/erû*, "eagle" in the protasis,[439] and PEŠ$_4$, *a/erû*, "to be pregnant" (CAD E 324–326) in the apodosis. The same assonance also lies behind the spelling of the "eagle-stone" (i.e. aetitis), a magical stone used during the Neo-Babylonian period (ca. 626–484 BC) as an amulet to help women during labour.[440] The stone is called *aban a/erê*, and can be written either ^{na4}PEŠ$_4$, or ^{na4}TI$_8$. Thompson (1936: 104-108) was the first to point out that a pun was behind this tradition.[441]

A.3.1.3. Morphologic Principle

The morphologic principle in cuneiform omens is based on paronomasia: in the present context, paronomasia means that the same signs, words, or verbal roots can be used in both protasis and apodosis of an omen, but with a different meaning, as well as a different morphology. Paronomasia has already been found in omens, for example in teratological omens of the series *Šumma izbu*, lit. "If the malformed new-born", by De Zorzi (2011: 67–68):

BAD MUNUS Ù.TU-*ma* GÌŠ-*šú* NU GÁL EN É *ul in-né-ši-ir* UŠ-*di* <A.ŠÁ>

If a woman gives birth and (the baby) has no penis, the lord of the house will not prosper, (there will be) the confiscation <of the field>.
(De Zorzi 2014: 422, 68)

In the omen quoted above, the association between the protasis and the apodosis is based on the reduplication of one sign with different readings, that is, GÌŠ, *išaru*, "penis" (CAD I/J 226–227), and UŠ, *rīdu*, "confiscation", or "persecution" (CAD R 324–325).

A short celestial omen, found in SAA 8, 36 (obv. 10–11), and in the tablet EAE 47, shows how puns might work on different levels at the same time, that is here the morphologic and the semantic level:

DIŠ ^{d}*e-ri-iš-ki-gal ik-kil-la-šá* GIM UR.MAḪ ŠUB KI BI ŠUB

If Ereškigal utters her roar like a lion, that region will be abandoned.
(Gehlken 2012: 159, 68')

439 See the readings in the Standard recension of Ura = *ḫubullu* 22 (Bloch-Horowitz 2015: 109, 308').
440 For the use of stones as amulets in Mesopotamia, see Schuster-Brandis (2008).
441 See also Reiner (1995: 123–124) and Noegel (2006: 99).

In this omen, the verb *nadû* (CAD N1 68–101) is written twice logographically (i.e. ŠUB) but the semantic context in which these logograms are used is different (i.e. *iddi*, "to roar (like a lion)", and *innaddi*, "to be abandoned"). A preterite is intended in the protasis (*iddi*), and a N-stem present or future in the apodosis (*innaddi*).[442]

Likewise, the above-mentioned case of *a/erû*, "eagle" and "to be pregnant" (see A.3.1.2.), could be considered an example of paronomasia. The wordplay with *nadû* works on a morphologic and semantic level, whereas the wordplay with *a/erû* works on a morphologic and phonetic level, because it is based on the similar spelling or pronunciation for two words with two different semantic fields.

A.3.1.4. Semantic Principle

On a semantic level, one can easily assume that analogies could create almost infinite combinations and possibilities because a loose association of ideas can hide behind every word and logogram. However, some of these associations are more frequent than others, so they are more easily identifiable to us. One of the most widespread analogies in celestial divination is based on the sign TÙR, *tarbaṣu*, "cattle-pen", which figuratively stays for a "halo" of a celestial body in divination or in an astronomical context, i.e. the cattle-pen of the Moon means its halo. The omens given below exemplify both the semantic fields in which TÙR is used, in the protasis and in the apodosis of omens:

> If the Moon is surrounded by a halo and Jupiter stands inside it, (there will be) an epidemic among the herd and the wild animals of the steppe.
> (see 2.2.3.2., p. 25)
>
> [If the O]ld Man is surrounded by a halo, in that year in the entire country (there will be) a downfall of people, (but the evil) [will] not [come close] to the livestock and the flock.
> (see 2.2.3.2., p. 25)

The way in which the word TÙR, *tarbaṣu*, covers two semantic fields has been discussed in chapter 2 (see 2.2.3.2.). In the present context, the quoted examples show how a semantic principle can be the semiotic link between protasis and apodosis, because the protases involving the "halo", predict events involving animals in herds.

The predictions can derive more directly from a sign or a word in the protasis – and so they derive from a semantic field. This is the case of the following omen, excerpted from one of the sources used for the reconstructed SIT 6:

> DIŠ ^{mul}KU$_6$ *ana* KIMIN (i.e. ^{mul}UGA^{mušen}) *i-mid* KU$_6$^{meš} *u* MUŠEN^{meš} *ud-[deš-šu-u...]*

442 The preterite of a conditional clause marks its anteriority with the apodosis, as the protasis precedes it logically and temporally, e.g. see Kouwenberg (2010: 94–95, 153), and Streck (2018: 128–132), who conveniently presents a survey on conditional clauses in Akkadian grammar, including earlier bibliographical references.

If the Fish comes in contact with ditto (i.e. the Raven), the fishes and the birds will the fishes and the birds will be[come abundant...].
(K 2138 l. 13', see SIT 6: 28 source H and parallels)

In the protasis of this omen, the Fish (^{mul}KU$_6$, *nūnu*) and Raven (^{mul}UGA^{mušen}, *āribu*) are linked, in a straightforward way, to a prediction for fishes (KU$_6$^{meš}, *nūnū*) and birds (MUŠEN^{meš}, *iṣṣūrū*).

A.3.1.5. Symbolic Principle

Analogies and cases of intertextuality – that link a protasis to an apodosis – between divination and mythology, religion, or the cultural framework of Mesopotamia can be considered symbolic principles of association.

Concerning the mythological and religious sphere, a symbolic principle behind the semiotic link of celestial omens is nothing other than the expression of the conception of the cosmos according to Mesopotamian people: celestial bodies are the representation of the anthropomorphic gods (see 2.2.3.1. § 2–3). The scribes of celestial omens sometimes explicitly related a celestial body in the protasis to the god with whom it is associated by mentioning him or her in the apodosis. That is precisely the case of several omens about the Pleiades:

[If in] Ajaru (i.e. Month II), the Pleiades, the divine [Seven], / the great [god]s, [rise] at [their] specified time, / the great gods will assemble [a]nd / give a good [counsel to the country], / good [winds] will blow.
(see 5.2.4., pp. 239)

(If) a planet comes close (and) reaches the Pleiades, the Sebettu will devour the country.
(see 5.1.8.4. § EAE 50–51, p. 230, 2b)

As summarised in the conclusions of this study (see 7.), the Pleiades mentioned in protases symbolise, or can be associated with, events brought about by two divine heptads in the apodoses, according to two parallel traditions: the Sebettu from the poem of "Erra and Išum", and a group of seven great gods appointed to the fates, the "divine Seven, the great gods" (see 5.3.1.–5.3.2.). Other celestial omens have several attestations of symbolic principles based on the religious sphere. For instance, in the following omen:

If Ereškigal utters her roar like a lion, that region will be abandoned.
(see A.3.1.3., p. 297)

In this omen, the roar of Ereškigal, the goddess of the underworld, symbolises (i.e. is a metaphor for) an earthquake. This is confirmed because the omen above is quoted with others about earthquakes (*rību*) in astrological and astronomical reports to Assyrian and Babylonian kings (e.g. in K 124 obv. 10–11, see SAA 8, 36). Another example from the religious sphere concerns the god Nergal/Erra:

DIŠ *ina* ^{iti}BÁR ^{mul}UD.KA.DUḪ.A IGI / MU.5.KAM* *ina* KUR URI^{ki} *ina* KA ^{d}*èr-ra* ÚŠ^{meš} / GÁL^{meš} *ana* MÁŠ.ANŠE NU T[E]

If in Nisannu (i.e. Month I) the Demon with the Gaping Mouth is visible, for 5 years in Akkad there will be a pestilence / by the command of Erra, (but) it will not appr[oach] the herd.
(Reiner-Pingree 1981: 67 XIII 5; 68 XIV 4)

We are aware that the Demon with the Gaping Mouth can be a substitute name for Mars, or be identified with Mars' divine counterpart Nergal/Erra, according to MUL.APIN, the Great Star List, and one Uranology text dating to the Late Babylonian period (ca. 484–30 BC).[443] Therefore, one can assume that, in the omen quoted above, the apodosis involving Nergal/Erra follows a protasis involving the Demon with the Gaping Mouth because the latter symbolises or is associated with the former.

Symbolic principles are also in place on a spatial level, that is, concerning the cultural sphere. As stars represent the gods, what happens in the sky reflects on earth. If there is an ominous sign in one of the four areas of the sky, its prediction is mirrored in the same area of the earth. For example, the entries EAE 52: 47–50 mirror a precise scheme of associations between areas of the sky (i.e. south-north-east-west), where the celestial phenomena are located, and areas of the earth (i.e. Akkad-Subartu-Elam-Amurru), or the countries which are affected by the prediction. This scheme of association was already discussed in 5.1.4. (see the commentary to the text of entries 46–60, pp. 167–168) and shown in Table 5. However, the order of the areas of earth affected in the apodoses is not always fixed. For instance, in the lunar eclipse omens dating to the first millennium BC, at least four different schemes, virtually adjusted on the four quadrants of the Moon, can be drawn from the omens (Rochberg-Halton 1988: 51–55); other schemes are also known, either drawn from omens or written in the form of rules.[444] Differences in the correspondences between the areas of the sky and the earth likely existed because the scribes adjusted the schemes within the written omens according to their needs; thus, all schemes were idiosyncratic, as first discussed by Rochberg-Halton (1988: 51–53), or they were used alongside celestial omens and expanded to explain new phenomena, as argued by Wainer (2016: 73).

Apart from the associations of areas of the sky and the world in EAE 52: 47–50, another scheme of association, in the form of a written rule involving the months and four areas of the world, is found in K 2254, a fragment of the series SIT:

10'. [^{iti}BÁR ^{iti}GU$_4$ ^{iti}SIG$_4$ KUR URI^{ki} ^{iti}ŠU ^{iti}NE ^{iti}KI]N KUR ELAM.MA^{ki} ^{iti}DU$_6$

---

443 More specifically, the Demon with the Gaping Mouth is a name of ^{d}U.GUR, "Nergal", in MUL.APIN (I i 28, see Hunger-Steele 2019: 35), and Uranology texts (Beaulieu-Frahm-Horowitz-Steele 2018: 36 D i 17), and a name of ^{mul}*ṣalbatānu*, "Mars", in the Great Star List (Koch-Westenholz 1995: 190 l. 108).

444 E.g. see the scheme in the Great Star List (Koch-Westenholz 1995: 202 ll. 293–294), in EAE 20, and in the tablet SIT 4 (K 3123 rev. 17', see App. B § 24); for other schemes of association drawn from lunar eclipse omens, see Rochberg-Halton (1988: 51–55, fig. 4–3, 4–5, 4–6).

^{iti}APIN ^{iti}GAN KUR MAR.TU^{ki}
11'. [^{iti}AB ^{iti}ZÍZ ^{iti}ŠE KUR SU.BIR$_4$^{ki} ITI]$^{⸢\text{meš}⸣}$ *šá mi-ši-iḫ* MUL^{meš}

10'. [Nisannu (i.e. Month I), Ajaru (i.e. Month II), Simanu (i.e. Month III) (are for) Akkad, Du'uzu (i.e. Month IV), Abu (i.e. Month V), Ulul]u (i.e. Month VI) (are for) Elam, Tešritu (i.e. Month VII), Araḫsamnu (i.e. Month VIII), Kislimu (i.e. Month IX) (are for) Amurru.
11'. [Ṭebetu (i.e. Month X), Šabaṭu (i.e. Month XI), Adaru (i.e. Month XII) (are for) Subartu. (These are) the month]s of the glow of the stars.
(K 2254 rev. 10'–11', see App. B § 25)

This scheme, which is restored based on the readable lands, associates a celestial event, a *mešḫu*, "glow", or "meteor" (see 5.1.4. commentary to the text of entries 37–43, p. 166) happening in specific months to predictions affecting the four lands;[445] more specifically, every land corresponds to three consecutive months each, as shown in the following list:

K 2254 obv. 10'–11'

Months I to III = Akkad
Months IV to VI = Elam
Months VII to IX = Amurru
Months to X to XII = Subartu

Many symbolic associations remain undetected or ambiguous to us, and only sometimes are they explicit. The scholars who wrote and worked with the omens considered themselves exponents of an erudite lore and were trained in every literary context. Hence, analogies and symbolisms were probably obvious to them. Symbolic associations are also the product of an overall conception of the cosmos, according to which divine agency is scattered and not necessarily conceived as unique (see A.1.–A.1.2.). Hence, they had different origins: they might find a match in the mythology (e.g., the Pleiades and the Sebettu or the divine Seven), religion (e.g. the earthquake as the roar of Ereškigal, the "Demon with the Gaping Mouth" associated with Nergal), or in a broader cultural and conceptual framework (e.g. the compass points and the areas of the world). Anyway, one must stress how the study of intertextuality in Mesopotamian literacy plays a considerable role: it helps us to illuminate possible interpretations of omens that would remain otherwise unclear, as was the case for the omens about the Pleiades.

A.3.1.6. Factual Principle

The so-called "impossible" protases (see fn. 268 and A.1.2.) are either based on a metaphorical language, or older belief on how the cosmos functioned, and equated, in

445 For parallels of this scheme of association, see Wainer (2016: 62–63) who presents a discussion about the consecutive-month scheme in relation to the *mešḫu*-phenomenon, including earlier bibliographical references.

commentaries, to literal or astronomically possible phenomena through analogies and associations of ideas. Nonetheless, this does not mean that Mesopotamian omens were oblivious to the cause-and-effect principles of physical nature. Some predictions find their roots in empirical evidence; in other words, practical experience is the only logic detectable. The prediction of the following celestial omen is then "obvious" for us:

> DIŠ *ina* UD *er-pi-šá* ŠÈG SUR ᵈTIR.AN.[NA G]IL ŠÈG NU SUR
> If on a cloudy day, (after) it had [ra]ined, a rainb[ow a]rches, it will not rain (again).
> (Reiner-Pingree 1981: 40 III 4a)

This omen refers to a rule of thumb or a proverbial knowledge because it is technically not exact. One can see two rainbows on the same day when two cloud banks producing rain showers move over the country with some time in between. Therefore, one should assume that this omen rather expresses a more or less high probability that, after it had rained, it will not rain again (or it will not rain again from the same cloud), a statement from which is possible to detect further anomalies.

Similarly, other omens are based on the probability of the consequences of this or that phenomenon. That is the case of other omens in the section of Adad in the series EAE:

> DIŠ KI *ina ka-la* UD-*mi i-nu-uš* BIR-*aḫ* KUR
>
> If the earth quakes all day long, (there will be) the scattering of the land
> (Gehlken 2011: 156, 55')
>
> DIŠ UD ŠÚ-*am-ma ina* IM.SI.SÁ ŠE.ER.ZI IM.DUGUD [...]
>
> If the day becomes cloudy and in the north (there is) brightness, (there will be) fog [...].
> (Gehlken 2012: 172, 6')

The predictions of the two omens quoted above are based on the more or less high probability of having soil cracking after a strong earthquake and fog after cloudiness.

A.4. Positive Versus Negative Predictions

As already mentioned (see A.2.1.2.), scholars have considered binary opposites one of the fundamental principles on which omens are based. For instance, the difference between the so-called *pars hostilis* and *pars familiaris* is the core of extispicy: this means that, in a liver, the same sign on the right side of the organ is considered positive, and on the left side, it is negative. Thus, a negative sign on a positive side produced a negative prediction and *vice versa*, but a negative sign on a negative side produced a positive prediction.[446] Following

446 For predictions according to binary opposites in Mesopotamian divination, see, among the others, Starr (1983: 15–24), Meyer (1987: 81–93), Jeyes (1991–1992: 35), Rochberg (2004: 55–58) and Maul (2018: 17–86). Winitzer (2017) explored the features of extispicy omens dating to the Old

this pattern, specific celestial bodies, directions, places, or colours (see A.2.1.1.2., A.2.1.1.4., A.3.1.5.) might have had a positive or negative symbolism. If one considers each element of an omen as an element of an equation – be it multiplication or addition – celestial omens could work as a sort of arithmetic rule: if something is positive, whenever it is combined or interacts with something negative, it produces a negative apodosis (Koch 2015: 12–15).

To assess if or why a celestial body (or one of its features) is either positive or negative, is not a straightforward task, but it can be deduced in some instances. For instance, the following entries from tablet EAE 52, already mentioned in previous subchapters (see A.2.1.1.5., A.3.1.2.), require a deeper analysis for their interpretation:

5'. DIŠ *ṣal-lum-[mu-ú* ^{mul}T]I$_8$^{mušen} IGI.DU$_8$ *ina* KUR DÙ.A.BI ^{munus}PEŠ$_4$^{meš} *šà* ŠÀ-*ši-na* NU SILIM^{meš}
6'. DIŠ *ṣal-lum-m[u]-ú* [^{mu}]^{l}UGA^{mušen} IGI.DU$_8$ *ina* KUR DÙ.A.BI ^{munus}PEŠ$_4$^{meš} [*šà* ŠÀ-*ši-n*]*a* SILIM^{meš}

5'. If a glo[w (coming from) the Ea]gle is visible, in the country altogether pregnant women will not bring (the children) in their wombs to term.
6'. If a gl[o]w (coming from) the Raven is visible, in the country altogether pregnant women will bring [(the children) in thei]r [wombs] to term.
(K 3099 + K 18689 rev. 5'–6' // K 3524 ll. 2, 4, see EAE 52: 40–41 sources E$_1$, I and parallels)

These two omens describe a glow (*ṣallummû*) coming out from two constellations, both named after two birds. On the vertical level, the Eagle (TI$_8$^{mušen}, *a/erû*) and the Raven (or Crow) (UGA^{mušen}, *āribu*) are thematically connected.[447] On the horizontal level, in the first omen (EAE 52: 40), there is a phonetic principle in place, and the Eagle had a positive meaning for women in labour. The second omen (EAE 52: 41) concerns the Raven, *āribu*: the Raven is conceived in the sky next to or over the Snake (^{mul}MUŠ, *nirāḫu*), that is associated with the god Ningišzida, the god of the underworld.[448] In the iconography, the Raven is pictured as a bird standing on the last part of the tail of a snake, i.e. the Snake (TCL 6, 12 rev.). The astral composition MUL.APIN (I ii 8–9, see Hunger-Steele 2019: 40) records the Snake (*Ningišzida bēl erṣēti*, "Ningišzida the lord of the underworld") and the Raven (*kakkab Adad*, "the star of Adad") as stars next to each other in the sky in the path of Anu. The Raven can also be written ^{mul}UG$_5$.GA, literally the "dead one": the writings

Babylonian period (ca. 2000–1500 BC). More specifically, he focused on working principles of additional protases and apodoses and working principles according to which the omens are ordered and connected.

447 For the names of the Eagle and the Raven written in Sumerian, see Salonen (1973: 80–81, 83, 104–106, 124–131), and Veldhuis (2004: 286–287, 299–301). See also Streck (2016) for further bibliography about birds.

448 DIŠ ^{mul}M[UŠ] x x x x x x *šú-ú kap-pi šá-kin* GÌR^{II} *ra-áš* UGA^{mušen} MUL ^{d}IŠKUR *ina muḫ-ḫi* KUN-*šú* "The S[nake ...] ... ... it is, who has wings, who has feet; the raven, the star of Adad (is) on top of its tail" (Beaulieu-Frahm-Horowitz-Steele 2018: 37 D iii 1–3).

^{mul}UG$_5$.GA and ^{mul}UGA are used freely in omens to indicate the same constellation (e.g. SIT 6: 24–25). This pun and the relationship with the Snake enclose the Raven in the realm of the dead.[449] Therefore, one can assume that the Raven had a negative meaning, at least in the sources mentioned above.

Guinan (2018) investigated bird omens in other divinatory contexts: the Akkadian word *āribu* is generally used for both crows and ravens (CAD A2 265–267), i.e. black birds, but it seems that the logogram UGA^{mušen} was used primarily for crows, which are smaller and more common birds. Guinan (2018: 23) suggested that the crow in diagnostic and terrestrial omens is used as the binary counterpart of the falcon (ŠUR.DÙ^{mušen}, *šurdû*). Following the working principle of the binary opposites, crows "turn an auspicious prediction into an inauspicious one and the reverse" (Guinan 2018: 15). In the two entries from tablet EAE 52 (40–41), the same reasoning might be in place: the Raven (^{mul}UGA^{mušen}) could be the negative counterpart of the Eagle (^{mul}TI$_8$^{mušen}).

Item	Writing	Symbolism
Light phenomenon	*ṣallummû*, "glow"	Negative ominous sign (–)
Eagle	TI$_8$^{mušen}, *a/erû*, "eagle"	Life and labor (+)
Raven	UGA^{mušen}, *āribu*, "raven", or UG$_5$.GA, "dead one"	Death (–)

Table 12. Interpretation of the individual elements in the apodoses of omens in EAE 52: 40–41.

In conclusion, the Eagle and the Raven are perceived as two animals and stars with different symbologies (see Table 12): life (i.e. pregnancy) and death. Considering the basic principles of *pars hostilis* versus *pars familiaris* in divination, the glow (*ṣallummû*) might have a negative meaning in this context, and together with the Raven, which is negative, it produced a positive apodosis (– × – = +). Whilst the comet produced a negative apodosis with the Eagle, which had a positive meaning for women in labour (– × + = –).[450]

449 The writing ^{mul}UG$_5$.GA, "dead one", is made explicit by the fragment BM 82923, a 36-star list in Late Babylonian ductus (see 4.1.3.4.). In BM 82923 obv. 17, the star ^{mul}UG$_5$.GA, otherwise written ^{mul}UGA^{mušen}, is identified with ^{d}EN ÚŠ, *bēl mūti*, "Lord of death" (Horowitz 2014: 141 l. 17). Such writing could be an example of Late Babylonian playful writing to underline the negative symbolism inherent in crows and ravens.

450 Nevertheless, this must remain a simple, abstracted rule which does not apply to every omen. For instance, in one weather omen mentioned by Tommaso Scarpelli during a personal conversation with the author, two negative elements par excellence in the protasis (i.e. an eclipse and a thunder) produce another negative protasis: DIŠ AN.GE$_6$ GAR-*ma* ^{d}IŠKUR GÙ-*šú* ŠUB ZÁḪ KUR *ina* KA DINGIR^{meš} DU$_{11}$.GA, "[If] an eclipse occurs and Adad thunders, (there will be) loss of the country by the command of the gods" (Rochberg-Halton 1988: 85 obv. 6'). It seems that, in this entry, two negative elements in the protasis have been added to each other (– + – = –) through the conjunction -*ma*, "and", and not combined; hence, the prediction caters for an (extremely) negative apodosis.

A.4.1. The Topicalisation of Planets

Identifying the symbolism according to which certain elements of celestial omens (i.e. stars, planets, phenomena, etc.) have a positive or a negative apodosis is crucial. In this respect, how planets in Mesopotamia were perceived is extremely relevant, not only to understand the predictions but also the rationale of factual explanations in commentaries. The planets were either perceived as positive or negative, and that resonated in their names and features. They were thought of as manifestations of the bad or good will of the great gods with whom they were associated, as often underlined in symbolic associations between protases and apodoses (see A.3.1.5). This is what Maul (2018: 195–203) has labelled "topicalisation", a concept that can be extended to stars as well.

Brown (2000: 53–74) collected and catalogued all the names of the five planets (i.e. Mars, Jupiter, Mercury, Saturn, and Venus) in Mesopotamia, analysing them according to the phenomena they portray in celestial omens.[451] The planet Mars ($^{d/mul}$*ṣalbatānu*) is the representation of Nergal/Erra in the night sky (see fn. 334). Like its anthropomorphic counterpart, Mars is related to plague and carnage (see 5.3.1.); therefore, Mars is an evil planet, known by many names, such as "Strange One" or "Other One" (^{mul}*aḫu* or ^{mul}MAN-*ma*, *šanûmma*), "Hostile One" (^{mul}*nakaru*), "Evil One" (^{mul}ḪUL or ^{mul}*lumnu*), and "Red One" (^{mul}SA$_5$, *sāmu*), all names that stress its evil characteristics (Koch-Westenholz 1995: 128–129; Reynolds 1998; Brown 2000: 70–72).

Jupiter ($^{d/mul}$SAG.ME.GAR) is the representation of Marduk, the king of the gods; it can be both a positive and negative planet, like the will of Marduk can. The planet is called "White One" (^{mul}BABBAR, *peṣû*), or "Heroic One" ($^{d/mul}$UD.AL.TAR, *dāpinu*), like the heroic Marduk in the cosmology of *Enūma eliš*. Jupiter was even given different names for its different positions during its path in the sky, underlying each name has specific characteristics: it is called "Šulpaea" ($^{d/mul}$ŠUL.PA.È), lit. "Lord of the Bright Rising", when it rises in the east, $^{d/mul}$SAG.ME.GAR when it is stationary in the east, "Ford" or "Crossing" ($^{d/mul}$*nēberu*) when it stands in the middle of the sky, and Heroic One when it is in the west (Koch-Westenholz 1995: 120–122).[452]

Despite conventionally being associated with Jupiter, Marduk can also be associated with Mercury ($^{mul.d}$AMAR.UTU, "Marduk-star", see SAA 8, 93: rev. 3). Mercury is called $^{d/mul}$(UDU.IDIM.)GU$_4$.UD, lit. "Calf of the Sun", and *šiḫṭu*, "Jumping One", a name that refers to the fast motion of the planet (Brown 2000: 67–68). Mercury was also associated either with Ninurta (MUL.APIN I ii 16, see Hunger-Steele 2019: 42) or Nabu (Gössmann 1950: 113–114 n. 290), so it can be either a positive or negative planet according to the context.[453]

---

451 Brown (2000: 63–64) identified five kinds of associations between the planets and their names: basic (a planet and a god), theological (features of gods), learned (made by scholars according to their expertise), observational (features of celestial phenomena), and symbolic (a planet and a divine symbol).

452 One report dating to the Neo-Assyrian period (ca. 911–612 BC) explains how these names for Jupiter are used for different positions in the sky (SAA 8, 147 obv. 7–rev. 1).

453 For Marduk's astral identity, see Rochberg (2009: 58–62).

Saturn is $^{\text{d/mul}}$(UDU.IDIM.)SAG.UŠ, *kajamānu*, "Steady", sometimes associated with Ninurta.[454] It was perceived as the "night Sun", as implied by another of its names as the star of the Sun god Šamaš (^{d}20 or ^{d}UTU). Šamaš is also the god of justice, *kittu* in Akkadian, a word which has the same root as *kajamānu*, the Semitic root *kun*, "to be steady" (Parpola 1983: 342–343, 3f.). It can be gathered that the names of Saturn could be based on a graphic pun with *kittu* as an attribute of Šamaš.

Venus ($^{\text{d/mul}}$nin.si$_4$.an.na, $^{\text{d/mul}}$*dilbat*) was the representation of Inanna/Ištar since early times. As both a morning and evening star, Venus embodies the gender-fluid features of the goddess Inanna/Ištar, that is, her masculine shape with warlike features as an evening star, and feminine shape as a morning star (see 3.1.1.3.1.).

The idea of topicalisation can be reflexively extended to specific stars and constellations when equated to planets. As noted in the reconstructed omens in chapter 5 (see 5.1.7., 5.3.–5.3.2.), stars and constellations in protases are substituted by – or they are considered alternative names of – planets in factual explanations. For instance, in commentaries, the Pleiades (MUL.MUL) are either called "Bristle" (^{mul}*zappu*), i.e. their Akkadian name, or "Mars", when they are said to be moving into or forward to celestial bodies:

> [DIŠ] $^{⸢\text{mul}⸣}$LU.LIM KI MUL.MUL *it-ten-tu id-ra-na-a-tu*$_4$ *ina* KUR GÁL^{meš} : $^{\text{mu}⸢\text{l}}$...] / ^{mul}*ṣal-bat-a-nu a-na* ^{mul}UDU.IDIM.GU$_4$.UD [TE-*ma*]
>
> [If] the Stag with the Pleiades go parallel to each other, there will be salinity in the land [...]. (It means that) the sta[r...], / (or) Mars [comes close] to Mercury.
> (LB 1321 rev. 7'–8', see SIT 6: 11 source B and parallels)

Considering the factual explanation of the entry given above, it is known that Mars is the astral representation of Nergal/Erra, and it is a negative planet. The link between the protasis (i.e. Pleiades) and the factual explanation (i.e. Mars) seems a symbolic principle (see A.3.1.5.) which resonates in mythology: in the poem of "Erra and Išum" (see 3.1.2.1.), where the Sebettu, the anthropomorphic counterparts of the Pleiades, are the companions of Nergal/Erra. Therefore, one can assume that the Stag represents Mercury and that the Pleiades and the Stag are topicalised, i.e. considered a negative and a positive asterism respectively, based on a factual explanation made by the scribes *a posteriori*.

---

454 E.g. see SAA 8, 154, where Saturn is not mentioned by name, but its phenomena are recorded by calling it "Sun" and "Ninurta".

Appendix B

Editions of Selected Sources

§ 1 Sm 319 = EAE 52 source B

Transliteration

Obverse

1–14. For the edition, see EAE 52: 1–7 (see pp. 147–148).
15. For the edition, see EAE 52: 12 (see pp. 149–150).

Reverse

1’. DIŠ [...]
(break of 3 lines)
5’’. DIŠ [...]
(break of 1 line)
7’’’. DIŠ/*ana* [...]
8’’’. LUG[AL...]
(end of the reverse)

Translation

Reverse

1’. If [...].
(break of 3 lines)
5’’. If [...].
(break of 1 line)
7’’’. If/To [...].
8’’’. Kin[g...].
(end of the reverse)

§ 2 K 2118 = EAE 52 source C; Commentary EAE 53 source U

Transliteration

Obverse

For the edition, see EAE 52: 5–13 (see pp. 148–150).

Reverse

1’. [DIŠ $^{m]ul}$[...]

2’. [DI]Š ^{d}EN.ME.ŠÁR.R[A...]
3’. DIŠ ^{d}EN.ME.ŠÁR.R[A...]
4’. DIŠ ^{d}EN.ME.ŠÁR.RA $^{m[ul}$...]
5’.   *um-šu*$_{14}$ *dan-*⸢*nu*⸣ [...]
6’. DIŠ ^{d}EN.ME.ŠÁR.RA $^{m[ul}$...]

7’. DIŠ ^{d}EN.ME.ŠÁR.RA [...]
8’. DIŠ ^{mul}MAN-*ma ana* ^{d}E[N.ME.ŠÁR.RA TE UN^{meš} DAGAL^{meš} ŠÀ KUR DÙG-*ab*]

9’. DIŠ MUL.MUL [^{mul}ŠUDUN KUR-*ud*...]
10’. DUB.51.K[AM* UD AN ^{d}*en-líl*...]
11’. *tup-pi* [...]
(remainder is missing)

Translation

Reverse

1’. [If the s]tar [...].

2’. [I]f Enmešarr[a...].
3’. If Enmešarr[a...].
4’. If Enmešarra [...] the s[tar...],
5’. strong heat [...].
6’. If Enmešarra [...] the s[tar...].
7’. If Enmešarra [...].
8’. If the Other One (i.e. Mars) [comes close to] E[nemšarra, people will spread, the country will rejoice].

9’. If the Pleiades [reaches the Yoke...].
10’. Tablet 51 [of *Enūma Anu Enlil*...].
11’. Tablet [of...].
(remainder is missing)

Comments

rev. 8’: For the restoration, see SAA 8, 503 obv. 6–7: DIŠ ^{mul}MAN-*ma ana* ^{mul}EN.ME.ŠÁR.RA TE / UN^{meš} DAGAL^{meš} ŠÀ KUR DÙG-*ab*. A shorter version of this omen is found in K 8000 rev. 16’ (ACh Ištar 24, 15): DIŠ ^{mul}MAN-*ma ana* ^{d}+EN.ME.ŠÁR.RA TE ŠÀ KUR DÙG-*ab* [...], “If the Other One comes close to Enmešarra, the country will rejoice [...]”.

rev. 9’: For the restoration, see Commentary EAE 53: 1 (see p. 178). This entry is also given in 5.1.3.2. § 3 (see p. 138).

§ 3 K 3099 + K 18689 (E$_1$) (+) Sm 259 (E$_2$) = EAE 52 sources E$_1$–E$_2$

Transliteration

Obverse

E$_1$

1’. [DIŠ MUL^{meš}-*šú* AN].⸢TA *nen*⸣-[*mu-du* KUR SU.BIR$_4$^{ki} KUR MAR.TU^{ki} MU.7.KAM* *i-šal-lal*]
2’. [DIŠ MU]L^{meš}-*šú* KI.TA *nen-mu-du* KUR ELAM.⸢MA^{ki}⸣ [KUR URI^{ki} KIMIN]

3’. [DIŠ M]UL^{meš}-*šú* AN.TA *rit-ku-su* KIMIN U$_5^{meš}$ KUR SU.⸢BIR$_4^{ki}$⸣ K[UR MAR.TU^{ki} KÚR^{me}-*ma*]
4’. [DIŠ M]UL^{meš}-*šú* KI.TA *rit-ku-su* KUR ELAM.MA^{ki} KUR URI^{ki} MU.5.KAM* *i-šal-lal* ⸢MU.5.KAM*⸣ [KUR URI^{ki} ZI-*ma*]
5’. [DIŠ M]UL^{meš}-*šú* AN.TA *da-ʾa-mu* MUL^{meš}-*šú* KI.TA *pa-nu-šú-<nu>* SAG.UŠ^{me} *ina* KUR SU.BIR$_4^{ki}$ *u* KUR MAR.T[U^{ki} KIMIN]
6’. KUR ELAM.MA^{ki} MU.5.KAM* ^{d}*èr-ra u* ^{d}IŠKUR UN^{meš}-*šú-nu* G[U$_7^{meš}$]
7’. [DIŠ] ⸢MUL^{meš}⸣-*šú* KI.TA *da-ʾa-mu* MUL^{meš}-*šú* AN.TA *pa-nu-šú-nu* SAG.UŠ^{me} *ina* KUR ELAM.MA^{ki} *u* KUR URI^{ki} M[U.5.KAM* KIMIN]
8’. [DIŠ MUL^{meš}-*š*]*ú* AN.TA SIG$_7^{meš}$-*ma* MUL^{meš}-*šú* KI.TA *pa-nu-šú-nu* SAG.UŠ^{me} *ina* KUR SU.BIR$_4^{ki}$ *u* K[UR MAR.TU^{ki} MU.3.KAM*]
9’. [ŠÈG^{me} *ina* A]N-*e* ILLU^{me} *ina* IDIM KUD^{me} GÁN.ZI NU SI.SÁ *ub-bu-ṭu* [GÁL-*ši*]
10’. [DIŠ MUL^{meš}-*šú* KI.T]A SIG$_7^{meš}$-*ma* MUL^{meš}-*šú* AN.TA *pa-nu-šú-nu* ⸢SAG.UŠ^{me} MU.5.KAM*⸣ [*ina* KUR ELAM.MA^{ki} *u* KUR URI^{ki} KIMIN]
11’. [DIŠ MUL^{meš}-*šú* AN.TA *ma-g*]*al* SA$_5^{meš}$ *ina* KUR SU.BIR$_4^{ki}$ *u* KUR MAR.TU$^{!k}$[^{i} GÁN.ZI SI.SÁ BURU$_{14}$ *ina-pu-uš* KUR ŠÀ...]
12’. [DIŠ MUL^{meš}-*šú* KI.TA *ma-gal* SA]$_5^{meš}$ *ina* KUR ELAM.MA^{ki} *u* KUR URI^{ki} ⸢GÁN⸣.[ZI SI.SÁ...]
13’. [DIŠ MUL^{meš}-*šú* AN.TA NU IGI.DU$_8^{meš}$-*ma*] ⸢KI.TA⸣ IGI.DU$_8^{meš}$ L[UGAL KUR SU.BIR$_4^{ki}$ *u* KUR MAR.TU^{ki} *i-šal-la-lu-ma* ŠÈG^{meš}...]
14’. [DIŠ MUL^{meš}-*šú* KI.TA NU IGI.DU$_8^{meš}$-*ma* AN.TA] ⸢IGI⸣.D[U$_8^{meš}$ LUGAL KUR ELAM.MA^{ki} *u* KUR URI^{ki} *i-šal-la-lu-ma* KUR SU.BIR$_4^{ki}$...]
(break of unknown length)
15’’– 20’’. For the edition, see EAE 52: 10–14 (see pp. 149–151).

E$_2$
1’–12’. For the edition, see EAE 52: 21–29 (see pp. 153–154).

Reverse
E$_2$ rev. 1–E$_1$ rev. 26’. For the edition, see EAE 52: 30–60 (see pp. 155–162).

Translation

Obverse
E$_1$
1’. [If its (i.e. the Field’s) up]per [stars] m[eet, Subartu will plunder Amurru for 7 years].
2’. [If] its lower [star]s meet, Elam [ditto (i.e. will plunder for 7 years) Akkad].
3’. [If] its upper [st]ars are conjoined ditto (i.e. Elam will plunder Akkad for 7 years). (If its upper stars) ride one on the other, Subartu (and) [Amurru will start hostilities].
4’. [If] its lower [st]ars are conjoined, Elam will plunder Akkad for 5 years, in the 5th year [Akkad will rise].
5’. [If] its upper [st]ars are dark, its lower stars look normal, in Subartu and

Amur[ru ditto (i.e. Subartu will plunder Amurru for 5 years, in the 5th year Amurru will rise)];

6'. (in) Elam for 5 years Erra and Adad will d[evour] their people.

7'. [If] its lower stars are dark, its upper stars look normal, in Elam and in Akkad [for 5] y[ears ditto (i.e. Erra and Adad will devour their people)].

8'. [If it]s upper [stars] are yellow/green and its upper stars look normal, in Subartu and [Amurru for 3 years]

9'. [rain from the s]ky (and) high floods from the springs will cease, the cultivated land will not prosper, [there will be] famine.

10'. [If its lowe]r [stars] are yellow/green and its upper stars look normal, for 5 years [in Elam and in Akkad ditto (i.e. rain from the sky and high floods from the springs will cease, the cultivated land will not prosper, there will be famine)].

11'. [If its upper stars are ve]ry red, in Subartu and Amurru [the cultivated land will prosper, the harvest will expand, the land the inside (lit. heart)...].

12'. [If its lower stars are very re]d, in Elam and Akkad the cultivated [land will prosper...].

13'. [If its upper stars are not visible, but] the lower are visible, the k[ing of Subartu and Amurru will plunder and rains...].

14'. [If its lower stars are not visible, but the upper] are visi[ble, the king of Elam and Akkad will plunder, and Subartu...].

(break of unknown length)

15''– 20''. See EAE 52: 10–14 (see pp. 149–151).

Comments

obv.: According to K 11096, source M_1 in Reiner-Pingree (1981: 64 XII 1–9), which duplicates K 3099 + K 18689 in obv. 1'–14' (see the comment given below), at the beginning of the obverse of E_1 there is a break of nine omens, which corresponds to circa ten lines.

obv. 1'–14': For the restoration, see Reiner-Pingree 1981: 58–59 IX 23–34; 64 XII 10–21.

§ 4 K 7214 = EAE 52 source G

Transliteration

1'. [...] ⸢UD^?meš *šú*? *ú*⸣ [...]

2'. [... b]i *it-ten-tu* x[...]

3'. [... dIŠKUR? R]A-*iṣ* KI.LAM TUR [...]

4'. [...] ⸢x x ⸣ [...]

5'. [...] U[D...]

6'. [...] AB.SÍN GUN-*sà* [LAL-*ṭa*...]

7'. [...]x ŠÀ KUR [...]

8'–21'. For the edition, see EAE 52: 44–56 (see pp. 157–162).

Translation

1’. [...] the days$^{?}$ ... [...].
2’. [...]... (they) go parallel ...[...].
3’. [... Adad$^{?}$ will d]evastate, the market rate will diminish [...].
4’. [...] ... [...].
5’. [...] the d[ay...].
6’ [...] the furrow [will diminish] its yield [...].
7’. [...]... the inside (lit. heart) of the country [...].

8’–21’. See EAE 52: 44–56 (see pp. 157–162).

Comments

1’–7’: The broken apodoses in these lines do not correspond to the reconstructed text EAE 52; perhaps they belong to different omens, which is indicated by the scribe with a horizontal line after l. 7’.

6’: For the restoration, see CAD M1 434 3b.

§ 5 K 11324 + K 12705 = EAE 52 source H

Transliteration

Obverse

1’–6’. For the edition, see EAE 52: 23–26 (see pp. 153–154).
7’–8’. For the edition, see EAE 52: 28 (see p. 154).

Reverse

1. DIŠ ^{mul}AŠ.GÁN *ina* KAL *u*$_{4}$*-mi* GUB-*ma* KI 20 *ir*-x[...]
2. ⸢SU.GU$_{7}$⸣ GÁL-*ši* Ú^{meš} UD^{meš} [...]
3. [^{d}UDU.IDI]M.GU$_{4}$.UD IGI-*ma* ŠÚ-*ma* GUR-*m*[*a*...]
4. [...] ⸢GU$_{4}$⸣ GÁL-*ma* x[...]
5. [...] ⸢IGI$^{?}$⸣-*šú*$^{o?}$-*ma ana* x[...]
6. [...] ⸢x⸣ [...]

(remainder is missing)

Translation

Reverse

1. If the Field stands still throughout the day and with the Sun it ...[...],
2. there will be famine, vegetation will dry up [...].
3. (It means that) [Merc]ury appears, sets, and returns a[nd...].
4. [...] there will be cattle (lit. head of cattle) ...[...].
5. [...] its$^{?}$ appearance$^{?}$ and towards ...[...].
6. [...] ... [...].

(remainder is missing)

§ 6 K 10845 = EAE 52 source J

Transliteration

1'. ⸢DIŠ ^{d}*dil-bat ana* ^{mul}AL⸣.LUL [...]

2'. KASKAL^{me} KÚR KUR *ú-šam-qa-a*[*t*$^{?}$...]

3'. DUMU-*šú* AŠ TE DIB MÁŠ°.ANŠE° TU[R...]

4'. É.GAL NUN *ina* ŠÀ-*šá* GÌR GU[B...]

5'. DIŠ *a-na* MIN ^{mul}KU$_6$ KU$_4$ [...]

6'. DIŠ ^{mul}KA$_5$.A *ana* ^{m}[^{ul}...]

7'–11'. For the edition, see EAE 52: 25–26 (see pp. 153–154).

Translation

1'. If Venus [...] towards the Crab [...],

2'. the roads of the enemy of the country will fal[l$^{?}$...],

3'. his son will usurp the throne, the herd will be littl[e...],

4'. the foot will sta[nd] inside the palace of the prince [...].

5'. If the Fish enters into ditto (i.e. the Crab) [...].

6'. If the Fox [...] into the s[tar...].

7'–11'. See EAE 52: 25–26 (see pp. 153–154).

Comments

In this fragment, which collects excerpted omens with factual explanations, not only the text that belongs with the preceding lines is indented (ll. 2'–4', 8', and 10') – as is customary in omens' fragments – but also the beginning of two omens (ll. 6'–7') is indented.

§ 7 Sm 1317 = EAE 52 source L; Commentary EAE 53 source N; SIT 6 source N

Transliteration

Obverse

1'–2'. For the edition, see EAE 52: 46 (see p. 158).

3'. For the edition, see EAE 52: 45 (see p. 158).

4'. DIŠ ^{mul}KAK.SI.SÁ *ana* IM.MER IGI-*šú* [GAR-*nu*]

5'. *ina u*$_4$-*mi* IGI-*ru* [IM.MER DU-*ma*]

6'. DIŠ ^{mul} KAK.SI.SÁ SA$_5$ KUR *ḫa-ru-bi-*[*e* GU$_7$]

7'. DIŠ *ina* ^{iti}ŠU ^{mul}KAK.SI.SÁ ^{mul}TI$_8^{mušen}$ KUR-*ud* [^{še}GIŠ.Ì SI.SÀ]

8'. DIŠ IŠ-BU TA ŠÀ IŠ-BU *ina* GÍR.TAB ŠUR-*ma* ⸢EN *ina*$^{?}$⸣ [...]

9'. DIŠ TA ŠÀ IŠ-BU *ina* GÍR.TAB 2 IŠ-BU^{me} ŠUR^{me}-*ma* ⸢EN⸣ *in*[*a*...]

10'. DIŠ TA ŠÀ IŠ-BU *ina* GÍR.TAB 3 IŠ-BU^{me} ŠUR^{me}-*ma* E[N...]

11'. DIŠ TA ŠÀ IŠ-BU <*ina*> MAR.GÍD.DA IŠ-BU ŠUR-*ma* [...]

12'. NUN *ina* BAD$_4$ DIB-*ma* NU È NÍG-*ga-šú* [...]

13'. DIŠ KIMIN-*ma* NU È NUN *ina* BAD$_4$ DIB-*ma* NU È

14’. For the edition, see Commentary EAE 53: 1 (see p. 178) and SIT 6: 2 (see pp. 205–206).
15’. For the edition, see Commentary EAE 53: 5 (see p. 180).
16’. For the edition, see Commentary EAE 53: 4 (see p. 179) and SIT 6: 3 (see p. 206).

Reverse
1–4. For the edition, see Commentary EAE 53: 3 (only rev. 3) (see p. 179) and SIT 6: 4–5 (see pp. 206–208).
5. DIŠ MUL.MUL ^{mul}U[GA$^{mušen?}$...]
6. DIŠ MUL.⌜MUL ^{mul}⌝[...]
7. DIŠ MUL.[MUL...]
8. DIŠ ^{m}[^{ul}...]
9. DIŠ [...]
(the rest of the reverse is severely damaged; remainder is missing)

Translation

Obverse
1’–2’. See EAE 52: 46 (see p. 158).
4’. If the face of the Arrow is [set up] to the north,
5’. when it appears during the day, [it will go northern].
6’. If the Arrow is red, the country [will eat] carobs.
7’. If in Du’uzu (i.e. Month IV) the Arrow reaches the Eagle, [sesame will be collected].

8’. If an IŠ-BU out of an IŠ-BU in the Scorpion flares up to in$^{?}$ [...].
9’. If out of an IŠ-BU in the Scorpion two IŠ-BU flare up to i[n...].
10’. If out of an IŠ-BU in the Scorpion three IŠ-BU flare up t[o ...].
11’. If out of an IŠ-BU in the Wagon an IŠ-BU flares up [...],
12’. the prince will be afflicted by hardship and will not escape (it), his belongings [...].
13’. If ditto (i.e. out of an IŠ-BU in the Wagon an IŠ-BU flares up) and does not go down, the prince will be afflicted by hardship and will not escape (it).

14’. See Commentary EAE 53: 1 (see p. 178) and SIT 6: 2 (see pp. 205–206).
15’. See Commentary EAE 53: 5 (see p. 180).
16’. See Commentary EAE 53: 4 (see p. 179) and SIT 6: 3 (see p. 206).

Reverse
1–4. See Commentary EAE 53: 3 (only rev. 3) (see p. 179) and SIT 6: 4–5 (see pp. 206–208).
5. If the Pleiades [...] the R[aven$^{?}$...].
6. If the Pleiades [...] the star [...].
7. If the Plei[ades...].
8. If the s[tar...].
9. If [...].

(the rest of the reverse is severely damaged; remainder is missing)

Comments

obv. 4'–7': For the restoration, see K 2894 + K 12290 obv. 18'–21' (ACh Ištar 28): DIŠ mulKAK.SI.SÁ *ana* IM.MER.RA IGI-*šú* GAR-*nu* / *ina u4-mi* IGI IM.MER.RA DU-*ma* / DIŠ mul⌜KAK.SI⌝.SÁ SA5 KUR *ḫa-ru-bi-e* GU7 / ⌜DIŠ⌝ *ina* itiŠU ⌜mul⌝KAK.SI.SÁ mulTI8mušen KUR-*ud* šeGIŠ.Ì SI.SÁ.

obv. 8'–13': These entries are about a luminous phenomenon that Reiner (1998: 244, 268) read as *iš-bu*.[455] The subjects of the protases are the Scorpion (GÍR.TAB, *zuqaqīpu*) and the Wagon (MAR.GÍD.DA, *ereqqu*), and even though the determinatives DINGIR or MUL are not written, the subjects are both likely addressed as celestial bodies. An *iš-bu* is possibly similar to a *mešḫu* or a *ṣallummû*, as suggested by the presence of the verb *ṣarāru* (ŠUR), "to flare up", which is used for *mešḫu* omens as well.[456] A logographic reading (IŠ-BU) for the signs *iš-bu* should be considered. First, because in Sm 1317 (obv. 8'–11') the two signs are preceded by TA ŠÀ (*ištu libbi*), "out of", hence a genitive case is expected to follow, and not an Akkadian nominative *iš-bu*. Second, in obv. 9'–10' *iš-bu* is followed by the plural determinative ME, which always follows logograms in divinatory texts.

§ 8 K 3558 = Commentary EAE 53 source A

Transliteration

Obverse

1–18a. For the edition, see Commentary EAE 53: 1–16 (see pp. 178–182).
19–33. For the edition, see Commentary EAE 53: 19–25 (see pp. 182–184).
34. For the edition, see Commentary EAE 53: 27 (see p. 184).
35. For the edition, see Commentary EAE 53: 30 (see p. 185).
36–42. For the edition, see Commentary EAE 53: 33–40 (see pp. 185–186).

Reverse

1'–5'. For the edition, see Commentary EAE 53: 52–55 (see pp. 187–189).
6'. [...] x[...]
7'–8'. For the edition, see Commentary EAE 53: 56–57 (see p. 189).
9'. [...]x TAG-*ma* [...]
10'–16'. For the edition, see Commentary EAE 53: 59–64 (see pp. 189–190).
17'. [...] x[...]
18'. [...]x AGA [...]

455 These entries are duplicated in the unpublished fragment K 8280 + K 11129 ll. 15'–18'.
456 See CAD M2 120–121 (*mišḫu* A), and CAD Ṣ 106–107 (*ṣarāru* B).

(break of unknown length)

19’’. [...] ⸢x⸣ [...]

20’’. [...] x[...]x IGI x[...]

21’’. [...] x[...$^{d/mul}$... *ina*] ⸢ŠÀ⸣ ^{mul}AB.SÍN GUB-*ma*

22’’. [DIŠ ^{mul}... ^{mu}]^{l}TI$_{8}^{mušen}$ KUR-*u*[*d*] KIMIN

23’’. [DIŠ ^{mul}... ^{mul}KA]K.⸢SI⸣.SÁ KUR-*ud* [ŠU.B]I.AŠ.ÀM

24’’. [DIŠ ^{mul}... *a*]-*dir* ^{d}⸢SAG.⸢UŠ⸣ *ina* ŠÀ ^{mul}AB.SÍN KUR-*ma*

25’’. [DIŠ...]x ⸢*meš-ḫu*⸣ [... *im-su*]*ḫ meš*-⸢*ḫu*⸣ TA ŠÀ-*šá* ŠUR-*ma*

26’’. [DIŠ...]x ⸢*meš-ḫu* [... *im-suḫ meš-ḫ*]*u šá ku-un-nu* GUB-*zu*

27’’. NÍG.PÀ[D.DA DUB.40]+⸢16.KAM*⸣ [DIŠ UD A]N ^{d}50

(the rest of the reverse is severely damaged until the edge)

Left edge

1. [...] ⸢^{d}⸣*ṣal-bat-a-nu* KI-*šú* GUB x[...]

2. SAG MU-*šú-nu* IT[I-*šú-nu*...]

Translation

Reverse

1’–5’. See Commentary EAE 53: 52–55 (see pp. 187–189).

6’. [...] ...[...].

7’–8’. See Commentary EAE 53: 56–57 (see p. 189).

9’. [...]... and touch [...].

10’–16’. See Commentary EAE 53: 59–64 (see pp. 189–190).

17’. [...] ...[...].

18’. [...]... tiara [...].

(break of unknown length)

19’’. [...] ...[...].

20’’. [...] ...[...]... appear ...[...].

21’’. [...] ...[... (It means that) the planet...] stands [in]side the Furrow.

22’’. [If the star...] reache[s] the Eagle, ditto (i.e. It means that the planet... stands inside the Furrow).

23’’. [If the star...] reaches the [Ar]row, ditto (i.e. It means that the planet... stands inside the Furrow).

24’’. [If the star... is da]rk. (It means that) Saturn rises inside the Furrow.

25’’. [If...]... the glow [... shines brightl]y. (It means that) the glow flares from its centre.

26’’. [If...]... the glow [... shines brightly. The glo]w (means) that it is firm (and) it stands still.

27’’. *Mukal*[*limtu*-commentary of tablet 5]6 of [*Enūma A*]*nu Enlil.*

(the rest of the reverse is severely damaged until the edge)

Left edge

1. [...] Mars stands (in) its position ...[...].

2. The beginning of their year, of [their] month[...].

Comments

rev. 27'': The number preserved in the subscript is damaged but based on the manuscript traditions of EAE previous scholars restored the number as ⸢54⸣ (e.g. Fincke 2001: 28; Frahm 2011: 149). The number 54 would "match" with an assumed version of EAE from Nineveh with Babylonian ductus (Fincke 2001: 37); nevertheless the units preserved are 6 and the number should be restored as [5]6.

§ 9 K 8744 = Commentary EAE 53 source B

Transliteration

1'–14'. For the edition, see Commentary EAE 53: 14–24d (see pp. 181–184).
15'. [... M]UL^{me} GUB^{me} [...]
16'. [...] ⸢^{d}⸣*dil-bat* GUB [...]
(remainder is missing)

Translation

1'–14'. See Commentary EAE 53: 14–24d (see pp. 181–184).
15'. [... the s]tars stand still [...].
16'. [...] Venus stands still [...].
(remainder is missing)

§ 10 Sm 197 = Commentary EAE 53 source E

Transliteration

1'–5'. For the edition, see Commentary EAE 53: 1–5 (see pp. 178–180).
6'. [DIŠ MU]L.MUL ^{mul}AŠ.⸢GÁN⸣ x[...]
7'. For the edition, see Commentary EAE 53: 6 (see p. 180).
8'. For the edition, see Commentary EAE 53: 8 (see p. 180).
9'. [DIŠ MUL.MU]L ⸢*u*⸣ ^{m}[^{ul}...]
(remainder is missing)

Translation

1'–5'. See Commentary EAE 53: 1–5 (see pp. 178–180).
6'. [If the Ple]iades ...[...] the Field, [...].
7'. See Commentary EAE 53: 6 (see p. 180).
8'. See Commentary EAE 53: 8 (see p. 180).
9'. [If the Pleiade]s and the s[tar...].
(remainder is missing)

§ 11 K 3923 + K 6140 + 81-7-27, 149 + 83-1-18, 479 = Commentary EAE 53 source G

Transliteration

Obverse

1'. DIŠ [MU]L.[MUL ^{mul}...]
2'. DIŠ MUL.⸢MUL⸣ ^{m}[^{ul}...]
3'. DIŠ MUL.MUL $^{⸢mul⸣}$[...]
4'. DIŠ MUL.MUL ^{mul}x[...]

5'. 9 MU^{meš} [...]

6'. DIŠ *ina* ^{iti}*a-da-ri* UD.25.K[AM*...]
7'. IM^{me} DÙG.GA^{meš} GUB^{meš} NU [...]
8'. [DIŠ] *ina* ^{iti}*še-rum*-BURU$_{14}$ UD.23.KAM* M[UL.MUL *u* ^{d}30...]
9'. [DIŠ *ina* ^{i}]^{ti}*pí-it*-KÁ UD.21.KAM* M[UL.MUL *u* ^{d}30...]
10'. [DIŠ *ina* ^{iti}DIN]GIR.MAḪ UD.19.KAM* M[UL.MUL *u* ^{d}30...]
11'. DIŠ *i*[*na* ^{iti}*a-b*]*i* UD.17.KAM* M[UL.MUL *u* ^{d}30...]
12'a. DIŠ *ina* ⸢^{iti}*la-lu*⸣-*bi-e* UD.⸢15⸣.[KAM* MUL.MUL *u* ^{d}30] ⸢IGI-*šú-nu*⸣-*t*[*i-ma* *iš-taq-lu*]
12'b. MU ⸢BI⸣ *eš-re-et* *ip-p*[*al-si-ḫu* TAG$_4$-*et*]
13'a. DIŠ *ina* ^{iti}*še-b*[*u*]-*ti* UD.13.K[AM* MUL.MUL] ⸢*u*⸣ ^{d}30 IGI-*šú-nu-ti*-[*ma*] ⸢*iš*⸣-*t*[*aq-lu*]
13'b. MU BI [*e*]*š-r*[*e-et*] *ip-pal*-⸢*si*⸣-*ḫu* T[AG$_4$-*et*]
14'a. DIŠ *ina* ^{iti}*še-er-ʾi*-[U]R[U$_4$ UD.11.KAM* MU]L.MUL *u* ^{d}30 IGI-*šú-nu-ti-ma* *iš-taq*-[*lu*]
14'b. MU BI *e*[*š-re-et*] *ip-pal-si-ḫu* TAG$_4$-[*et*]
15'a. DIŠ *ina* ^{iti}*tam-ḫi-ri* U[D.9.KAM* MU]L.MUL *u* ^{d}30 IGI-*šú-nu-ti*-⸢*ma*⸣ *iš-taq*-⸢*lu*⸣
15'b. MU BI *eš-re-et* *ip-pal-si*-⸢*ḫu*⸣ TAG$_4$-*et*
16'a. DIŠ *ina* ^{iti}*si-li-li-ti* UD.7.KAM* MUL.MUL *u* ^{d}30 IGI-*šú-nu-ti-ma* *i*[*š*]-*taq-lu*
16'b. MU BI *eš-re-et* *ip*-⸢*pal*⸣-*si*-[*ḫ*]*u* TAG$_4$-*et*
17'a. DIŠ *ina* ^{iti}*ḫul-dúb-bi-e* UD.5.KAM* MUL.MUL *u* ^{d}30 IGI-*šú-nu-t*[*i*]-*ma* ⸢*iš*⸣-*taq-lu*
17'b. MU BI *eš-re-et* *ip-pa*[*l-s*]*i-ḫu* TAG$_4$-*et*
18'a. DIŠ *ina* ^{iti}*ša-ba-ṭi* UD.3.KAM* ⸢MUL.MUL⸣ *u* ^{d}30 IGI-*šú-nu-t*[*i-m*]*a* *iš-taq-lu*
18'b. MU BI *eš-re-et* *i*[*p-pal-s*]*i-ḫu* TAG$_4$-*et*

(end of the obverse)

Reverse

1–4. For the edition, see 4.2.2. § 2a (see pp. 116–118).

---

5. DIŠ DUB *ni-ṣir-ti* AD.ḪAL [...]x ⸢SUM$^{?}$⸣ *ana* É.⸢ZI.DA$^{?}$⸣ [... g]i$^{?}$
6. IBILA-*ka* MIN$^{?}$ *ta-š*[*em*...] ⸢tùm mi x⸣ [...]

7–21. For the edition, see Commentary EAE 53: 26–41 (see pp. 184–186).

Translation

Obverse

1'. If the [Pl]e[iades, the star...].
2'. If the Pleiades, the s[tar...].
3'. If the Pleiades, the star [...].
4'. If the Pleiades, the star ...[...].

5'. 9 entries [...].

6'. If on the 25t[h] day of Adari (i.e. Nisannu) [...],
7'. good winds will blow, not [...].
8'. [If] on the 23rd day of Šer'u-ebūri (i.e. Ajaru) the P[leiades and the Moon...].
9'. [If on] the 21st day of Pīt-bābi (i.e. Simanu) the P[leiades and the Moon...].
10'. [If on] the 19th day [of Bēl]et-ilī (i.e. Du'uzu) the P[leiades and the Moon...].
11'. If o[n] the 17th day [of Ab]u (i.e. Abu) the P[leiades and the Moon...].
12'a. If on the 15th day of Lalubû (i.e. Ululu) you observe [the Pleiades and the Moon and they are balancing],
12'b. that year is normal, (if) they are a[part (lit. they fall down), it is left behind (i.e. will be intercalary)].
13'a. If on the 13t[h] day of Šeb[ū]tu (i.e. Tešritu) you observe [the Pleiades] and the Moon [and they] are b[alancing],
13'b. that year is [n]or[mal], (if) they ar[e a]part (lit. they fall down), it is [left behind (i.e. will be intercalary)].
14'a. If on [the 11t]h day of Šer'i-[er]ēš[i] (i.e. Araḫsamnu) you observe [the Ple]iades and the Moon and they are balanc[ing],
14'b. that year is n[ormal], (if) they are apart (lit. they fall down), it is left behi[nd] (i.e. will be intercalary).
15'a. If on the [9th] d[ay of Tamḫīru (i.e. Kislimu) you observe the [Pl]eiades and the Moon and they are balancing,
15'b. that year is normal, (if) they are apart (lit. they fall down), it is left behind (i.e. will be intercalary).
16'a. If on the 7th day of Sililītu (i.e. Ṭebetu) you observe the Pleiades and the Moon and they are b[a]lancing,
16'b. that year is normal, (if) they are a[p]art (lit. they fall down), it is left behind (i.e. will be intercalary).
17'a. If on the 5th day of Ḫuldubbû (i.e. Šabaṭu) you observe the Pleiades and the Moon and they are balancing,
17'b. that year is normal, (if) they are [ap]art (lit. they fall down), it is left behind (i.e. will be intercalary).
18'a. If on the 3rd day of Šabaṭu (i.e. Adaru) you observe the Pleiades and the Moon [and] they are balancing,

18’b. that year is normal, (if) they a[re ap]art (lit. they fall down), it is left behind (i.e. will be intercalary).

(end of the obverse)

Reverse

1–4. See 4.2.2. § 2a (see pp. 116–118).

5. If the tablet of the arcana, the secret of [...]... give?, to the Ezida? [...]...
6. your heir... , you li[sten...] ... [...].

7–21. See Commentary EAE 53: 26–41 (see pp. 184–186).

Comments

obv. 1’–18b’: For the restoration, see Hunger-Reiner (1975); for a discussion on the meaning of these entries for the intercalation, see 4.2.2. § 2c (see pp. 120–124).

obv. 6’–18b’: The names of the months are taken from the Elamite calendar (Hunger 1980: 302). In the translation, the Elamite names are followed in brackets by the corresponding Akkadian names. The equations between the names of the Elamite calendar and the Babylonian calendar are given in an article by Reiner (1975), who also took this fragment into consideration.

rev. 5–6: These lines are damaged and difficult to reconstruct, yet they likely represent a rubric describing the origin of the preceding entries (rev. 1–4). A similar rubric can be found in BM 42282 + BM 42294 obv. 1 (Brack-Bernsen-Hunger 2008: 4): [D]UB *ni-ṣir-tu*₄ AN-*e pi-riš-tú* DINGIR^meš GAL^meš, “[T]ablet of the arcana of the sky, the secret of the great gods”.

§ 12 Sm 1054 = Commentary EAE 53 source H

Transliteration

Obverse

1’. [... KUR] ⸢MAR⸣.TU^k[^i? ...]
2’. [...]-⸢*ti*⸣-*i* ⸢bul⸣ x[...]
3’. For the edition, see Commentary EAE 53: 20 (see pp. 182–183).
4’. [*ṣa-ra-r*]*u* *ba-ʾa-*[*lu*...]
5’. For the edition, see Commentary EAE 53: 19 (see p. 182).

6’–7’. For the edition, see Commentary EAE 53: 1 (see p. 178).

8’. [LIBIR.RA.BI.GIM A]B.SAR BA.AN.È ⸢x x⸣[...]
(end of the obverse)

Translation

Obverse

1’. [...] Amurru [...].
2’. [...]... ...[...].
3’. See Commentary EAE 53: 20 (see pp. 182–183).
4’. [... to flas]h (means) to be br[ight...].
5’. See Commentary EAE 53: 19 (see p. 182).

6’–7’. See Commentary EAE 53: 1 (see p. 178).
8’. [Like its ancient exemplar, writ]ten (and) collated ...[...].
(end of the obverse)

Comments

obv. 4’: Restored after AO 6464 (TCL 6, 17) rev. 11: *ṣa-ra-ri* : *ra-bu-u šá ba-a-lu*, “to flash (means) to be abnormally bright”. For the edition and earlier bibliography, see CCP 3.1.8.A.a (https://ccp.yale.edu/P363690 accessed 26.01.2021).

§ 13 Sm 247 = Commentary EAE 53 source I

Transliteration

Obverse

1’. [... ^{mul}S]AG.UŠ ^{mul}⸢x⸣ x[...]
2’. [...] [...]
3’. [... ^{d}*di*]*l*$^{?}$*-bat* [...]
4’. [... *b*]*i-ri-šu-nu* ^{d}IŠKUR K[A-*šú* ŠUB...]
5’. [...]x ŠUB-*ma* ^{d}30 *ina* KI.TA x[...]
6’. [... $^{d/mul}$... ^{m}]^{ul}ÙZ TE-*ma ina* ŠÀ 5 GU_4.UD^{meš} [...]
7’. [...]x-*ma* [...]
8’. For the edition, see Commentary EAE 53: 20 (see pp. 182–183).
9’. [...]x *ina* EN.NUN.UD.ZAL.LE *ša* AN.GE_6 ^{d}⸢30⸣ [...]
10’. [...]x *ki-mi-*⸢*i*⸣ 20 <IGI>*-mar še-e-ru-um-ma* [...]
11’. [... *ina ri-bi i*]*na a-sur-rak-ki* ZAG GÙB [...]
12’. [...]*-di u na-ma-ri* [...]
13’. [... *a-ra-d*]*u a-la-ku* [*na-ma-ru a-ma-ru*...]
14’. [...]x-*ma* DIRI^{meš}*-ma* KI.$DURU_5$ SI.SÁ x[...]
15’. [...] I[M...]
16’. [...] ⸢x x x⸣ [...]
(remainder is missing)

Reverse

1’. [...] x[...]
(break of one line)
3’’. [...] ⸢x pa⸣ [x x] ⸢x⸣ [...]
4’’. [...] ⸢x⸣ [...]

(break of one line)
6’’. [...] ⸢x⸣ [...]
7’’. [...] ⸢x x x⸣ [...]
(remainder is missing)

Translation

Obverse

1’. [... S]aturn, the star... ...[...].
2’. [...] [...].
3’. [... Ve]nus$^{?}$ [...].
4’. [... b]etween them, Adad [will] t[hunder...].
5’. [...]... (it) will fall. (It means that) the Moon [...] below ...[...].
6’. [... the planet...] and comes close to the She-Goat, inside five ... [...].
7’. [...]... and [...].
8’. See Commentary EAE 53: 20 (see p. 182–183).
9’. [...]... during the morning watch of an eclipse of the Moon [...].
10’. [...]... when the Sun is visible and it is morning [...].
11’. [... when it sets i]n the right depth to [...].
12’. [...] and dawn [...].
13’. [... to go dow]n (means) to go, [to dawn (means) to be visible...].
14’. [...]... and they will exceed and the flooded ground will prosper ...[...].
15’. [...] the wi[nd...].
16’. [...] ... [...].
(remainder is missing)

Comments

obv. 6’: The line is fragmentarily preserved, and it is unclear whether its second half is a continuation of the protasis, an apodosis, or an explanation. The logograms GU$_4$.UD, *šahaṭu*, “to jump” (CAD Š1 88–92), is also the name of the planet Mercury, either written $^{d/mul}$(UDU.IDIM.)GU$_4$.UD, or simply ^{mul}UDU.IDIM, *bibbu*, “wild sheep” or the word for “planet” (see CAD B 217–219). Hence, in the present context, one could propose the following interpretations for the second half of the line: *ina* ŠÀ 5-<*šu*> <^{d}>GU$_4$.UD *meš*-[*ḫa*$^{?}$ *im-šu-uḫ*$^{?}$], “therein five times Mercury [produces$^{?}$ a] gl[ow$^{?}$]”, with *ina libbi*, “inside”, in its adverbial use and referred to the previously mentioned celestial bodies (see CAD x); *ina* ŠÀ 5 GU$_4$.UD^{meš}, “therein five *attacks* or *risings* (of celestial bodies)”, interpreting GU$_4$.UD^{meš} as *šiḫṭu* (see CAD Š2 416–417); *ina* ŠÀ 5 GU$_4$.UD^{meš}, “inside the five planets”, with GU$_4$.UD as an alternative and playful writing for *bibbū*, “wild sheep” or “planets”.

obv. 11’–13’: For the restoration of obv. 11’ and 13’, see K 3123 rev. 2’–3’ (App. B § 24, see p. 346). Line 13’ is a commentary to line 12’.

§ 14 BM 38301 (80-11-12, 183) = Commentary EAE 53 source K; SIT 6 source HH

Transliteration

Obverse

For the edition, see Commentary EAE 53: 40–55 (see pp. 186–189).

Reverse

1. [DIŠ ^{mu}]^{l}⸢MAR.GÍD.DA⸣ *ina* MAŠ SÌLA ^{d}30 GUB [...] šu x[x x]
2. [DIŠ K]IMIN *ina* SI ZAG ^{d}30 GUB *ina* KUR MAR.TU[^{ki} x x]
3. [DIŠ K]IMIN *ina* SI GÙB ^{d}30 GUB *ina* KUR URI^{ki} [x x]
4. [DIŠ KIMI]N Á IM.U$_{18}$.LU ^{mul}GE$_6$ *i-n*[*a n*]*a-ma-ri ina* EN.NU.UN x[x]
5. [DIŠ KIMIN] Á IM.MER ^{mul}GE$_6$ *i-*⸢*na na*⸣*-ma-ri* AN.GE$_6$ [GAR-*an*]
6. [DIŠ KI]MIN Á IM.KUR.RA ^{mul}GE$_6$ *i-*⸢*na na-ma*⸣*-ri i-na* TUR-*tú* [x x x]
7. [DIŠ] ⸢KIMIN⸣ Á IM.MAR.TU ^{mul}⸢GE$_6$ *i*⸣-[*na na-m*]*a-ri a-ba-ku* [x x x]
8. ⸢DIŠ⸣ KIMIN *ana* ^{mul}ŠUDUN *pa-ni-šá us-*[*saḫ-ḫi-ir*] ŠE *ina* AB.SÍ[N...]
9. ⸢DIŠ⸣ KIMIN *ib-riq-ma* GUB ZI-*ut* ERÍN-⸢*man*⸣-[*da ina* K]UR ELAM.MA$^{<ki>}$ MU.1[+x.KAM*]
10. [DIŠ] ⸢KIMIN⸣ *a-dir* ERÍN-*man-da* ZI-*am-ma* MU.⸢2⸣.KAM* KUR ELAM.[MA^{ki}...]
11. [...] *ina* KUR ^{munus}[KÚR...]
12. [DIŠ K]IMIN *meš-ḫa im-šu-uḫ* *ina* KUR SU.BIR$_4$$^{<ki>}$ *u* KUR M[AR$^{?}$.TU^{ki}...]
13. [DIŠ M]UL *tak-tak-ku š*[*a*] ^{mul}MAR.GÍD.DA [*iṣ-ru-uḫ*...]
14. [...] ^{d}+*en-líl* K[ÚR...] ⸢x x⸣ [...]

---

15. [...]x *u* x[...]
(remainder is missing)

Translation

Reverse

1. [If] the Wagon stands in the shoulder of the Moon [...] ...[...].
2. [If d]itto (i.e. the Wagon) stands in the right horn of the Moon, in Amurru [...].
3. [If d]itto (i.e. the Wagon) stands in the left horn of the Moon, in Akkad [...].
4. [If ditt]o (i.e. the Wagon) (is at) the south side of the Black One a[t d]awn, in the watch ...[...].
5. [If ditto] (i.e. the Wagon) (is at) the north side of the Black One at dawn, an eclipse [will occur].
6. [If di]tto (i.e. the Wagon) (is at) the east side of the Black One at dawn, for the small [...].
7. [If] ditto (i.e. the Wagon) (is at) the west side of the Black One a[t da]wn, the dispatching [...].
8. If ditto (i.e. the Wagon) turns a[way] towards the Yoke, the barley in the furro[w...].
9. If ditto (i.e. the Wagon) flashes and stand still, (there will be) the attack of the

enemy ho[rde in] Elam the 1[+xth] year.
10. [If] ditto (i.e. the Wagon) is black, the enemy horde will rise, the 2nd year the Ela[m...].
11. [...] in the country [(there will be) hostilities...].
12. [If d]itto (i.e. the Wagon) produces a glow, in Subartu and A[murru$^{?}$...].
13. [If the s]tar of the *taktaku* (i.e. a part of the chariot) of the Wagon [flares up...].
14. [...] Enlil will [...] the e[nemy...].

---

15. [...]... and ...[...].
(remainder is missing)

Comments

rev.: The reverse of BM 38301 duplicates the reverse of K 6102, a fragment dating to the Neo-Babylonian period (ca. 626–484 BC) which has the following subscript: [DUB.x]+25.KAM* DIŠ UD AN ^{d}+EN.LÍL.LÁ x [...], "[Tablet x]+25 of *Enūma Anu Enlil*" (K 6102 rev. 15'–16'). K 6102 represents the tablet EAE 55 (Reiner-Pingree 1981: 71 text XVIII) because its incipit is listed in the Aššur catalogue (Rochberg 2018: 122 fn. 5; 125 obv. i. 10').

rev. 4': The same omen is given in SIT 6: 23 (see pp. 215–216).

Rev. 4'–7': For the "Black One" (^{mul}GE$_6$, *ṣalmu*) see 5.1.6. commentary to the text 23, p. 221.

rev. 8': For the restoration, see K 6102 rev. 10' (AAT 80): [DIŠ ^{mu}]^{l}MAR.GÍD.DA *ana* ^{mul}ŠUDUN *pa-ni-šá us-sa*[*ḫ-ḫi-ir*...].

rev. 13': For the restoration, see K 6102 rev. 14' (AAT 80): [... *ta*]*k*-⌜*tak*⌝-*ku šá* ^{mul}MAR.GÍD.DA *iṣ-ru-uḫ* ^{d}+*en-líl* KÚR 21 ⌜x x⌝. For *taktaku*, or *takšakku*, a part of a chariot, see CAD T 89.

§ 15 79-7-8, 271 = Commentary EAE 53 source L; SIT 6 source II

Transliteration

1'. [...]x ⌜^{d}⌝[...]
2'. [... ^{mul}]*ma*-⌜*ak-ru*⌝ [...]
3'. [... ^{mul}UDU.IDIM].⌜GU$_4$⌝.UD *ana* UGU ⌜KI⌝ ^{m}[$^{ul?}$...]
4'. [...]x-*ma u ina šu-ut* $^{d+}$⌜*en*⌝-[*líl*...]
5'–6'. For the edition, see Commentary EAE 53: 19–20 (see pp. 182–183).
7'. [DIŠ *ina* SAG MU MUL.M]UL *pa-ni-ma* [30 *ka-rit*...]
8'. For the edition, see SIT 6: 33 (see p. 218).
9'. [...]x [...]
10'. [...]x-*šú u* x[...]
11'. [...^{d}]30 [...]
(remainder is missing)

Translation

1'. [...]... the star [...].

2’. [...] the Red One (i.e. Mars) [...].
3’. [... Mer]cury over the top of the s[tar$^?$ ...].
4’. [...]... and in the path of En[lil...].
5’–6’. See Commentary EAE 53: 19–20 (see pp. 182–183).
7’. [If at the beginning of the year the Pleiad]es are in front [and the Moon is behind...].
8’. See SIT 6: 33 (see p. 218).
9’. [...]... [...].
10’. [...]... and ...[...].
11’. [...] the Moon [...]
(remainder is missing)

Comments

2’: For *makrû* as a name for Mars, see MSL 11: 40 l. 28.

§ 16 K 5713 + K 7129 + Rm 2, 114 = SIT 6 source A

Transliteration

Obverse

1’–20’. For the edition, see SIT 6: 1–17 (see pp. 205–213).
21’. DIŠ ^{mul}UG$_5$.GA KASKAL ^{d}UTU KUR-*ud* KI.LAM TUR ^{d}UDU.IDIM.SAG.UŠ : ^{d}UDU.IDIM.GU$_4$.UD ^{d}ŠUL.PA.⸢È⸣ KUR-*ma*
22’. ⸢DIŠ⸣ ^{mul}UG$_5$.GA *ana* ^{d}ŠUL.PA.È TE SU.⸢GU$_7$⸣ GÁL-*ši*
(end of the obverse)

Reverse

1. [DIŠ] ⸢^{mul}UG$_5$.GA *ú*⸣-x[...$^{d/mul}$... ^{d}*ṣal-bat*]-*a-nu* KUR-*ma*
2. DIŠ ^{mul}UG$_5$.⸢GA ^{m}[^{ul} ... SI].SÀ

---

3. DIŠ ⸢^{mul}⸣ [x x] x[...$^{d/mul}$... ^{d}UDU.IDIM.G]U$_4$.UD KUR-*ma*
4. DIŠ ⸢^{mul}⸣ [...$^{d/mul}$... ^{d}UDU.IDIM].GU$_4$.UD KUR-*ma*
5. DIŠ ⸢^{mul}⸣ [... ^{m}]^{ul}SIM.MAḪ
6. *lu* [...] GUB-*ma*
7. [DIŠ] ⸢^{mul} x⸣ [...] ⸢NU⸣ SI.SÀ
8. [DIŠ ^{mul}]⸢UR.BAR⸣.[RA ...$^{d/mul}$... ^{d}]⸢ŠUL⸣.PA.È KUR-*ma*
9. [DIŠ] ⸢x *ina* x x⸣ [...] *ana* KUR KÚR
10. [DIŠ ^{mu}]^{l}APIN KASKAL 20 KUR-*ud* ⸢la⸣ [...$^{d/mul}$... ^{d}UDU.IDIM].SAG.UŠ KUR-*ma*

---

11. [DIŠ ^{mu}]^l KA.⸢MUŠ.Ì.KÚ.E⸣ *ana* ^{mu}[^{l}...]x ŠUR-*nun* ^{d}UDU.IDIM.SAG.UŠ ^{d}UDU.IDIM KUR-*ma*
12. [x x] ⸢*šá* x ^{d}*é-a*$^?$ GUB$^?$-*zu*$^?$ ^{mul}⸣[...] BAD-*ma* KALAM NUN SIG$_7$
13. [x] ⸢x x x x⸣ x[...] ^{d}*èr-ra* *ana* ÚŠ^{meš}

---

14. [...] AN.G[E$_6$]

---

15. [...] ⌜x x x x x GUB-*iz* d?⌝[...]
(remainder is missing)

Translation

Obverse

1'–20'. See SIT 6: 1–17 (see pp. 205–213).

21'. If the Raven reaches the path of the Sun, the market rate will diminish. (It means that) Saturn – (var.) Mercury reaches Šulpaea (i.e. Jupiter).

22'. If the Raven comes close to Šulpaea (i.e. Jupiter), there will be famine.
(end of the obverse)

Reverse

1. [If] the Raven ... ...[... (It means that) the planet...] reaches [Ma]rs.
2. If the Raven [...] the s[tar..., it will] prosper.

---

3. If the star [...] ...[... (It means that) the planet...] reaches [Merc]ury.
4. If the star [... (It means that) the planet...] reaches [Merc]ury.
5. If the star [...] the Swallow,
6. or [...] (it) stands still.
7. [If] the star ...[...] (it) will not prosper.
8. [If] the Wol[f... (It means that) the planet...] reaches Šulpaea (i.e. Jupiter).
9. [If] ... in ... [...] will start hostilities against the country.
10. [If] the Plow reaches the path of the Sun ...[... (It means that) the planet...] reaches [Sat]urn.

---

11. [If] Pāšittu [...] towards the sta[r...,]... it will rain. (It means that) Saturn reaches a planet.
12. [...]... ... of Ea? (they) stand still?, the star [...]. If the country is not green.
13. [...] ... [...] Erra (is) for the pestilence.

---

14. [...] Eclip[se].

---

15. [...] ... ... (it) stands still, the planet? [...].
(remainder is missing)

Comments

obv. 21': The same omen is given in SIT 6: 24 (see p. 216). A variant for the apodosis of this omen is: *šá-ni-iš ri-ig-mu* GAR-*an*, "or: there will be noise" (SAA 8, 82 obv. 6).

rev. 12: In Akkadian BAD-*ma*, *šumma*, "if", introduces a variant apodosis (see 5.1.5.1. commentary to the text of entry 2, p. 191).

§ 17 K 2177 + K 7869 + Rm 473 = SIT 6 source C

Transliteration

Obverse

For the edition, see SIT 6: 1–26 (see p. 205–217).

Reverse

1'. [...]x
2'. [...]x
3'. [...]x
4'. [... t]ir$^{?}$
5'. [...] ⸢ŠUB$^{?}$⸣-*ma*
6'. [...] ⸢KUR⸣-[*ma*]
(break of unknown length)
7''. [...]x
8''. [...]x ud
(break of unknown length)
9'''. DUB.6.KAM*-*ma* DIŠ 30 *ina ta-mar-t*[*i-šu*] *mu-kal-lim-tu*$_{4}$
10'''. KUR ^{m}AN.ŠÁR.[DU$_{3}$.A] ⸢MAN ŠÚ⸣ M[AN] KUR AN.ŠÁR^{ki}
(end of the reverse)

Translation

Reverse

1'. [...]...
2'. [...]...
3'. [...]...
4'. [...]...
5'. [...] (it) will fall$^{?}$.
6'. [...] (it) will reach.
(break of unknown length)
7''. [...]...
8''. [...]...
(break of unknown length)
9'''. Tablet 6 of *Šumma Sîn ina tāmart*[*išu*], *mukallimtu*-commentary.
10'''. Land of Ashur[banipal], king of all, k[ing] of Aššur.
(end of the reverse)

§ 18 K 2170 + K 3629 = SIT 6 source F

Transliteration

Obverse

1–11. For the edition, see SIT 6: 1–7, 9 (see pp. 205–210).
12. [...]x : KÚR-*ri* KUR-*su*
13. For the edition, see SIT 6: 10 (see p. 210).
14. [... UD/MU.x]+⸢3.KAM*⸣ KUR-*ma* : ^{d}[UDU].IDIM

15. [...]x ⸢KUR⸣ [...] ⸢TE⸣
(remainder is missing)

Reverse

1'. [...] [...]

2'. [... a]n$^?$ DUMU^{meš} LUGAL GI.⸢NA^{meš}⸣ [...]
3'. [... ^{d}*ṣa*]*l-bat-a-nu ana* ^{mul}AL.LU[L TE-*ma*]

4'. [... *ina* ÍD GÁL^{meš} KU$_6$^{meš} Í]D *ma-la ba-šu-ú* GU$_7$^{meš} ka x[...]
5'. [... x-*im-ma* GU$_7$ *ina* KUR GAR-*an* BAL.GI.KU$_6$ NÍG.BÚN.NA.KU$_6$ KU$_6$^{me} ÍD *ana na-b*]*a-li i-šal-li* ^{d}*ṣal-bat-a-nu ana* ^{mul}A[L.LUL TE-*ma*]

6'. [...] ⸢DAM⸣-*sa ana* GAZ SUM-*i*[*n*]
7'. [...]x sag LIBIR^{meš} KUR DIRI^m[^{eš}]
8'. [... D]IB-*su-ma ina* ŠÀ-*šú* : *ina* MU BI NU TI-*uṭ* [x]
9'. [...a]s$^?$ ^{d}*dil-bat ana* ^{mul}AB.SÍN TE-*ma*

10'. [DIŠ ^{mul}LUGAL *ana* ^{mul}UR.BAR.RA TE *ina* UD NU NAM-*šú* ÚŠ É.GAL NUN KAR-']*a* ZI.GA ÌR *u* GÉME *ina* É.GAL GÁL-*ši*
11'. [KI.LAM KUR KI.LAM LAL-*a* GU$_7$ ^{d}]*ṣal-bat-a-nu ana* ^{mul}LUGAL TE-*ma*

12'. [...]a IGI^{meš} BURU$_5$^{meš} IGI^{meš} DINGIR^{meš} A.RÁ KUR
13'. [...] ⸢ZÁLAG$^?$⸣-*ir* BURU$_5$$^{ḫi.a}$ *ina* IGI.DU$_8$ ZI-*ma*
14'. [...]x ^{d}*dil-bat ana* ^{mul} ^{d}AMAR.UTU TE-*ma*

15'. [...] ⸢ig ad$^?$⸣ iš BURU$_5$$^{ḫi.a}$ ZI-*ma nam-maš-še-e*
16'. [...]⸢^{d}⸣*dil-bat ana* ^{d}GU$_4$.UD TE-*ma*

17'. [...]x ERÍN *la kit-ti* KUR DIB-*bat* DUMU ^{lú}DAM.GÀR
18'. [...]x AŠ.TE DIB-*bat* KUR EN MAN-*ma* TUKU-*ši*
19'. [...]x⸢^{meš} SIG$_5$ *ana* NÍG.GA LUGAL IZI ŠUB-*ut*⸣
20'. [...] nu ^{d}*ṣal-bat-a-nu ana* ^{mul}GU$_4$.AN.NA TE-*ma*

21'. [...]x A.MAḪ DIB-*bat* TAG-*ma*
22'. [...]x ^{mul}UDU.IDIM *ana* ^{mul}AGA.AN.NA TE-*ma*

23'. [... g]a ^{d}*ṣal-bat-a-nu* KI ^{mul}GAM

24'. [...] KUR NINDA *nap-šá* GU$_7$ ŠÈG^{meš} *ṭaḫ-du-tu*$_4$
25'. [... $^{d/mul}$...] ⸢*ana*⸣ ^{d}GU$_4$.UD TE

26'. [...] ⸢*ina*$^?$ MU$^?$⸣^{meš} *ina* ITI BI ^{d}30 AN.GE$_6$ GAR-*an*
27'. [... ^{d}*ṣal-bat*]-⸢*a*⸣-*nu* ⸢*ana*⸣ ^{d}AŠ.GÁN TE-*ma*

28’. [... *ina* Š]À ⸢MÁŠ⸣.ANŠE $^{giš\circ}$ TUKUL$^{?\circ}$ GÁL-*ú* TÉŠ.BI ŠÚR ^{d}*ṣal-bat-a-nu ana* $^{mul.d}$AMAR.UTU TE-*ma*

29’. [DIŠ ^{mul}ḪÉ.GÁL-*la-a* SUKKAL ^{d}*nin-lil ana* ^{mul}ÙZ TE... *m*]*a-la* DÙG.GA *ina* KUR GÁL ^{d}GÌR *u* ^{d}NIDABA *ina* KUR GÁL^{meš}-*ma* KUR DAGAL-*iš* ^{d}UDU.IDIM.GU$_4$.UD *ana* ^{d}⸢*ṣal-bat*⸣-[*a-nu* TE-*ma*]

30’. [... r]i *u su-a-lu*$_4$ KUR DIB-*bat* ^{d}*ṣal-bat-a-nu ana* ^{d}*dil-bat* TE-*ma*

31’. [... ZÍ]Z$^{?}$ *ina* KUR GÁL^{meš}

32’. [... $^{d/mul}$... ^{mul}SIP]A.ZI.AN.NA TE-*ma*

33’. [...]^{meš} ^{d}*dil-bat ana* ^{mul}ŠUL.PA.È TE-*ma*

34’. [... *i*$^{?}$-*ṭ*]*e*$^{?}$-*eḫ-ḫu-ú ana* ^{mul4}*ú-ṣur-ti ta-nam-bi*

35’. [...] ^{mul}ELAM.MA MU GAR

36’. [...] MU-*šú*$^{?}$ *ša kal* MUL$_4$^{meš} SUKUD.GIM

37’. [DUB.x.KAM* DIŠ 30 *ina ta-ma*]*r-ti-šú* NÍG.PÀD.DA UD AN EN.LÍL

38’. [... *kī pî lē’i* GABA.R]I TIN.TIR^{ki} BÁRA.SIPA^{ki}

Upper edge

39’. [... AB.SAR.ÀM] BA.AN.È

40’. [*tup-pi* ^{md}AG-*zu-qu-up*-GI.NA DUMU ^{md}AMAR.UT]U.MU.BA-*šá* ^{lú}DUB.SAR

41’. [ŠÀ.BAL.BAL ^{m}*gab-bi*-DINGIR^{meš}-*ni*-KAM*]-⸢*eš*⸣ ^{lú}GAL DUB.SAR^{meš}
(end of the tablet)

Translation

Obverse

1–11. See SIT 6: 1–7, 9 (see pp. 205–210).

12. [...]... – (var.) enemies of his country.

13. See SIT 6: 10 (see p. 210).

14. [... on the x]+3rd [day/month] (it) reaches – (var.) [a pla]net.

15. [...]... (it) reaches [...] (it) comes close.
(remainder is missing)

Reverse

1’. [...] [...].

2’. [...]... the sons of the king [will...] regular offerings [...],

3’. [... (It means that) M]ars [comes close] to the Cra[b].

4’. [... there will be in the river, the fishes] which are [(in) the ri]ver will be eaten ... ...[...],

5’. [... and the consumption in the country will occur, the river] will toss [the *raqqu*-turtle, the *šeleppu*-turtle (and) fishes onto the sh]ore. (It means that) Mars [comes close] to the C[rab].

6’. [...] will giv[e] her husband for the battle,
7’. [...] the old ... of the country will exceed,
8’. [... he/it s]eizes him/it and in his/its heart – (var.) in that year (he/it) will not get well [...],
9’. [...]... (It means that) Venus comes close to the Furrow.

10’. [If the King-star comes close to the Wolf, he will die before his time, the palace of the prince will be robbe]d, there will be a loss of slaves in the palace,
11’. [the business of the country will devour the smaller business.] (It means that) Mars comes close to the King-star.

12’. [...]... (they) appear, the locusts will appear, the gods, the course of the country
13’. [...] (it) shines bright[?], the locusts will attack when (it is) visible,
14’. [...]... (It means that) Venus comes close to the Marduk-star.

15’. [...]... ... ... locusts will attack and the herd (of wild animals),
16’. [... (It means that)] Venus comes close to Mercury.

17’. [...]... an unreliable army will seize the country, the son of a merchant
18’. [...]... will usurp the throne, the country will have another ruler,
19’. [...] ... good, the fire will fall on the property of the king,
20’. [...]... (It means that) Mars comes close to the Bull of Heaven.

21’. [...]... the flood will seize and affect (badly),
22’. [...]... (It means that) a planet comes close to the Crown of Anu.

23’. [...] (It means that) Mars (is) with the Crook.

24’. [...] the country will have much food to eat, rain will be abundant,
25’. [... (It means that) the planet...] comes close to Mercury.

26’. [...] in[?] the years[?], in that month an eclipse of the Moon will occur.
27’. [... (It means that) Mar]s comes close to the Field.

28’. [... insid]e the city there will be weapon(s)[?], and at the same time (the city) will get furious. (It means that) Mars comes close to the Marduk-star.

29’. [If the star of Abundance, the vizier of Ninlil, comes close to the She-Goat, ... e]verything good there will be in the country, there will be Sumukan and Nisaba in the country, the country will grow. (It means that) Mercury [comes close] to Ma[rs].

30’. [...] and the phlegm will seize the country. (It means that) Mars comes close to Venus.

31’. [... emme]r$^{?}$ will be in the country,
32’. [... (It means that) the planet...] comes close to [the Tru]e Shepherd of Anu.

33’. [...]... (It means that) Venus comes close to Šulpaea (i.e. Jupiter).

34’. [...(they) co]me close$^{?}$, you call the star (as) “(divine) design” (lit. “drawing”).
35’. [...] the star of Elam will establish the year.
36’. [...] its year$^{?}$ for all the stars, as (given) above.

37’. [Tablet x of *Šumma Sîn ina tāma*]*rtišu*, *mukallimtu*-commentary of *Enūma Anu Enlil.*
38’. [... according to a wooden board, origin]al from Babylon (and) Borsippa

Upper edge
39’. [... written (and)] collated.
40’. [A tablet of Nabû-zuqup-kēnu, son of Mardu]k-šumu-iqīša, the scribe
41’. [descendant of Gabbi-ilāni-ēr]eš, the chief of the scribes.
(end of the tablet)

Comments

rev. 4’–5’: For the restoration, see the unpublished K 6534 ll. 1–2 [... ^{m}]^{eš}-*ma* KU$_{6}^{meš}$ ÍD *ma-l*[*a*...] / [...]⸢KU$_{6}$ NÍG.BÚN⸣.NA.KU$_{6}$ KU$_{6}^{me}$ ÍD *ana na-ba-l*[*i* ...]; Sm 1510 obv. 4’–6’ [...]x^{meš} *ina* ÍD GÁL^{meš}-*ma* KU$_{6}^{meš}$ ÍD *ma-la ba-š*[*u*...] / [...]x-*im-ma* GU$_{7}$ *ina* KUR ⸢GAR⸣-*an* BAL.GI.KU$_{6}$ NÍG.BÚN.NA.KU$_{6}$ [...] / [...] ⸢*i*⸣-*šal-la* ^{d}*ṣal-bat-a-nu ana* ^{mul}AL.LUL TE-*m*[*a*]. In this broken entry is possible to detect a semantic working principle behind the parts of the omen (see A.3.1.4.): the factual explanation tells us that the planet Mars, a negative planet (see A.4.1.), affects the Crab (^{mul}AL.LUL, *alluttu*), which very likely symbolises the turtles and the fishes in the apodosis.

rev. 10’–11’: For the restoration, see K 1522 + K 3594 obv. 11’–12’ (App. B § 22 P$_{3}$, see p. 337).

rev. 23’: The first sign of the line, visible only by its second half, could also be [i]g.

rev. 28’: Given the context, it is likely that MÁŠ.ANSE, *būlu*, “herd”, is here meant as a synonym of *nammaššû*, “herd (of wild animals)”, but also “population, city” (CAD N1 234b). That is rarely attested in apodoses (George 2013a: 34, 12 §6).

rev. 29’: For the restoration, see K 3780 obv. ii 7’–8’ (App. B § 23, see p. 343).

rev. 34'–36':	In rev. 34', the Akkadian *uṣurtu*, lit. "drawing", or "(divine) design" (CAD U/W 290–295) is attested as a name for a celestial body in SAA 8, 124 obv. 8, i.e. as a name of ^{mul}MAŠ.TAB.BA.GAL.GAL, *tū'amū rabûtu*, "Great Twins", the zodiacal sign Gemini (see CAD U/W 293b for further bibliography). In rev. 35', for "star of Elam" (^{mul}ELAM.MA) as a name for Mars, see the Great Star List (Koch-Westenholz 1995: 198, 200 ll. 237–240). For the scribal abbreviation SUKUD.GIM, "as (given) above", see Fincke (2022). These three lines (rev. 34'–36') are likely an explanatory section of the scribe, as attested by *ta-nam-bi*, "you call", from *nabû*, "to name" (CAD N1 32–39) in rev. 34', a verb used in hermeneutical contexts to explain how to call or understand something (Gabbay 2016: 170–171), and by the use of SUKUD.GIM. Similar to rev. 34', the stars of the GIŠ.ḪUR, *uṣurtu*, are mentioned in K 2254 rev. 6' (App. B § 25, see p. 349), again in an explanatory context. Based on this comparison, one could assume that a group of stars, called "stars of the (divine) designs", were somehow individuated or used by ancient scholars for specific divinatory purposes.
rev. 37'–41':	The colophon is restored based on the colophons of Nabû-zuqup-kēnu (Hunger: 1968: 92 n. 296–311).

§ 19 K 5277 = SIT 6 source G

Transliteration

Obverse

For the edition, see SIT 6: 1–10 (see pp. 205–210).

Reverse

1'. [... *ki-i pî* ... *šaṭir*(*ma*) B]A.AN.È

2'. [*tup-pi* ^{md}AG-*zu-qu-up*-GI.NA DUMU ^{md}AMAR.UTU.MU.BA-*šá* ^{lú}DUB.SAR ŠÀ.BAL.BAL ^{m}*gab-bi*-DINGIR^{meš}-*ni*-KA]M*-*eš* ^{lú}GAL DUB.SAR^{meš}

3'. [^{uru}*kal-ḫa* ^{iti}X UD.X.KÁM *li-mu* ^{m}*man-nu-ki*-^{d}*aš+šur*-ZU ^{lú}GAR.KUR ^{uru}*t*]*il-le-e*

4'. [MU.13.KAM ^{m}LUGAL.GI.NA EGIR-*ú* LUGAL KUR *aššur*]⌜^{ki} *ù* MU.1.KAM* LUGAL⌝ KÁ.DINGIR.RA^{ki}

(end of the reverse)

Translation

Reverse

1'. [According to... written (and) c]ollated.

2'. [A tablet of Nabû-zuqup-kēnu, son of Marduk-šumu-iqīša, the scribe, descendant of Gabbi-ilāni-ere]š, the chief of the scribes.

3'. [Kalḫu, month x, xth day, eponym of Mannu-kī-Aššur-lē'i, governor of T]illê

4'. [13th year, Sargon the Second, king of Assyria], and 1st year (as) king of Babylon.

(end of the reverse)

§ 20 BM 44005 (81-7-1, 1766) = SIT 6 source L

Transliteration

Obverse

1–2. For the edition, see SIT 6: 12 (see p. 211).
3–5. For the edition, see SIT 6: 8–9 (see pp. 209–210).
6. [DIŠ ^{mul}].⸢^{d}⸣AMAR.UTU *meš-ḫa im-*⸢*šuḫ*⸣ [...]
7. [...] *ina* ^{mul4}UD.⸢ZAL$^{?}$.LE$^{?}$ GUB⸣ [...]
8. [...]x ⸢x ru⸣ [...]
(remainder is missing)

Reverse

1'. [...] ⸢x x x⸣ [...]
2'. [...] BURU$_{14}$ ⸢GUB$^{?}$-*ma* mi⸣ [...]
3'. [... MU]L$^{?}$ *el-*{x}*-mu-šú* ⸢x⸣ x[...]

Upper edge

4'. [...]x MU^{meš} *mu-kal-lim-da* x[...]
5'. [... *ṣa-a-tú u*] *šu-ut* KA *šá* K[A *ummâni ša...*]

(end of the reverse)

Translation

Obverse

1–2. See SIT 6: 12 (see p. 211).
3–5. See SIT 6: 8–9 (see pp. 209–210).
6. [If] the Marduk-[star] produces a glow, [...].
7. [...] stands in the star of the morning$^{?}$ watch$^{?}$ [...].
8. [...]... .... [...].
(remainder is missing)

Reverse

1'. [...] ... [...].
2'. [...] and the harvest will stand$^{?}$... [...].
3'. [...] the amber$^{?}$ [sta]r$^{?}$ ...[...].

Upper edge

4'. [...]... entries from a *mukallimtu*-commentary ...[...].
5'. [... lemmata and] oral explanations following the sayi[ngs of a master scholar relating to ...].

(end of the reverse)

Comments

rev. 3’: The *elmušu*- or *elmešu*-stone is a brilliant precious stone unattested in economic texts but referred to because of its colourful and bright reflection, associated with the sky and the celestial bodies (AHw 205a; CAD E 107–108). The shade of colour of the *elmušu*-stone might have been similar to amber (Thavapalan 2020: 270 fn. 1013 for earlier bibliography on this topic). The fact that a star is compared to a stone is not unique, as in Mesopotamia there was the widespread belief that heaven was made of precious stones (Horowitz 1998: 263).

edge 4’-5’: According to the labels given by Frahm (2011: 53), the subscript could correspond to a *ṣâtu* commentary type 6b: *ṣâtu u šūt pî ša pî ummâni ša* (tablet incipit), “lemmata and oral explanations following the sayings of a master scholar relating to (tablet incipit)”, but the entries were taken from another *mukallimtu*-commentary which is a characteristic attested only in another EAE commentary (Frahm 2011: 56). In line 4’, the writing *mu-kal-lim-da* could be an example of playful writing based on the root of *lamādu*, “to learn”. It is attested in two commentaries of EAE from the Ashurbanipal library in Nineveh (Frahm 2011: 42, fn. 158). However, Johannes Hackl in a private conversation with the author noted that the spelling *mu-kal-lim-da* might simply derive from a labial and dental shift (*mt* < *nd*) since there was no cuneiform sign for the sound *lin*. He pointed out the example of *tašlimtu*/*tašlindu*, “malicious talk”, from *šalāmu*, with the prefix *ta*- and feminine *-t*, attested as *taš-li-in-du* or *taš-lim-da* (AHw 1338–1339 “tašlimtu(m)”).

§ 21 Rm 192 = SIT 6 soruce O

Transliteration

Upper edge

1. [DIŠ ^{m}]^{ul}UDU.IDIM *ana* ^{m}[^{ul}UD]U.IDIM TE AB.SÍN IGI-*šá ip-te-te kiš-šu-tú* GÁL-[*ši*]
2. [*ṭ*]*ar-du* AŠ.TE D[IB-*bat* N]IDBA DINGIR^{meš} *i-šak-kan* É.KUR^{meš} TÉŠ.BI *i-za-an-*[*na-an*]

Obverse

3. ⸢DIŠ ^{mul}⸣UDU.IDIM *ana* ^{mu}[^{l}UDU].IDIM TE DUMU LUGAL *šá ina* URU ZAG.MU *áš-bu ana* AD-*šú* ḪI.GAR ⸢GAR-*an*$^{?}$-*ma*$^{?}$⸣
4. AŠ.TE NU DIB-*bat* DUMU *ma-a*[*m*]-*ma-na-ma* E$_{11}$-*ma* AŠ.TE DIB-*bat* É^{meš} DINGIR^{meš} GAL^{meš}
5. *ana* KI-*ši-na* GUR⸢^{meš} SÁ⸣.DUG$_{4}$ DINGIR^{meš} GAL^{meš} *ú-kan* É DINGIR^{meš} GAL^{meš} 1-*niš i-za-an-*⸢*na-an*⸣
6. DIŠ ^{mul}*ṣal-bat-a-nu u* ^{d}UDU.⸢IDIM⸣ *im-taḫ-ru-ma* GUB^{meš} ZI-*ut* ELAM.MA^{ki}
7. DIŠ ^{mul}KU$_{6}$ *a-dir* KU$_{6}$ *ina* ÍD ⸢*e*⸣-*ru-ta*$_{5}$ NU DÙ MUŠEN *ina* AN-*e* NUNUZ NU ŠUB

8. ^{mul}KU$_6$ ^{d}*ṣal-bat-a-nu* *a-dir* ^{d}SAG.UŠ
9. DIŠ ^{mul}APIN KASKAL ^{d}UTU KUR-*ud* SU.GU$_7$ *ina* KUR GÁ[L]
10. DIŠ ^{mul}UDU.IDIM *u* ^{d}*ṣal-bat-a-nu im-daḫ-ru-ma* ⸢GUB^{meš}⸣ ZI-*ut* ELAM.⸢MA⸣^{k}[^{i}]
11. DIŠ ^{d}*sar*$_6$*-ru u* ^{d}UDU.IDIM *im-daḫ-ru-ma* GUB^{meš} *ina* MU BI ⸢ZI⸣-*ut* ELAM.MA^{k}[^{i}]
12–14. For the edition, see SIT 6: 2–3 (see pp. 205–206).
15. DIŠ ^{mul}*ṣal-bat-a-nu* NÍGIN-*ma* ŠE.ER.ZI ÍL-*ši* LUGAL ⸢ELAM⸣.[MA^{ki}]
16. KAL-*ma* NÍG.TUK x[...]

Lower edge
17. [DIŠ] ^{mul}*ṣal-bat-a-nu* ŠE.ER.ZI ÍL-*ši* LUGAL ELAM.MA^{ki} x[...]
18. ^{d}U.GUR *ina šub-ti-šú* ZI.IR ERÍN KÚR ⸢KUR S[U$^{?}$.BIR$_4$$^{?ki?}$...]
19. [x]x 6 ITI[$^{meš?}$...]

Reverse
1. DIŠ ^{d}U.GUR *ina* AN-*e ana* ^{mu}[$^{l.d}$AMAR.UTU...]
2. ⸢GIN$_7$ ⸣ $^{mul.d}$AMAR.UTU-*ma* [... *ina* GAB-*šú* GUB-*iz si-ḫu si-id-ru*]
3. LUGAL ERÍN-*šú* TUR-*ár* LUGAL ḪU[L^{meš}-*šú ú-nap-pa-aṣ* LUGAL NÍ.G]I [*ina* KUR GÁL-*ma*]
4. AŠ.TE NU DIB-*bat ana* SILIM LUGAL-*šú* ⸢NUN^{meš}-*šú* x uḫ ga$^{?}$⸣ ti x[x]
5. *šá* ^{d}U.GUR KI $^{mul.d}$AMAR.UTU *ul-ta-pak-ma* GUB-[*ma*$^{?}$]
(end of the reverse)

Translation

Upper edge
1. [If] a planet comes close to (another) [pl]anet (and) the furrow has opened its surface, there will be illegitimate power,
2. an [e]xile w[ill usurp] the throne, he will institute [fo]od offerings for the gods, he will e[ndow] all the temples altogether.

Obverse
3. If a planet comes close to (another) [pla]net, the son of the king who resides in (another) town at the beginning of the year will start a revolt against his father,
4. (but) he will not usurp the throne, the son of som[e]one else will ascend and will usurp the throne, the temples of the great gods
5. will return to their place, he will place the regular offering to the great gods, he will endow the temples of the great gods together.
6. If Mars and a planet face each other (i.e. are in opposition) and stand still, (there will be) the revolt of Elam.
7. If the Fish is dark, the fish in the river will not drop the spawn, the bird in the sky will not drop the egg.
8. The Fish (means that) Mars is obscured, (or) Saturn.
9. If the Plow reaches the path of the Sun, there will [be] famine in the country.
10. If a planet and Mars face each other and stand still, (there will be) the revolt of Elam.

11. If the Liar (i.e. Mars) and a planet face each other and stand still, in that year (there will be) the revolt of Elam.
12–14. See SIT 6: 2–3 (see pp. 205–206).
15. If Mars is surrounded and has a brilliant sheen, the king of Elam
16. will be strong and rich ...[...].

Lower edge
17. [If] Mars has a brilliant sheen, the king of Elam ...[...],
18. Nergal will make (him) grieve on his throne, the army of the enemy S[ubartu[?]...].
19. [...]... six month[s[?]...].

Reverse
1. If Nergal in the sky [...] towards [the Marduk]-st[ar...],
2. like the Marduk-star [... stands in front of it, (there will be) the revolt (and) battle line],
3. the king will reduce his army, the king [will smite his] en[emies, there will be an usurpe]r [king in the country],
4. (but) he will not usurp the throne, for the well-being of his king (and) his princes... ...[...].
5. (It means) that Nergal (i.e. Mars) is piled up with the Marduk-star and stands still.
(end of the reverse)

Comments

This commentary collects excerpted omens with Mars as the main subject, mainly indicated by substitute names (i.e. "planet", "Fish", "Plow", "Pleiades", "Nergal").

obv. 1–5: The same omens are given in tablet EAE 56 (TCL 6, 16 obv. 34–36, see Largement 1957: 242). In obv. 3, the traces of the last two signs of the lines point at GAR-*an-ma*, *išakkan-ma*, lit. "he will set up", rather than DÙ-*uš*, *ippuš*, "he will make", as in TCL 6, 16 obv. 35.

obv. 3–5: The long apodoses of these entries are so detailed that they are probably a description of a historical event that happened in the past or was handed down as such. This fact is not unique, as other so-called "historical apodoses" or "historical omens" are attested, for instance, in extispicy. The idea behind such apodoses was probably to tie the future to the past, and so they validated the divination, which was always based on a case-by-case investigation. In simple words, if an important event for the royal house has happened once when any specific planetary phenomena were happening, there is the possibility that it could happen again (see Koch-Westenholz 1995: 13–19; Koch 2015: 14–15 and fn. 36).

obv. 11: For "Liar" (^{d}*sar*$_{6}$-*ru*) as a name for Mars, see the Great Star List (Koch-Westenholz 1995: 198, 200 ll. 237–240).

rev. 1–4: For the restoration, see K 10932 obv. 1–4 (ACh Suppl. 2, 70, 20–23) ⸢^{d}⸣U.GUR *ina* AN-*e ana* $^{mul4.d}$A[MAR.UTU...] / *ina* GAB-*šú* GUB-*iz si-ḫu si-*⸢*id-ru*⸣ x[...] / LUGAL NÍ.GI *ina* KUR GÁL-*ma* AŠ TE NU [...] / *ša* ^{d}U.GUR KI $^{mul4.d}$AMAR.UTU *u*[*l-ta-pak-ma* GUB...]; K 3780 obv. i 1'–4' (App. B § 23, see p. 342). See also the comments to obv. 3–5 for the meaning of the apodosis in this context.

§ 22 K 1494a (P_1) (+) K 1494b (P_2) (+) K 1522 + K 3594 (P_3) = SIT 6 source P_1–P_3

Transliteration

Obverse

P_1

1'. ⸢DIŠ ^{mul} x⸣ [...]

2'. *ana* KUR 20 *ina lu*[*m-ni*...]

3'. ŠÈG^{meš} [...]

4'. DIŠ ^{mul}MAN-*ma ana* [...]

5'. DIŠ ^{mul}ŠU.GI *ana* [...]

6'. DIŠ ^{mul}MAN-*ma a-n*[*a*...]

7'. [DIŠ] ^{mul}UDU.IDIM *ana* [...]

8'. [DIŠ] ^{mul}UDU.IDIM *ana* x[...]

9'. DIŠ ^{mul}MAN-*ma ana* ^{mu}[^{l} ...]

10'. DIŠ ^{d}PA *u* ^{d}LUGAL x[...]

11'. DIŠ ^{mul}MAN-*ma ana* ^{mul}[...]

12'. DIŠ ^{mul} AL.LUL ^{mu}[^{l}...]

13'. DIŠ ^{mul} AL.LUL MUL^{me}[^{š}...]

14'. DIŠ ^{mul}⸢AL⸣.LUL *ana* ^{mul}Š[U.GI...]

(break of unknown length)

P_2

1'. [...]⸢^{meš}⸣

2'. [...] be

3'. [...] ⸢la$^{?}$⸣

4'. [...]x^{meš}

5'. [... ^{me}]^{š}

6'. [...]-*ma*

7'. [... -*m*]*a*

8'. [...]x

(break of unknown length)

P$_3$

1'. KU$_6$⸢^{meš} x x⸣ [...]

2'. DIŠ ^{mul}*ṣal-bat-a-nu ana* ^{mul}[...]
3'. DIŠ ^{mul}*ṣal-bat-a-nu ana* ŠÀ ^{mu}[^{l}...]
4'. DIŠ ^{mul}*ṣal-bat-a-nu ana* ŠÀ ^{mu}[^{l}...]
5'. DIŠ ^{mul}*ṣal-bat-a-nu u* ^{mul}UR.[MAḪ/BAR.RA...]
6'. DIŠ ^{mul}*ṣal-bat-a-nu ana* 15 ^{mul}[...]
7'. DIŠ ^{mul}*ṣal-bat-a-nu ana* 2,30 ^{mul}[...]
8'. DIŠ ^{mul}*ṣal-bat-a-nu ana* SAG ^{mu}[^{l} ...]x ⸢LUGAL⸣ [...]
9'. DIŠ ^{mul}*ṣal-bat-a-nu ana* EGIR ^{mu}[^{l} ... EG]IR ⸢LUGAL⸣ [...]
10'. For the edition, see SIT 6: 30 (see p. 218).
11'. DIŠ ^{mul}LUGAL *ana* ^{mul}UR.B[AR.RA TE *ina*] UD NU NAM.TAR-*šú* ÚŠ É.GAL NUN KAR-*ʾa*
12'. ZI.GA ÌR *u* GÉME *ina* ⸢É⸣.[GAL *ka-ru-u*]*r-tu*$_4$ GÁL^{me} KI.LAM KUR KI.LAM LAL-*a* GU$_7$ ^{d}*ṣal-bat-a-nu* TE-*ma*
13'. DIŠ ^{mul}*ṣal-bat-a-nu ina* ^{d}UTU.ŠÚ.A [...] ⸢*ú*⸣-*qer-rim-ma* GUB *ub-bu-ṭu ina* KUR URI^{ki} GÁL-*ma*
14'. TÙR x[... ...]^{meš} TUR-⸢*ni*⸣

15'. DIŠ ^{mul}UG$_5$.GA *ana* ^{mul}[... u]r BURU$_5$$^{ḫi.a}$ ZI-*ma* KUR [x]
16'. DIŠ ⸢^{d}*ṣal*⸣-*bat*-⸢*a*⸣-[*nu*] *ina* ŠÀ ^{mul}AB.SÍN GUB-[*ma*]
17'. DIŠ ^{mul}UG$_5$.[GA ...]-*šu-ur* ŠE *ina* KI^{didli} UR$_4$
18'. For the edition, see SIT 6: 14 (see pp. 211–212).
^{d}⸢*ṣal-bat-a-nu ana* ^{mul}⸣[SIM.MAḪ TE-*ma*]
19'. [...] DINGIR A.ŠÀ ^{d}*ba*-⸢*ba*$_6$$^{?}$⸣ [...]
20'. [...] ⸢A⸣.ŠÀ ^{mul}AŠ.GÁN ^{d}[...]
(end of the obverse)

Reverse

P$_3$

1. DIŠ ^{mul}AŠ.GÁN *ana* ^{mul}APIN ⸢TE UN⸣^{meš} SÙḪ-*ma* ŠEŠ ŠEŠ-*šú* GU$_7$ URU KI URU$^{?}$ [KÚ]R-⸢*ir*⸣
2. ^{d}IŠKUR RA-*iṣ* GIG^{meš} *ina* KUR GÁL^{meš} KA ÍD^{meš} *is-sek-kir*
3. ^{d}*ṣal-bat-a-nu ana* ^{mul}AB.SÍN *ú-lu ana* ^{mul}AŠ.GÁN TE-⸢*ma*⸣
4. DIŠ ^{mul}UDU.IDIM *ana* A.ŠÀ TE ⸢A$^{?}$.ŠÀ⸣ 5 ⸢GU$_4$$^{?}$⸣.UD MUL *ana* ^{mul}ÙZ TE-*ma*
5. *ina* KUR SU.BIR$_4$⸢^{ki} LÚ.NA⸣.ME NU TAG$_4$
6. ⸢DIŠ ^{mul}UG$_5$⸣.[GA x] ⸢x x x⸣ KUR-*u*[*d*] KÙ.BABBAR *iq-qir šá ina la a-dan-ni-šú* IGI-*mar*

---

7. [...]x ⸢UR⸣ *ina* KUR GÁL-*ši* ^{d}*ṣal-bat-a-nu*
8. [... Á] ⸢:⸣ *i-di* : Á : *qa-nu*
9. [...] SU.GU$_7$ ⸢KUR⸣ DIB-*bat*
10. [... UR]I^{ki} x[...]x ŠU$^{?meš}$-*šú*
11. [...]x [... b]i

12. DIŠ ^{m}[^{ul}...]
13. l[u... ... KU]R-⸢*ma*⸣ x[...]
14. ^{d}[*ṣal-bat-a-n*]*u ana* ^{mu}[^{l}...]

---

15. DIŠ ^{d}ŠÁR.⸢UR$_{4}$⸣ [*u* ^{d}ŠÁ]R.GAZ *šá* KUN [^{mul}GÍR.TAB *it-ta-na-an-bi-ṭu* ^{giš}TUKUL KUR URI^{ki} ZI-*ú*]

---

16. DIŠ ^{mul}KU$_{6}$ ^{mul}SUḪUR.MÁŠ^{ku6} TAG$_{4}$ i[b...]
17. ⸢DIŠ⸣ [^{mul} L]U[GAL$^{?}$...] x[...]
(break of unknown length)

P$_{2}$

1’. [...]⸢$^{meš?}$⸣
2’. [...]⸢^{meš}⸣-*ma*
3’. [...] *ina* UR BI *i-še-et*
4’. [... *i*]-*še-et*

5’. [...] ⸢KÚR$^{?}$⸣^{meš}

6’. [...]x ni ú
7’. [... KU]R URI^{ki}
8’. [...]x ir

9’. [...]x
(break of unknown length)

P$_{1}$

1’. ⸢x ŠÀ⸣ [...] ⸢x⸣ [...]⸢$^{meš?}$⸣ [...] ⸢x x x⸣ [...]

2’. DIŠ $^{mul.lú}$ḪUN.GÁ I[G]I LUGAL MAR.TU^{ki} *ina* ^{giš}TUKUL ŠUB-*ut*

3’. DIŠ ^{mul}*e$_{4}$-ru$_{6}$* MUL.MUL KUR-[*u*]*d* GU$_{7}$-*ti* ^{d}IMIN.BI ^{d}UDU.IDIM.SAG.UŠ ^{d}*zap-pa* KUR-*ma*
4’–5’. For the edition, see SIT 6: 10–11 (see pp. 210–211).
6’. DIŠ MUL$_{4}$.MUL$_{4}$ *šat-ḫu-ma u* [GE$_{6}$^{me}]^{š} ÚŠ^{meš} *ina* KUR GÁL^{meš}
7’. DIŠ MUL$_{4}$.MUL$_{4}$ *ni-iḫ-su-*[*ma* U]D.DA-*su-nu* NU GÁL-*ši* ^{d}*èr-ra* ZI-*ma* UN^{meš} KUR *ú-šam-qat*
8’. DIŠ MUL.MUL ⸢*a*⸣-*d*[*ir*$^{?}$ *ša-aḫ*]-*lu-uq-ti* UN^{meš} *ḫu-šaḫ-ḫu*

9’. ⸢x⸣ x[... M]UL ^{d}UTU *šá ki-na-a-ti*

(end of the reverse)

Translation

Obverse

P_1

1'. If the star... [...],
2'. to the country of the king in ha[rm...],
3'. rain [...].
4'. If the Other One (i.e. Mars) [...] towards [...].
5'. If the Old Man [...] towards [...].
6'. If the Other One [...] toward[s...].
7'. [If] a planet [...] towards [...].
8'. [If] a planet [...] towards ...[...].

9'. If the Other One [...] towards the sta[r...].
10'. If Šullat and Ḫaniš ...[...].

11'. If the Other One [...] towards the star [...].
12'. If the Crab [...] the sta[r...].
13'. If the Crab [...] the stars [...].
14'. If the Crab [...] towards the O[ld Man...].
(break of unknown length)

P_2

Only the last signs of 8 lines are visible, the rest is broken off; break of unknown length.

P_3

1'. fishes ...[....].

2'. If Mars [...] towards the star [...].
3'. If Mars [...] into the sta[r...].
4'. If Mars [...] into the sta[r...].
5'. If Mars and the L[ion]/W[olf...].
6'. If Mars [...] towards the right of the star [...].
7'. If Mars [...] towards the left of the star [...].
8'. If Mars [...] towards the head of the sta[r ...,]... the king [...].
9'. If Mars [...] towards the back of the sta[r..., beh]ind the king [...].
10'. See SIT 6: 30 (see p. 218).
11'. If the King-star [comes close to] the W[olf], (he) will die [before] his time, the palace of the prince will be robbed,
12'. (there will be) loss of slaves in the p[alace,] there will be [voracious hun]ger, the business of the country will devour the smaller business. (It means that) Mars is coming close.
13'. If Mars comes close to [...] in the west (lit. sunset) and stands still, there will be famine in Akkad,
14'. the cattle-pen ...[...]... will be diminished.

15’. If the Raven [...] towards the star [...,]... the locusts will attack and the country [will...].
16’. (It means) If Mar[s] stands inside the Field.
17’. If the Ra[ven...,]... the barley on earth will be cursed.
18’. See SIT 6: 14 (see pp. 211–212).
19’. [...] the god of the field (is) Baba? [...].
20’. [...] the field of the Field, the god [...].
(end of the obverse)

Reverse
P3
1. If the Field comes close to the Plow, people will be confused and (one) brother will eat the (other) brother, a city will [be at w]ar with (another) city,
2. Adad will devastate, there will be sick people in the country, the mouth of the (twin) rivers will be blocked.
3. (It means that) Mars comes close to the Furrow or the Field.
4. If a planet comes close to the area, the *area*? *rises*? five (times), (and) a star comes close to the She-Goat,
5. in Subartu no one will be saved.
6. If the Ra[ven] reache[s...], the silver will be scarce. (It means) that it appears in its period.

---

7. [...]... there will be in the country. (It means) Mars.
8. [... Á] (means) side – (var.) Á (means) horn.
9. [...] the famine will seize the country.
10. [... Akka]d ...[...]... his hands?.
11. [...]... [...].
12. If the s[tar...],
13. o[r... (it) rea]ches and ...[...].
14. (It means that) [Mar]s [...] towards the sta[r...].

---

15. If Šarur [and Ša]rgaz of the Tail [of the Scorpion are constantly bright, the weapons of Akkad will be raised].

---

16. If the Fish leaves behind the Goat Fish ...[...]
17. If [the K]i[ng-star?...] ...[...].
(break of unknown length)

P2
1’. [...]...
2’. [...] and ...
3’. [...] he will escape in that city.
4’. [... he] will escape.

5’. [...] the enemies?.

6’. [...] ...

7’. [... the lan]d of Akkad.

8’. [...] ...

9’. [...]...

(break of unknown length)

P_1

1’. ... the inside (lit. heart) [...] ... [...] ...[...] ... [...].

2’. If the Hired man ri[se]s, the king of Amurru will fall under the weapon(s).

3’. If the Frond reaches the Pleiades, (there will be) consumption by the Sebettu. (It means that) Saturn reaches the Bristle.

4’–5’. See SIT 6: 10–11 (see pp. 210–211).

6’. If the Pleiades are elongated and [become dark], there will be a pestilence in the country.

7’. If the Pleiades are contracted [and] do not [h]ave their glooming, Erra will rise and kill the people of the country.

8’. If the Pleiades da[rken$^{?}$, (there will be) the dest]ruction of people (and) famine.

9’. ... ...[... the s]tar of Šamaš of the justice.

(end of the reverse)

Comments

Reiner (1998: 227, 228) considered these fragments as part of EAE 53, and also related to the unedited fragments K 3147, K 6145, K 10512, K 3223, and K 2894 + K 12290. However, these sources are only similar content-wise; they likely represent one of the unedited tablets of EAE, or the alleged EAE tablet about Mars (Koch 2015: 175). Rm 192 (App. B § 21), K 2170+ (App. B § 18) and K 2177+ (App. B § 17) (which duplicates tablet SIT 6 on the reverse) have excerpts from that unidentified section. Reiner (2006: 314) wrote about the alleged EAE Mars tablet: “I would not consider it impossible that the original version of Tablet 69 dealt with Mars omens, especially Mars’s ‘coming close’ to other celestial bodies and that the phenomenon ‘coming close’, whatever its original meaning may have been, would have been predicated of other celestial bodies too.”

P_3 obv. 11’–12’: For the restoration of obv. 11’, see LBAT 1543 obv. 10’ (Biggs 1967a: 129), and 81-2-4, 429 ll. 6’–7’ [...]x ^{mul}LUGAL *ana* ^{mul}UR.BAR.RA TE-*ḫ*[*u*$^{?}$...] / [... *k*]*a*-⸢*ru*⸣-*ur*-tu_4 GÁL-*ši* me$^{?}$ e ⸢ki⸣ [...]. A parallel omen is in K 2170+ rev. 10’–11’ (App. B § 18, see p. 327).

P_3 obv. 19’: The goddess Baba, or Bau, is a goddess of fertility and healing mainly worshipped in Lagaš (see, e.g., Richter 2004: 101–102, 455–456, 514–519).

P$_3$ rev. 1: For references to cannibalism in Mesopotamia, see fn. 156.

P$_3$ rev. 4: The reading of the second part of the protasis (⸢A$^?$.ŠÀ⸣ 5 ⸢GU$_4$$^?$⸣.UD, "the *area*$^?$ *rises*$^?$ five times) is unclear. In protases of celestial omens, the logogram A.ŠÀ, *eqlu*, "field, area", is used to indicate Jupiter when it moves towards celestial directions (e.g. the "area" of Akkad, Elam, Guti, Amurru, i.e. north, south, east, and west, see Reiner-Pingree 2005: 20–21).

P$_3$ rev. 15: This omen is quoted in many reports together with various explanatory lines, see SAA 8, 51 obv. 1–3; SAA 8, 52 obv. 1–6; SAA 8, 185 obv. 3–5; SAA 8, 370 rev. 3–7; SAA 8, 502 obv. 7–10.

P$_1$ rev. 3'–8': The section of entries about the Pleiades – separated from the others by two horizontal rulings – epitomise the perception of this asterism in divination. The scribe selected four omens: the first is about a conjunction, and the other three are about the luminosity of the Pleiades. The apodoses and factual explanations show the symbolic principle of the semiotic link (see A.3.1.5.): the Pleiades cater of consumption (GU$_7$, *ukultu*), pestilence (ÚŠ^{meš}, *mūtānu*), or famine (SU.GU$_7$, *ḫusaḫḫu*) brought by the Sebettu and Erra.

P$_1$ rev. 6'–7': For the restoration, see Reiner-Pingree (1981: 72 XV 28–29).

§ 23 K 3780 = SIT 6 source U

Transliteration

Obverse i

1'. ⸢*si-ḫu si-id-ru*⸣ L[UGAL ERÍN-*šú* TUR-*ár*...]

2'. LUGAL ḪUL^{meš}-*šú ú*-⸢*nap-pa*⸣-*a*[*ṣ* LUGAL NÍ.GI *ina* KUR GÁL-*ma*]

3'. AŠ.TE NU DIB-*bat ana* SILIM LUGAL *u* ERÍN-*ni-šú* x[...]

4'. ⸢*ša*⸣ ^{d}U.GUR KI $^{mul.d}$AMAR.UTU *ul-t*[*a-pak-ma* GUB-*ma*$^?$]

5'. BAD *lu-um-nu ina* IGI ^{d}ŠUL.PA.È GUB SU.GU[$_7$...]

6'. BAD *lu-um-nu* EGIR ^{d}ŠUL.PA.È GUB MU BI ⸢*dam-qat*⸣ [...]

7'. DIŠ ^{mul}*e-tu-ra-me* SUKKAL ^{mul}*a-nu-ni-tum ana* ^{mul}ŠU.PA TE

8'. *ina* MU BI *ina* TIL MU LUGAL ÚŠ BURU$_{14}$ *nap-ša* KUR GU$_7$ BURU$_{14}$ ŠE *u* ^{še}GIŠ.Ì ⸢GÁL⸣

9'. DIŠ ^{mul}ZUBI *lum-mun* BALA ZÁḪ

10'. DIŠ ^{mul}*ṣal-bat-a-nu ana* ^{d}SAG.ME.GAR TE *mí-iq-tu dan-nu ina* KUR GÁL

11'. DIŠ ^{d}SAG.ME.GAR ^{mul}MAN-*ma* TE-*šú ina* MU BI LUGAL URI^{ki} BAD-*ma* BURU$_{14}$ KUR SI.SÁ

12'. DIŠ ^{mul}UDU.IDIM *ana* ^{d}ŠUL.PA.È TE KI.LAM LAL A.RI.A *ana* BAD$_4$ NIGIN^{meš}

13'. ^{d}*èr-ra* KUR GU$_7$ *bi-ib-lu*$_4$ A^{meš} KUR *ub-bal*

14'. LUGAL *ana* LUGAL *ze-ra-a-ti* KIN *mi-iq-tú dan-nu ina* KUR GÁL

15'. *ina* ⸢BURU$_{14}$⸣ *um-šu*$_{14}$ *dan-nu* GÁL *ina* EN.TE.NA EN.TE.NA *dan-nu ina* KUR GÁL

16'. ⸢DIŠ ^{mul}x x⸣ ^{d}SAG.ME.⸢GAR⸣ *is-niq* ÚŠ^{meš} *qú-bu-ri ina* KUR GÁL^{meš}

17'. [...] *u* DÙG-*bi* ⸢^{giš}⸣TUKUL KUR MAR:TU<ki> UGU ^{giš}TUKUL KUR URU^{ki} GAR^{meš}

Obverse ii

1'. [...]x [...]
2'. [DIŠ] ^{mul}MAR.GÍD.DA *ana* MUL.⸢MUL TE⸣ [...]
3'. ⸢x x x x x x x x⸣ x[...]
4'. SI.SÀ BURU$_{14}$ ⸢x pa x⸣ ^{d}⸢NIDABA⸣ *ana* x[...]
5'. BURU$_{14}$ KUR NU GU$_{7}$-*kam* ^{d}IŠKUR GÙ ⸢ŠUB⸣ x[...]
6'. DIŠ ^{mul}MAN-*ma ana* ^{d}⸢UR$^{?}$⸣.MAḪ TE *ta-lit-tú* NAM.LÚ.U$_{18}$.⸢LU⸣ [...]
7'. DIŠ ^{mul}ḪÉ.GÁL-*la-a* SUKKAL ^{d}*nin-lil ana* ^{mul}ÙZ TE x[...]
8'. ⸢*ba*$^{?}$⸣-*a-šu-šu* DÙG.GA *ina* K[UR G]ÁL ^{d}GÌR ^{d}NIDABA *ina* KUR G[ÁL^{meš}-*ma* KUR DAGAL-*iš*]
9'. For the edition, see SIT 6: 27 (see p. 217).
10'. DIŠ ⸢^{mul}⸣UDU.IDIM *ana* A.ŠÀ TE *ina* KUR SU.BIR$_{4}$ NIN x[x]
11'. DIŠ ⸢KI-*šu*⸣-*nu* ^{mul}MAR.GÍD.DA *ina* AN-*e* GE$_{6}$ AN.GE$_{6}$ GAR

12'–13'. For the edition, see SIT 6: 2–3 (see pp. 205–206).
14'. DIŠ MIN *ana* ⸢^{mul}⸣MUŠ KUR-*ud ina* MU BI GÁN.BA TUR
15'. DIŠ ^{mul}SIPA.⸢ZI⸣.AN.NA *ana* MUL.MUL TE SU.GU$_{7}$ *i-mad*
16'. For the edition, see SIT 6: 24 (see p. 216).
17'. DIŠ ^{mul}UD.KA.D[UḪ.A GÌ]R$^{?}$ 15 [^{m}]^{ul}LU.LIM *er-ḫi* RI.RI.<GA> ⸢*bu-li*⸣ GALGA ⸢BIR⸣-*aḫ*

(end of the obverse)

Reverse i

1. [DIŠ ^{d}S]AG.ME.GAR *u* ^{mul}ŠUDUN *it-tén-me-du ina* MU BI ⸢DINGIR⸣[^{me} GU$_{7}$^{me}]
2. [KUR *i*]*š-šá-lil-ma šá* SAḪAR.ŠUB.BA-*a* ⸢*ina*⸣ [KUR GÁL^{meš}]
3. [DIŠ ^{d}S]AG.ME.GAR KI ^{mul}ŠUDUN IGI-*ma ir-bi* UN^{meš} NÍG.[GA-*ši-na ana* KÙ.BABBAR BÚR^{meš}]

---

4. ⸢DIŠ⸣ ^{mul}UR.MAḪ MUL^{meš}-*šú* {x} *il-ta-pu*-[*ú* LUGAL *a-šar* DU-*ku li-is-su*]
5. DIŠ ^{mul}LUGAL ŠE.ER.ZI *na-ši* LUGA[L URI^{ki} *ga-me-ru-tú* DÙ-*uš*]
6. DIŠ ^{d}SAG.ME.GAR KI *ni-ṣ*[*ir-ti*...]

---

7. ⸢DIŠ ^{d}⸣SAG.ME.GAR *ana* DAL.[BA.AN.NA...]
8. KUR ELAM.MA[^{ki}...]
9. DIŠ ^{d}MIN *ina* ^{iti}DU$_{6}$ *ana ši-ir-* ʾ[*i*$^{?}$...]
10. DIŠ ^{d}MIN *ana* IGI ^{mul}BAN [...]
11. ⸢DIŠ ^{d}MIN *ana*⸣ EGIR ^{mul}BAN [...]
12. ⸢DIŠ⸣ [^{d}MIN] *ana* LI.DUR ^{mu}[^{l}BAN...]
13. [DIŠ ^{d}MI]N *ina* ZAG ^{mul}BAN IGI [...]

---

14. [DIŠ ^{d}MIN] ⸢*ana*⸣ SAG ^{mul}x[...]

15. [...]x x x x[...]
(remainder is missing)

Reverse ii

1. [... *ina*] ⸢Á-*šú* ^{mul}GE$_6$⸣ GUB ^{d}IŠKUR RA
2. [...]x-*šú* ⸢^{mul}⸣GE$_6$ GUB AN.GE$_6$
3. [...]x ⸢ŠUB$^{?}$⸣-*ma* GUB^{meš} ZI-*ut* ELAM.MA^{ki}
4. [... ^{še}GI]Š.Ì NIM SIG$_5$
5. [...]x ⸢iš$^{?}$⸣ *ú-šaq-qà-ma*
6. [...]x *la ba-šú u* KÙR *ana* ⸢KUR⸣ RA
7. [...]x ⸢ŠUR$^{?}$⸣ KÚR URU ⸢ZAG⸣ *ina* KUR mu x[...]
8. [...^{še}GI]Š.⸢Ì NIM SIG$_5$⸣
(remainder is missing)

Translation

Obverse i

1’. Revolt (and) battle line, [the army of]the k[ing will be reduced...],
2’. the king will smite his enemies, [there will be a usurper king in the country],
3’. (but) he will not usurp the throne, for the well-being of the king and his army ...[...].
4’. (It means that) Nergal is hea[ped up] with the Marduk-star [and$^{?}$ stands].
5’. If the Evil One (i.e. Mars) stands before Šulpaea (i.e. Jupiter), (there will be) famin[e...].
6’. If the Evil One stands behind Šulpaea, that year will be good [...].
7’. If the Cattle-pen, the vizier of Anunitu, comes close to the Resplendent One,
8’. in that year, at the end of the year, the king will die, the harvest will be abundant, the country will eat, there will be harvest, barley, and sesame.
9’. If the Crook is poorly visible, the reign will perish.
10’. If Mars comes close to Jupiter, there will be a strong disease in the country.
11’. If Jupiter comes close to the Other One (i.e. Mars), in that year the king of Akkad will die and the harvest of the country will prosper.
12’. If a planet comes close to Šulpaea (i.e. Jupiter), the market rate will diminish, the offspring will gather in a fortress,
13’. Erra will devour the country, a devastating flood will carry off the country,
14’. the king will send hostile messages to (another) king, there will be a strong disease in the country,
15’. in the summer there will be a strong heat, in winter there will be a strong frost in the country.
16’. If the star... approaches Jupiter, there will be a pestilence (requiring) graves in the country,
17’. [...] and good thing(s), the weapon of Amurru will be held against the weapon of Akkad.

Obverse ii

1'. [...]... [...].
2'. [If] the Wagon comes close to the Pleiades, [...].
3'. (only traces of signs)
4'. (it) will prosper, the harvest ... ... Nisaba ...[...],
5'. the harvest of the country will not be eaten, Adad will thunder... [...].
6'. If the Other One (i.e. Mars) comes close to the Lion?, the offspring of mankind [...].
7'. If the Abundance, the vizier of Ninlil, comes close to the She-Goat, ...[...]
8'. will be? for him, there [will] be good thing(s) in the la[nd], Sumukan (and) Nisaba w[ill be] in the country, [the country will grow].
9'. See SIT 6: 27 (see p. 217).
10'. If a planet comes close to the field, in Subartu the queen ...[...].
11'. If with them the Wagon in the sky becomes dark, an eclipse will occur.

12'–13'. See SIT 6: 2–3 (see pp. 205–206).
14'. If ditto (i.e. the Pleiades) reaches the Snake, in that year the market rate will diminish.
15'. If the True Shepherd of Anu comes close to the Pleiades, the famine will be severe.
16'. See SIT 6: 24 (see p. 216).
17'. If the Demon with the Gaping Mo[uth], the right [fee]t of the Stag, is quick, (there will be) an epidemic among the herd, the order will be thwarted.

(end of the obverse)

Reverse i

1. [If J]upiter and the Yoke keep meeting, in that year the god[s will devour],
2. [the land w]ill be plundered, and [there will be] leprosy in [the country].
3. [If J]upiter with the Yoke is visible and sets, the people [will sell their] prop[erty for silver].

---

4. If the stars of the Lion *continually flar*[*e*, the king be victorious wherever he goes.]
5. If the King-star has a brilliant sheen, the kin[g of Akkad will show overpowering strength].
6. If Jupiter with the sec[ret...].

---

7. If Jupiter [...] the distance between [...],
8. Elam [...].
9. If the ditto-god (i.e. Jupiter) in Tešritu (i.e. Month VII) [...] towards the furr[ow?...].
10. If the ditto-god (i.e. Jupiter) [...] to the front of the Bow [...].
11. If the ditto-god (i.e. Jupiter) [...] to the back of the Bow [...].
12. If [the ditto-god (i.e. Jupiter)] [...] to the navel of the [Bow...].

13. [If the ditt]o-god (i.e. Jupiter) is visible at the right of the Bow [...].

---

14. [If the ditto-god (i.e. Jupiter)] [...] to the top of the star ...[...].
15. [...]... ... ...[...].
(remainder is missing)

Reverse ii
1. [... at] its side the Black One stands still, Adad will devastate.
2. [...] its ... the Black One stands still, (there will be) an eclipse.
3. [...]... (it) fall$^{?}$, and (they) stand still, (there will be) an attack of Elam.
4. [...] the early [sesa]me will be good.
5. [...]... ... (he) will raise.
6. [...]... it will not exist and the enemy will smite the land.
7. [...]... (it) flares$^{?}$, the enemy [...] the border town in the country ... ...[...].
8. [...] early [sesa]me will be good.
(remainder is missing)

Comments

obv. i 1'–4':	For the restoration, see Rm 192 obv. 3–5 (App. B § 21, see p. 333).
obv. i 5'–6':	The unpublished fragment K 7020 ll. 5'–6' duplicates these lines: DIŠ *lu-um-nu ina* IGI ⸢^{mul}⸣[ŠUL.PA.È GUB...] / DIŠ *lu-um-nu ina* {x} EGIR ^{mu}[^{l}ŠUL.PA.È GUB...]. Jeanette C. Fincke identified K 7020 and passed the information to the author in April 2020.
obv. i 7':	The star ^{mul}*e-tu-ra-me* is a pseudo-loanword from (È.)TÚR, *tarbaṣu*, "cattle-pen", also attested elsewhere (e.g. see Reiner 2006: 319, and BM 62741 rev. 9, see CCP 6.7.A (https://ccp.yale.edu/P461274 accessed 29.06.2022).
obv. ii 8':	For the restoration, see K 2170+ rev. 29' (App. B § 18, see p. 328).
rev. i 4:	The verb *iltappû* is a Gtn-stem preterite from *šapû* (CAD Š1 487–490), "to flare", and it has been attested in Assyrian and Babylonian reports (SAA 8, 2 obv. 6–7; SAA 8, 9 obv. 13; SAA 8, 289 obv. 3–5) and MUL.APIN (II iii 30, see Hunger-Steele 2019: 103) which also allowed the restoration of the apodosis of the entry. The meaning has been left uncertain (e.g. Reiner-Pingree 2005: 181; CAD Š1 489 2'c), and it could refer to a peculiar luminous phenomenon. The variant *ultappû*, a Dt-stem present/future, or Dtn-stem preterite, is also attested (SAA 8, 437 obv. 3–5).

§ 24 K 3123

Transliteration

Reverse

1'.	[	...] ⸢x x x⸣ [...]
2'.	[	... GU]B$^{?}$-*ma ina ri-bi ina* ⸢*a-šur-rak*⸣-[*ki* ZAG]
3'.	[	*a-r*]*a*-⸢*du a*⸣-*la-ku* ⸢*na-ma-ru*⸣ *a-ma-r*[*u*]

4’. [... *a-r*]*a-di-ma* KI ^{d}UTU IGI-*mar*
5’. [... *a-r*]*a*$^{?}$*-di-ma* KI ^{d}UTU NU IGI-*mar*
6’. [...] AN.GE$_{6}$ 30 NU GAR-*an*
7’. [DIŠ *ina* IGI.DU$_{8}$.A ^{d}30 *u*$_{4}$*-mu er-pu* G]ÁL-*ku li-ti-ik-šú* DUG
8’. [... *li*]-*ti-ik-šú maš-qu-ú* KIMIN ^{na4}*aš-pú-ú* ⸢*šá*⸣ U$_{4}$.SAKAR *šá* ŠÀ ^{d}30 ^{d}PA *šá* ŠÀ ^{d}UTU ⸢*šá*$^{?}$⸣ [^{d}]⸢LUGAL⸣
9’. [DIŠ AN.GE$_{6}$ EN.NUN.AN.USÁ]N ⸢*a*⸣-*na* ⸢NAM.ÚŠ⸣^{meš} ⸢: DIŠ AN.GE$_{6}$⸣ EN.NUN.MÚRU.BA *a-na* KI.LAM TUR.RA
10’. [DIŠ AN.GE$_{6}$ EN.NUN.UD.Z]AL.⸢LE⸣ *a-na* ⸢GIG⸣ AN.⸢TI.LA :⸣ DIŠ UD.⸢DUG$_{4}$⸣.GA EN.NUN.AN.USÁN *a-na* 3 ITI UD.10.KÁM
11’. [DIŠ UD.DUG$_{4}$.GA E]N.NUN.⸢MÚRU.BA *a-na*⸣ 6 ITI UD.20.KÁM ⸢:⸣ DIŠ ⸢UD⸣.DUG$_{4}$.GA EN.NUN.UD.ZAL.LE *a-na* 10 ITI^{meš}
12’. [DIŠ AN.GE$_{6}$ EN.NU]N.AN.⸢USÁN *a-na* KUR URI^{ki} :⸣ DIŠ AN.GE$_{6}$ EN.NUN.MÚRU.BA *a-na* KUR SU.BIR$_{4}^{ki}$
13’. [DIŠ AN.GE$_{6}$ EN.NUN.U]D.⸢ZAL.LE *a-na* KUR ELAM.MA^{ki} :⸣ DIŠ ^{iti}BÁR ^{iti}NE ^{iti}GAN KUR URI^{ki}
14’. [DIŠ ^{iti}GU$_{4}$ ^{iti}KIN ^{iti}A]B ⸢KUR ELAM.MA^{ki}⸣ [:] DIŠ ^{iti}SIG$_{4}$ ^{iti}DU$_{6}$ ^{iti}ZÍZ KUR MAR.TU^{ki}
15’. [DIŠ ^{iti}ŠU ^{iti}APIN ^{iti}ŠE KUR] ⸢SU⸣.BIR$_{4}^{ki}$ ITI^{meš} *šá* AN.TA.LÙ ^{d}30
16’. [DIŠ UD.13.KÁM KUR URI^{ki} UD.14.KÁ]M ⸢KUR ELAM⸣.MA^{ki} UD.15.KÁM KUR MAR.TU^{ki} UD.16.KÁM KUR SU.BIR$_{4}^{ki}$ UD^{meš} *šá* AN.TA.LÙ ^{d}30
17’. [IM.GÀL.LU KUR ELAM.MA^{ki} I]M ⸢SI.SÁ⸣ KUR URI^{ki} IM.KUR.RA KUR SU.BIR$_{4}^{ki}$ *u gu-ti-i* IM.MAR.TU KUR MAR.TU^{ki}
18’. [DIŠ KASKAL *šu-u*]*t* ⸢^{d}+*en*⸣-[*líl* KUR URI^{k}]^{i} ⸢KASKAL⸣ *šu-ut* ^{d}*a-nim* KUR ELAM.MA^{ki} KASKAL *šu-ut* ^{d}*é-a* KUR MAR.TU^{ki} *u* SU.BIR$_{4}^{ki}$
19’. [*enūma*$^{?}$] ⸢^{d}⸣30 AN.⸢GE$_{6}$ *iš-tak*⸣-*nu* ITI *u*$_{4}$*-ma* EN.NUN IM KASKAL *u* KI^{meš} MUL^{meš} *šá ina* ŠÀ-*bi* AN.GE$_{6}$ GAR-*nu* ḪI.ḪI-*ma*
20’. [EŠ.BA]R *a-na šá* ITI-*šú* ⸢<<U$_{4}$>>-*um*⸣-*šú* EN.NUN-*šú* IM-*šú* KASKAL-*šú u* MUL-*šú* SUM-*in*
21’–24’. For the edition, see 4.2.2. § 2a (see pp. 116–118).

25’. [DIŠ MAN *in*]*a* EŠ.BAR ^{d}*a-nun-na-ki i-bak-ki* [...]
26’. [DUB].⸢4⸣.KÁM DIŠ 30 *ina* IGI.DU$_{8}$.A-*šú* [...]
27’. [...]x x[...]
(remainder is missing)

Translation

Reverse
1’. [...] ... [...].
2’. [... (it) stand]s$^{?}$; when it sets (it means that it is) in the [right] dep[th].
3’. [... to go d]own (means) to go, to dawn (means) to be visibl[e].
4’. [... to go] down. (It means that) it appears with the Sun.
5’. [... to go] down. (It means that) it does not appear with the Sun.
6’. [...] (It means that) the eclipse of the Moon will not occur.

7'. [If at the appearance of the Moon] you have [a cloudy day], its checking device is a vessel,

8'. [...] its [ch]ecking device is a *mašqû*-vessel, ditto (?). The jasper of the crescent (is) the inside (lit. heart) of the Moon, Nabu at sun[s]et (is) the inside of the Sun, Ḫaniš?.

9'. [An eclipse of the evening watc]h (is) for the deaths; an eclipse of the middle watch (is) for small price(s);

10'. [an eclipse of the morning w]atch (is) for the sick people to recover. The period of the evening watch (is) 3 months and 10 days;

11'. [the period of the m]iddle watch (is) 6 months and 20 days; the period of the morning watch (is) 10 months.

12'. [An eclipse of the even]ing watch (is) for Akkad; an eclipse of the middle watch (is) for Subartu;

13'. [an eclipse for the morni]ng watch (is) for Elam. Nisannu (i.e. Month I), Abu (i.e. Month V), Kislimu (i.e. Month IX) (are for) Akkad;

14'. [Ajaru (i.e. Month II), Ululu (i.e. Month VI), Ṭeb]etu (i.e. Month X) (are for) Elam; Simanu (i.e. Month III), Tešritu (i.e. Month VII), Šabaṭu (i.e. Month XI) (are for) Amurru;

15'. [Du'uzu (i.e. Month IV), Araḫsamnu (i.e. Month VIII), Adaru (i.e. Month XII) (are for)] Subartu. (These are) the months of the eclipse of the Moon.

16'. [The 13th day (is for) Akkad, the 14t]h day (is for) Elam, the 15th day (is for) Amurru, the 16th day (is for) Subartu. (These are) the days of the eclipse of the Moon.

17'. [The south wind (is for) Elam,] the north [wi]nd (is for) Akkad, the east wind (is for) Subartu and Gutium, the west wind (is for) Amurru.

18'. [The pat]h of En[lil (is for) Akkad], the path of Anu (is for) Elam, the path of Ea (is for) Amurru and Subartu.

19'. [When?] the Moon makes an eclipse, you consider the month, the day, the watch, the wind, the path and earth, the stars that set during (or: inside) the eclipse,

20'. and you give [the decisio]n for its month, its day, its morning watch, its wind, its path and its star.

21'–24'. See 4.2.2. § 2a (see pp. 116–118, 1–8).

25'. [If the Sun] weeps [a]t the decision of the Anunnaki [...]

26'. [Tablet] 4 of *Šumma Sîn ina tāmartišu* [...]

27'. [...]... ...[...].

(remainder is missing)

Comments

rev. 2'–3': For the restoration, see Sm 247 obv. 11'–13' (App. B. § 13 see p. 320).

rev. 7'–8': For the restoration, see Oppenheim (1974: 200, 64–65). For references to time measurement through checking devices, such as the *mašqû*-

vessel, or water clock, Brown, Fermor and Walker (1999–2000), and Brack-Bernsen (2005).

rev. 9’–18’: For the restoration and a discussion on the schemes of association in these lines, with their parallels and further bibliography, see Wainer (2016: 57–58, 76–79).

rev. 14’–15’. In these lines, the months associated with the lands are different from those given K 2254 obv. 10 (see A.3.1.5., pp. 300–301), because the scheme in K 3123 fit the lunar eclipse omens, while the scheme in K 2254 obv. 10–11 is based on another celestial event, a *mešḫu*, “glow”, or “meteor”.

rev. 25’: For the restoration, see K 4026 l. 1 (van Soldt 1995: 36 III 65).

§ 25 K 2254

Transliteration

Obverse

1. [...]x *šá-ta-ḫu a-ra-ku*$_{13}$
2–5. For the edition, see 4.2.2. § 2a (see pp. 116–117, 1–4).
6. [... ŠE]Š$^{?}$-*ma* ME GAR
7. [...]x *ina* IGI.LÁ 30 IM.U$_{18}$.LU DU
8. [...] *ina* ITI-*šú*
9. For the edition, see 4.2.2. § 2c (see p. 122, 2–3).
10. [...] ⸢*i*⸣-*šal-lim*-{x}-*ma*
11–12. For the edition, see 4.2.2. § 2c (see p. 122, 1–2).
13. [...] LAL^{meš}-*ma*
14. [...] LAL^{meš}-*ma*
15. [... L]AL^{meš}-*ma*
16. [...]x-*ma*
17. [... -*m*]*a*

(remainder is missing)

Reverse

1’. [...]x x[...]
2’. [... I]M.MAR DU *ina* ⸢ITI BI 1 ŠÈG⸣[^{meš} ŠUR^{meš}]
3’. [... G]ÙB BAD IM.MAR DU^{meš} *ina* ITI BI 2 ŠÈG^{meš} ŠUR^{meš}
4’. [... G]ÙB BAD IM.MAR DU^{meš} *ina* ITI BI 3 ŠÈG^{meš} ŠUR^{meš}
5’. [...]x *ma-gar-ri* LUGAL *ina* MÈ ŠUB-*ut*
6’. [...]x MUL^{meš} GIŠ.ḪUR
7’. [...] *šá* KUR SU.BIR$_{4}^{ki}$
8’. [...] *šá i-dir-ti* MUL^{meš} [M]UL^{meš} ⸢MÁŠ⸣
9’. [...] *ana* ^{d}UTU.ŠÚ.A GUB-*ma bar-tu*$_{4}$
10’–11’. For the edition, see A.3.1.5. (see pp. 300–301).

12’. [DUB.x.KÁM DIŠ 30 *ina*] ⸢*ta-mar*⸣-*ti-šú* NÍG.PÀD.DA UD AN $^{d+}$*en-líl*
13’. [... *kī pî* ^{giš}D]A-*ʾu*$_{5}$ GABA.⸢RI⸣ KUR:TIN.TIR^{ki} BÁRA.SIPA$^{⸢ki⸣}$

14’. [AB.SAR.ÀM BA.AN.È *ṭup-pi* ^{md}AG]-⸢*zu*⸣-*qup*-GI.NA DUMU ^{m}[^{d}AMA]R.UTU.⸢MU⸣.[B]A-*šá*
15’. [...] ^{lú}⸢DUB⸣.SAR
16’. [ŠÀ.BAL.BAL ^{m}*gab-bi*-DINGIR^{meš}-*ni*-KA]M-*eš* ^{lú}GAL.DUB.SAR^{meš}
(end of the reverse)

Translation

Obverse

1. [...]... to be elongated (means) to be behind.
2–5. See 4.2.2. § 2a (see pp. 116–117, 1–4).
6. [... will prevai]l$^{?}$, and you make a prediction.
7. [...]... (It means that) at the appearance of the Moon the south wind blows.
8. [...] in its month.
9. See 4.2.2. § 2c (see p. 122, 2–3).
10. [...] and it will be well.
11–12. See 4.2.2. § 2c (see p. 122, 1–2).
13. [...] they are balancing.
14. [... they are balancing.
15. [... they are b]alancing.
16. [...]...
17. [...]...
(remainder is missing)

Reverse

1’. [...]... ...[...].
2’. [...] (he/it) goes [w]est, in that month it [will rain] once.
3’. [...] (he/it) is open (on the) [l]eft, they go west, in that month it will rain twice.
4’. [...] (he/it) is open (on the) [l]eft, they go west, in that month it will rain three times.
5’. [...]... of the wheel, the king will fall in the battle.
6’. [...]... the stars of the “(divine) design” (lit. “drawing”).
7’. [...] of Subartu.
8’. [...] of the obscurement of the stars, [the s]tars of divination.
9’. [...] (it) stands to the west (lit. towards sunset), (there will be) rebellion.
10’–11’. See A.3.1.5. (see pp. 300–301).

12’. [Tablet x of *Šumma Sîn ina*] *tāmartišu*, *mukallimtu*-commentary of *Enūma Anu Enlil.*
13’. [... according to a wooden] board from Babylon (and) Borsippa,
14’. [written (and) collated. Tablet of Nabû]-zuqup-kēnu, son of [Mar]duk-šumu-[iqī]ša,
15’. [...] the scribe,
16’. [descendant of Gabbi-ilāni-e]reš, the chief of the scribes.
(end of the reverse)

Comments

obv. 1: Akkadian *šatāḫu*, "to be elongated", is a verb used in other protases,[457] and in Commentary EAE 53: 20 it is used to indicate a specific configuration of the Pleiades for intercalation, i.e. when they are ahead (*panû*, "to be ahead") of the Moon, in order to define what was considered a "normal" year (see 4.2.2. § 2b, pp. 118–120). Obv. 1' explains the verb *šatāḫu* as *arāqu,* "to become pale or yellow/green". As an alternative, one can read *a-ra-kum*, from *arāku*, "to become long", or "to be delayed", which would be a synonym of *šatāḫu*, but it would be opposite to the meaning of Commentary EAE 53: 20 and, therefore, to the following explanations in obv. 2–5 (see below).

obv. 6, 8: For a discussion on GIŠ.ḪUR, *uṣurtu*, "(divine) design" (lit. "drawing"), see the comments for K 2170+ rev. 34'–36' (App. B § 18, see p. 328). The difference between "the stars of divination" and "the stars of the (divine) design" in this context is unclear.

rev. 12'–16': These lines are partially duplicated by K 6686 + K 9234 rev. 4'–5' (App. B § 26; see 4.2.2. § 2c, see pp. 120–124). Both K 2254 and K 6686+ were written by the Assyrian scribe Nabû-zuqup-kēnu (May 2018: 110–124),[458] and whereas K 2254 belongs to the tradition of *Šumma Sîn ina tāmartišu*. K 6686+ was written and collated from an unknown-to-us ancient exemplar (rev. 10': [... LIBIR.RA.BI.GI]M$^{?}$ AB.SAR.ÀM ⸢BA⸣.AN.È, "[... lik]e$^{?}$ [its ancient exemplar], written and collated"). The colophon of K 2254 is restored based on the colophons of Nabû-zuqup-kēnu (Hunger: 1968: 92 n. 296–311).

§ 26 K 6686 + K 9234

Transliteration

Obverse

1'. [...] ⸢maḫ⸣ [...]x

2'. [...]x {*ina*} 1-*šu*$^{?}$ UD^{meš}

3'. [...]x ⸢AN⸣-*ú šu-la* È-*ni* IGI

4'. [... MU]L$^{?meš}$ *ina* ^{d}UTU.È *ne-mu-ru*

5'. [...] ^{d}UDU.IDIM.GU$_4$.UD ^{d}UDU.IDIM.SAG.UŠ

6'. [... ^{d}]*ṣal-bat-a-nu* ^{mul}SAG.ME.GAR

7'. [...] ⸢^{d}UTU⸣.È GUB^{meš}-*ma* SIG$_5$

---

457 Commentary EAE 53: 20; see also SAA 8, 351 rev. 4, see 5.2.1.; K 1494a+ rev. 6', see App. B § 22 P_1; Sm 247 obv. 8', see App. B § 13.

458 To Nabû-zuqup-kēnu belonged also App. B § 19 and 28.

8’. [...] ⸢^{d}⸣UTU.⸢ŠÚ⸣.A GUB^{meš}-*ma bar-t*[*u*$_{4}$]

9’. [...]x ⸢AN⸣-*e* ⸢NU⸣ x[...]
10’. [...]x [...]
(remainder is missing)

Reverse
1’. [...] ⸢LAL^{meš}⸣-[*ma*]
2’. [...] ⸢ú⸣ 12 [x x KI] ⸢30⸣ LAL^{meš}-*ma*
3’. [...] LAL^{meš}-*ma*
4’. [...]x-*us-su* [x x]x *i-šal-lim-ma*
5’. For the edition, see 4.2.2. § 2c (see p. 122, 2).

6’. [...]x *a-bu-ub* [KU]Š$_{7}$-*tì* GAR-*an*
7’. [... *ina*$^{?}$ AN$^{?}$]-*e* GÁL^{meš} *i*[*t*]-*bal*-⸢*ma*⸣
8’. [...]x *u*$_{4}$-*mi* IGI.DU *it-bal*-[*m*]*a* NU IGI
9’. [...]x *ka-ra-ru*-⸢*ú*⸣ AN.BA[R$_{7}$]

10’. [... LIBIR.RA]-⸢*ú*⸣ AB.SAR.ÀM ⸢BA⸣.AN.È
11’. [*ṭup-pi* ^{md}AG-*zu-qu-up*-GI.NA DUMU ^{md}AMAR.UTU.MU].BA-*šá* ^{lú}DUB.SAR ŠÀ.BAL.BAL ^{m}⸢*gab-bi*-DINGIR⸣^{m}[^{eš}-*n*]*i*-⸢KAM*⸣-*eš* ⸢^{lú}⸣GA[L DUB.SAR^{meš}]
(end of the reverse)

Translation

Obverse
1’. [...] ... [...]...
2’. [...]... for the first$^{?}$ of the days.
3’. [...]... the sky brings forth the ..., (and) it is visible.

4’. [...the sta]rs$^{?}$ appear in the east (lit. sunrise).
5’. [...] Mercury, Saturn.
6’. [...] Mars, Jupiter.
7’. [...] (they) stand in the east (lit. sunrise), (there will be) prosperity.
8’. [...] (they) stand in the west (lit. sunset), (there will be) a revolt.

9’. [...] the sky will not ...[...]
10’. [...]... [...].
(remainder is missing)

Reverse
1’. [...] they are balancing.
2’. [...]... twelve [... with] the Moon they are balancing.
3’. [...] they are balancing.
4’. [...]... [...]... and it will be well.
5’. See 4.2.2. § 2c (see p. 122, 1).

6’. […]… a [devastati]ng flood will occur.
7’. […] (they) are [in$^{?}$ the sk]y$^{?}$. (It means that) it d[i]sappears.
8’. […]… (It means that) it disappears the day before [a]nd it is not visible.
9’. […]… the brilliance of the Sun at midday (means) afternoo[n].

10’. [… ancient exempla]r, written and collated.
11’. [Tablet of Nabû-zuqup-kēnu, son of Marduk-šumu]-iqīša, the scribe, descendant of Gabbi-il[ān]i-ēreš, the chie[f of the scribes].
(end of the reverse)

Comments

obv. 3’: Assuming that AN-*ú*, *šamû*, “sky”, is the subject of this broken line, then the spelling *šu-la*, is unclear. Perhaps it is an accusative from *šulû*, “road” (CAD S 370–371), or see *šūlu* C in CAD Š 259b.

rev. 10’–11’: The colophon is restored based on the colophons of Nabû-zuqup-kēnu (Hunger: 1968: 92 n. 296–311; May 2018: 122).

§ 27 VAT 7850 (+) AO 6486 = EAE 52 source Q; Commentary EAE 53 source T

Transliteration

Reverse

1’. […]^{meš} ⸢NÍGIN$^{?}$⸣ DINGIR ⸢x⸣ […]
2’. […]-*ma šá-*⸢*ru-ru*⸣ ⸢ÍL⸣-*ma* ⸢:⸣ DIŠ MIN ⸢TE⸣ […]
3’–4’. For the edition, see EAE 52: 40–41 (see pp. 156–157).
5’. […] AN : ^{mul}MAR.GÍD.DA : ^{mul}AB.SÍN : ^{mu}[^{l}…]
6’–9’. For the edition, see EAE 52: 44–47 (see pp. 157–159).
10’. [DIŠ ŠU.G]I TÙR NÍGIN-⸢*ma*⸣ : ^{mul}ŠU.GI *a-*⸢*dir*⸣ : ^{d}GENNA *ina* ŠÀ IGI […]
11’. […] ⸢*ina* ŠÀ : ⸣ DIŠ ⸢MIN⸣ *meš-ḫa im-šuḫ* : GU$_{4}$.UD *ina* ^{mul}ŠU.GI *šá-ru-*[*ru* ÍL-*ma*]
12’. […] ⸢*a*⸣*-dan-šú* DIB-*ma* : DIŠ MIN *meš-ḫa* GAR-*an ina* ^{mul}ŠU.GI IGI-*m*[*a*…]
13’. […] ⸢x⸣ *ana* ^{mul}ŠU.GI TE : AN *i-na* ^{mul}ŠU.GI UŠ-*ma* […]
14’. […] ⸢x⸣-*ma* : EN.ME.ŠÁR.RA *ma-diš* SA$_{5}$: GU$_{4}$.UD *ina* ŠU.GI NÍGIN-*m*[*a*…]
15’. [… ^{mul}ŠU].⸢GI⸣ TA *kin-ṣi-šú* EN *a-si-di-šú* : EN.ME.ŠÁR.RA *šum-*[*šu*…]
16’. […] ⸢BAD⸣ : *né-su-ú* : EN.ME.ŠÁR.RA MUL IGI-*šú* : GIM *kar-ra-ri-*⸢*e*⸣ […]
17’. […] ⸢x⸣ MUL *ma-diš* SA$_{5}$: AN *ina* ÍL *šá-ru-ru* KI *dil-bat ina* ^{mul}[…]
18’. […] DIŠ MIN MUL IGI-*šú ma-diš e-kil* : GU$_{4}$.UD *ina* ^{mul}ŠU.GI IGI-*m*[*a*…]
19’. [… DIŠ MIN *a-d*]*ir* : AN.GE$_{6}$ *ina* ^{mul}GIGIR GAR-*ma* : DIŠ ^{mul}MAN-*ma ana* EN.ME.ŠÁR.RA [TE]
20’. [*ina* ŠÀ KUR DÙ]G-*ab* UN^{meš} DAGAL^{meš} : ^{mul}ZUBI ŠE.ER.ZI *na-ši* […]
21’. […] ⸢x x x x⸣ ^{mul}ŠU.GI *šá-ru-ru* ÍL-*ma*$^{?}$:⸣ ^{mu}[^{l}…]
22’. [DIŠ ^{mul}GÀM *lum-mu-u*]*n* BALA ZÁḪ : ^{d}PA.ME.GAR *ina* ^{mul}ŠU.GI *ma-diš* TUR-*ma* : GAR : AN […]
23’. […] ⸢x-*ma*$^{?}$⸣ : ^{mul}GÀM *zi-mu-šú* : *us-sà-na-la-mu* : GENNA *ina* ŠU.GI *ana*

^{d}PA.ME.GAR TE-[*ma*$^{?}$]

24’. [DIŠ ^{d}PA.ME].⸢GAR *ina* ^{mul}⸣ŠU.GI TUR-*ma* : IM^{meš} KI ^{d}UDU.IDIM^{meš} KIN.KIN-*ma* ME.A GAR-*an*

25’. [*ṣa-a-tú u šu-ut* K]A *mál-sú-ut* ÉŠ.GÀR *šá* DIŠ UD $^{d+}$60 ⸢$^{d+}$EN⸣.LÍL.LÁ

26’. [*ša* ŠÀ DIŠ *ina* ^{iti}B]ÁR ^{mul}AŠ.GÁN *u* MUL.MUL IGI^{meš} AL.TIL

27’. [DIŠ MUL.MU]L ^{mul}⸢ŠUDUN⸣ KUR-*ud* : AN *ina* UR.A KI ^{d}PA.ME.GAR GUB-*ma*

28’. [GIM LIBIR-*šú* SA]R-*ma ba-rù up-puš*$_4$ GABA.RI-*e* ^{giš}DA SUMUN-*bar* IM $^{md+}$60-ŠEŠ-GÁL-⸢*ši*⸣

29’. [A *šá* ^{m}*ki-di*]*n*-⸢60⸣ ŠÀ.BAL.BAL ^{m}*é-kur-za-kir* ^{lú}MAŠ.MAŠ ^{d}60 *u an-tu*$_4$ [x$^{?}$]

30’. [^{lú}ŠEŠ.GU.LA *šá*] ⸢É *re*⸣-*eš* ^{lú}DUB UD $^{d+}$60 $^{d+}$EN.LÍL.LÁ ⸢UNUG⸣^{ki}-*ú qàt* ^{m}*ina-qí-bit*-[^{d}60]

31’. [*ana a-ḫa-a-zi-šú* GÍ]D.DA UD^{meš}-*šú* TIN ZI-*tì-šú u* ⸢*kun*⸣-*nu* SUHUŠ^{meš}-⸢*šú*⸣ [x$^{?}$]

32’. [NU GÁL-*e* GIG^{m}]^{eš}-*šú ù* MUD *be-lu-ti-šú* SAR-*ma ina* ⸢UNUG⸣^{ki} ⸢*u*⸣ [É *re-eš*]

33’. [É EN-*ú-ti-šú-nu ú-k*]*in pa-lìḫ* $^{d+}$60 *u* ⸢*an*⸣-*t*[*u*$_4$] ⸢*li*⸣-*iṣ-ṣur l*[*i-šá-qir*]

34’. [UNUG^{ki} ^{iti}x UD].⸢28.KAM*⸣ MU.77.⸢KAM* ^{m}*si-lu*⸣-*ku* LUGAL (end of the reverse)

Translation

Reverse

1’. [...]... (they are) surrounded$^{?}$, the god ...[...].

2’. [...] and to have a brilliant sheen. If ditto (?) comes close [...].

3’–4’. See EAE 52: 40–41 (see pp. 156–157).

5’. [...] Mars – (var.) the Wagon – (var.) the Furrow – (var.) the sta[r...].

6’–9’. See EAE 52: 44–47 (see pp. 157–159).

10’. [If the Old Ma]n is surrounded by a halo – (var.) The Old Man is dark. (It means that) Saturn is visible inside [...].

11’. [...] inside. If ditto (i.e. the Old Man) produces a glow. (It means that) Mercury [has] a brilliant she[en].

12’. [...] and (it) passes by (at) its proper time. If ditto (i.e. the Old Man) produces a glow(and) it is visible in the Old Man [...].

13’. [...]... (it) comes close to the Old Man. (It means that) Mars approaches the Old Man [...].

14’. [...]... and – (var.) Enmešarra is intensively red. (It means that) Mercury circles around the Old Man [...].

15’. [... the Old] Man from his knee to his heel, [his] name is Enmešarra [...].

16’. [...] to open (means) to be far. Enmešarra (is) the star before it. (It is) like the brilliance of the Sun at midday [...].

17’. [...] ... the star is intensively red. (It means that) Mars has a brilliant sheen with Venus in the star [...].

18'. […] If ditto (i.e. Enmešarra), the star before it is very dark. (It means that) Mercury is visible in the Old Man […].

19'. [… If ditto (i.e. Enmešarra) is da]rk. (It means that) an eclipse in the Wagon occurs. If the Other One (i.e. Mars) [comes close] to Enmešarra,

20'. [inside the country (there will be) happin]ess of the widespread people. (It means that) the Crook have a brilliant sheen […].

21'. […] … … the Old Man has a brilliant sheen. (It means that) the sta[r…].

22'. [If the Crook if is poorly visibl]e, the reign will perish. (It means that) Jupiter in the Old Man (is) very small – GAR – AN […].

23'. […] … and[?] – (var.) The glow of the Crook lays flat. (It means that) Saturn in the Old Man comes close to Jupiter.

24'. [If Jupi]ter in the Old Man is small. You keep investigating the tablets with the planets and make a prediction.

25'. [Lemmata and oral explan]ations relating to a "reading" of *Enūma Anu Enlil*,

26'. [(referring to the entries) from "If in N]isannu the Field and the Pleiades are visible" – completed.

27'. [If the Pleiade]s reach the Yoke. (It means that) Mars stands in the Lion with Jupiter.

28'. [Writte]n, collated and made [according to its ancient exemplar] from a wooden board. Tablet of Anu-aḫu-ušabši,

29'. [son of Kidi]n-Anu, descendant of Ekur-zākir, the priest of Anu and Antu […[?]],

30'. [high priest of] the Bīt rēš, scribe of *Enūma Anu Enlil* from Uruk. Hand of Ina-qibīt-[Anu],

31'. [for his education, the len]gth of his days, the life of his soul and to establish his foundations […],

32'. [the non-existence of] his [illness]es and kin, he wrote his rulership, and in Uruk and [Bīt rēš]

33'. [he establish]ed [the house of his rulership]. May who reveres Anu and Ant[u] preserve and v[alue (it)].

34'. [Uruk, month x, day] 28, year 77 of Seleukus, the king.
(end of the reverse)

Comments

In VAT 7850 (+) AO 6486, *Glossenkeile* are not only used to mark variants and explanations, but also to separate each sentence and/or entry if it begins in the middle of a line. For the restoration, see Table 13, which contains the parallel omens from the reconstructed tablet EAE 52, other unidentified EAE fragments, and reports to Neo-Assyrian kings.

rev. 4': For ÁB as a logogram for *kakkabu*, "star", see Borger (2004: 396–397 n. 672).

rev. 5': For the logogram AN as a name for Mars, see Brown (2000: 56).

rev. 15’: See also 5.1.4. further comments on the text (see pp. 170–171).

rev. 24’: The second half of the line (“You keep investigating the tablets with the planets and make a prediction”) show that the tablet was meant to be not only a *ṣâtu*, “lemmata”, commentary (see rev. 25’–26’ below) but also a guide to scholars to interpret the signs of the sky by referring to older omens and more updated ones like, for instance, the “tablets with the planets” mentioned by the scribe.

rev. 25’–26’: According to Frahm (2011: 52 fn. 225), VAT 7850 (+) AO 6486 is a *ṣâtu* commentary type 4a: *ṣâtu u šūt pî malsût iškār enūma anu enlil ša libbi* (tablet incipit), “Lemmata and oral explanations (relating to) a “reading” of the series *Enūma Anu Enlil* (and referring to entries) from (tablet incipit)”.

rev. 26’–27’: See also 5.1.3.2. § 2 (see p. 138), EAE 52: 1 (see p. 147), and Commentary EAE 53: 1 (see p. 178).

rev. 27’–34’: The tablet was written by Ina-qibīt-Anu, son of Anu-aḫu-ušabši from the Ekur-zākir family based in Uruk, and it dates to the Hellenistic period (235/[?]/28 BC) (Frahm 2011: 150 o). For the restoration of the colophon, see Hunger (1968: 37–38 n. 87, 43–44 n. 103), and VAT 7830 rev. 13’–19’, whose edition is provided online (CCP 3.1.55.G, https://ccp.yale.edu/P461321 accessed 05.03.2021).

EAE omens	**VAT 7850 (+) AO 6486**
EAE 52: 40: DIŠ *ṣal-lum-[mu-ú]* ^{mul}TI$_8^{\text{mušen}}$ IGI.DU$_8$	3': [...] DIŠ MIN ^{mul}TI$_8^{\text{mušen}}$ IGI ⸢:⸣ AN *ina* AB.SÍN NÍGIN-*m*[*a*$^?$...]
EAE 52: 41: DIŠ *ṣal-lum-m*[*u*]*-ú* ^{mul}UGA^{mušen} IGI.DU$_8$	4': [...] DIŠ MIN ^{mul}⸢UGA$^?$⸣ IGI : GU$_4$.UD *ina* ^{áb}AB.S[ÍN...]
EAE 52: 44: DIŠ ^{mul}ŠU.GI MUL^{meš}-*šú bi-rit-su-nu ma-gal* BAD-*at*	6': [DIŠ ^{mul}ŠU.GI MUL^{meš}-*šú*] *bi-rit-*⸢*su*⸣*-nu ma-gal* BAD-*at*
EAE 52: 45: DIŠ ^{mul}ŠU.GI MUL^{meš}-*šú nen-mu-du*	7': [DIŠ] MIN MUL^{meš}-*šú* [*ne*]*n-mu-du* <:> *ṭe*$_4$*-ḫu-tú ina* ŠÀ GÁL [...] 8'. [... *e*]*-mi-*⸢*du*⸣ : *sa-na-qu*
EAE 52: 46: DIŠ ^{mul}ŠU.GI TÙR NÍGIN	8': DIŠ MIN TÙR NÍGIN : ^{d}30 *ina* ŠU.GI TÙR [NÍGIN...]
EAE 52: 47: DIŠ ^{mul}ŠU.GI TÙR NÍGIN-*ma* KÁ-*šú ana* IM.U$_{18}$.LU BAD	9': DIŠ MIN TÙR NÍGIN-*ma* ⸢KÁ⸣-*šú ana* U$_{18}$ BAD : ^{d}30 DAGAL-*šú*
83-1-18, 322 obv. 1–2 (SAA 8, 549): DIŠ ^{mul}EN.ME.ŠÁR.[RA] / MUL-*šú ma-ʾa-diš* [SA$_5$]	14': EN.ME.ŠÁR.RA *ma-diš* SA$_5$
K 3632 rev. 15 (Reiner-Pingree 1998: 104, 24): [TA... ^{mul}]ŠU.GI EN *a-si-di* [^{d}]EN.ME.ŠÁR.RA]	15': ^{mul}ŠU.GI TA *kin-ṣi-šú* EN *a-si-di-šú* : EN.ME.ŠÁR.RA *šum-*[*šu*...]
K 2118 rev. 8' (App. B § 2): DIŠ ^{mul}MAN-*ma ana* ^{d}E[N.ME.ŠÁR.RA TE UN^{meš} DAGAL^{meš} ŠÀ KUR DÙG-*ab*] **K 8000 rev. 16 (ACh Ištar 24, 15)**: DIŠ ^{mul}MAN-*ma ana* $^{\text{d+}}$EN.ME.ŠÁR.RA *ina* ŠÀ KUR DÙG-*ab* [UN^{meš} DAGAL^{meš}] **K 759 obv. 6–7 (SAA 8, 503)**: DIŠ ^{mul}MAN-*ma ana* ^{mul}EN.ME.ŠÁR.RA TE / UN^{meš} DAGAL^{meš} ŠÀ KUR DÙG-*ab*	19': DIŠ ^{mul}MAN-*ma ana* EN.ME.ŠÁR.RA [TE] 20'. [*ina* ŠÀ KUR DÙ]G-*ab* UN^{meš} DAGAL^{meš}
K 871 rev. 1 (SAA 8, 170): [DIŠ ^{d}]⸢GÀM⸣ ŠE.ER.ZI *na-ši* **K 742 rev. 4 (SAA 8, 115)**: DIŠ ^{d}GÀM [ŠE.ER].⸢ZI⸣ [ÍL]	20': ^{mul}ZUBI ŠE.ER.ZI *na-ši*
K 3780 obv. i 9' (App. B § 23): DIŠ ^{mul}ZUBI *lum-mun* BALA ZÁḪ	22': [DIŠ ^{mul}ZUBI *lum-mu-u*]*n* BALA ZÁḪ
Sm 1267 l. 9' (Reiner-Pingree 1981: 48 VI 4b): [^{mul}ZUBI *zi-mu-šú* : *uṣ-ṣa-na-l*]*a-mu* BALA ZÁḪ-*ma* MAN-*ma* DU$_6$+GUB-*a*	23': ^{mul}ZUBI *zi-mu-šú* : *uṣ-ṣa-na-la-mu*

Table 13. Correspondences between the omens from the series EAE and the so-called astronomical and astrological reports, and the omens in VAT 7850 (+) AO 6486 (App. B § 27).

Bibliography

Aaboe, A. 2001. *Episodes from the Early History of Astronomy*. New York: Springer.
Abusch, T. 1987. "Alaktu and Halakhah. Oracular Decision, Divine Revelation". *HThR* 80, pp. 15–42.
—. 2015. *The Witchcraft Series Maqlû*. WAW 37. Atlanta: SBL Press.
Abusch, T. Schwemer, D. 2011. *Corpus of Mesopotamian Anti-witchcraft Rituals, Volume One*. AMD 8/1. Leiden-Boston: Brill.
—. 2016. *Corpus of Mesopotamian Anti-witchcraft Rituals, Volume Two*. AMD 8/2. Leiden-Boston: Brill.
Abusch, T., Schwemer, D., Luukko, M., Van Buylaere, G. 2020. *Corpus of Mesopotamian Anti-witchcraft Rituals, Volume Three*. AMD 8/3. Leiden-Boston: Brill.
Al-Rawi, A. R., Black, J. A. 1989. "The Second Tablet of 'Išum and Erra'". *Iraq* 51, pp. 111–122.
Al-Rawi, A. R., George, A. 1991–1992. "Enūma Anu Enlil XIV and Other Early Astronomical Tablets". *AfO* 38/39, pp. 52–73.
—. 2006. "Tablets from the Sippar Library, XIII. *Enūma Anu Ellil* XX". *Iraq* 68, pp. 23–57.
—. 2014. "Back to the Cedar Forest: The Beginning and the End of Tablet V of the Standard Babylonian Epic of Gilgameš". *JCS* 66, pp. 69–90.
Allen, G. 2000. *Intertextuality*. London-New York: Routledge.
Alster, B. 1976. "On the Earliest Sumerian Literary Tradition". *JCS* 28/2, pp. 109–126.
—. 1985. "Geštinanna as Singer and the Chorus of Uruk and Zabalam: UET 6/1 22". *JCS* 37/2, pp. 219–228.
Ambos, C. 2013. *Der König im Gefängnis und das Neujahrsfest im Herbst*. Dresden: ISLET.
—. 2013a. "Rites of Passage in Ancient Mesopotamia: Changing Status by Moving through Space: *Bīt Rimki* and the Ritual of the Substitute King" in Ambos, C., Verderame, L. (eds.) *Approaching Rituals in Ancient Cultures. Proceedings of the Conference, November 28-30, 2011, Roma.* Pisa: Fabrizio Serra, pp. 39–54.
Annus, A. 2008. "The Soul's Journey and Tauroctony: On Babylonian Sediment in the Syncretic Religious Doctrines of Late Antiquity" in Dietrich, M. L. G., Kulmar, T. (eds.) *Body and Soul in Conceptions of the Religions*. Münster: Ugarit-Verlag, pp. 1–46.
—. 2014. "Seeing Otherwise: On the Rules of Comparison in Historical Humanities" in Geller, M. J. (ed.) *Melammu. The Ancient World in an Age of Globalization*. Berlin: epubli, pp. 359–372.
—. 2016. *The Overturned Boat. Intertextuality of the Adapa Myth and the Exorcistic Literature*. SAAS 24. Winona Lake: Eisenbrauns.
Archi, A. 2010. "The Heptad in Anatolia". *Hethitica XVI*, pp. 21–34.
Arnaud, D. 1985. *Recherches au Pays d'Aštata. Emar VI Tome 1.* Paris: Editions Recherche sur le Civilisations.

Assmann, A. 2003. "Etymographie: Zeichen im Jenseits der Sprache" in Assmann, A., Assmann, J. (eds.) *Hieroglyphen: Stationen einer anderen abendländischen Grammatologie*. München: Wilhelm Fink, pp. 37–63.
Attinger, P. 1998. "Inanna et Ebih". *ZA* 88/2, pp. 164–195.
Attinger, P., Krebernik, M. 2005. "L'Hymne à Ḫendursaĝa (Ḫendursaĝa A)" in Rollinger, R. (ed.) *Von Sumer bis Homer. Festschrift für Manfred Schretter zum 60. Geburtstag am 25. Februar 2004*. Münster: Ugarit-Verlag, pp. 21–104.
Aveni, A. F. 1996. "Astronomy in the Americas" in Walker, C. (ed.) *Astronomy before the Telescope*. London: British Museum Press, pp. 269–303.
Bácskay, A. 2013. "Asakkû: Demons and Illness in Ancient Mesopotamia" in Jacobus, H., de Hemmer Gudme, A. K., Guillaume, P. (eds.) *Studies in Magic and Divination in the Biblical World*. Piscataway: Gorgias Press, pp. 1–7
—. 2015. "Magical-Medical Prescriptions against Fever: An Edition of the Tablet BM 42272". *JMC* 26, pp. 1–32.
Bartelmus, A. S. 2016. *Fragmente einer grossen Sprache*. UAVA 12/2. Boston-Berlin: De Gruyter.
Beaulieu, P. 1999. "The Babylonian Man in the Moon". *JCS* 51, pp. 91–99.
Beaulieu, P., Frahm, E., Horowitz, W., Steele, J. 2018. *The Cuneiform Uranology Texts. Drawing the Constellations*. Philadelphia: American Philosophical Society Press.
Beckman, G. 2007. "A Hittite Ritual for Depression (CTH 432)". *DBH* 25, pp. 69–81.
Behrens, H. 1978. *Enlil und Ninlil. Ein sumerischer Mythos aus Nippur*. StPohl SM 8. Roma: Pontificio Istituto Biblico.
Bennett, E. A. 2021. *The Meaning of Sacred Names and Babylonian Scholarship. The Gula Hymn and Other Works*. Dubsar 25. Münster: Zaphon.
Biggs, R. D. 1967. *ŠÀ.ZI.GA, Ancient Mesopotamian Potency Incantations*. TCS 2. Locust Valley: J. J. Augustin Publisher.
—. 1967a. "More Babylonian Prophecies". *Iraq* 29/2, pp. 117–132.
Bilić, T. 2007. "A Note on the Celestial Orientation: Was Gilgamesh Guided to the Cedar Forest by the Pleiades?". *Vjesnik Arheološkog muzeja u Zagrebu* 40/1, pp. 11–14.
Bjorkman, J. 1973. "Meteors and Meteorites in Ancient Near East". *Meteoritics* 8, pp. 91–132.
Bloch, Y., Horowitz, W. 2015. "Ura = *ḫubullu* XXII: The Standard Recension". *JCS* 67, pp. 71–125.
Böck, B. 1995. "Sumerisch a.rá und Divination in Mesopotamien". *AulaOr.* 13, pp. 151–159.
—. 2010. "Physiognomy in Ancient Mesopotamia and Beyond. From Practice to Handbook" in Annus, A. (ed.) *Divination and Interpretation of Signs in the Ancient World*. Chicago: Oriental Institute of Chicago, pp. 199–219.
Borger, R. 1973. "Keilschrifttexte verschiedenen Inhalts", in Beek, M. A., Kampman, A. A., Nijland, C., Ryckmans, J. (eds.) *Symbolae Biblicae et Mesopotamicae Francisco Mario Theodoro de Liagre Bohl Dedicatae*. Leiden: Brill, pp. 36–55.
—. 2004. *Mesopotamische Zeichenlexikon*. AOAT 305. Münster: Ugarit-Verlag.
Bottéro, J. 1978. "Antiquités assyro-babyloniennes". *AEPHE (IVème Section)*, pp. 107–164.
Brack-Bernsen, L. 2005. "The 'Days in Excess' from MUL.APIN: On the 'First Intercalation' and 'Water Clock' Schemes from MUL.APIN". *Centaurus* 47, pp. 1–29.

—. 2007. "The 360-Days Year in Mesopotamia" in Steele, J. M. (ed.) *Calendars and Years: Astronomy and Time in the Ancient Near East*. Oxford: Oxbow Books, pp. 83–100.
—. 2020. "The Observational Foundations of Babylonian Astronomy" in A.C. Bowen, A. C., Rochberg, F. (eds.) *Hellenistic Astronomy. The Science in Its Context*. Leiden-Boston: Brill, pp. 171–189.
Brack-Bernsen, L., Hunger, H. 2008. "BM 42282+42294 and the Goal-Year Method". *SCIAMVS* 9, pp. 3–23.
Brack-Bernsen, L., Steele, J. M. 2004. "Babylonian Mathemagics: Two Mathematical Astronomical-Astrological Texts" in Burnett, C., Hogendjik, J. P., Plofker, K., Yano, M, (eds.) *Studies in the History of the Exact Sciences in Honour of David Pingree*. Leiden-Boston: Brill, pp. 95–125.
Briquel-Chatonnet, F. (ed.) 1996. *Mosaique de langues, mosaique culturelle: Le bilinguisme dans le Proche-Orient ancien*. Paris: Maisonneuve.
Britton, J. 2010. "Studies in Babylonian Lunar Theory: Part III. The Introduction of the Uniform Zodiac". *Archive for History of Exact Sciences* 64/6, pp. 617–663.
Britton, J., Walker, C. 1996. "Astronomy and Astrology in Mesopotamia" in Walker, C. (ed.) *Astronomy before the Telescope*. London: British Museum Press, pp. 42–67.
Brown, D. R. 2000. *Mesopotamian Planetary Astronomy-Astrology*. CM 18. Groningen: Styx.
—. 2010. "What Shaped Our Corpuses of Astronomical and Mathematical Cuneiform Texts" in Bretelle-Establet, F. (ed.) *Looking at It from Asia: The Processes that Shaped the Sources of History and Science*. Paris: Springer, pp. 277–304.
—. (ed.) 2018. *The Interactions of Astral Science*. Bremen: Hempen Verlag.
Brown, D., Fermor, J., Walker, C. 1999–2000. "The Water Clock in Mesopotamia". *AfO* 46/47, pp. 130–148.
van Buren, E. D. 1939–1941. "The Seven Dots in Mesopotamian Art and Their Meaning". *AfO* 13, pp. 277–289.
Cagni, L. 1969. *L'epopea di Erra*. StSem. 34. Roma: Istituto di Studi del Vicino Oriente.
—. 1977. "The Poem of Erra". *SANE* 1, pp. 63–119.
Campion, N. 2012. *Astrology and Cosmology in the World's Religions*. New York-London: New York University Press.
Caplice, R. 2002. *Introduction to Akkadian. Fourth Edition*. StPohl SM 9. Roma: Pontificio Istituto Biblico.
Capomacchia, A. M. G., Verderame, L. 2011. "Some Considerations about Demons in Mesopotamia". Verderame, L. (ed.) *Demoni Mesopotamici*. Brescia: Morcelliana, pp. 291–297.
Casaburi, M. C. 2003. *Tre-stelle-per-ciascun(-mese): l'Astrolabio B: edizione filologica*. Napoli: Università degli studi di Napoli "L'Orientale".
Cavigneaux, A. 1976. *Die sumerisch-akkadischen Zeichenlisten. Überlieferungsprobleme*. PhD dissertation, Ludwig Maximillians Universität.
—. 1983. "Lexicalische Listen". *RlA* 6, pp. 609–641.
Cavigneaux, A., Krebernik, M. "Numušda". *RlA* 9, pp. 611–614.
Civil, M. 1974. "Medical Commentaries from Nippur". *JNES* 33, pp. 329–338.

—. 1994. *The Farmer's Instructions. A Sumerian Agricultural Manual*. AulaOr. Suppl. 5. Barcelona: Editorial Ausa.

Cohen, M. E. 1981. *Sumerian Hymnology: The Ersemma*. HUCA Suppl. 2. USA: KTAV Publishing House.

—. 1988. *The Canonical Lamentations of Ancient Mesopotamia. Volume 1*. Potomac: Capital Decisions.

Cohen, Y. 2007. "Akkadian Omens from Hattuša and Emar: The *šumma immeru* and *šumma ālu* Omens". *ZA* 97/2, pp. 233–251.

Cooley, J. 2008. "'I Want to Dim the Brilliance of Šulpae!' Mesopotamian Celestial Divination and the Poem of Erra and Išum". *Iraq* 70, pp. 179–188.

—. 2013. *Poetic Astronomy in the Ancient Near East*. History, Archaeology, and Culture of the Levant 5. Winona Lake: Eisenbrauns.

Cooper, J. 2001. "Literature and History: The Historical and Political Referents of Sumerian Literary Texts." in Abusch, T. (ed.) *Historiography in the Cuneiform World*. Bethesda: CDL Press, pp. 131–138.

Craig, J. A. 1899. *Astrological-Astronomical Texts*. Leipzig: J.C. Hinrichs'sche Buchhandlung.

Cryer, F. H. 1994. *Divination in Ancient Israel and Its Near Eastern Environment. A Socio-Historical Investigation*. JSOT Suppl.142. Sheffield: JSOT Press.

De Zorzi, N. 2009. "Bird Divination in Mesopotamia: New Evidence From BM 108874". *KASKAL* 6, pp. 85–136.

—. 2011. "The Omen Series *Šumma Izbu*: Internal Structure and Hermeneutic Strategies". *KASKAL* 8, pp. 43–75.

—. 2014. *La serie teratomantica Šumma Izbu. Testo, tradizione, orizzonti culturali*. Padova: Sargon Editrice e Libreria.

Delnero, P. 2006. *Variation in Sumerian Literary Compositions: A Case Study based on the Decad*. PhD dissertation, University of Pennsylvania.

—. 2016. "Divination and Religion as a Cultural System" in Fincke, J. C. (ed.) *Divination as Science. A Workshop Conducted during the 60th Rencontre Assyriologique Internationale, Warsaw, 2014*. Winona Lake: Eisenbrauns, pp. 147–166.

van Dijk, J. 1983. *Lugal ud me-lám-bi Nir-ĝál: le récit épique et didactique des Travaux de Ninurta, de Déluge et de la Nouvelle Création*. Leiden: Brill.

Dossin, G. 1935. "Prières aux 'Dieux de la nuit' (AO 6769)". *RA* 32, pp. 179–187.

Dunbar, K. 2001. "The Analogical Paradox: Why Analogy Is So Easy in Naturalistic Setting Yet So Difficult in the Psychological Laboratory" in Gentner, D., Holyoak, K. J., Kokinov, B. N. (eds.) *The Analogical Mind. Perspectives from Cognitive Science*. Cambridge-London: MIT Press, pp. 313–334.

Ebeling, E. 1931. *Tod und Leben nach den Vorstellungen der Babylonier*. Berlin: Walter de Gruyter & Company.

—. 1938. "Enmešarra". *RlA* 2, pp. 396–397.

Eco, U. 1971. *Le forme del contenuto*. Milano: Bompiani.

—. 1979. *Lector in fabula: la cooperazione interpretativa nei testi narrativi*. Milano: Bompiani.

—. 1998. *Trattato di semiotica generale*. Milano: Bompiani.

Edzard, D. O. 1972–1975. "Ḫendursanga". *RlA* 4, pp. 324–325.

—. 1980. "*Kannibalismus". *RlA* 5, pp. 389–390.
—. 1990. "Gilgameš und Huwawa A. I. Teil". *ZA* 80, pp. 165–203.
—. 1991. "Gilgameš und Huwawa A. II. Teil". *ZA* 81, pp. 165–233.
—. 1993. *Gilgameš und Huwawa: Zwei Versionen der sumerischen Zedernwaldepisode nebst einer Edition von Version B*. SbMünchen 4. München: Verlag der Bayerischen Akademie der Wissenschaften.
—. 1997. *Gudea and His Dynasty*. RIME 3/1. Toronto-Buffalo-London: University of Toronto Press.
Edzard, D. O., Wilcke, C. 1976. "Die [H]endursanga-Hymne" in Barry, L. Eichler, B. L., Heimerdinger, J. W., Sjöberg, A. W. (eds.) *Kramer Anniversary Volume. Cuneiform Studies in Honor of Samuel Noah Kramer*. Neukirchen-Vluyn: Butzon & Bercker, pp. 139–176.
Englund, R. K. 1988. "Administrative Timekeeping in Ancient Mesopotamia". *JESHO* 31, pp. 121–185.
Falkenstein, A. 1959. "Zur ersten Tafel des Erra-Mythos". *ZA* 53, pp. 200–208.
Farber, W. 1987–1990. "Lilû, Lilītu, Ardat-lilî A". *RlA* 7, pp. 23–24.
—. 1989. "(W)ardat-lilî(m)". *ZA* 79, pp. 14–35.
—. 1990. "*Mannam lušpur ana Enkidu*: Some New Thoughts about an Old Motif". *JNES* 49/4, pp. 299–321.
Fincke, J. C. 2001. "Der Assur-Katalog der Serie *enūma anu enlil* (EAE)". *OrNS* 70, pp. 19–39.
—. 2003–2004. "The Babylonian Texts of Nineveh". *AfO* 50, pp. 111–149.
—. 2004. "The British Museum's Ashurbanipal Library Project". *Iraq* 66, pp. 55–60.
—. 2006. "Omina, die göttlichen 'Gesetzte' der Divination". *JEOL* 40, pp. 131–47.
—. 2009. "Ist die mesopotamische Opferschau ein nächtliches Ritual?". *BiOr.* 66, 519–558.
—. 2013. "The Solar Eclipse Omen Text from *Enūma Anu Enlil*". *BiOr.* 70, pp. 582–608.
—. 2013a. "'If a Star Changes into Ashes...' A Sequence of Unusual Celestial Omens". *Iraq* 75, pp. 171–196.
—. 2013b. "Additions to Already Edited *enūma anu enlil* (*EAE*) Published in BPO 3 as Group F". *KASKAL* 10, pp. 89–110.
—. 2014. "Additions to Already Edited *enūma anu enlil* (*EAE*) Tablets, Part II: The Tablets Concerning the Appearance of the Sun Published in PHIANS 73, part I". *KASKAL* 11, pp. 103–139.
—. 2014a. "Divination im Alten Orient: Ein Überblick" in Fincke, J. C. (ed.) *Divination in Ancient Near East. A Workshop on Divination Conducted during the 54th Rencontre Assyriologique Internationale, Würzburg, 2008*. Winona Lake: Eisenbrauns, pp. 1–20.
—. 2015. "Additions to Already Edited *enūma anu enlil* (*EAE*) Tablets, Part III: A New Copy from Babylonia for the Tablet on Planets (MUL.UDU.IDIM) of the Omen Series". *KASKAL* 12, pp. 267–279.
—. 2016. "The Oldest Mesopotamian Astronomical Treatise: enūma anu enlil" in Fincke, J. C. (ed.) *Divination as Science. A Workshop Conducted during the 60th Rencontre Assyriologique Internationale, Warsaw, 2014*. Winona Lake: Eisenbrauns, pp. 107–146.

—. 2016a. "Additions to Already Edited *enūma anu enlil* (*EAE*) Tablets, Part IV: The Lunar Eclipse Omens from Tablets 15–19 Published by Rochberg-Halton in AfO Beih 22". *KASKAL* 13, pp. 89–119.

—. 2016b. "*ākilu*, a Pest, lit. "Eater, Devourer", in Omen Apodoses and Other Texts". *NABU* 2016 n. 102.

—. 2017. "Additions to Already Edited *enūma anu enlil* (*EAE*) Tablets, Part V: The Lunar Eclipse Omens from Tablet 20 Published by Rochberg-Halton in AfO Beih 22 with an Excursus on *šurinnu* (ŠU.NIR)". *KASKAL* 14, pp. 55–74.

—. 2018. "Of *tirku*, Moles and Other Spots on the Skin according to the Physiognomic Omens" in Panayotov, S. V., Vacín, L. (eds.) *Mesopotamian Medicine and Magic. Studies in Honor of Markham J. Geller*. Leiden-Boston: Brill, pp. 203–231.

—. 2019. "Additions to Already Edited *enūma anu enlil* (*EAE*) Tablets, Part VI: The Lunar Eclipse Omens from Tablets 21–22 Published by Rochberg-Halton in AfO Beih 22". *KASKAL* 16, pp. 95–132.

—. 2021. "BM 39279, Another "Astral Magic" Text, also Called the "Exorcist's Almanac" (BRM IV 19 and 20, STT 300, LBAT 1626, SpTU IV 243 and CBS 562)". *Akkadica* 142, pp. 79–92.

—. 2022. "SUKUD.GIM, "as above", and Intertextual Reference in the Great Star List Used as an Editorial Note in the Colophon of the Diagnostic Omen Series SA.GIG". *NABU* 2022 n. 26.

Finkelstein, J. J. 1963. "The Antediluvian Kings: A University of California Tablet". *JCS* 17/2, pp. 39–51.

Foster, B. R. 2005. *Before the Muses: An Anthology of Akkadian Literature. Third Edition*. Bethesda: CDL Press.

Frahm, E. 2011. *Babylonian and Assyrian Text Commentaries. Origins of Interpretation*. GMTR 5. Münster: Ugarit-Verlag.

Frame, G. 2020. *The Royal Inscriptions of Sargon II, King of Assyria (721–705 BC)*. RINAP 2. Pennsylvania: Eisenbrauns.

Frankena, R. 1959–1962. "Die Worte der Sibitti in der I. Tafel des Irra-Epos". *JEOL* 16, pp. 40–47.

—. 1961. "New Materials for the Tākultu Ritual: Additions and Corrections". *BiOr*. 18, pp. 199–207.

Frankfort, H., Frankfort, H. A., Wilson, J. A., Jacobsen, T. 1946. *The Intellectual Adventure of Ancient Man*. Chicago-London: University of Chicago Press.

Frechette, C. G. 2012. *Mesopotamian Ritual-Prayers of Hand-lifting*. AOAT 379. Münster: Ugarit-Verlag.

Freedman, S. M. 1998. *If a City Is Set on a High: The Akkadian Omen Series Šumma Alu ina Mēlê Šakin. Volume 1: Tablets 1–21* OccPubl. S. N. Kramer Fund 17. Philadelphia: The University of Pennsylvania Museum.

—. 2017. *If a City Is Set on a High: The Akkadian Omen Series Šumma Alu ina Mēlê Šakin. Volume 3: Tablets 41–63*. Winona Lake: Eisenbrauns.

Friberg, J. 1987–1990. "Mathematik". *RlA* 7, pp. 531–585.

Gabbay, U. 2016. *The Exegetical Terminology of Akkadian Commentaries*. CHANE 82. Leiden-Boston: Brill.

Gehlken, E. 2005. "Die Adad-Tafeln der Omenserie *Enūma Anu Enlil*. Teil 1: Einführung". *BagM* 36, pp. 235–273.

—. 2007. "Die Serie DIŠ *Sîn ina tāmartīšu* im Überblick". *NABU* 2007 n. 4.

—. 2008. "Die Adad-Tafeln der Omenserie *Enūma Anu Enlil*. Teil 2: Die ersten beiden Donnertafeln (EAE 42 und EAE 43)". *ZOrA* 1, pp. 256–314.

—. 2012. *Weather Omens of Enūma Anu Enlil*. CM 43. Leiden-Boston: Brill.

Geller, M. J. 1985. *Forerunners to Udug-Ḫul: Sumerian Exorcistic Incantations*. FAOS 12. Stuttgart: Franz Steiner.

—. 2011. "Review *Divinatorische Texte I: Terrestrische, teratologische, physiognomische und oneiromantische Texte* by N. P. Heeßel". *WO* 41, pp. 118–121.

—. 2014. *Melothesia in Babylonia: Medicine, Magic, and Astrology in the Ancient Near East*. Berlin: STMAC.

—. 2016. *Healing Magic and Evil Demons*. BAM 8. Boston-Berlin: De Gruyter.

Gentner, D., Holyoak, K. J., Kokinov, B. N. (eds.) 2001. *The Analogical Mind. Perspectives from Cognitive Science*. Cambridge-London: MIT Press.

George, A. R. 1992. *Babylonian Topographical Texts*. OLA 40. Leuven: Peeters.

—. 1993. *House Most High. The Temples of Ancient Mesopotamia*. MesCiv. 5. Winona Lake: Eisenbrauns.

—. 1999. *The Epic of Gilgamesh: A New Translation*. London: Penguin Books.

—. 2003. *The Babylonian Gilgamesh Epic. Volume 1 and 2*. Oxford: Oxford University Press.

—. 2007. "The Epic of Gilgamesh: Thoughts on Genre and Meaning" in Azize, J., Weeks, N. (eds.) *Gilgamesh and the World of Assyria. Proceedings of the Conference Held at the Mandelbaum House, the University of Sydney, 21–23 July 2004*. Leuven: Peeters, pp. 37–66.

—. 2013. "The Poem of Erra and Ishum: A Babylonian Poet's View of War" in Kennedy, H. (ed.) *Warfare and Poetry in the Middle East*. New York: I.B. Tauris, pp. 39–71.

—. 2013a. *Babylonian Divinatory Texts Chiefly in the Schøyen Collection*. CUSAS 18. Bethesda: CDL Press.

—. 2015. "The Gods Išum and Ḫendursanga: Night Watchmen and Street-Lighting in Babylonia". *JNES* 74/1, pp. 1–8.

Glassner, J. J. 1984. "Pour un lexique des termes et figures analogiques en usage dans la divination mésopotamienne". *JA* 272, pp. 15–46.

Goody, J. 1977. *The Domestication of the Savage Mind*. Cambridge: Cambridge University Press.

—. 1986. *The Logic of Writing and the Organization of Society*. Cambridge: Cambridge University Press.

Gössmann, P. F. 1950. *Planetarium Babylonicum, oder die sumerisch-babylonischen Stern-Namen*. ŠL 4/2. Rom: Verlag des Päpstl. Bibelinstituts.

—. 1955. *Das Era-Epos*. Würzburg: Augustinus-Verlag.

Graßhoff, G. 2011. "Babylonian Meteorological Observations and the Empirical Basis of Ancient Science" in Gebhard J. Selz, G. J., Wagensonner, K. (eds.) *The Empirical Dimension of Ancient Near Eastern Studies*. Wien: LIT, pp. 33–48.

Grayson, A. K. 1987. *Assyrian Rulers of the Third and Second Millennia BC (to 1115 BC)*. RIMA 1. Toronto-Buffalo-London: University of Toronto Press.

—. 1991. *Assyrian Rulers of the Early First Millennium BC I (1114–859 BC)*. RIMA 2. Toronto-Buffalo-London: University of Toronto Press.
—. 1996. *Assyrian Rulers of the Early First Millennium BC II (858–745 BC)*. RIMA 3. Toronto-Buffalo-London: University of Toronto Press.
Grayson, A. K., Novotny, J. 2012. *The Royal Inscriptions of Sennacherib, King of Assyria (704–681 BC) Part 1*. RINAP 3/1. Winona Lake: Eisenbrauns.
—. 2014. *The Royal Inscriptions of Sennacherib, King of Assyria (704–681 BC) Part 2*. RINAP 3/2. Winona Lake: Eisenbrauns.
Graziani, S. 1979. "Note sui Sibitti". *AIUON* 39, pp. 673–690.
Greaves, S. W. 2000. "Ominous Homophony and Portentous Puns in Akkadian Omens" in Noegel, S. B. (ed.) *Puns and Pundits: Wordplay in the Hebrew Bible and Ancient Near Eastern Literature*. Bethesda: CDL Press, pp. 103–113.
Groneberg, B. 1986. "Die sumerisch-akkadische Inanna/Ištar: Hermaphroditos?". *WO* 17, pp. 25–46.
Guinan, A. K. 1996. "Left/Right Symbolism in Mesopotamian Divination". *SAAB* 10, pp. 5–10.
—. 2018. "Crow Omens in Mesopotamia" in Crisostomo, C. J., Escobar, E. A., Tanaka, T., Veldhuis, N. (eds.) *The Scaffolding of Our Thoughts. Essays on Assyriology and the History of Science in Honor of Francesca Rochberg*. Leiden-Boston: Brill, pp. 15–25.
Hall, M. G. 1985. *A Study of the Sumerian Moon-God, Nanna/Suen*. PhD dissertation, University of Pennsylvania.
—. 1986. "A Hymn to the Moon-God, Nanna". *JCS* 38/2, pp. 152–166.
Hallo, W. W. 1996. "Bilingualism and the Beginnings of Translation" in Fox, M. et alia (eds.) *Texts, Temples, and Traditions, A Tribute to Menahem Haran*. Winona Lake: Eisenbrauns, pp. 345-358.
—. 2010. *The World's Oldest Literature. Studies in Sumerian Belles-Lettres*. CHANE 35. Leiden-Boston: Brill.
Halloran, J. H. 2006. *Sumerian Lexicon. A Dictionary Guide to the Ancient Sumerian Language*. Los Angeles: Logogram Publishing.
Hallowell, A. I. 1960. "Ojibwa Ontology, Behavior and World View" in Diamond, S. (ed.) *Culture in History: Essays in Honor of Paul Radin*. New York: Columbia University Press, pp. 19–52.
Hätinen, A. 2021. *The Moon God Sîn in Neo-Assyrian and Neo-Babylonian Times*. Dubsar 20. Münster: Zaphon
Haubold, J., Steele, J., Stevens, K. 2019. "The Astronomical Diaries: Content, Structure, Style" in Haubold, J., Steele, J., Stevens, K. (eds.) *Keeping Watch in Babylon. The Astronomical Diaries in Context*. Leiden-Boston: Brill, pp. 1–18.
Hautamäki, A. 2020. *Viewpoint Relativism, A New Approach to Epistemological Relativism Based on the Concept of Points of View*. Synthese Library 419. Switzerland: Springer.
Heeßel, N. P. 2000. *Babylonisch-assyrische Diagnostik*. AOAT 43. Münster: Ugarit-Verlag.
—. 2005. "Stein, Pflanze und Holz. Ein neuer Text zur 'medizinischen Astrologie." *OrNS* 74, pp. 1–22.

—. 2008. "Astrological Medicine in Babylonia." In Akasoy, A., Burnett, C. and Yoeli-Talin, R. (eds.) *Astro-Medicine. Astrology and Medicine, East and West*. Firenze: Sismel, pp. 1–16.
—. 2021. *Divinatorische Texte III. Astrologische Omina*. KAL 13. WVDOG 160. Wiesbaden: Harrassowitz.
Heimpel, W. 1982. "A Catalogue of Near Eastern Venus Deities". *SMS* 4/3, pp. 59–72.
Hilgert, M. 2009. "Von 'Listenwissenschaft' und 'epistemischen Dingen'. Konzeptuelle Annäherungen an altorientalische Wissenpraktiken". *Journal for General Philosophy of Science* 40, pp. 277–309.
Horowitz, W. 1998. *Mesopotamian Cosmic Geography*. MesCiv. 8. Winona Lake: Eisenbrauns.
—. 2000. "Astral Tablets in the Hermitage, Saint Petersburg". *ZA* 90, pp. 194–206.
—. 2005. "Some Thoughts on Sumerian Star-Names and Sumerian Astronomy" in Sefati, Y., Artzi, P., Cohen, C., Eichler, B. L., Hurowitz, V. A. (eds.) *An Experienced Scribe Who Neglects Nothing, Ancient Near Eastern Studies in Honor of Jacob Klein*. Bethesda: CDL Press, pp. 163–178.
—. 2014. *The Three Stars Each: The Astrolabes and Related Texts*. AfO Beih. 33. Wien: Institut für Orientalistik der Universtät Wien.
—. 2015. "Mesopotamian Star Lists" in Ruggles, C. L. N. (ed.) *Handbook of Archaeoastronomy and Ethnoastronomy*. Wien-New York: Springer, pp. 1829–1833.
Horowitz, W., Andre, A. Kritsch, I. 2018. "The Gwich'in Boy in the Moon and Babylonian Astronomy". *Arctic Anthropology* 55/1, pp. 91–104.
Hugh-Jones, S. 1979. *The Palm and the Pleiades: Initiation and Cosmology in Northwest Amazonia*. Cambridge Studies in Social and Cultural Anthropology 24. Cambridge: Cambridge University Press.
Hunger, H. 1968. *Babylonische und assyrische Kolophone*. AOAT 2. Kevelaer-Neukirchen-Vluyn: Butzon & Bercker.
—. 1976. "Astrologische Wettervorhersagen". *ZA* 66, pp. 234–260.
—. 1980. "Kalender". *RlA* 5, pp. 297–302.
—. 1992. *Astrological Reports to Assyrian Kings*. SAA 8. Helsinki: Helsinki University Press.
—. 2001. *Astronomical Diaries and Related Texts from Babylonia, Volume V: Lunar and Planetary Texts*. DÖAW 299. Wien: Österreichische Akademie der Wissenschaften.
—. 2004. "Stars, Cities and Predictions" in Burnett, C., Hogendjik, J. P., Plofker, K., Yano, M. (eds.) *Studies in the History of the Exact Sciences in Honour of David Pingree*. Leiden-Boston: Brill, pp. 16–32.
—. 2005. "Plejaden". *RlA* 10, p. 592.
—. 2005a. "Planeten". *RlA* 10, pp. 589–591.
—. 2006. *Astronomical Diaries and Related Texts from Babylonia, Volume VI: Goal Year Texts*. DÖAW 346. Wien: Österreichische Akademie der Wissenschaften.
—. 2011. "The Relation of Babylonian Astronomy to Its Culture and Society". in Valls-Gabaud, D., Boksenberg, A. (eds.) *Proceedings IAU Symposium no. 260, 2009*. Cambridge: Cambridge University Press, pp. 62–73.

—. 2014. *Astronomical Diaries and Related Texts from Babylonia, Volume VII: Almanacs and Normal Star Almanacs*. DÖAW 466. Wien: Österreichische Akademie der Wissenschaften.
—. 2019. "Astrological Texts from Late Babylonian Uruk" in Proust, C., Steele, J. (eds.) *Scholars and Scholarship in Late Babylonian Uruk*. Wien-New York: Springer, pp. 171–185.
—. 2020. "The Texts and Aims of Babylonian Astronomy" in A.C. Bowen, A. C., Rochberg, F. (eds.) *Hellenistic Astronomy. The Science in Its Context*. Leiden-Boston: Brill, pp. 272–283.
Hunger, H., Brack-Bernsen, L. 2002. "TU 11: A Collection of Rules for the Prediction of Lunar Phases and of Month Length". *SCIAMVS* 3, pp. 3–90.
Hunger, H., Pingree, D. 1989. *MUL.APIN. An Astronomical Compendium in Cuneiform*. AfO Beih. 24. Horn: Berger & Söhne.
—. 1999. *Astral Sciences in Mesopotamia*. HdO 44. Leiden-Boston-Köln: Brill.
Hunger, H., Reiner. E. 1975. "A Scheme for Intercalary Months from Babylonia". *WZKM* 67, pp. 21–28.
Hunger, H., Steele, J. 2019. *The Babylonian Astronomical Compendium MUL.APIN*. London-New York: Routledge.
Jacobsen, T. 1939. *The Sumerian King List*. AS 11. Chicago: University of Chicago Press.
—. 1977. "*Inuma Ilu awīlum*" in de Jong Ellis, M. (ed.) *Essays on the Ancient Near East in Memory of Jacob Joel Finkelstein*. Hamden: Archon Books, pp. 113–117.
Jean, C. F. 1924. "^{d}VII-bi". *RA* 21, pp. 93–104.
Jensen, P. 1900. *Assyrisch-babylonische Mythen und Epen*. Berlin: Reuther & Reichard.
—. 1928. "Astralmythen". *RlA* 1, pp. 305–309.
Jeyes, U. 1980. "Death and Divination in the Old Babylonian Period" in Alster, B. (ed.) *Death in Mesopotamia. Papers Read at the XXVIe Rencontre Assyriologique Internationale*. Copenhagen: Akademisk Forlag, pp. 107–122.
—. 1989. *Old Babylonian Extispicy: Omen Texts in the British Museum*. PIHANS 64. Leiden: Nederlands Historisch-Archaeologisch Instituut te Istanbul.
—. 1991–1992. "Divination as Science in Ancient Mesopotamia". *JEOL* 32, pp. 23–41.
Jiménez, E. 2014. "New Fragments of Gilgameš and Other Literary Texts from Kuyunjik". *Iraq* 76, pp. 99–121.
—. 2018. "Highway to Hell: The Winds as Cosmic Conveyors in Mesopotamian Incantation Texts" in van Buylaere, G., Luukko, M., Schwemer, D., Mertens-Wagschal, A. (eds.) *Sources of Evil: Studies in Mesopotamian Exorcistic Lore*. Leiden-Boston: Brill, pp. 316–350.
Jones, A. 2004. "A Study of Babylonian Observations of Planets Near Normal Stars". *Archive for History of Exact Sciences* 58/6, pp. 475–536.
Jursa, M. 2001–2002. "Göttliche Gärtner? Eine bemerkenswerte Liste". *AfO* 48/49, pp. 76–89.
Kapelrud, A. S. 1968. "The Number Seven in Ugaritic Texts". *VT* 18/4, pp. 494–499.
Kelley, D. H., Milone, E. F. 2005. *Exploring Ancient Skies. An Encyclopedic Survey of Archaeoastronomy*. New York: Springer.

Khait, I. 2014. "New Readings in YOS 10" in Fincke, J. (ed.) *Divination in Ancient Near East. A Workshop on Divination Conducted during the 54th Rencontre Assyriologique Internationale, Würzburg, 2008*. Winona Lake: Eisenbrauns, pp. 77–89.

Klein, J., Sefati, Y. 2020. *From the Workshop of the Mesopotamian Scribe: Literary and Scholarly Texts from the Old Babylonian Period*. Pennsylvania: Eisenbrauns.

Koch, J. 1997. "Zur bedeutung von LÁL in den „Astronomical Diaries" und in der Plejaden-Schaltregel". *JCS* 49, pp. 83–101.

—. 2003. "Neues vom Beschwörungstext BA 10/1, 81 No. 7 rev. 1–8". *WO* 33, pp. 89–99.

Koch(-Westenholz), U. 1995. *Mesopotamian Astrology: An Introduction to Babylonian and Assyrian Celestial Divination*. CNIP 19. Copenhagen: Museum Tusculanum Press.

—. 1999. "The Astrological Commentary *Šumma Sîn ina tāmartīšu* Tablet 1" in Gyselen, R. (ed.) *La science des cieux. Sages, mages, astrologues*. Bures-sur-Yvette: Groupe pour l'étude de la civilisation du Moyen-Orient, pp. 149–165.

—. 2000. *Babylonian Liver Omens. The Chapters Manzāzu, Padānu and Pān tākalti of the Babylonian Extispicy Series Mainly from Aššurbanipal's Library*. CNIP 25. Copenhagen: Museum Tusculanum.

—. 2015. *Mesopotamian Divination Texts: Conversing with the Gods; Sources from the First Millennium BCE*. GMTR 7. Münster: Ugarit-Verlag.

—. 2019. "Principles of Astrological Omen Compositions. Some Challenges of Reserve Engineering the Astrological Hermeneutic". *KASKAL* 16, pp. 221–235.

Kolev, R. 2013. *The Babylonian Astrolabe: The Calendar of Creation*. SAAS 22. Winona Lake: Eisenbrauns.

Konstantopoulos, G. V. 2015. *They are Seven: Demons and Monsters in the Mesopotamian Textual and Artistic Tradition*. PhD dissertation, University of Michigan.

Kouwenberg, N. J. C. 2010. *The Akkadian Verb and Its' Semitic Background*. LANE 2. Winona Lake: Eisenbrauns.

Kramer, S. N. 1963. *The Sumerians: Their History, Culture, and Character*. Chicago: University of Chicago Press.

Krebernik, M. 1997. "Mondgott A I". *RlA* 8, pp. 360–369.

—. 2001. *Tall Bi'a Tuttul-II. Die Altorientalischen Schriftfunde*. WVDOG 96/2. Saarbrücker: Saarbrücker Druckerei und Verlag.

—. 2011. "Šar-ur und Šar-gaz". *RlA* 12, pp. 84–86.

Kristeva, J. 1986. *The Kristeva Reader. Edited by Toril Moi*. Columbia University Press.

Kroll, G. 1903. *Codices Vindobonenses*. Catalogus Codicum Astrologorum Graecorum VI. Bruxelles: Henrici Lamertin.

Kugler, F. X. 1909. "Auf den Trümmern des Panbabylonismus". *Anthropos* 4, pp. 477–499.

Kuhn, T. 1996. *The Structure of Scientific Revolutions, 3rd ed.* Chicago: University of Chicago Press.

Kunstmann, W. G. 1932. *Die babylonische Gebetsbeschworung*. Leipzig: J. C. Hinrichs.

Kurtik, G. E. 2007. *The Star Heaven of Ancient Mesopotamia: The Sumero-Akkadian Names of Constellations and Other Heavenly Bodies* [in Russian]. Saint Petersburg: ALETHEIA.

—. 2016. "Observations of the Planet Venus in Archaic Uruk: the Problem and Researches". *NABU* 2016 n. 84.

Labat, R. 1951. *Traité akkadien de diagnostics et pronostics médicaux. I. Transcription et traduction*. Collection de travaux de l'académie internationale d'histoire des sciences 7. Paris-Leiden: Académie internationale d'histoire des sciences.

—. 1965. *Un calendrier babylonien des travaux, des signes et des mois (séries iqqur īpuš)*. Paris-Genf: Bibliothèque de l'École des Hautes Études, Sciences historiques et philologiques.

—. 1972–1975. "Hemerologien". *RlA* 4, pp. 317–323.

Læssøe, J. 1955. *Studies on the Assyrian Ritual and Series* bît rimki. Copenhagen: Ejnar Munksgaard.

Lakoff, G., Johnson, M. 2003. *Metaphors We Live by. Second Edition*. Chicago-London: University of Chicago Press.

Lambert, W. G. 1957. "Ancestors, Authors, and Canonicity". *JCS* 11, pp. 1–14.

—. 1957–1958. "Review *Das Era-Epos* by F. Gössmann". *AfO* 18, pp. 395–401.

—. 1957–1958a. "A Part of the Ritual for the Substitute King". *AfO* 18, pp. 109–112.

—. 1959–1960. "An Address of Marduk to Demons. New Fragments". *AfO* 19, pp. 114–119.

—. 1962. "A Catalogue of Texts and Authors". *JCS* 16, pp. 59–77.

—. 1974–1977. "Review *Untersuchungen zur Formensprache der babylonischen Gebetsbeschworungen* by W. R. Mayer". *AfO* 25, pp. 197–199.

—. 1990. "The Name of Nergal Again". *ZA* 80, pp. 40–52.

—. 1996. "The Etymology and Meaning of Ṣalbatānu". *NABU* 1996 n. 123.

—. 1997. "Procession to the Akītu House". *RA* 91/1, pp. 49–80.

—. 1998. "The Qualifications of Babylonian Diviners" in Maul, S. (ed.) *Festschrift für Rykle Borger zu seinem 65. Geburtstag am 24. Mai 1994*. Groningen: Styx, pp. 141–158.

—. 2007. *Babylonian Oracle Questions*. MesCiv. 13. Winona Lake: Eisenbrauns.

—. 2013. *Babylonian Creation Myths*. MesCiv. 16. Winona Lake: Eisenbrauns.

Lambert, W. G., Millard, A. R. 1968. *Atra-ḫasīs. The Babylonian Story of the Flood*. Oxford: Claredon Press.

Largement, R. 1957. "Contribution à l'étude des astres errants dans l'astrologie chaldéenne I". *ZA* 52, pp. 235–264.

Larsen, M. T. 1987. "The Mesopotamian Lukewarm Mind: Reflections on Science, Divination and Literacy" in Rochberg-Halton, F. (ed.) *Language, Literature, and History: Philological and Historical Studies Presented to Erica Reiner*. New Haven: American Oriental Society, pp. 203–225.

Lawson, J. N. 1994. *The Concept of Fate in Ancient Mesopotamia of the First Millennium. Towards an Understanding of Šīmtu*. OBC 7. Wiesbaden: Harrassowitz.

Lawson, K. 2012. "Another Look at an Aramaic Astral Bowl". *JNES* 71/2, pp. 209–230.

Lehoux, D. 2004. "Observation and Prediction in Ancient Astrology". *Studies in History and Philosophy of Science* 35, pp. 227–246.

Leibovici, M. 1956. "Sur l'astrologie médicale Néo-Babylonienne". *JA* 244, pp. 275–280.

Leichty, E. 1970. *The Omen Series Šumma Izbu*. TCS 4. Locust Valley: J. J. Augustin Publisher.

—. 1993. "The Origins of Scholarship" in Galter, H. D. (ed.) *Die Rolle der Astronomie in den Kulturen Mesopotamiens. Beiträge zum 3. Grazer Morgenländische Symposion (23.–27. September 1991)*. Graz: Grazer Morgenländische Studien, pp. 21–29.
—. 2011. *The Royal Inscriptions of Esarhaddon, King of Assyria (680–669 BC)*. RINAP 4. Winona Lake: Eisenbrauns.
Lenzi, A. 2011. *Reading Akkadian Prayers and Hymns: An Introduction*. Ancient Near East Monographs 3. Atalanta: SBL Press.
Lévi-Strauss, C. 1969. *The Raw and the Cooked*. Introduction to a Science of Mythology Volume 1. New York: Harper & Row.
Linssen, M. J. H. 2004. *The Cults of Uruk and Babylon: The Temple Ritual Texts as Evidence for Hellenistic Cult Practises*. CM 25. Leiden: Brill.
Litke, R. 1998. *A Reconstruction of the Assyro-Babylonian God Lists, ^{d}An : ^{d}A-nu-um and An : Anu ša amēli*. TBC 3. New Heaven: Yale Babylonian Collection.
Liverani, M. 2014. *The Ancient Near East. History, Society and Economy*. London-New York: Routledge.
Livingstone A. 1986. *Mystical and Mythological Explanatory Works of Assyrian and Babylonian Scholars*. Oxford: Oxford University Press.
—. 1997. "Menologie". *RlA* 8, pp. 59–60.
—. 1999. "The Magic of Time" in Abusch, T., van der Toorn, K. (eds.) *Mesopotamian Magic: Textual, Historical, and Interpretative Perspectives*. Groningen: Styx, pp. 131–137.
—. 2013. *Hemerologies of Assyrian and Babylonian Scholars*. CUSAS 25. Bethesda: CDL Press.
Lloyd, G. E. R. (1966) 1992. *Polarity and Analogy: Two Types of Argumentations in Early Greek Thought*. Cambridge: Cambridge University Press.
Machinist, P. 2005. "Order and Disorder: Some Mesopotamian Reflections" in Shaked, S. (ed.), *Genesis and Regeneration: Essays on Conceptions of Origins*. Jerusalem: Israel Academy of Sciences and Humanities, pp. 31–61.
Machinist, P., Sasson, J. M. 1983. "Rest and Violence in the Poem of Erra". *JAOS* 103/1, pp. 221–226.
Machinist, P., Tadmor, H. 1993. "Heavenly Wisdom" in Cohen, M., Snell, D. & Weisberg, D. (eds.) *The Tablet and the Scroll: Near Eastern Studies in Honor of William W. Hallo*. Bethesda: CDL Press, pp. 146–151.
Mannikka, E. 1996. *Angkor Wat: Time, Space, and Kingship*. Honolulu: University of Hawaii Press.
Maul, S. M. 1994. *Zukunftsbewältigung. Eine Untersuchung altorientalischen Denkens anhand der babylonisch-assyrischen Löserituale (Namburbi)*. BagF 18. Mainz: Philipp von Zabern.
—. 1999. "Das Wort im Worte. Orthographie und Etymologie als hermeneutische Verfahren babylonischer Gelehrter" in Most, G. W. (ed.) *Commentaries-Kommentare*. Göttingen: Vandenhoeck und Ruprecht, pp. 1–18.
—. 2005. "Omina und Orakel A". *RlA* 10, pp. 45–88.
—. 2018. *The Art of Divination in the Ancient Near East: Reading the Signs of Heaven and Earth*. Texas: Baylor University Press.

May, N. M. 2018. "The Scholar and Politics: Nabû-zuqup-kēnu, His Colophons and the Ideology of Sargon II", in Koslova, N. V. (ed.) *Proceedings of the International Conference Dedicated to the Centenary of Igor Mikhailovich Diakonoff (1915–1999)*. St. Petersburg: The State Hermitage Publishers, pp. 110–164.

Mayer, W. R. 1976. *Untersuchungen zur Formensprache der babylonischen Gebetsbeschwörungen*. StPohl SM 5. Roma: Pontificio Istituto Biblico.

—. 2018. "Das Gebet an die Götter der Nacht in KUB 4, 47". *OrNS* 87, pp. 265–274.

McNamara, P. 2009. *The Neuroscience of Religious Experience*. Cambridge: Cambridge University Press.

van der Meer, P. E. 1939. "Tablets of the *ḪAR-ra = ḫubullu* Series in the Ashmolean Museum". *Iraq* 6/2, pp. 144–179.

Messier, C. 1784. "Catalogue des nébuleuses et d'amas d'étoiles". *Connaissance des Temps pour l'Année 1784*, pp. 227–267.

Meyer, J. 1987. *Untersuchungen zu den Tonlebermodellen aus dem Alten Orient*. AOAT 39. Kevelaer-Neukirchen-Vluyn: Butzon & Bercker.

Miller, G. A. 1955. "The Magical Number Seven, Plus or Minus Two. Some Limits on Our Capacity for Processing Information". *Psychological Review* 101/2, pp. 343–352.

Mohr, H. 2006. "Light / Enlightenment". *The Brill Dictionary of Religion. Volumes I, II, III, and IV*, pp. 1103–1108.

Monroe, M. W. 2016. *Advice from the Stars: The Micro-Zodiac in Seleucid Babylonia*. PhD dissertation, Brown University.

—. 2016a. "The Micro-Zodiac in Babylon and Uruk: Seleucid Zodiacal Astrology" in Steele, J. M. (ed.) *The Circulation of Astronomical Knowledge in the Ancient World*. Leiden-Boston: Brill, pp. 119–138.

Moortgat, A. 1940. *Vorderasiatische Rollsiegel*. Berlin: Gebr. Mann.

Muroi, K. 2014. "The Origin of the Mystical Number Seven in Mesopotamian Culture; Division by Seven in the Sexagesimal Number System". Issued by arXiv, https://doi.org/10.48550/arXiv.1407.6246 accessed 13.05.2022.

Nadali, D., Polcaro, A. 2016. "The Sky from the High Terrace: Study on the Orientation of the Ziqqurat in Ancient Mesopotamia". *Mediterranean Archaeology and Archaeometry* 16/4, pp. 103–108.

Negretti, N. 1973. *Il Settimo Giorno*. Analecta Biblica Dissertationes 55. Roma: Pontificio Istituto Biblico.

Neugebauer, O. 1955. *Astronomical Cuneiform Texts: Babylonian Ephemerides of the Seleucid Period for the Motion of the Sun, the Moon, and the Planets. Volumes 1–3*. London: The Institute for Advanced Study.

—. 1969. *The Exact Sciences in Antiquity. Second Edition*. New York: Dover Publications.

—. 1975. *A History of Ancient Mathematical Astronomy*. Berlin: Springer.

Neugebauer, O., Sachs, A. J. 1967. "Some Atypical Astronomical Cuneiform Texts". *JCS* 21, pp. 183–218.

Niederreiter, Z. 2008. "Le role des symbols figurés attribués aux membres de la cour de Sargon II: des emblèmes créés par les lettrés du palais au service de l'idéologie royale". *Iraq* 70, pp. 51–86.

Noegel, S. B. 2006. "On Puns and Divination: Egyptian Dream Exegesis from a Comparative Perspective" in Szpakowska, F. (ed.) *Through a Glass Darkly: Magic, Dreams, and Prophecy in Ancient Egypt*. Swansea: The Classical Press of Wales, pp. 95–119.

—. 2011. "'Wordplay' in the Song of Erra". Heimpel, W., Frantz, G. (eds). *Strings and threads. A Celebration of the Work of Anne Draffkorn Kilmer*. Winona Lake: Eisenbrauns, pp. 161–193.

O'Meara, S. J. 1998. *Deep-Sky Companions: The Messier Objects*. Cambridge: Cambridge University Press.

Oelsner, J. 1986. *Materialien zur babylonischen Gesellschaft und Kultur in hellenistischer Zeit*. Assyriologia 7. Budapest: ELTE Sokszorosítóüzemében.

—. 2005–2006. "Der 'Hilprecht-Text': die Jenaer astronomisch-mathematische Tafel HS 245 (früher HS 229) und die Paralleltexte Sm 162 (CT 33, 11) Rs. sowie Sm 1113 (AfO 18, 393f.)". *AfO* 51, pp. 108–124.

Oelsner, J., Horowitz, W. 1997–1998. "The 30-Star-Catalogue HS 1897 and the Late Parallel BM 55502". *AfO* 44/45, pp. 176–185.

Oppenheim, L. 1956. "The Interpretation of Dreams in the Ancient Near East. With a Translation of an Assyrian Dream-Book". *TAPS* 46/3, pp. 179–373.

—. 1959. "A New Prayer to the 'Gods of the Night'". *AnBi.* 12, pp. 282–301.

—. 1964. *Ancient Mesopotamia: A Portrait of a Dead Civilization*. Chicago: University of Chicago Press.

—. 1974. "A Babylonian Diviner's Manual". *JNES* 33/2, pp. 197–220.

Orchiston, W. 1996. "Australian Aboriginal, Polynesian and Maori Astronomy" in Walker, C. (ed.) *Astronomy before the Telescope*. London: British Museum Press, pp. 318–303.

Ornan, T. 2009. "In the Likeness of Man. Reflections on the Anthropocentric Perception of the Divine in Mesopotamian Art" in Porter, B. (ed.) *What Is a God?: Anthropomorphic and Non-Anthropomorphic Aspects of Deity in Ancient Mesopotamia*. Winona Lake: Eisenbrauns, pp. 93–152.

Ortony, A. 1993. *Metaphor and Thought (2nd ed)*. Cambridge: Cambridge University Press.

Oshima, T. 2011. *Babylonian Prayers to Marduk*. ORA 7. Tübingen: Mohr Siebeck.

—. 2013. "Review *Mesopotamian Ritual-Prayers of 'Hand-lifting' (Akkadian Šuillas): An Investigation of Function in Light of the Idiomatic Meaning of the Rubric* by C. G. Frechette.". *BSOAS* 76/1, pp. 111–112.

—. 2019. "Legends of Sargon. History to His Story: Forming the Warrior King Archetype" in Da Riva, R., Lang, M., Fink, S. (eds.) *Literary Change in Mesopotamia and Beyond. Proceedings of the 2nd and 3rd Melammu Workshops*. Münster: Zaphon, pp. 43–56.

Ossendrijver, M. 2012. *Babylonian Mathematical Astronomy: Procedure Texts*. Sources and Studies in the History of Mathematics and Physical Sciences. Berlin: Springer.

—. 2016. "Conceptions of the Body in Mesopotamian Cosmology and Astral Science" in Buchheim, T., Meissner, D., Wachsmann, N. (eds.) *Soma: Körperkonzepte und körperliche Existenz in der antiken Philosophie und Literatur*. Hamburg: Felix Meiner, pp. 143–158.

—. 2018. "Babylonian Scholarship and the Calendar during the Reign of Xerxes" in Waerzeggers, C., Seire, M. (eds.), *Xerxes and Babylonia. The Cuneiform Evidence*. Leuven-Paris-Bristol: Peeters, pp. 135–163.
—. 2019. "Babylonian Market Predictions" in Haubold, J., Steele, J., Stevens, K. (eds.) *Keeping Watch in Babylon. The Astronomical Diaries in Context*. Leiden-Boston: Brill, pp. 53–78.
—. 2021. "Weather Prediction in Babylon". *JANEH* 8/2, pp. 223–258.
von der Osten, H. H. 1934. *Ancient Oriental Seals in the Collection of Mr. Edward T. Newell*. Chicago: University of Chicago Press.
Papke, W. 1984. "Zwei Plejaden-Schaltregeln aus dem 3. Jahrtausend". *AfO* 31, pp. 67–70.
Parpola, S. 1983. *Letters from Assyrian Scholars to the Kings Esarhaddon and Assurbanipal, Part II: Commentary and Appendices*. AOAT 5/2. Kevelaer-Neukirchen-Vluyn: Ugarit-Verlag.
—. 1993. *Letters from Assyrian and Babylonian Scholars*. SAA 10. Helsinki: Helsinki University Press.
—. 2017. *Assyrian Royal Rituals and Cultic Texts*. SAA 20. Helsinki: Helsinki University Press.
Peirce, C. S. 1991. *Peirce on Signs: Writings on Semiotic by Charles Sanders Peirce. Edited by Hoopes James*. Chapel Hill-London: University of North Carolina Press.
Perdibon, A. 2019. *Mountains and Trees, Rivers and Springs: Animistic Beliefs and Practices in Ancient Mesopotamian Religion*. LAOS 11. Wiesbaden: Harrassowitz.
Peterson, J. 2019. *The Literary Sumerian of Old Babylonian Ur: UET 6/1–3 in Transliteration and Translation with Select Commentary. Part I: UET 6/1*. Cuneiform Digital Library Preprints 15.
Pettinato, G. 1982. *Testi lessicali bilingui della biblioteca L. 2769. Parte I*. MEE 4. Napoli: Istituto Universitario Orientale di Napoli.
Pingree, D. 1987. "Venus Omens in India and Babylon" in Rochberg-Halton, F. (ed.) *Language, Literature, and History: Philological and Historical Studies Presented to Erica Reiner*. New Haven: American Oriental Society, pp. 293–316.
—. 1996. "Astronomy in India" in Walker, C. (ed.) *Astronomy before the Telescope*. London: British Museum Press, pp. 123–142.
Pingree, D., Walker, C. 1988. "A Babylonian Star-Catalogue: BM 78161" in Leichty, E., de J. Ellis, M., Gerardi, P. (eds.) *A Scientific Humanist. Studies in Memory of Abraham Sachs*. Philadelphia: University of Pennsylvania Museum, pp. 313–322.
Pirngruber, R. 2013. "The Historical Sections of the Astronomical Diaries in Context: Developments in a Late Babylonian Scientific Text Corpus". *Iraq* 75, pp. 197–210.
Plett, H. F. (ed.) 1991. *Intertextuality*. Berlin-New York: De Gruyter.
Pohl, A. 2022. *Die akkadischen Hymnen der altbabylonischen Zeit. Grammatik, Stilistik, Editionen*. LAOS 13. Wiesbaden: Harrassowitz.
Polvani, A. M. 2005. "The Deity IMIN.IMIN.BI in Hittite Texts". *OrNS* 74/3, pp. 181–194.
Ponchia, S. 2013–2014. "Hermeneutical Strategies and Innovative Interpretation in Assyro-Babylonian Texts: The Case of Erra and Išum". *SAAB* 20, pp. 61–72.
Pongratz-Leisten, B. 1994. *Ina šulmi īrub. Die kulttopographische und ideologische Programmatik der akītu-Prozession in Babylonien und Assyrien im 1. Jahrtausend v. Chr*. BagF 16. Mainz am Rehin: Philipp von Zabern.

—. 2011. "Divine Agency and Astralization of the Gods in Ancient Mesopotamia" in Pongratz-Leisten, B. (ed.) *Reconsidering the Concept of Revolutionary Monotheism*. Winona Lake: Eisenbrauns, pp. 137–187.
Porada, E. 1938. "Die Siegel aus der Sammlung des Franziskanerklosters Flagellatio in Jerusalem". *Berytus* V, pp. 1–26.
Porter, B. N. 2009. "Blessings from a Crown, Offerings to a Drum: Were There Non-Anthropomorphic Deities in Ancient Mesopotamia?" in Porter, B. N. (ed.) *What Is a God?: Anthropomorphic and Non-Anthropomorphic Aspects of Deity in Ancient Mesopotamia*. Transaction of the Casco Bay Assyriological Institute Volume 2. Winona Lake: Eisenbrauns, pp. 153–194.
—. (ed.) 2009a. *What Is a God?: Anthropomorphic and Non-Anthropomorphic Aspects of Deity in Ancient Mesopotamia*. Transaction of the Casco Bay Assyriological Institute Volume 2. Winona Lake: Eisenbrauns.
Puhvel, J. 1991. "Names and Numbers of the Pleiad" in Leslau, W., Kaye, A. S. (eds.) *Semitic Studies in Honor of Wolf Leslau on the Occasion of His Eighty-Fifth Birthday, November 14th, 1991*. Wiesbaden: Harrassowitz, pp. 1243–1247.
Radner, K. 1995. "Format and Content in Neo-Assyrian Texts", in Mattila, R. (ed.) *Nineveh, 612 BC: The Glory and Fall of the Assyrian Empire. Catalogue of the 10th Anniversary Exhibition of the Neo-Assyrian Text Corpus Project*. Helsinki: Helsinki University Press, pp. 63–78.
Rappenglück, M. 1997. "The Pleiades in the "Salle des Taureaux", Grotte de Lascaux. Does a Rock Picture in the Cave of Lascaux Show the Open Star Cluster of the Pleiades at the Magdalénien Era (ca 15.300 BC)?" in Jaschek, C., Atrio Barandela, F. (eds.) *Actas del IV Congreso de la Seac "Astronomia en la Cultura". Salamanca: Universidad de Salamanca*, pp. 217–225.
—. 2008. "The Pleiades and Hyades as Celestial Spatiotemporal Indicators in the Astronomy of Archaic and Indigenous Cultures" in Wolfschmidt, G. (ed.) *Prähistorische Astronomie und Ethnoastronomie. Proceedings des Kolloquiums des Arbeitskreises Astronomiegeschichte in der Astronomischen Gesellschaft am 24. September 2007 in Würzburg*. Hamburg: Universität Hamburg.
Ratzon, E. 2016. "Early Mesopotamian Intercalation Schemes and the Sidereal Month" *Mediterranean Archaeology and Archaeometry* 16, pp. 143–151.
Reiner, E. 1958. *Šurpu: A Collection of Sumerian and Akkadian Incantations*. AfO Beih. 11. Graz: Selbstverlag des Herausgebers.
—. 1960. "Plague Amulets and House Blessings". *JNES* 2, pp. 148–155.
—. 1960a. "Fortune-Telling in Mesopotamia". *JNES* 19/1, pp. 23–35.
—. 1961. "The Etiological Myth of the 'Seven Sages'". *Or.* 30, pp. 1–11.
—. 1975. "Inscription from a Royal Elamite Tomb". *AfO* 24, pp. 87–102.
—. 1993. "Two Babylonian Precursors of Astrology". *NABU* 1993 n. 26.
—. 1995. *Astral Magic in Babylonia*. TAPS 85/4. Philadelphia: American Philosophical Society.
—. 1998. "Celestial Omen Tablets and Fragments in the British Museum" in Maul, S. (ed.) *Festschrift für Rykle Borger zu seinem 65. Geburtstag am 24 Mai. 1994*. Groningen: Styx, pp. 215–302.

—. 1999. "Babylonian Celestial Divination" in Swerdlow, N. M. (ed.) *Ancient Astronomy and Celestial Divination*. Cambridge-London: MIT Press, pp. 22–37.

—. 2006. "If Mars Comes Close to Pegasus..." in Guinan, A. K., de J. Ellis, M., Ferrara, A. J., Freedman, S. M., Rutz, M. T., Sassmannshausen, L., Tinney, S., Waters, M. W. (eds.) *If a Man Builds a Joyful House: Assyriological Studies in Honor of Erle Verdun Leichty*. Leiden: Brill, pp. 313–323.

Reiner, E., Pingree, D. 1975. *Babylonian Planetary Omens, I*. BiMes. 2/1, Malibu: Undena Publications.

—. 1981. *Babylonian Planetary Omens, II*. BiMes. 2/2, Malibu: Undena Publications.

—. 1998. *Babylonian Planetary Omens, III*. CM 11. Groningen: Styx.

—. 2005. *Babylonian Planetary Omens, IV*. CM 30. Leiden-Boston: Brill-Styx.

Reinhold, G.G.G., Golinets, V. (eds.). 2008. *Die Zahl Sieben im Alten Orient. Studien zur Zahlensymbolik in der Bibel und ihrer altorientalischen Umwelt*. Frankfurt am Main-Berlin-Bern-Bruxelles-New York-Oxford-Wien: Peter Lang.

Reisman, D. 1973. "Iddin-Dagan's Sacred Marriage Hymn". *JCS* 25/4, pp. 185–202.

Renshaw, S. L. 2012. "The Inspiration of Subaru as a Symbol of Cultural Values and Traditions in Japan" in Campion, N., Sinclair, R. (eds.) *Culture and Cosmos, Vol. 16 nos. 1 and 2*. England: Culture and Cosmos and Sophia Centre Press, pp. 175–191.

Renzi-Sepe, M. T. 2021. "A Note on the Series *Šumma Sîn ina tāmartīšu*". *OrNS* 90/1, pp. 113–117.

Reynolds, F. 1998. "Unpropitious Titles of Mars in Mesopotamian Scholarly Tradition" in Prosecky, J. (ed.) *Intellectual Life of the Ancient Near East: Papers Presented at the 43rd Rencontre Assyriologique International, Prague, July 1–5, 1996*. Prague: Oriental Institute, pp. 347–358.

Richter, T. 2004. *Untersuchungen zu den lokalen Panthea Süd- und Mittelbabyloniens in altbabylonischer Zeit. 2., verbesserte und erweiterte Auflage*. AOAT 257. Münster: Ugarit-Verlag.

Roberts, J. J. M. 1971. "Erra: Scorched Earth". *JCS* 24, n. 1/2, 11–16.

Robertson, J. F. 1981. *Redistributive Economy in Ancient Mesopotamian Societies: A Case Study from Isin-Larsa Period Nippur*. PhD dissertation, University of Pennsylvania.

Rochberg(-Halton), F. 1984. "Canonicity in Cuneiform Texts". *JCS* 36/2, pp. 127–144.

—. 1988. *Aspects of Babylonian Celestial Divination: The Lunar Eclipse Tablets of Enūma Anu Enlil*. AfO Beih. 22. Horn: Berger & Söhne.

—. 1996. "Personifications and Metaphors in Babylonian Celestial Omina". *JAOS* 116/3, pp. 475–485.

—. 1998. *Babylonian Horoscopes*. TAPS 88/1. Philadelphia: American Philosophical Society.

—. 1999. "Continuity and Change in Omen Literature". in Böck, B., Cancick-Kirschbaum, E., Richter, T. (eds.). *Munuscula Mesopotamica: Festschrift für Johannes Renger*. Münster: Ugarit-Verlag, pp. 415–425.

—. 1999a. "Review *The Concept of Fate in Ancient Mesopotamia of the First Millennium: Toward an Understanding of Šīmtu* by J. N. Lawson". *JNES* 58, pp. 54–58.

—. 2000. "Scribes and Scholars: The *ṭupšar Enūma Anu Enlil*" in Marzahn, J., Neumann, H. (eds.) *Assyriologica et Semitica. Festschrift für Joachim Oelsner anläßlich seines 65. Geburtstages am 18. Februar 1997*. Münster: Ugarit-Verlag, pp. 359–376.

—. 2004. *The Heavenly Writing: Divination, Horoscopy, and Astronomy in Mesopotamian Culture*. Cambridge: Cambridge University Press.

—. 2009. "'The Stars Their Likenesses': Perspectives on the Relation between Celestial Bodies and Gods in Ancient Mesopotamia" in Porter, B. (ed.) *What Is a God?: Anthropomorphic and Non-Anthropomorphic Aspects of Deity in Ancient Mesopotamia*. Winona Lake: Eisenbrauns, pp. 41–91.

—. 2010. *In the Path of the Moon. Babylonian Celestial Divination and Its Legacy*. AMD 6. Leiden-Boston: Brill.

—. 2010a. "'If P then Q': Form and Reasoning in Babylonian Divination" in Annus, A. (ed.) *Divination and Interpretation of Signs in the Ancient World*. Chicago: Oriental Institute of Chicago, pp. 19–28.

—. 2011. "The Heavens and the Gods in Ancient Mesopotamia: The View from a Polytheistic Cosmology" in Pongratz-Leisten, B. (ed.) *Reconsidering the Concept of Revolutionary Monotheism*. Winona Lake: Eisenbrauns, 117–136.

—. 2015. "The Babylonians and the Rational" in Johnson, C. (ed.) *In the Wake of the Compendia. Infrastructural Contexts and the Licensing of Empiricism in Ancient and Medieval Mesopotamia*. Boston-Berlin: De Gruyter, pp. 209–246.

—. 2016. *Before Nature: Cuneiform Knowledge and the History of Science*. Chicago-London: University of Chicago Press.

—. 2018. "The Catalogues of *Enūma Anu Enlil*: Medicine, Magic and Divination". in Steinert, U. (ed.) *Assyrian and Babylonian Scholarly Text Catalogues*. Boston-Berlin: De Gruyter, pp. 121–136.

—. 2018a. "*Ina lumun attalî Sîn*: On Evil and Lunar Eclipses" in van Buylaere, G., Luukko, M., Schwemer, D., Mertens-Wagschal, A. (eds.) *Sources of Evil: Studies in Mesopotamian Exorcistic Lore*. Leiden-Boston: Brill, pp. 285–315.

—. 2020. "The Babylonian Contribution to Greco-Roman Astronomy" in Bowen, A. C., Rochberg, F. (eds.) *Hellenistic Astronomy. The Science in Its Context*. Leiden-Boston: Brill, pp. 147–159.

—. 2020a. "Hellenistic Babylonian Astral Divination and Nativities" in Bowen, A. C., Rochberg, F. (eds.) *Hellenistic Astronomy. The Science in Its Context*. Leiden-Boston: Brill, pp. 472–489.

Röllig, W. 1987–1990. "Luḫuššu". *RlA* 7, p. 159.

Roth, M. T. 1997. *Law Collections from Mesopotamia and Asia Minor. Second Edition*. WAW 6. Atlanta: SBL Press.

Rutz, M. 2016. "Astral Knowledge in an International Age: Transmission of the Cuneiform Tradition, ca. 1500–1000 B.C." in Steele, J. M. (ed.) *The Circulation of Astronomical Knowledge in the Ancient World*. Leiden-Boston: Brill, pp. 18–54.

—. 2018. "A Late Babylonian Compilation Concerning Ritual Timing and Materia Medica" in Crisostomo, C. J., Escobar, E. A., Tanaka, T., Veldhuis, N. (eds.) *The Scaffolding of Our Thoughts. Essays on Assyriology and the History of Science in Honor of Francesca Rochberg*. Leiden-Boston: Brill, pp. 97–112.

Sachs, A. 1948. "A Classification of the Babylonian Astronomical Tablets of the Seleucid Period". *JCS* 2, pp. 271–290.

Sachs, A. J., Hunger, H. 1988. *Astronomical Diaries and Related Texts from Babylonia, Vol. I. Diaries from 652 B.C. to 262 B.C.* DÖAW 195. Wien: Verlag der Österreichischen Akademie der Wissenschaften.

—. 1989. *Astronomical Diaries and Related Texts from Babylonia, Vol. II. Diaries from 261 B.C. to 165 B.C.* DÖAW 201. Wien: Verlag der Österreichischen Akademie der Wissenschaften.

—. 1996. *Astronomical Diaries and Related Texts from Babylonia. Volume III: Diaries from 164 B.C. to 61 B.C.* DÖAW 247. Wien: Verlag der Österreichischen Akademie der Wissenschaften.

Sakuma, Y. 2014. "Analyse hethitischer Vogelflugorakel" in Fincke, J. (ed.) *Divination in Ancient Near East. A Workshop on Divination Conducted during the 54th Rencontre Assyriologique Internationale, Würzburg, 2008.* Winona Lake: Eisenbrauns, pp. 37–52.

Salonen, A. 1973. *Vögel und Vogelfang im alten Mesopotamien.* AASF (B) 180. Helsinki: Academia Scientiarum Fennica.

de Saussure, F. (1916) 2011. *Course in General Linguistics. Trans. W. Baskin.* New York: Columbia University Press.

Schaudig, H. 2001. *Die Inschriften Nabonids von Babylon und Kyros' des Großen samt den in ihrem Umfeld entstandenen Tendenzschriften.* AOAT 256. Münster: Ugarit-Verlag.

Schaumberger, J. 1935. *Sternkunde und Sterndienst in Babel. Assyriologische, astronomische und astralmythologische Untersuchungen.* Münster: Aschendorffsche Verlagsbuchhandlung.

Schmidtchen, E. 2021. *Mesopotamische Diagnostik: Untersuchungen zu Rekonstruktion, Terminologie und Systematik des babylonisch-assyrischen Diagnosehandbuches und eine Neubearbeitung der Tafeln 3–14.* BAM 13. Berlin-Boston: De Gruyter.

Schreiber, M. 2018. "Astrologische Wettervorhersagen und Kometenbeobachtungen" in Kleber, K., Neumann, G., Paulus, S. (eds.) *Grenzüberschreitungen Studien zur Kulturgeschichte des Alten Orients. Festschrift für Hans Neumann zum 65. Geburtstag am 9. Mai 2018.* Münster: Zaphon, pp. 739–756.

—. 2019. "Calendrics and Pharmacology Combined". *NABU* 2019 n. 51.

—. 2020. "Late Babylonian Astrological Physiognomy" in Johnson, J. C., Stavru, A. (eds.) *Visualizing the Invisible with the Human Body: Physiognomy and Ekphrasis in the Ancient World.* Berlin-Boston: De Gruyter, pp. 119–140.

—. 2020a. "Egalkura and Late Astrology" in Johnson, J. C. (ed.) *Patients and Performative Identities.* University Park: Penn State University Press, pp. 35–48.

Schuster-Brandis, A. 2008. *Steine Als Schutz- Und Heilmittel: Untersuchung zu ihrer Verwendung in der Beschworungskunst Mesopotamiens im 1. Jt. v. Chr.* AOAT 46. Münster: Ugarit-Verlag.

Schuster, H. S. 1938. "Die nach Zeichen geordneten sumerisch-akkadischen Vokabulare". *ZA* 44, pp. 217–270.

Schwemer, D. 2011. "Magic Rituals: Conceptualizations and Performance." in Radner, K., Robson, E. (eds.) *The Oxford Handbook of Cuneiform Culture.* Oxford: Oxford University Press, pp. 418–442.

Scurlock, J. 2005. "Sorcery in the Stars: STT 300, BRM 4, 19–20 and the Mandaic Book of the Zodiac". *AfO* 51, pp. 125–146.

—. 2014. *Sourcebook for Ancient Mesopotamian Medicine*. WAW 36. Atlanta: SBL Press.
—. 2016. "Divination between Religion and Science" in Fincke, J. (ed.) *Divination as Science. A Workshop Conducted during the 60th Rencontre Assyriologique Internationale, Warsaw, 2014*. Winona Lake: Eisenbrauns, pp. 1–10.
Scurlock, J., Andersen, B. 2005. *Diagnoses in Assyrian and Babylonian Medicine: Ancient Sources, Translations, and Modern Medical Analyses*. Urbana-Chicago: University of Illinois Press.
Sefati, Y. 1998. *Love Songs in Sumerian Literature: Critical Edition of the Dumuzi-Inanna Songs*. Ramat-Gan: Bar-Ilan University Press.
Seidl, U. 1989. *Die babylonischen Kudurru-reliefs: Symbole mesopotamischer Gottheiten*. OBO 87. Schweiz-Gottingen: Universitätsverlag Freiburg-Vandenhoeck & Ruprecht.
Sjöberg, Å. 1973. "Hymn to Numusda with a Prayer for King Siniqisam of Larsa and a Hymn to Ninurta". *OrS* 22, pp. 107–121.
Sjöberg, Å., Bergmann, E., Gragg, G. B. 1969. *The Collection of the Sumerian Temple Hymns*. TCS 1. Locust Valley: J. J. Augustin Publisher.
Sladek, W. R. 1974. *Inanna's Descent to the Netherworld*. PhD dissertation, The Johns Hopkins University.
Smith, G. 1875. *Assyrian Discoveries. An Account of Explorations and Discoveries on the Site of Nineveh, During 1873 and 1874*. London: Sampso Low, Marston Low and Searle.
von Soden, W. 1931. "Der hymnisch-epische Dialekt des Akkadischen". *ZA* 40, pp. 163–227.
—. 1933. "Der hymnisch-epische Dialekt des Akkadischen". *ZA* 41, pp. 90–183.
—. 1965. "Leistung und Grenze sumerische und babylonischer Wissenschaft" in Landsberger, B. (ed.) *Die Eigenbegrifflichkeit der babylonischen Welt*. Darmstadt: Sonderausgabe Wissenschaftliche Buchgesellschaft, pp. 21-124 (Orig. 1936 *WO* 2, pp. 411–464, 509–557).
—. 1969. "'Als die Götter (auch noch) Mensch waren'. Einige Grundgedanken des altbabylonischen Atramḫasīs-Mythus". *Or*, 38/3, pp. 415–432.
van Soldt, W. H. 1995. *Solar Omens of Enūma Anu Enlil: Tablets 23(24) - 29(30)*. PIHANS 73. Leiden: Nederlands Historisch-Archaeologisch Instituut te Istanbul.
van der Spek, R. J. 1993. "The Astronomical Diaries as a Source for Achaemenid and Seleucid History". *BiOr*. 50, pp. 91–101.
Starr, I. 1983. *The Rituals of the Diviner*. BiMes. 12. Malibu: Undena Publications.
Steele, J. M. 2011. "Making Sense of Time: Observational and Theoretical Calendars" in Radner, K., Robson, E. (eds.) *The Oxford Handbook of Cuneiform Culture*. Oxford: Oxford University Press, pp. 470–485.
—. 2011a. "Astronomy in Late Babylonian Uruk" in Ruggles, C. L. N. (ed.) *"Oxford IX" International Symposium on Archaeoastronomy Proceedings IAU Symposium No. 278, 2011*. Cambridge: Cambridge University Press, pp. 331–334.
—. 2015. "Mesopotamian Astrological Geography" in Barthel, P., van Kooten, G. (eds.) *The Star of Bethlehem and the Magi. Interdisciplinary Perspectives from Experts on the Ancient Near East, the Greco-Roman World, and Modern Astronomy*. Leiden-Boston: Brill, pp. 201–216.
—. 2017. *Rising Time Schemes in Babylonian Astronomy*. Switzerland: Springer.

—. 2019. "The Early History of the Astronomical Diaries" in Haubold, J., Steele, J., Stevens, K. (eds.) *Keeping Watch in Babylon. The Astronomical Diaries in Context*. Leiden-Boston: Brill, pp. 19–52.

Stephenson, F. R., Walker, C. B. F. (eds.) 1985. *Halley's Comet in History*. London: Trustees of the British Museum.

Steinert, U. 2017. "Cows, Women and Wombs: Interrelations Between Texts and Images from the Ancient Near East" in Kertai, D., Nieuwenhuyse, O. (eds.) *From the Four Corners of the Earth. Studies in Iconography and Cultures of the Ancient Near East in Honour of F. A. M. Wiggermann*. Münster: Ugarit-Verlag, pp. 205–258.

Stern, S. 2012. *Calendars in Antiquity: Empires, States and Societies*. Oxford: Oxford University Press.

Stevens, K. 2019. "From Babylon to Baḫtar: the Geography of the Astronomical Diaries" in Haubold, J., Steele, J., Stevens, K. (eds.) *Keeping Watch in Babylon. The Astronomical Diaries in Context*. Leiden-Boston: Brill, pp. 198–236.

Stockhusen, M. 2019. *Studien zum Transfer astralwissenschaftlicher Konzepte zwischen Ägypten und Mesopotamien in spätpharaonischer Zeit. Eine kulturhistorische Analyse mit einem Ausblick in die griechisch-römische Epoche*. PhD dissertation, Universität Leipzig.

Stol, M. 1992. "The Moon as Seen by the Babylonians" in Meijer, D. J. W. (ed.) *Natural Phenomena: Their Meaning, Depiction, and Description in the Ancient Near East*. Amsterdam: Royal Netherlands Academy of Arts and Sciences, pp. 245–277.

Streck, M. P. 1995. *Zahl und Zeit. Grammatik der Numeralia und des Verbalsystems im Spätbabylonischen*. CM 5. Groningen: Styx.

—. 2001. "Review *If a City Is Set on a Height. The Akkadian Omen Series Šumma Alu ina Mēlê Šakin. Volume 1: Tablets 1–21* by S. M. Freedman". *OLZ* 96/2, pp. 216–222.

—. 2016. "Vogel A". *RlA* 14, pp. 577–580.

—. 2018. *Altbabylonisches Lehrbuch. Dritte, überarbeitete Auflage*. Porta Linguarum Orientalium 23. Wiesbaden: Harrassowitz.

Streck, M. P., Wasserman, N. 2018. "The Man Is Like a Woman, The Maiden Is a Young Man. A New Edition of Ištar-Louvre". *OrNS* 87, pp. 1–38.

Stucken, E. 1896. *Astralmythen der Hebraeer, Babylonier und Ägypte*. Leipzig: E. Pfeiffer.

Swerdlow, N. M. 1998. *The Babylonian Theory on Planets*. Princeton: Princeton University Press.

Szarzyńska, K. 1993. "Offerings for the Goddess Inana in Archaic Uruk". *RA* 87/1, pp. 7–28.

Tadmor, H., Yamada, S. 2011. *The Royal Inscriptions of Tiglat-Pileser III (744–727 BC) and Shalmaneser V (726–722 BC)*. RINAP 1. Winona Lake: Eisenbrauns.

Tendahl, M., Gibbs, R. W. 2008. "Complementary Perspectives on Metaphor: Cognitive Linguistics and Relevance Theory. *Journal of Pragmatics* 40, pp. 1823–1864.

Thavapalan, S. 2020. *The Meaning of Color in Ancient Mesopotamia*. CHANE 104. Leiden-Boston: Brill.

Thavapalan, S., Stenger, J., Snow, C. 2016. "Color and Meaning in Ancient Mesopotamia: The Case of Egyptian Blue". *ZA* 106/2, pp. 198–214.

Thompson, R. G. 1900. *The Reports of the Magicians and Astrologers of Nineveh and Babylon in the British Museum*. London: Luzac & Co.

—. 1936. *A Dictionary of Assyrian Chemistry and Geology*. Oxford: Clarendon Press.
Toomer, G. J. 1996. "Ptolemy and His Greek Predecessors" in Walker, C. (ed.) *Astronomy before the Telescope*. London: British Museum Press, pp. 68–91.
Tuplin, C. 2019. "Logging History in Achaemenid, Hellenistic and Parthian Babylonia: Historical Entries in Dated Astronomical Diaries" in Haubold, J., Steele, J., Stevens, K. (eds.) *Keeping Watch in Babylon. The Astronomical Diaries in Context*. Leiden-Boston: Brill, pp. 79–119.
Ungnad, A. 1941. "Besprechungskunst und Astrologie in Babylonien". *AfO* 14, pp. 251–284.
Urton, G. 1987–2005. "Ethnoastronomy". in Johnson, L. (ed.) *Encyclopedia of Religion (2nd ed.) vol. 5*. Detroit: Thomson Gale, pp. 2862–2866.
Vanstiphout, H. 1995. "The Matter of Aratta: An Overview". *OLP* 26, pp. 5–20.
—. 2003. *Epics of Sumerian Kings: The Matter of Aratta*. WAW 20. Atlanta: SBL Press.
Veldhuis, N. 1991. *A Cow of Sîn*. LOT 2. Groeningen: Styx.
—. 1999. "Continuity and Change in the Mesopotamian Lexical Tradition" in Roest, B., Vantisphout, H. (eds.) *Aspect of Genre and Type in Pre-Modern Literary Cultures*. Groeningen: Styx, pp. 101–118.
—. 2004. *Religion, Literature and Scholarship: The Sumerian Composition Nanše and the Birds, with a Catalogue of Sumerian Bird Names*. CM 22. Leiden-Boston: Brill-Styx.
—. 2006. "How to Classify Pigs: Old Babylonian and Middle Babylonian Lexical Texts" in Lion, B., Michel, C. (eds.) *De la domestication au tabou: Le cas des suidés dans le Proche-Orient ancien*. Paris: De Boccard, pp. 25–29.
—. 2010. "The Theory of Knowledge and the Practice of Celestial Divination" in Annus, A. (ed.) *Divination and Interpretation of Signs in the Ancient World*. Chicago: Oriental Institute of Chicago, pp. 77–91.
—. 2014. *History of the Cuneiform Lexical Tradition*. GMTR 6. Münster: Ugarit-Verlag.
Verderame, L. 2002. *Le tavole I-VI della serie astrologica Enūma Anu Enlil*. Nisaba 2. Roma: Di.Sc.A.M.
—. 2002a. "*Enūma Anu Enlil* tablets 1-13" in Steele, J. M., Imhausen, A. (eds.) *Under One Sky. Astronomy and Mathematics in Ancient Near East*. Münster: Ugarit-Verlag, pp. 447–455.
—. 2014. "The Halo of the Moon" in Fincke, J. C. (ed.) *Divination in Ancient Near East. A Workshop on Divination Conducted during the 54th Rencontre Assyriologique Internationale, Würzburg, 2008*. Winona Lake: Eisenbrauns, pp. 91–104.
—. 2016. "Pleiades in Ancient Mesopotamia". *Mediterranean Archaeology and Archaeometry* 16/4, pp. 109–117.
—. 2017. "On the Early History of the Seven Demons (Sebettu)" in Kertai, D., Nieuwenhuyse, O. (eds.) *From the Four Corners of the Earth. Studies in Iconography and Cultures of the Ancient Near East in Honour of F. A. M. Wiggermann*. Münster: Ugarit-Verlag, pp. 283–296.
—. 2017a. "The Seven Attendants of Hendursaĝa: A Study of Animal Symbolism in Mesopotamian Cultures". *SANER* 12, pp. 396–415.
Virolleaud, C. 1908–1912. *L'astrologie chaldéenne. Le livre intitulé «enuma <Anu> iluBêl»*. Paris: Paul Geuthner.

Volk, K. 1995. *Inanna und Šukaletuda: Zur historisch-politischen Deutung einers sumerischen Literaturwerkes*. SANTAG 3. Wiesbaden: Harrassowitz.
Wainer, Z. 2016. "Traditions of Mesopotamian Celestial-Divinatory Schemes and the 4th Tablet of *Šumma Sin ina Tāmartišu*", in Steele, J. M. (ed.) *The Circulation of Astronomical Knowledge in the Ancient World*. Leiden-Boston: Brill, pp. 55–82.
Walker, C. 1983. "The Myth of Girra and Elamatum". *AnSt.* 33, pp. 145–152.
—. (ed.) 1996. *Astronomy before the Telescope*. London: British Museum Press.
Walker, C. B., Bromhead, F., Hunger, H. 1977. "Zwölfmaldrei". *MDOG* 109, pp. 27–34.
Ward, W. H. 1910. *The Seal Cylinders of Western Asia*. Washington: Carnegie Institution of Washington.
Wardle, D. (ed.) 2006. *Cicero on Divination. Book 1*. Clarendon Ancient History Series 1. Oxford: Clarendon Press.
Warner, B. 1996. "Traditional Astronomical Knowledge in Africa" in Walker, C. (ed.) *Astronomy before the Telescope*. London: British Museum Press, pp. 318–328.
Wasserman, N. 1999. "An Allusion to the Epic of Gilgamesh in a Ritual to Ištar". *NABU* 1999 n. 81.
Watson, R., Horowitz, W. 2011. *Writing Science before the Greeks*. CHANE 48. Leiden-Boston: Brill.
Wee, J. Z. 2012. *The Practice of Diagnosis in Mesopotamian Medicine: With Editions of Commentaries on the Diagnostic Series Sa-gig*. PhD dissertation, Yale University.
—. 2014. "Grieving with the Moon: Pantheon and Politics in The Lunar Eclipse". *JANER* 14, pp. 29–67.
—. 2014a. "Lugalbanda Under the Night Sky: Scenes of Celestial Healing in Ancient Mesopotamia". *JNES* 73/1, pp. 23–42.
—. 2016. "Virtual Moons over Babylonia: The Calendar Text System, Its Micro-Zodiac of 13, and the Making of Medical Zodiology" in Steele, J. M. (ed.) *The Circulation of Astronomical Knowledge in the Ancient World*. Leiden-Boston: Brill, pp. 139–229.
Weichenhan, M. 2016. *Der Panbabylonismus. Die Faszination des himmelischen Buches im Zeitalter der Zivilisation*. Berlin: Frank & Timme.
Weidner, E. F. 1915. *Handbuch der babylonischen Astronomie. Erster Band*. Leipzig: J.C. Hinrichs'sche Buchhandlung.
—. 1919. "Babylonische Hypsomatabilder". *OLZ* 22, pp. 10–16.
—. 1925. "Ein astrologischer Kommentar aus Uruk". *StOr.* 1, pp. 347–358.
—. 1936–1937. "Die 84 Tafel der Serie *šumma âlu ina mêlê šakin*". *AfO* 11, pp. 358–360.
—. 1938. "Enmešarra am Himmel". *RlA* 2, pp. 397–398.
—. 1941–1944. "Die astrologische Serie Enûma Anu Enlil". *AfO* 14, pp. 172–195, 308–318.
—. 1954–1956. "Die astrologische Serie Enûma Anu Enlil". *AfO* 17, pp. 71–89.
—. 1959–1960. "Ein astrologischer Sammeltext aus der Sargonidenzeit". *AfO* 19, pp. 105–113.
—. 1967. *Gestirn-Darstellungen auf babylonischen Tontafeln*. SbWien 254/2. Graz: Hermann Böhlaus Nachf.
—. 1968–1969. "Die astrologische Serie Enûma Anu Enlil". *AfO* 22, pp. 65–75.
von Weiher, E. 1971. *Der babylonische Gott Nergal*. AOAT 11. Kevelaer-Neukirchen-Vluyn: Butzon & Bercker.

West, M. L. 1988. *Hesiod Theogony and Works and Days. Translated with an Introduction and Notes by M. L. West*. Oxford-New York: Oxford University Press.
Wiggermann, F. A. M. 1992. *Mesopotamian Protective Spirits: The Ritual Texts*. CM 1. Leiden-Boston: Brill.
—. 1998–2001. "Nergal A. Philologisch". *RlA* 9, pp. 215–223.
—. 2005. "Pašittu". *RlA* 10, pp. 363–364.
—. 2011. "The Mesopotamian Pandemonium: A Provisional Census" in Verderame, L., Capomacchia, A. M. G. (eds.) *Demoni Mesopotamici*. Brescia: Morcelliana, pp. 298–322.
—. 2011a. "Siebengötter A". *RlA* 12, pp. 459–466.
—. 2013. "Sumuqan A". *RlA* 13, pp. 308–309.
Wilcke, C. 1969. *Das Lugalbandaepos*. Wiesbaden: Harrassowitz.
—. 1980. "Inanna/Ištar A". *RlA* 5, pp. 74–87.
Wilhelm, G. 1997. "Menschenfresser". *RlA* 8, p. 60.
Williams, C. 2002. "Signs from the Sky, Signs from the Earth" in Steele, J. M., Imhausen, A. (eds.) *Under One Sky. Astronomy and Mathematics in Ancient Near East*. Münster: Ugarit-Verlag, pp. 473–495.
Winitzer, A. 2017. *Early Mesopotamian Divination Literature: Its Organizational Framework and Generative and Paradigmatic Characteristics*. AMD 12. Boston: Brill.
Wiseman, D. J. 1969. "A Lipšur Litany from Nimrud". *Iraq* 31/2, pp. 24–44.
Wisnom, S. 2020. *Weapons of Words: Intertextual Competition in Babylonian Poetry*. CHANE 106. Leiden-Boston: Brill.
Woods, C. 2006. "Bilingualism, scribal learning, and the death of Sumerian" in Sanders, S. (ed.) *Margins of Writing. Oriental Institute Seminars 2*. Chicago: University of Chicago Press, pp. 95–124.
Yildiz, F., Gomi, T. 1988. *Die Puzriš-Dagan-Texte der Istanbuler Archäologischen Museen. Teil II: nr. 726-1379*. FAOS 16. Wiesbaden: Franz Steiner.
Young, S. 1987. "Stars" in Jones, L. (ed.) *Encyclopedia of Religion (2nd ed.) vol. 13*. Detroit: Thomson Gale, pp. 8733–8736.

Plates

Plate 1

K 3918 + K 6239

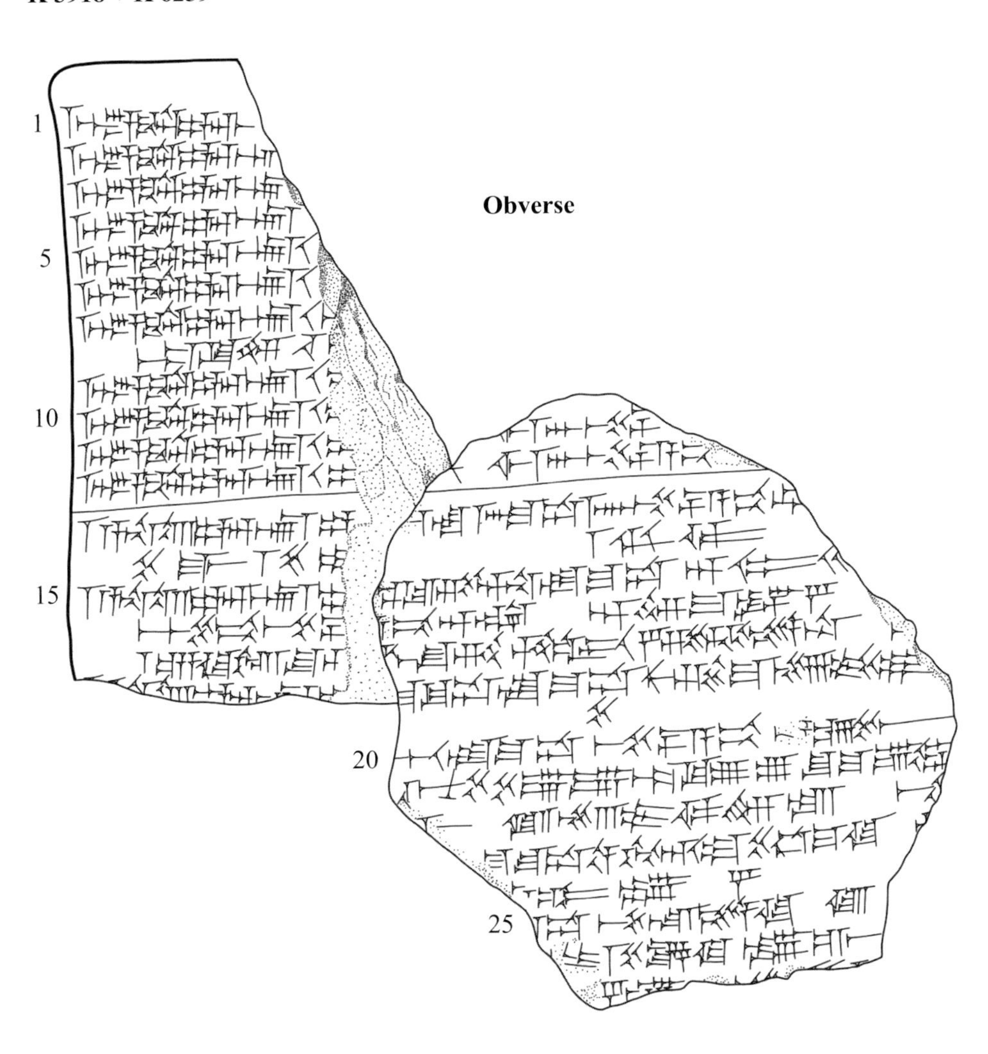

cm

Plate 2

Sm 319

Obverse

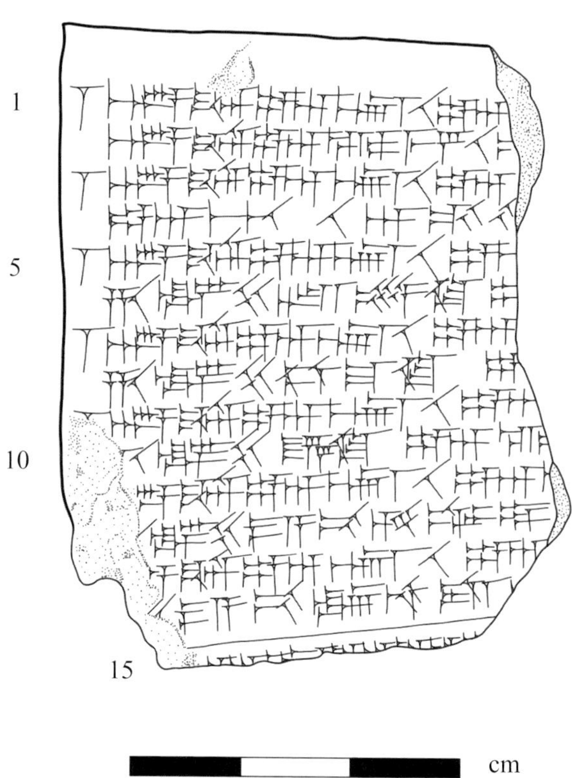

Plate 3

Sm 319

Reverse

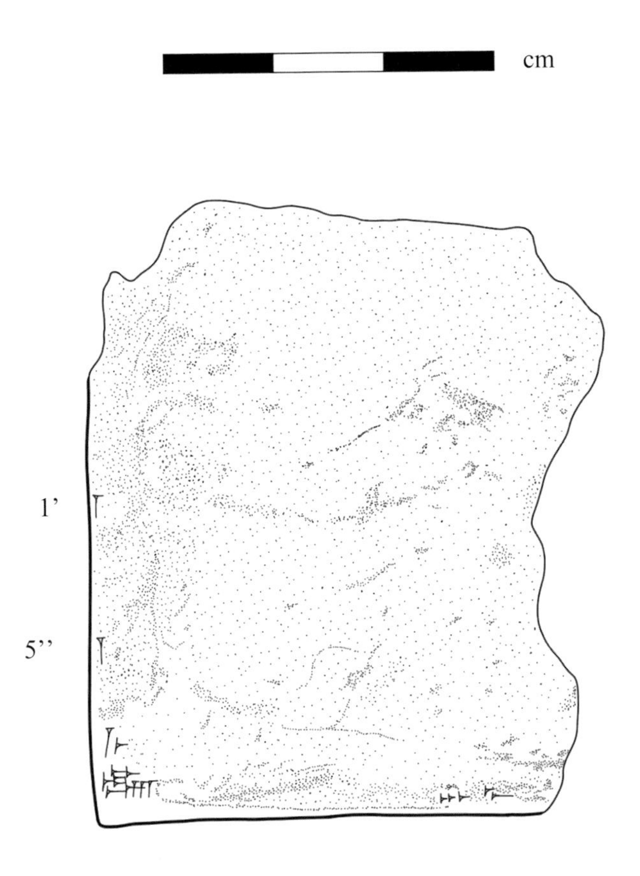

Plate 4

K 2118

Obverse

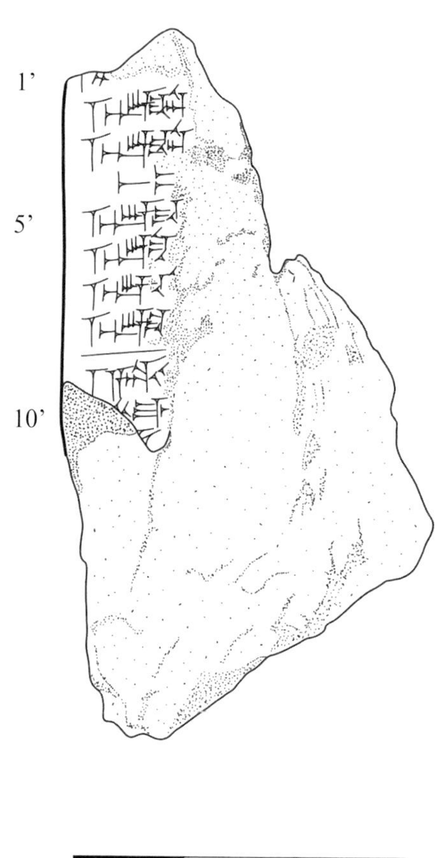

Plate 5

K 2118

Reverse

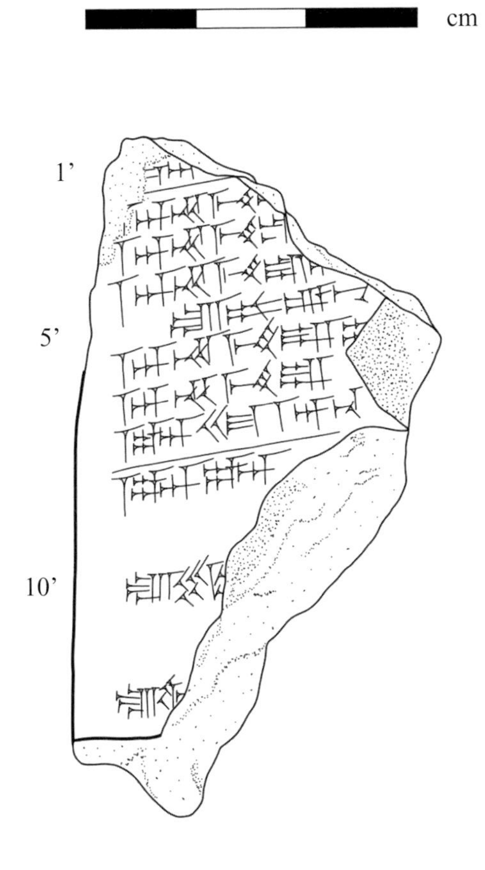

Plate 6

K 11632

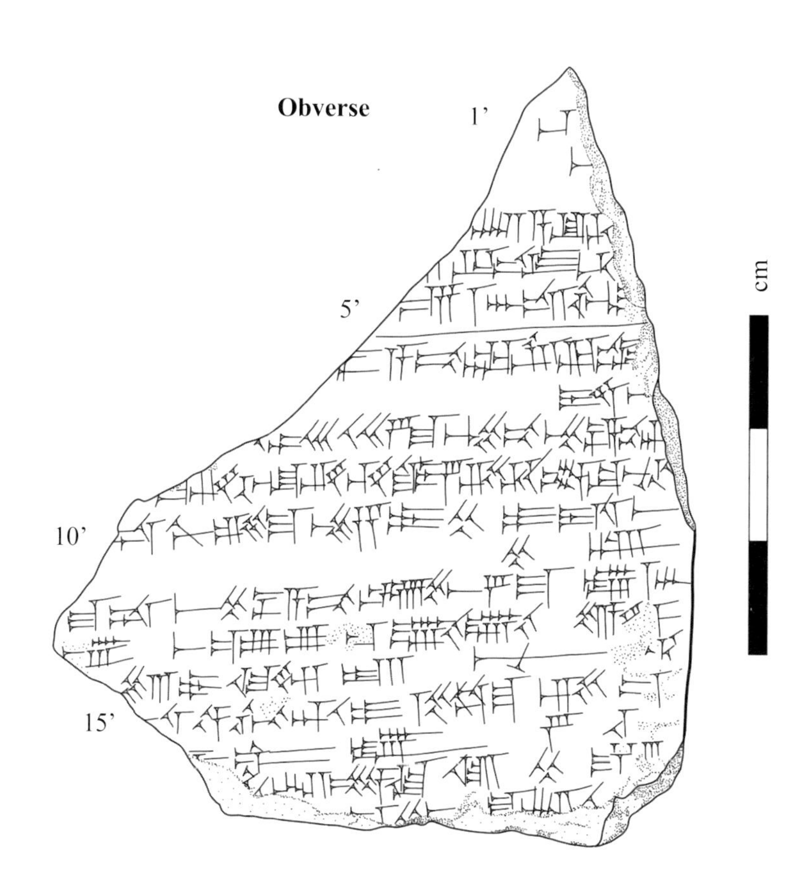

Plate 7

K 3099 + K 18689 (+) Sm 259 (1: 1)

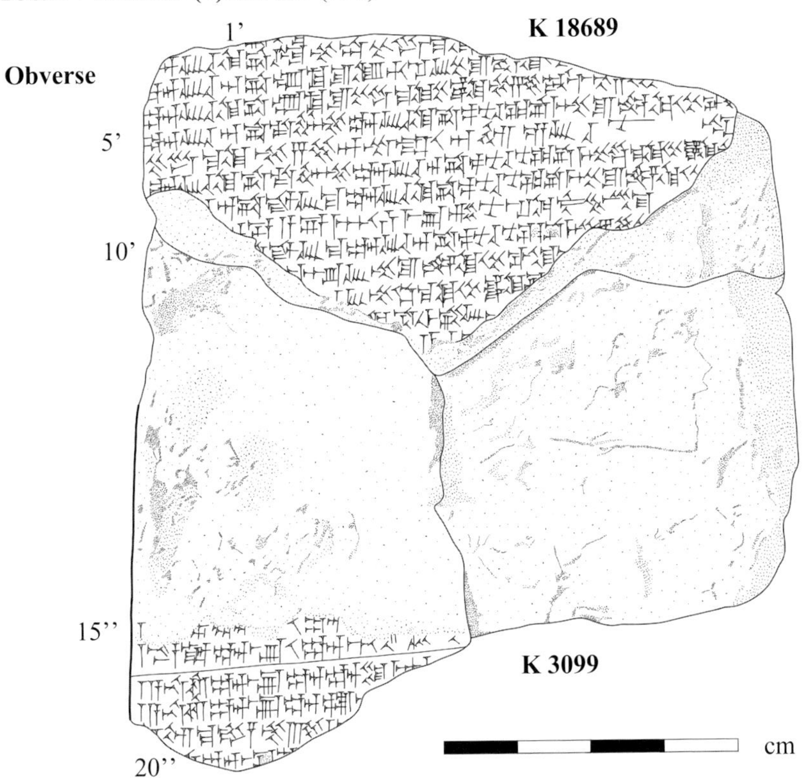

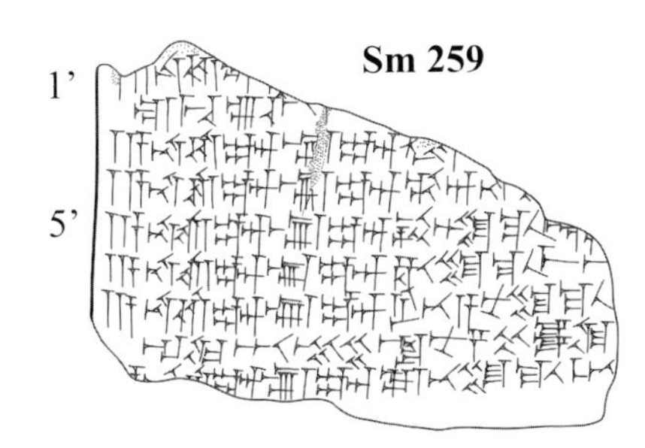

Lower edge

Plate 8

K 3099 + K 18689 (+) Sm 259 (1: 1)

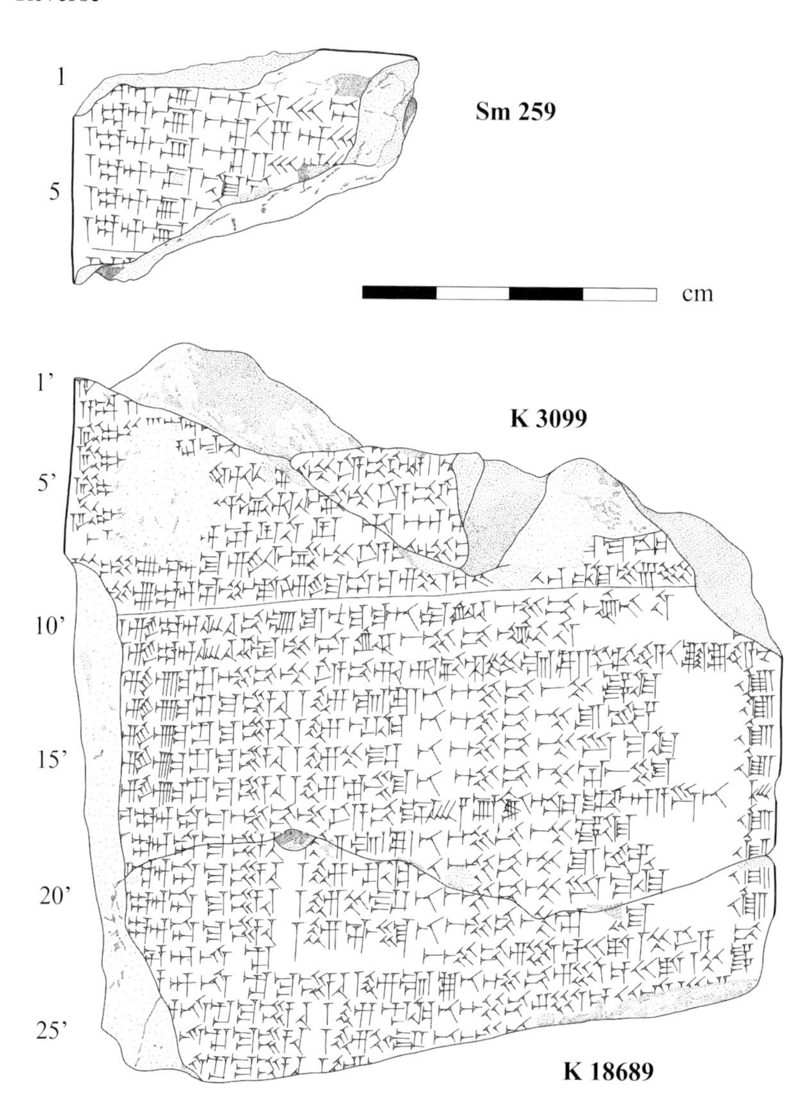

Plate 9

Rm 100

Reverse (?)

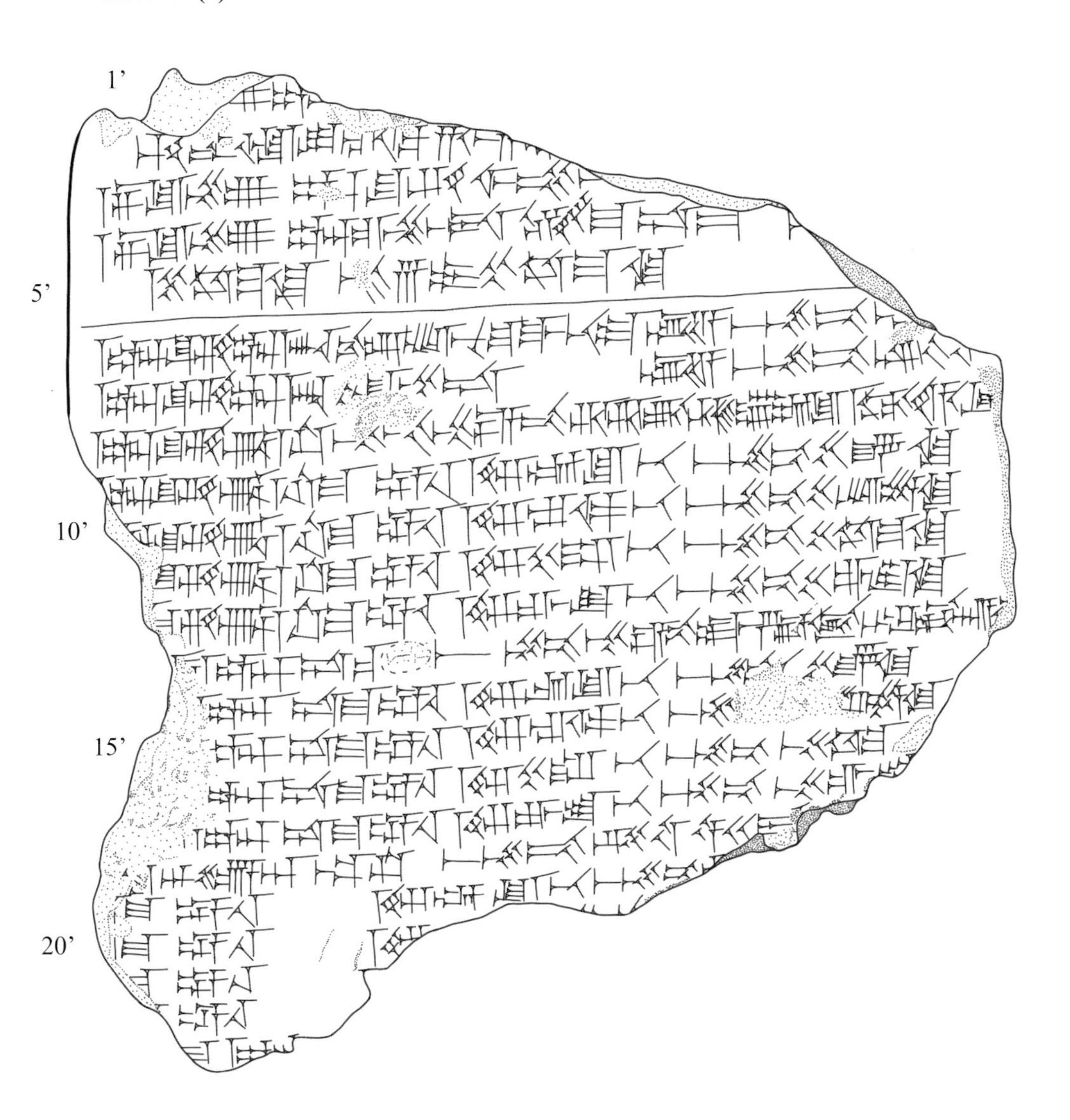

Plate 10

K 7214

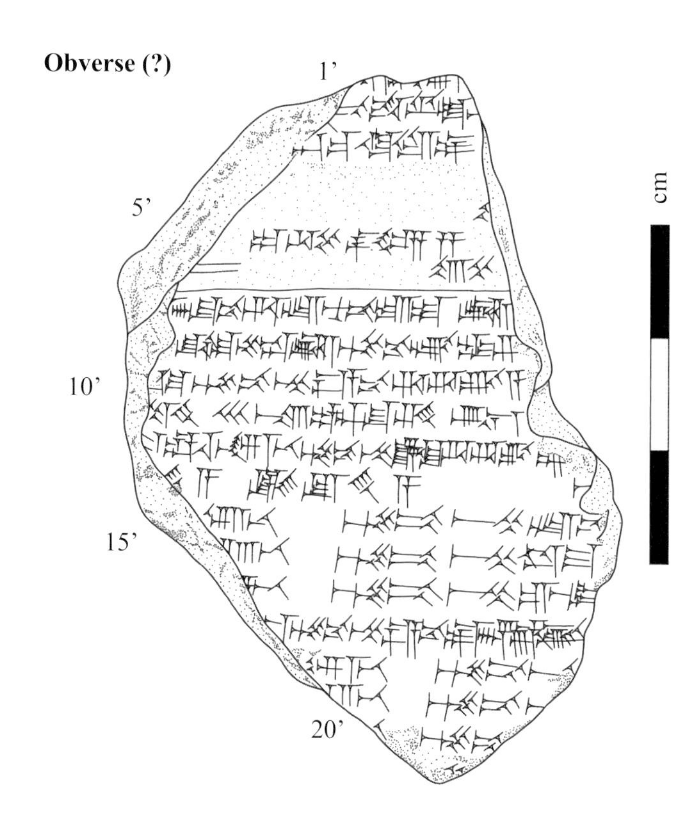

K 10845

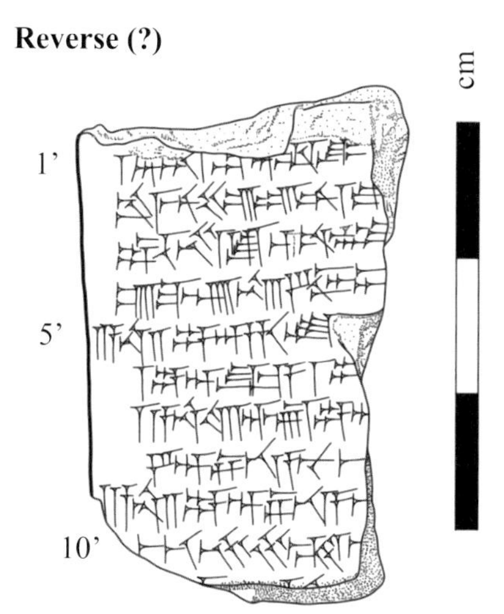

Plate 11

K 11324 + K 12705

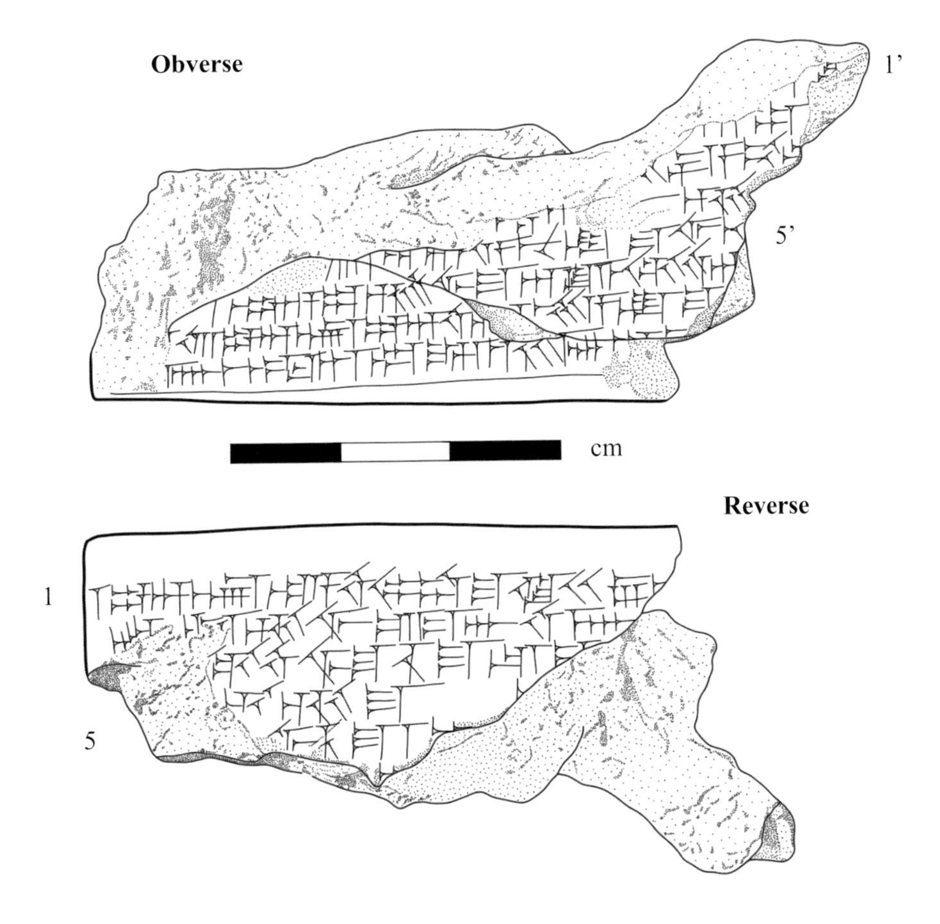

Plate 12

K 3524

Reverse (?)

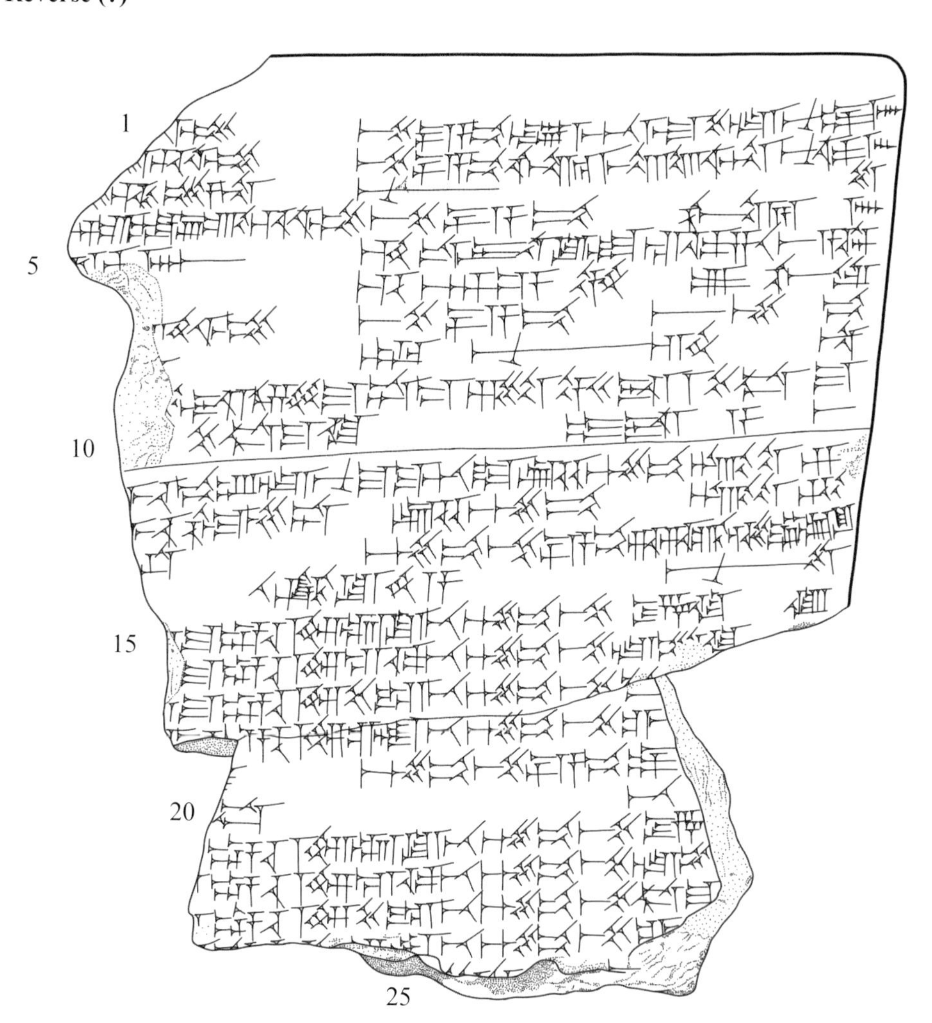

Plate 13

K 7986

Obverse (?)

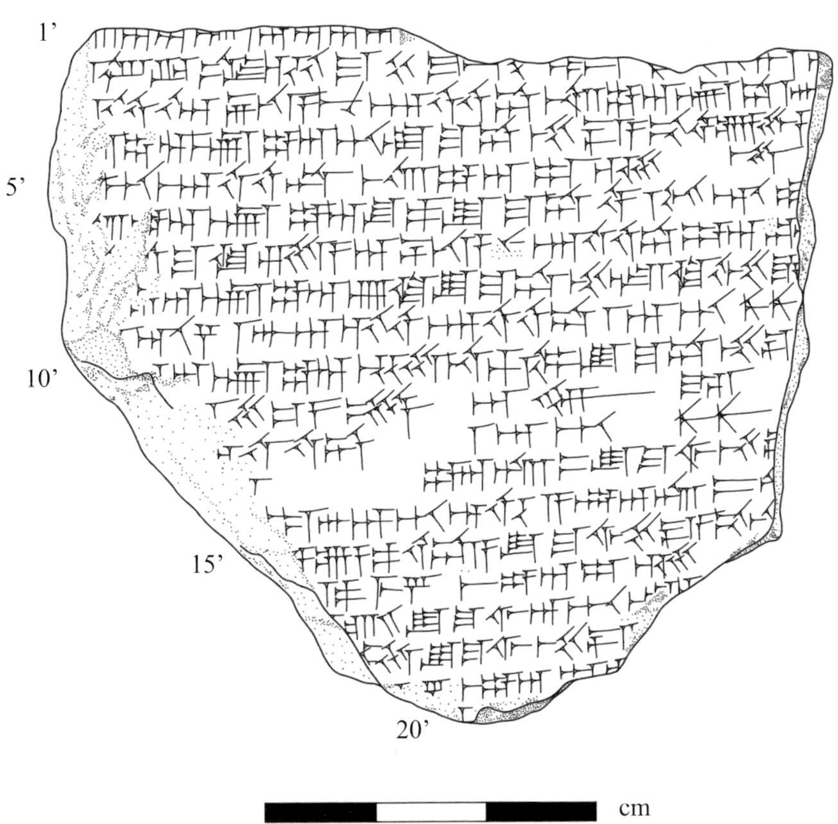

Plate 14

Sm 1317

Obverse

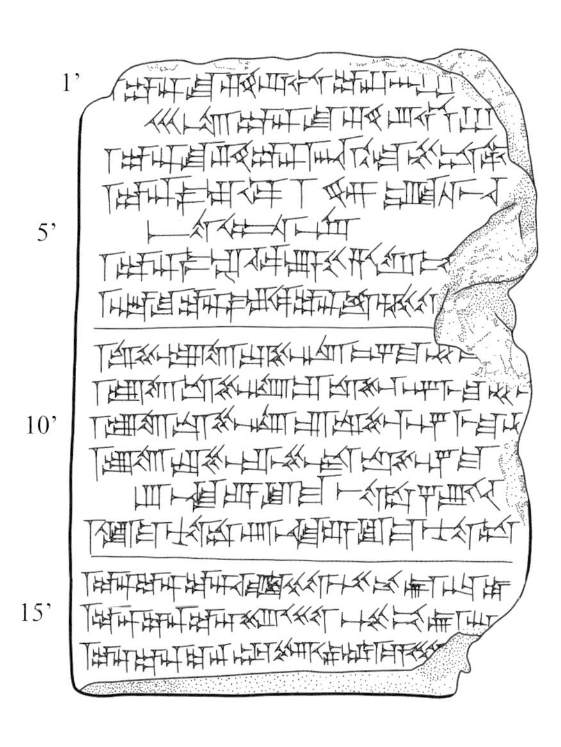

Plate 15

Sm 1317

Reverse

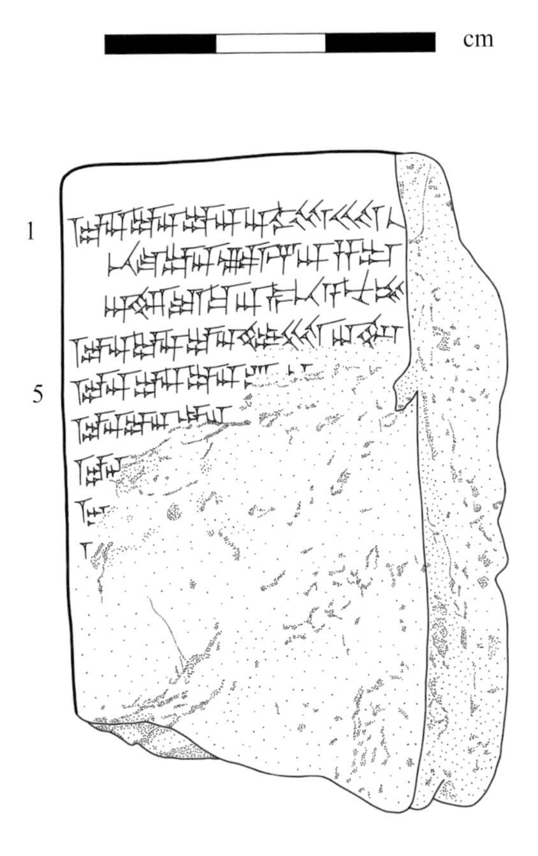

Plate 16

K 3558

Obverse

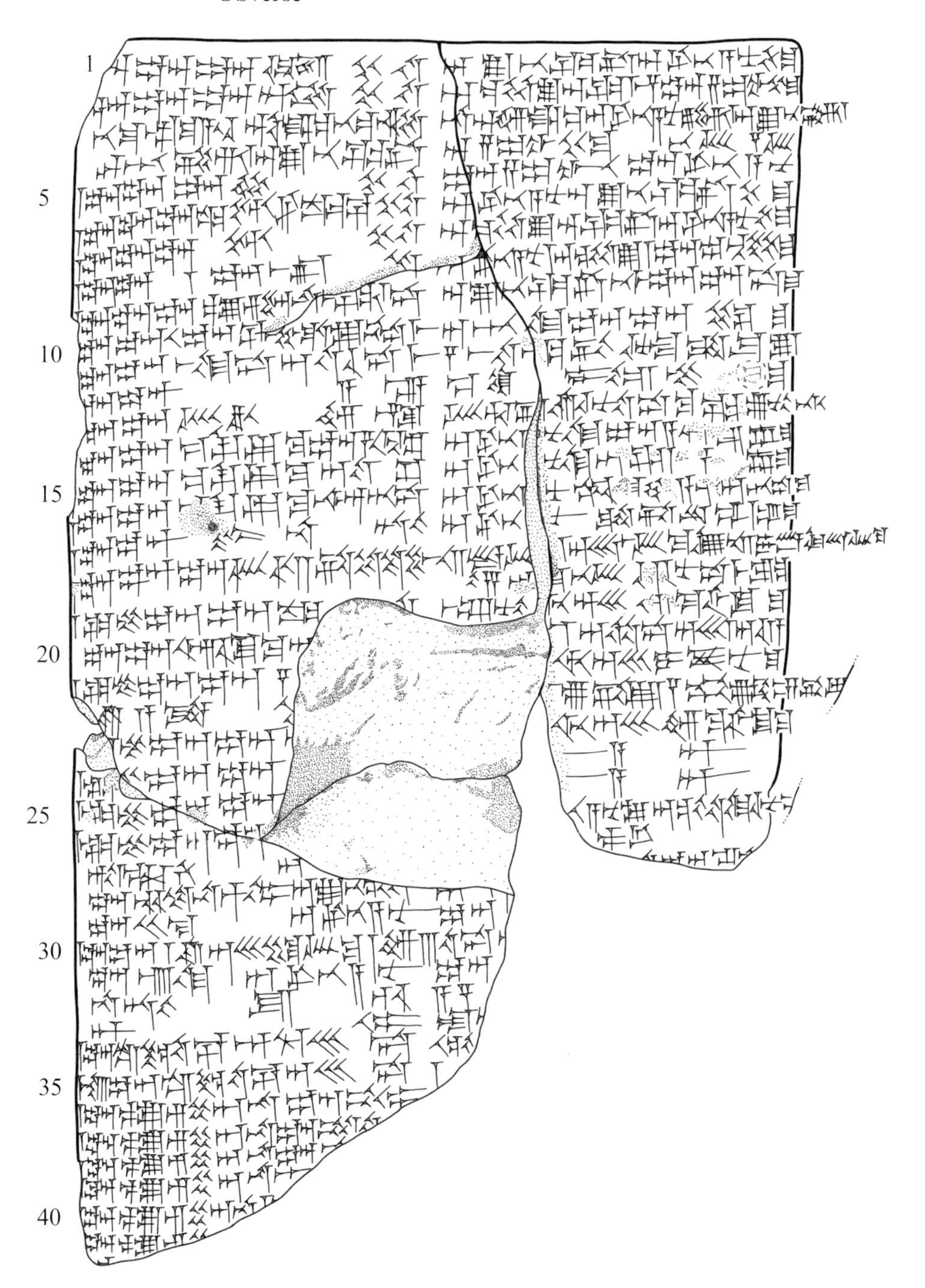

Plate 17

K 3558

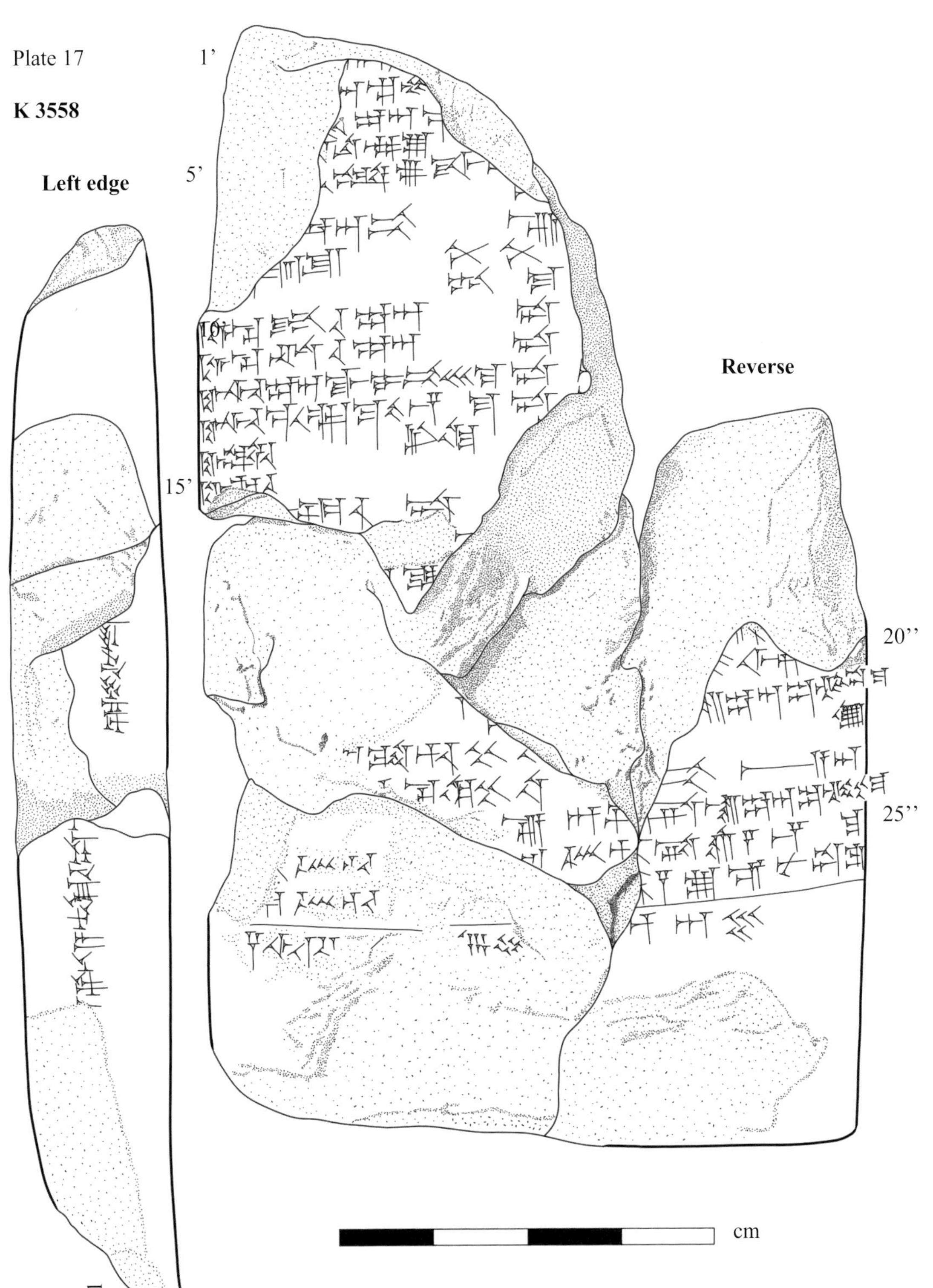

Plate 18

K 8744

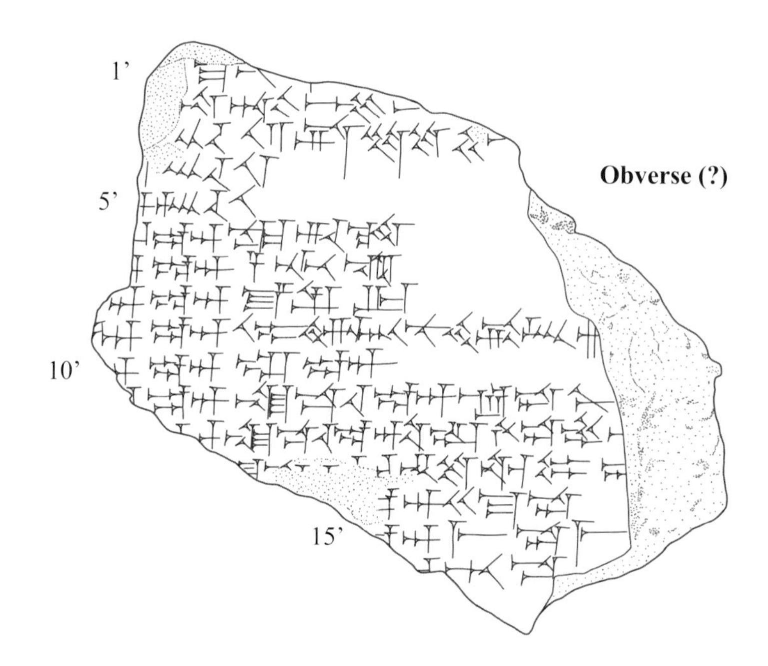

Sm 1946

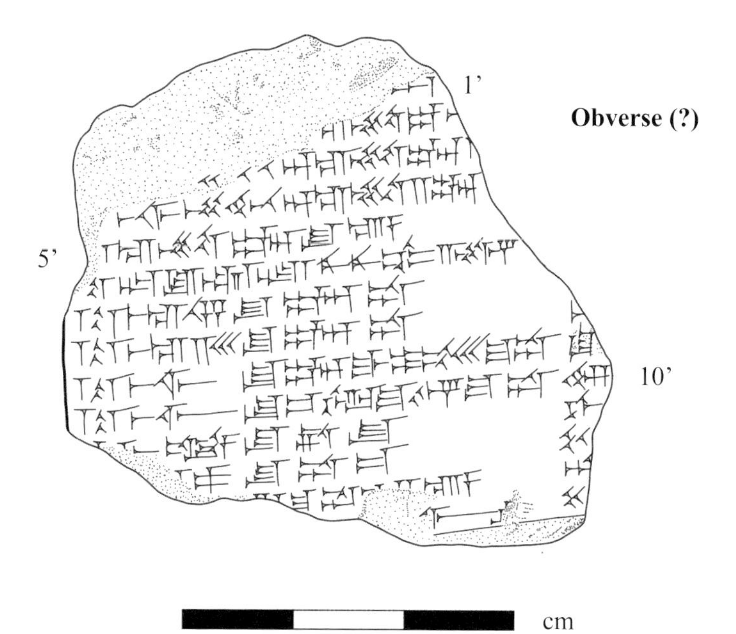

Plate 19

Sm 1349

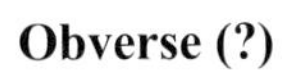

Obverse (?)

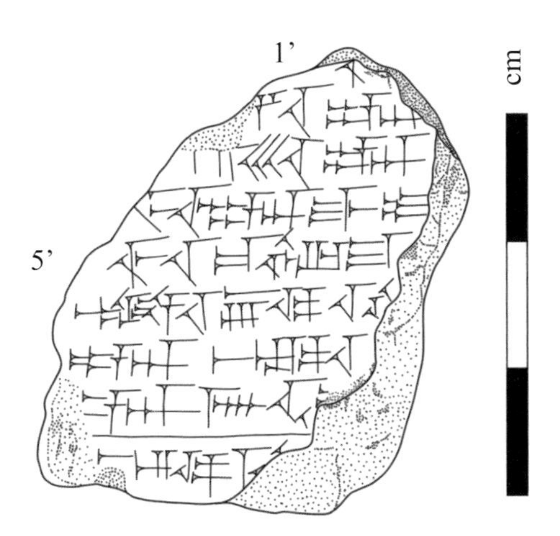

Sm 197

Obverse (?)

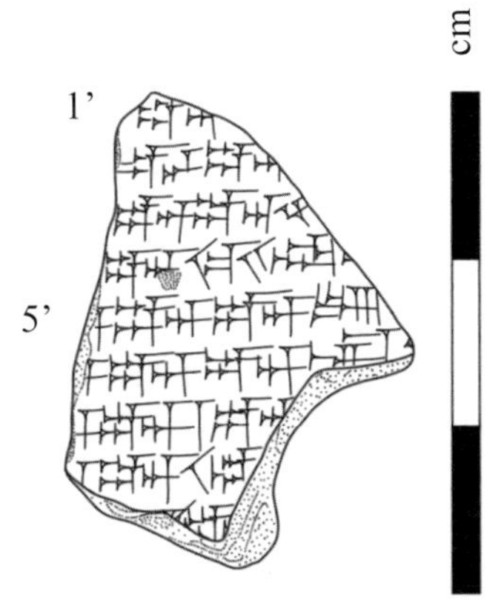

Plate 20

K 11001 + K 15541

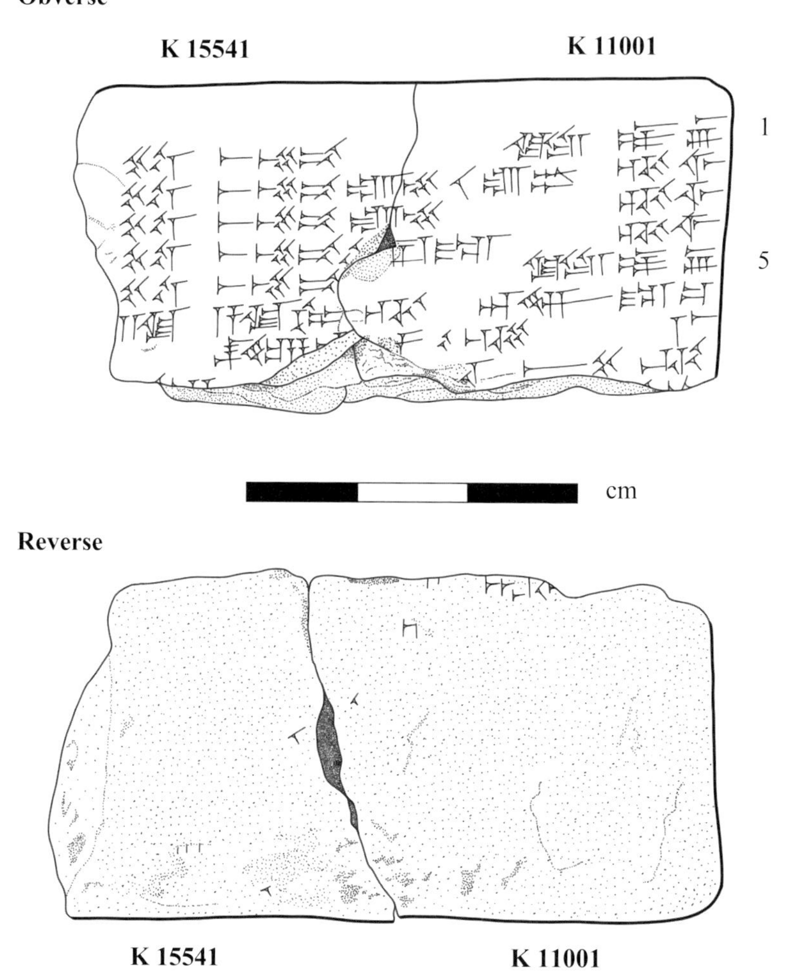

Plate 21

K 3923 + K 6140 + 81-7-27, 149 + 83-1-18, 479

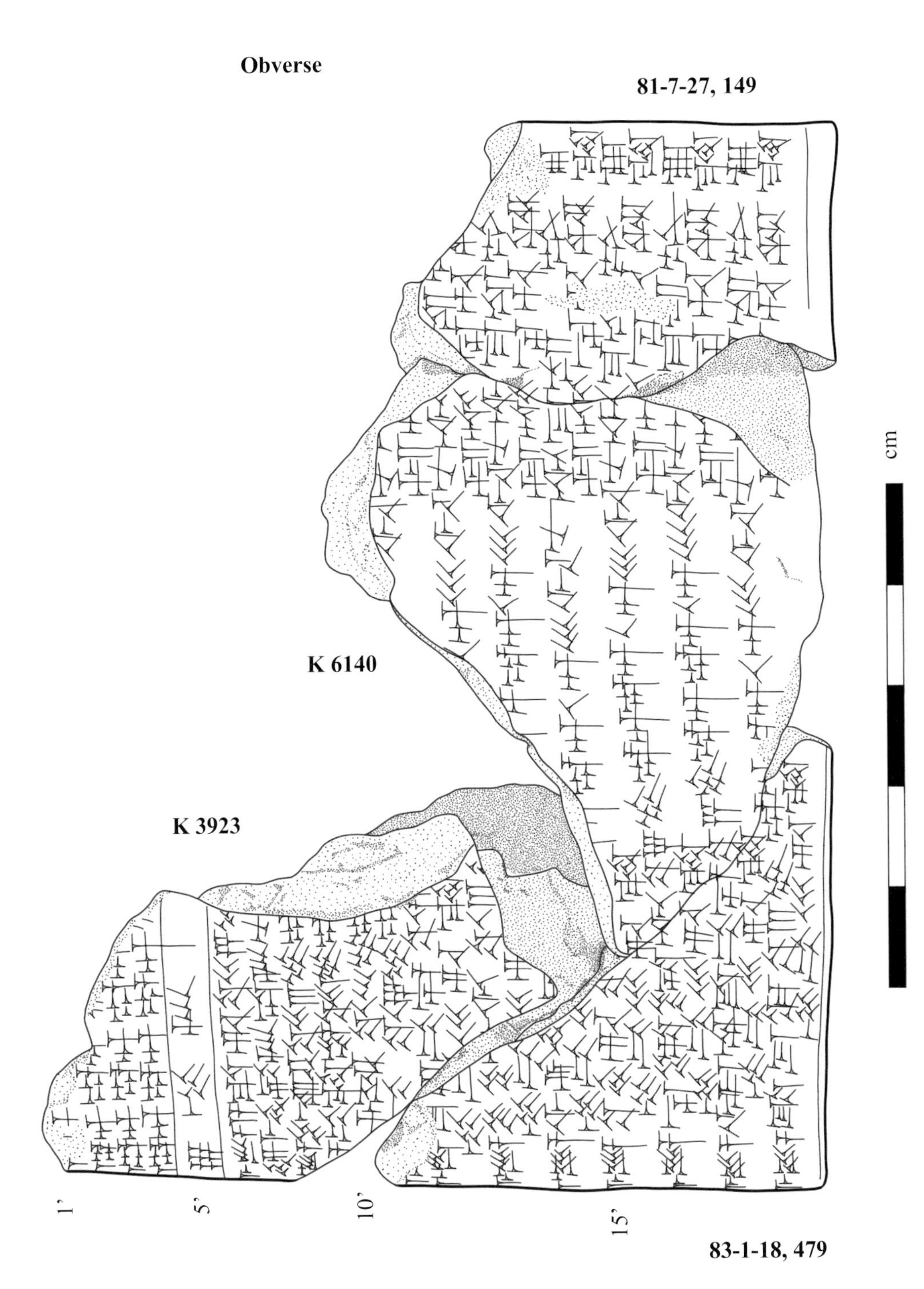

Plate 22

K 3923 + K 6140 + 81-7-27, 149 + 83-1-18, 479

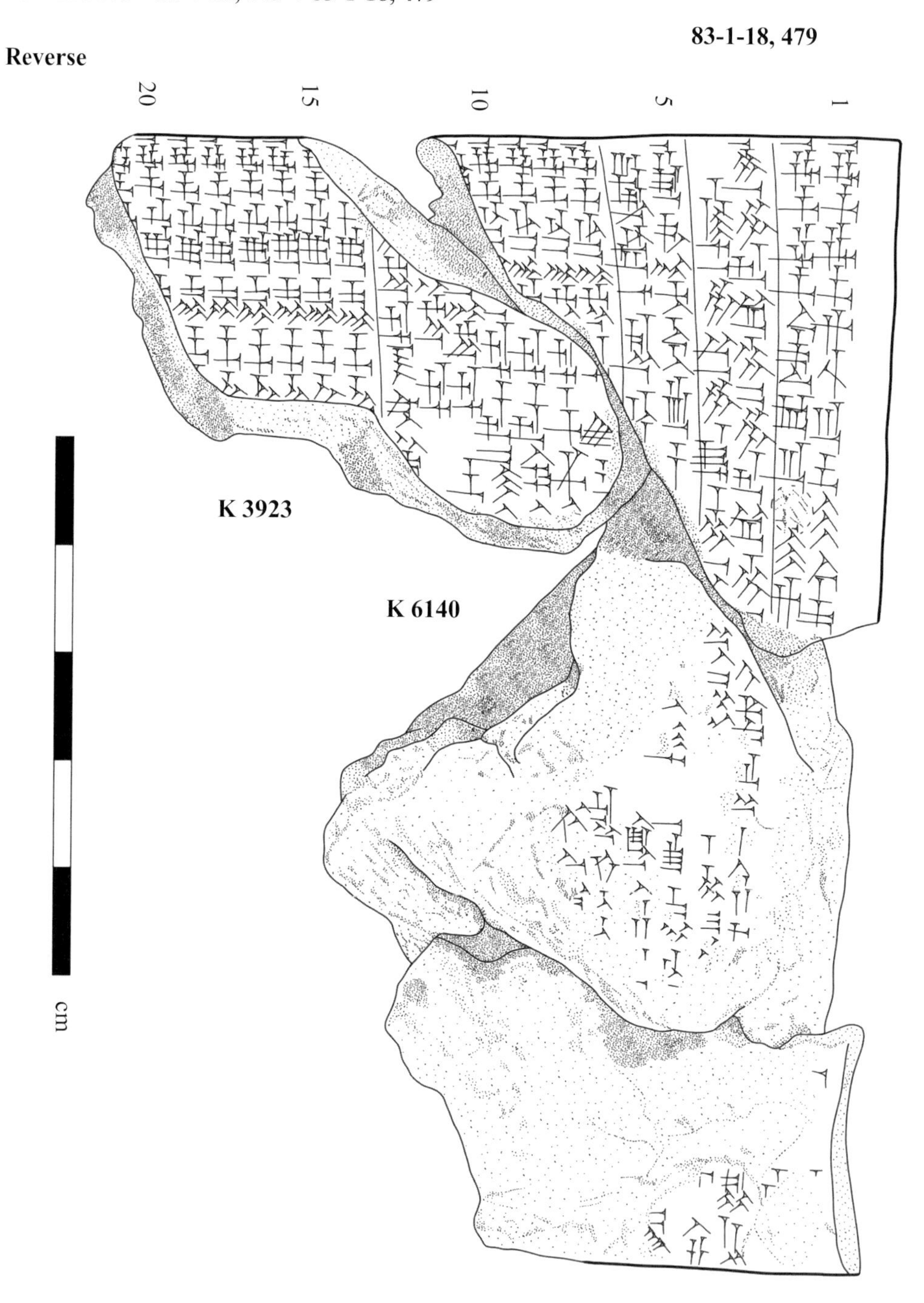

Plate 23

Sm 1054

Obverse

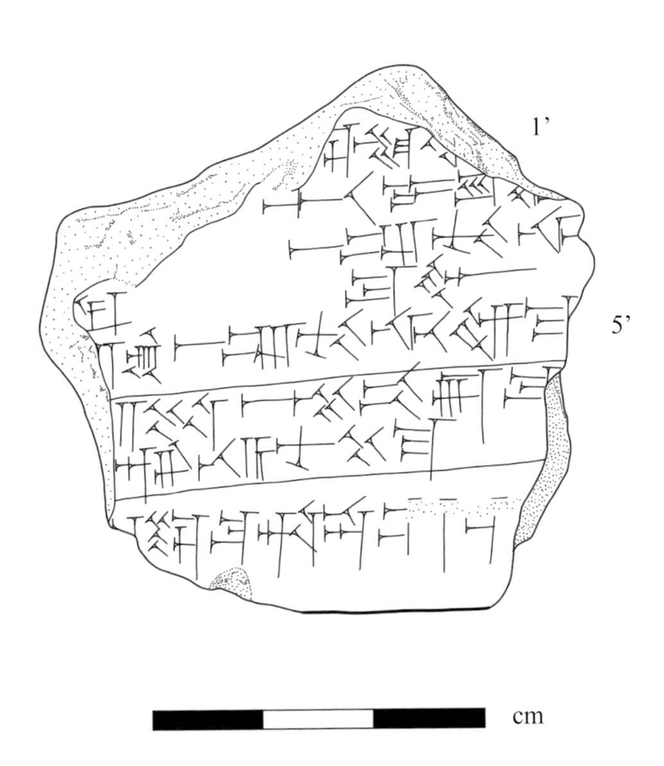

Plate 24

Sm 247

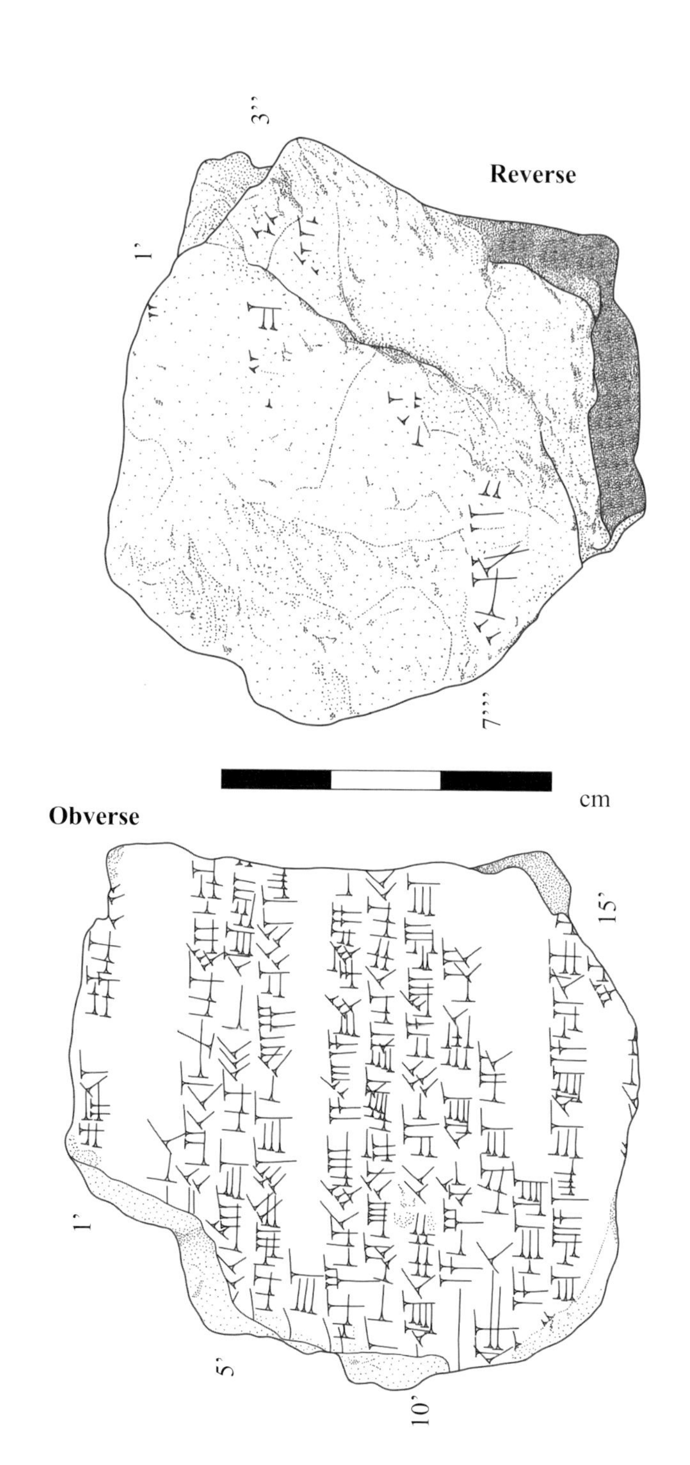

Plate 25

BM 38301

Obverse

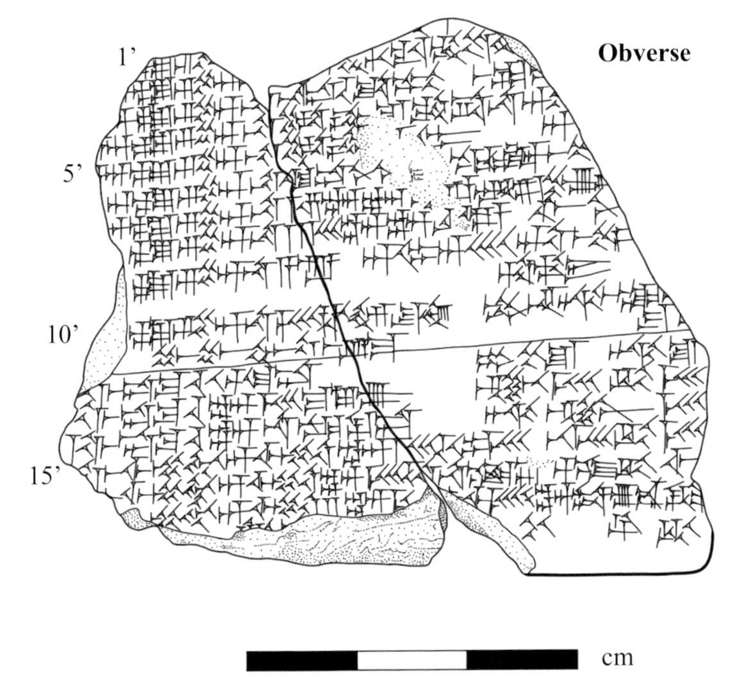

cm

Reverse

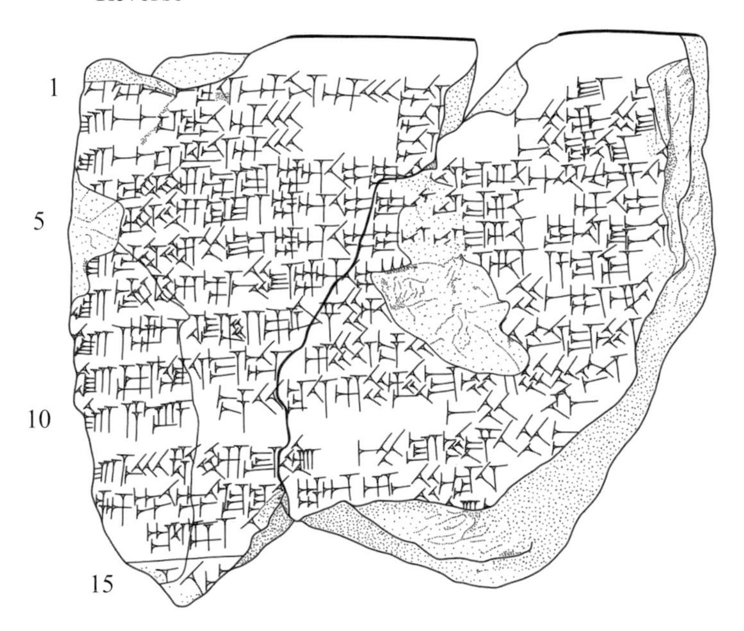

Plate 26

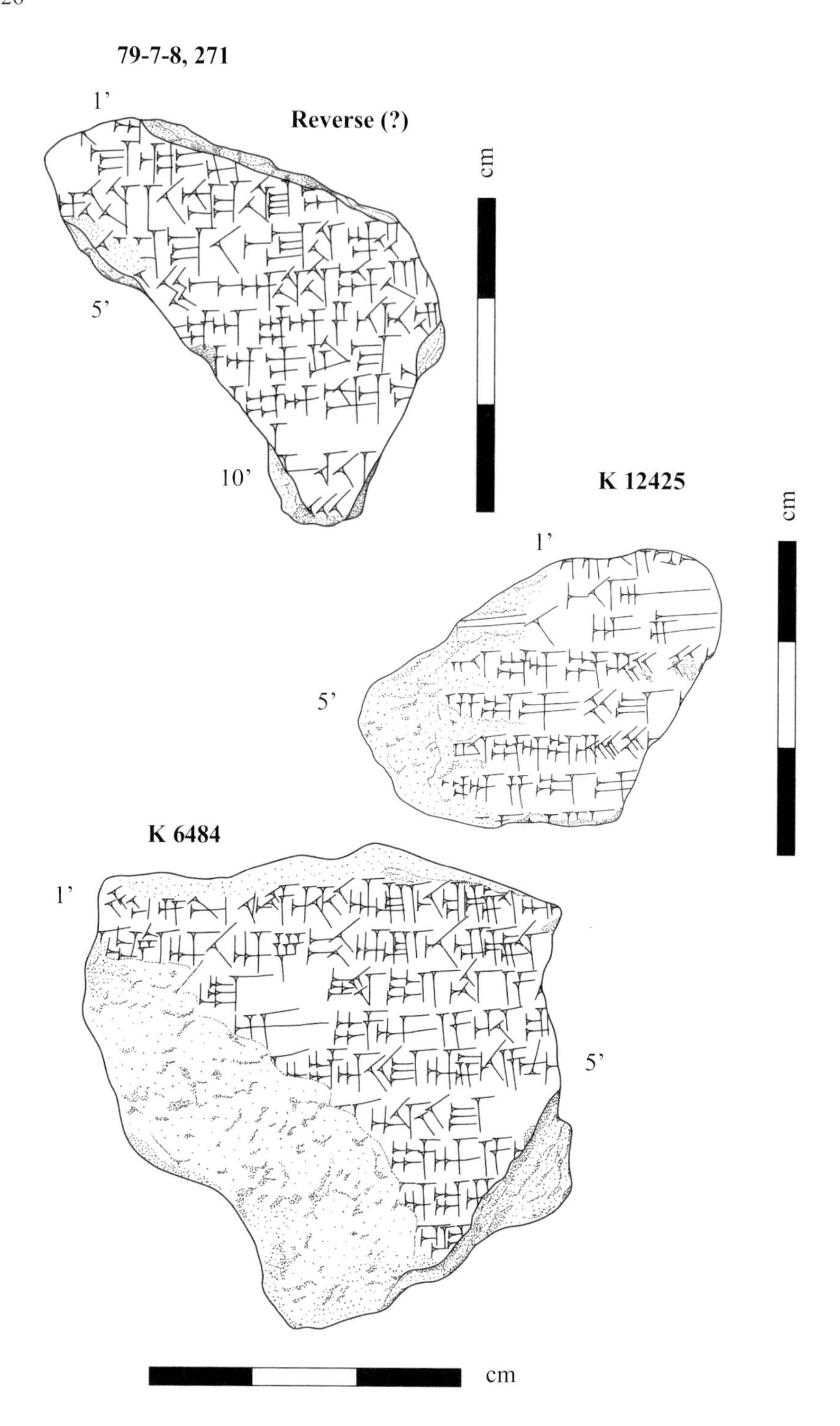
79-7-8, 271
Reverse (?)
1'
5'
10'
cm
K 12425
1'
5'
cm
K 6484
1'
5'
cm

Plate 27

K 5713 + K 7129 + Rm 2, 114 (1: 1)

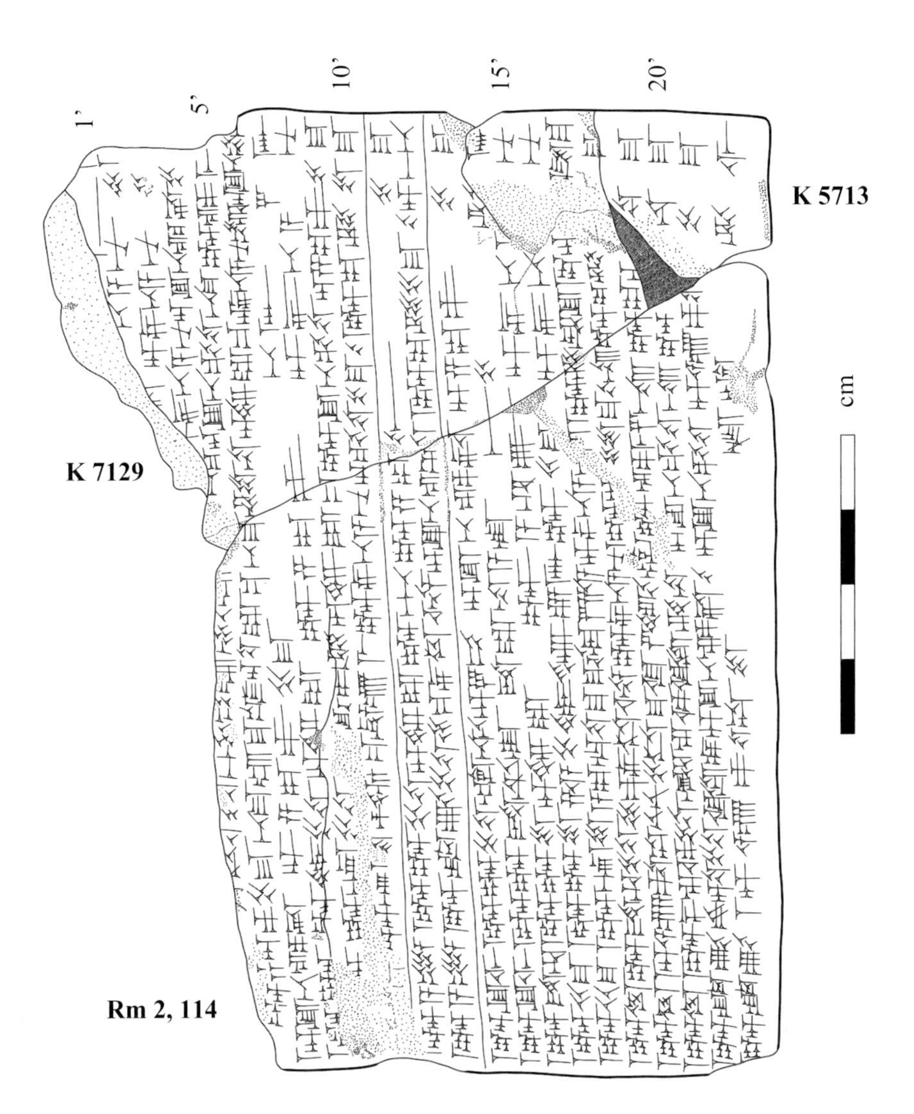

Plate 28

K 5713 + K 7129 + Rm 2, 114 (1: 1)

Reverse

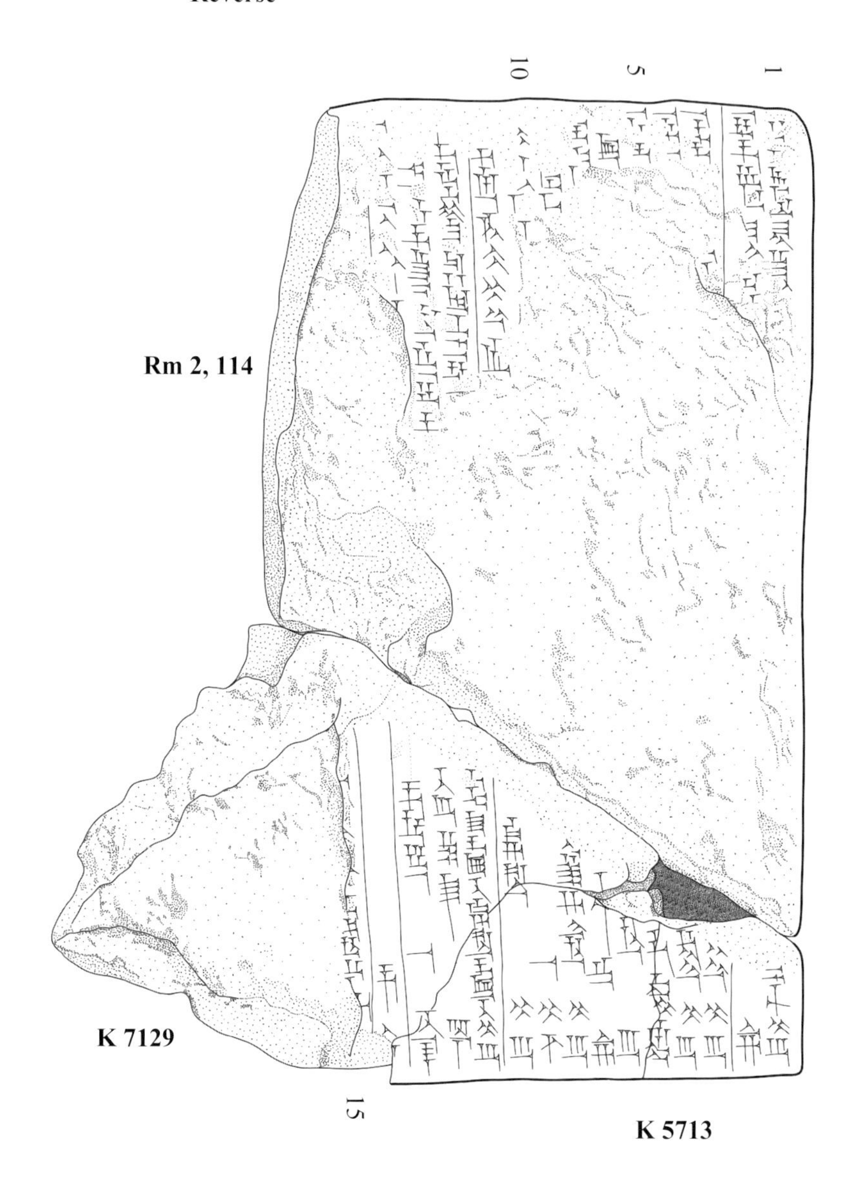

Plate 29

K 2177 + K 7869 + Rm 473 (1: 1)

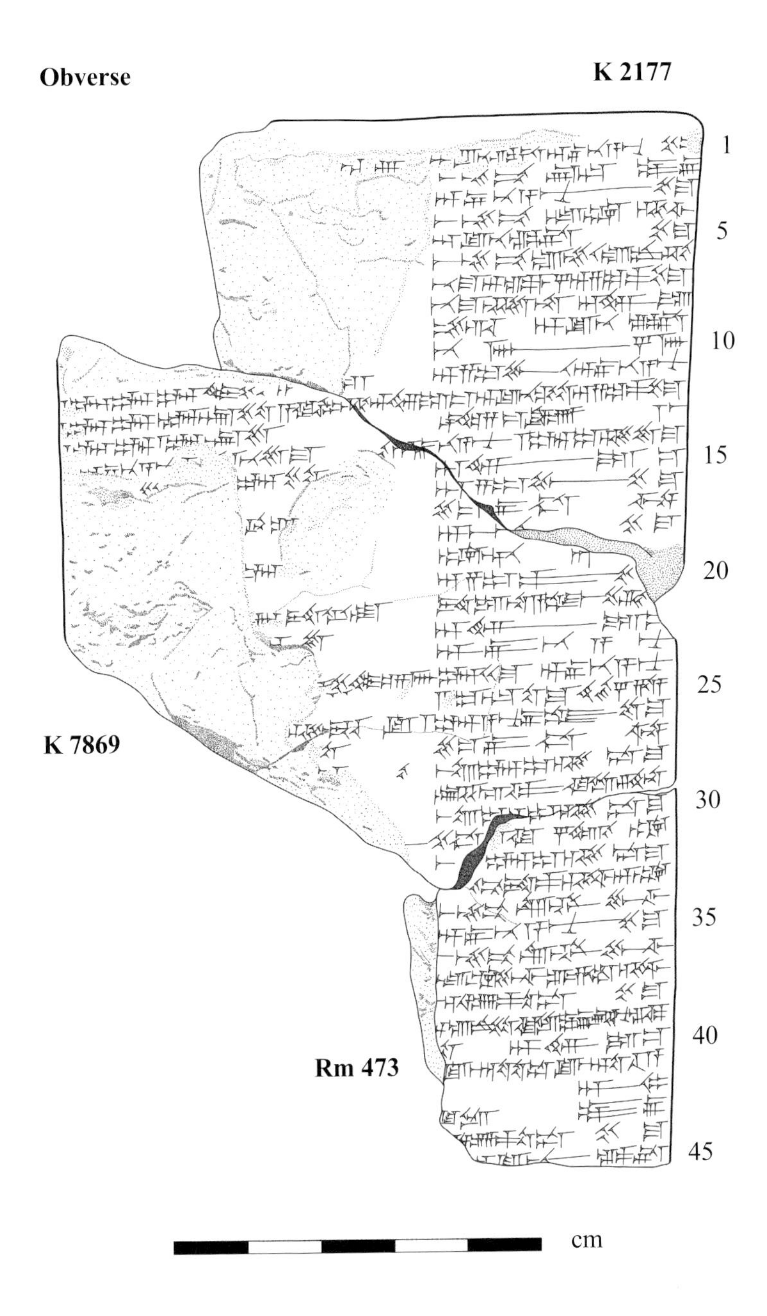

cm

Plate 30

K 2177 + K 7869 + Rm 473 (1: 1)

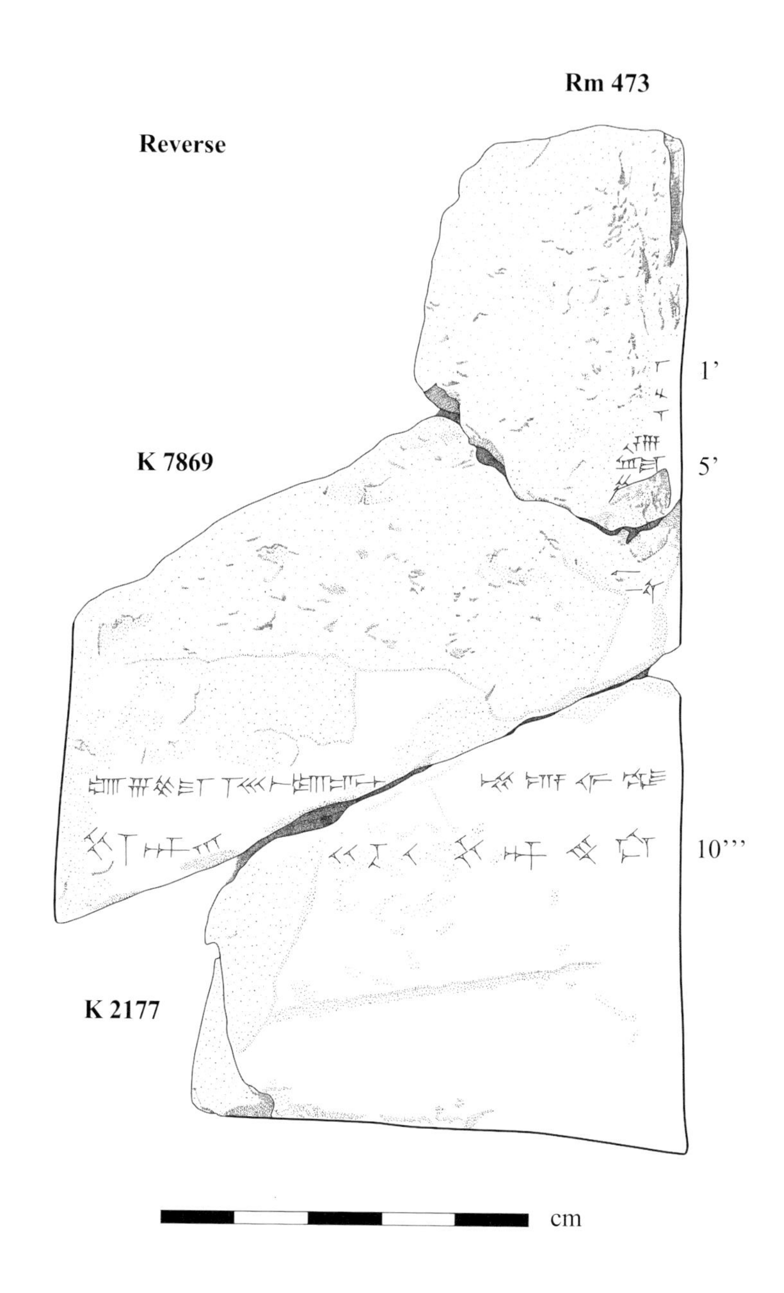

Plate 31

K 2170 +K 3629 (1: 1)

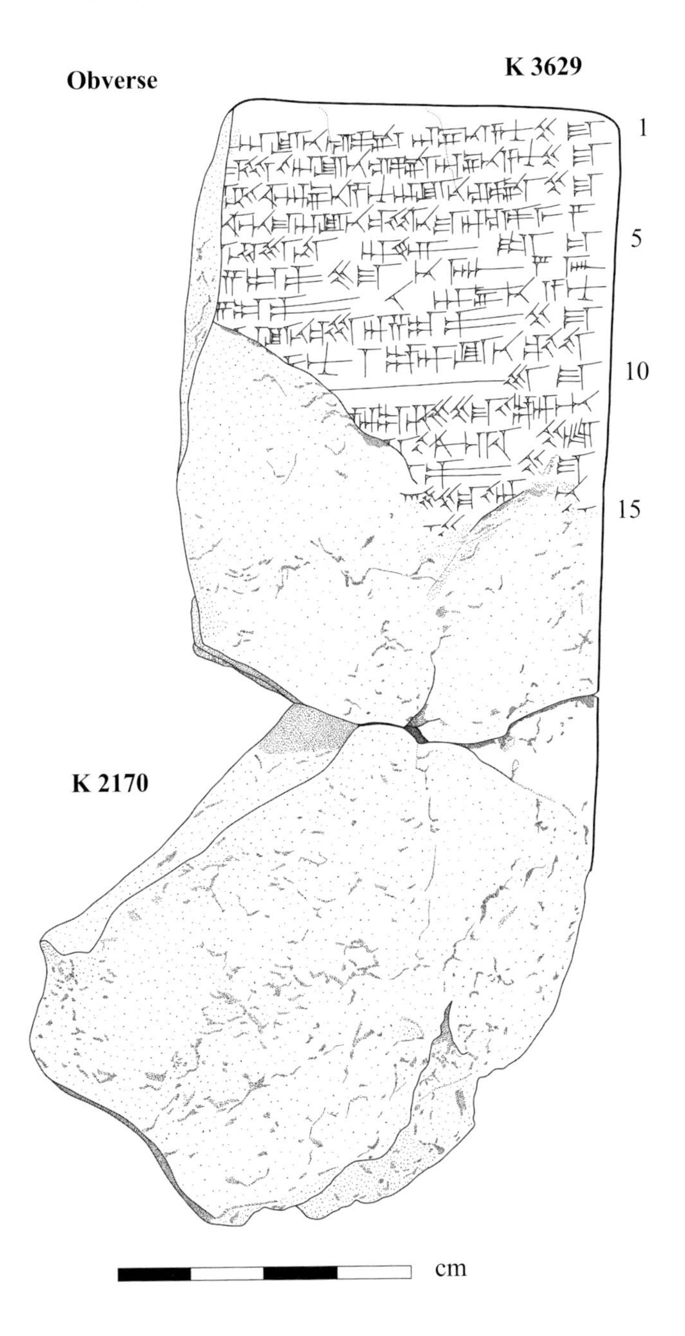

Plate 32

K 2170 + K 3629 (1: 1)

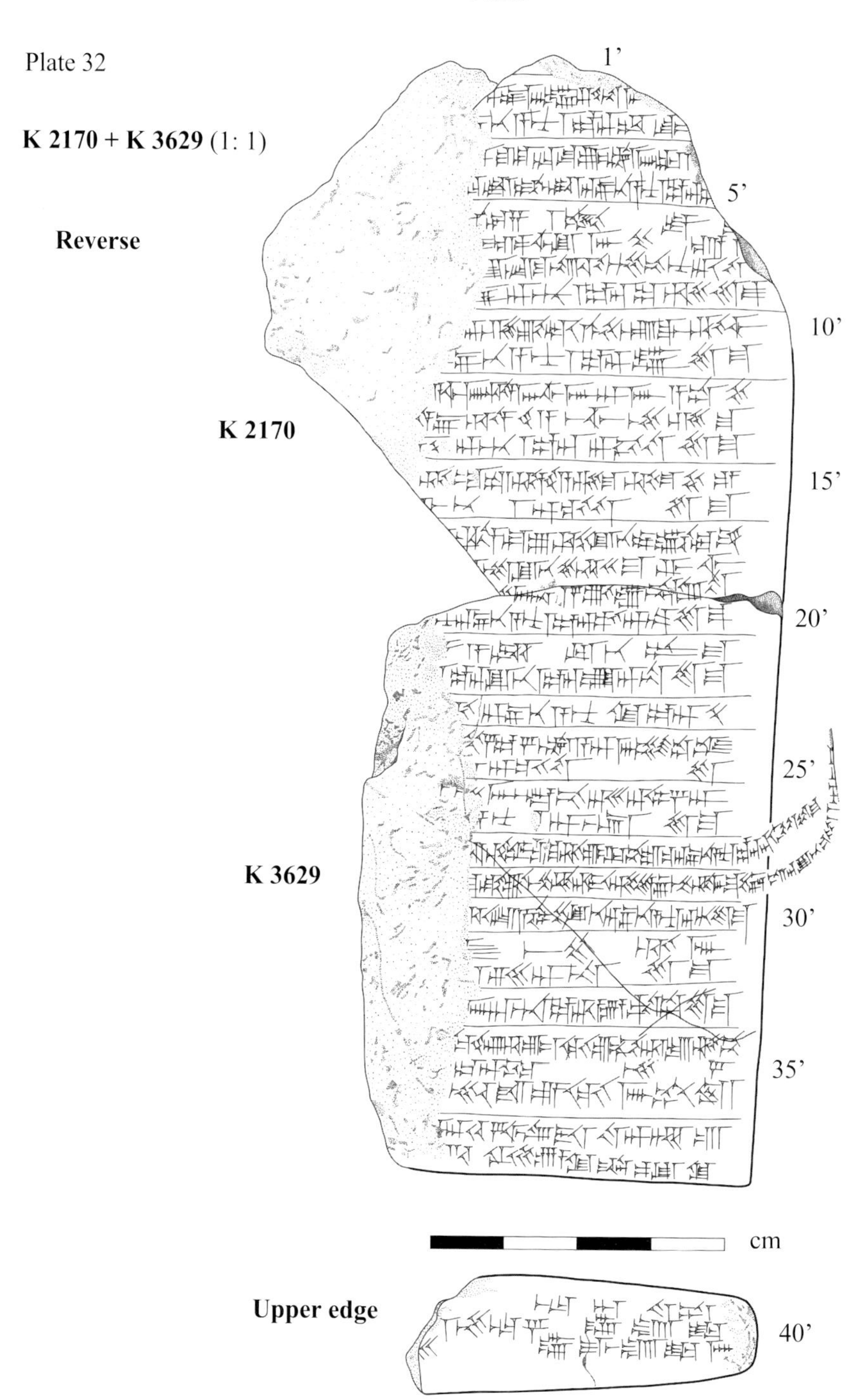

Plate 33

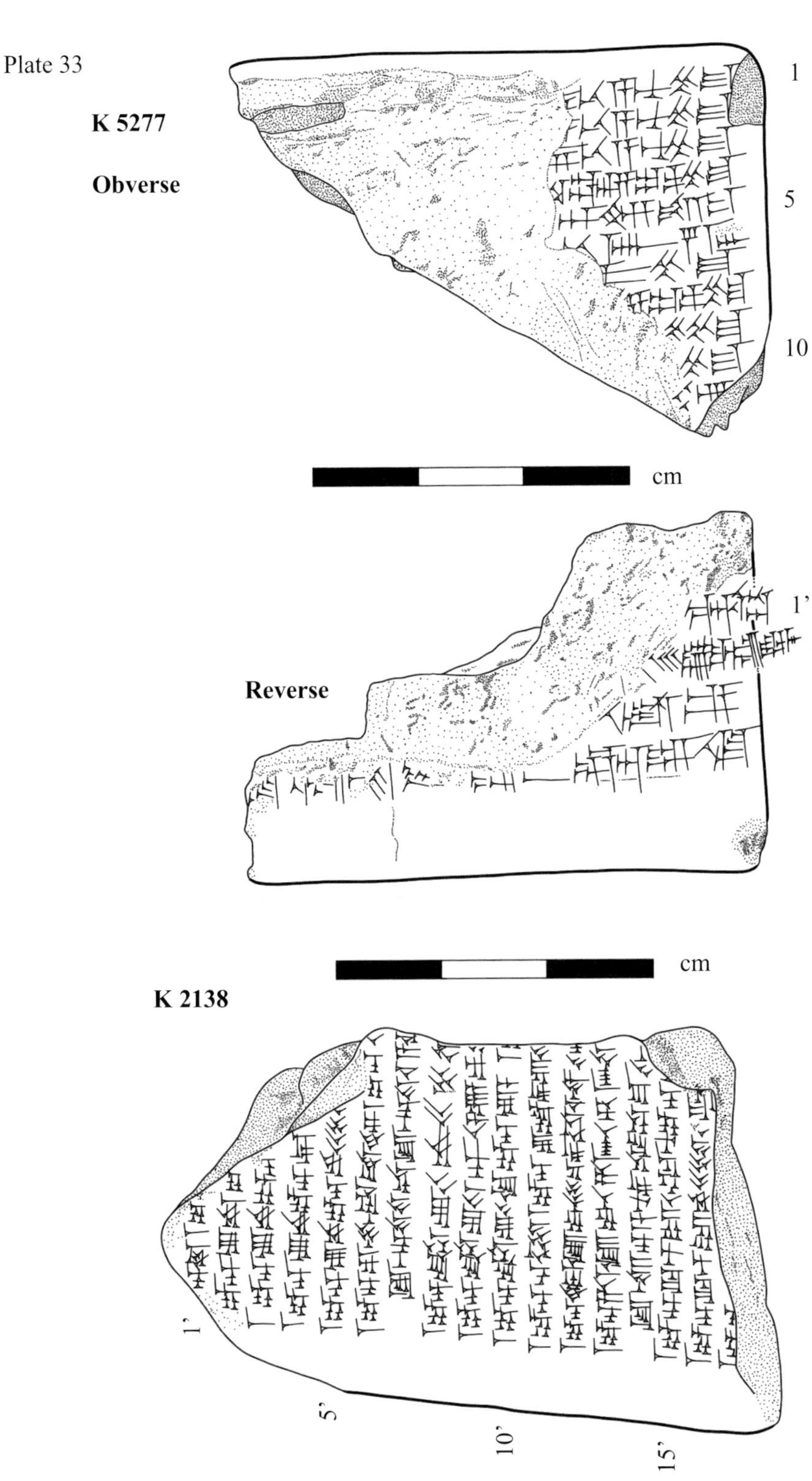

Plate 34

K 2301

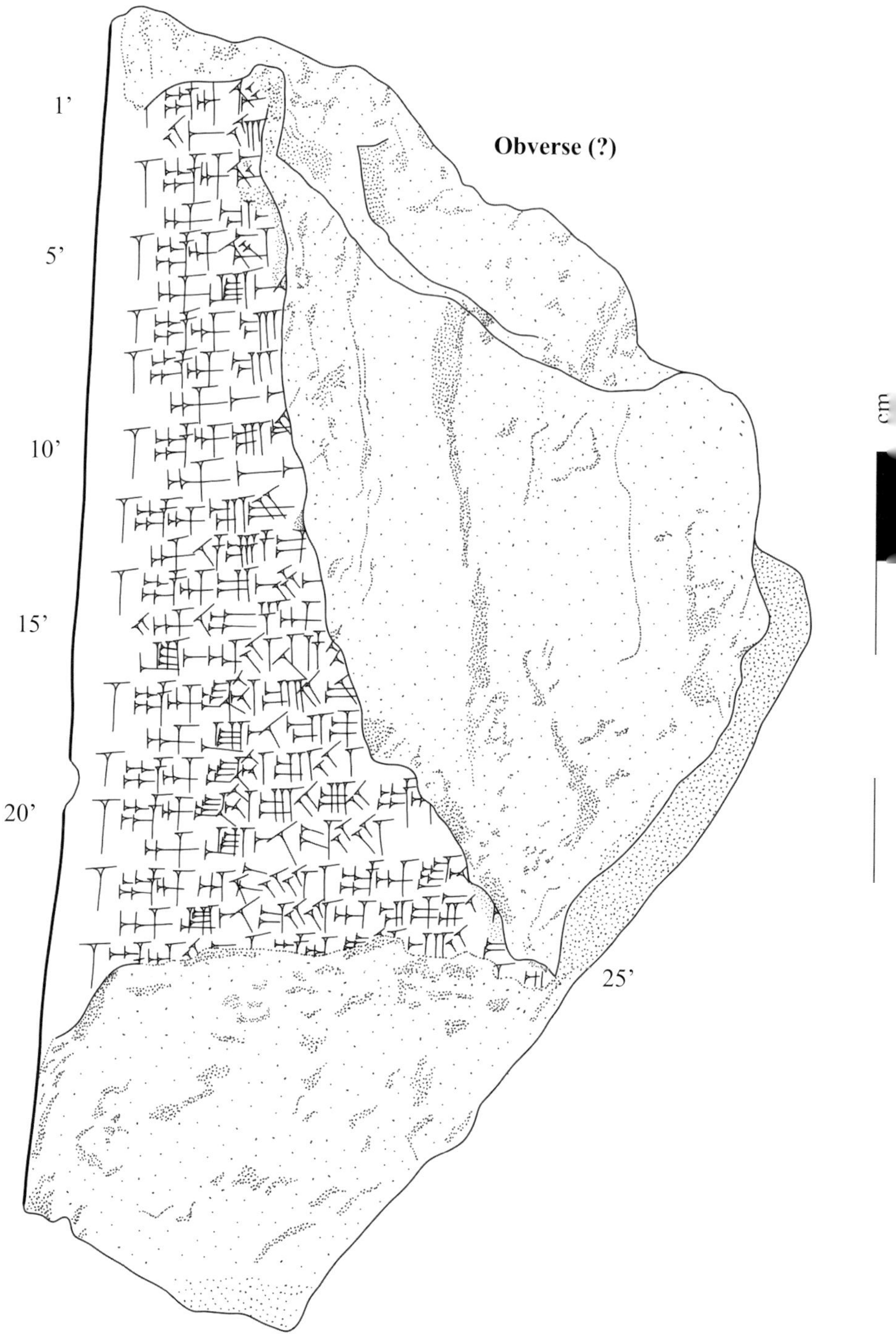

Obverse (?)

1’

5’

10’

15’

20’

25’

cm

Plate 35

Rm 477

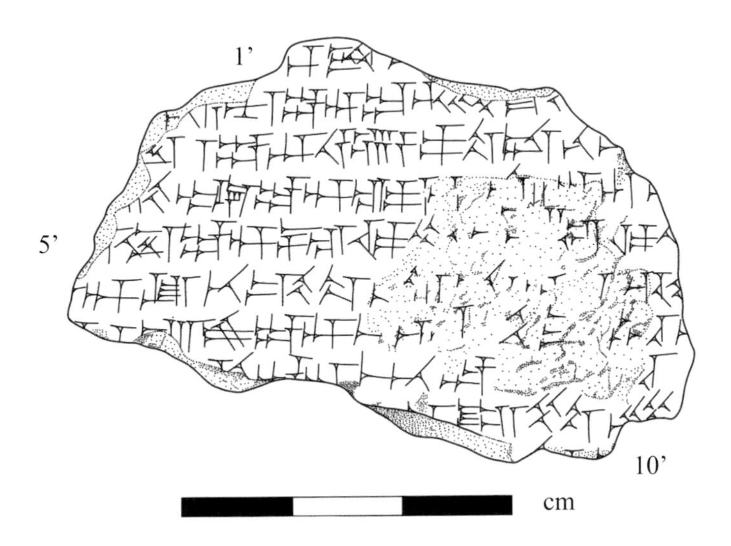

K 12149

Obverse (?)

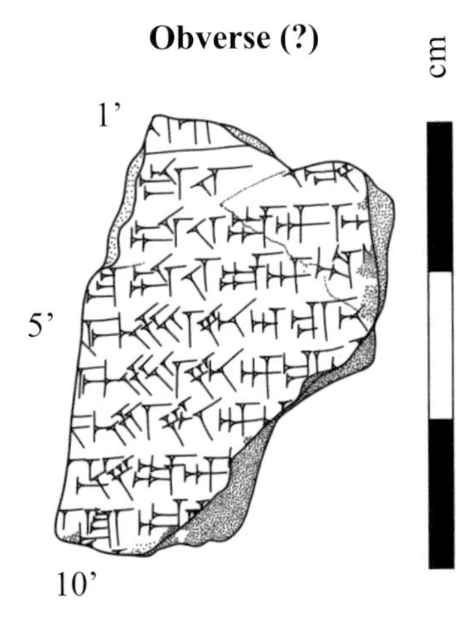

Plate 36

BM 44005

Obverse

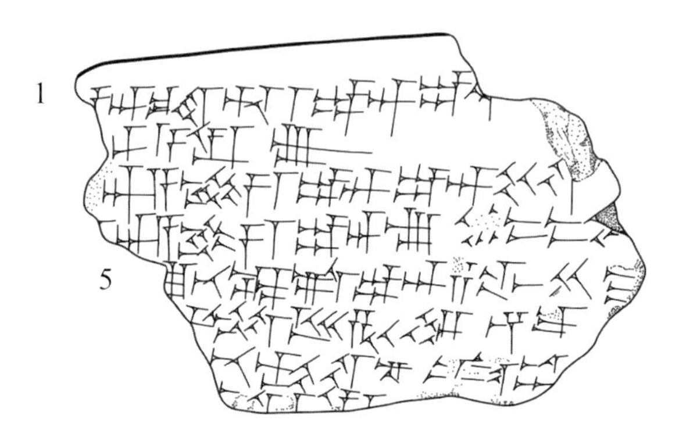

Reverse

Upper edge

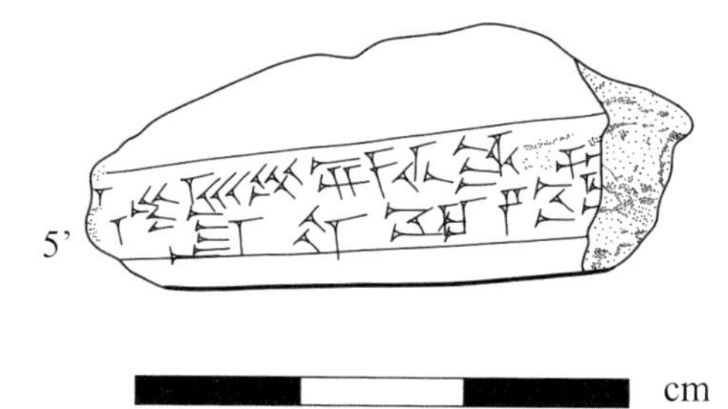

Plate 37

Rm 192 (1: 1)

Upper edge

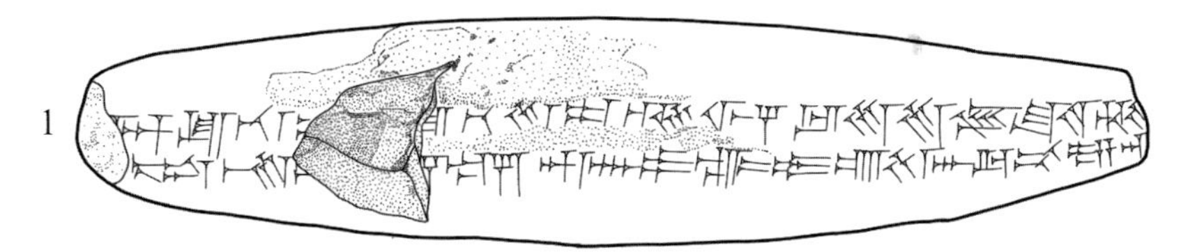

Obverse

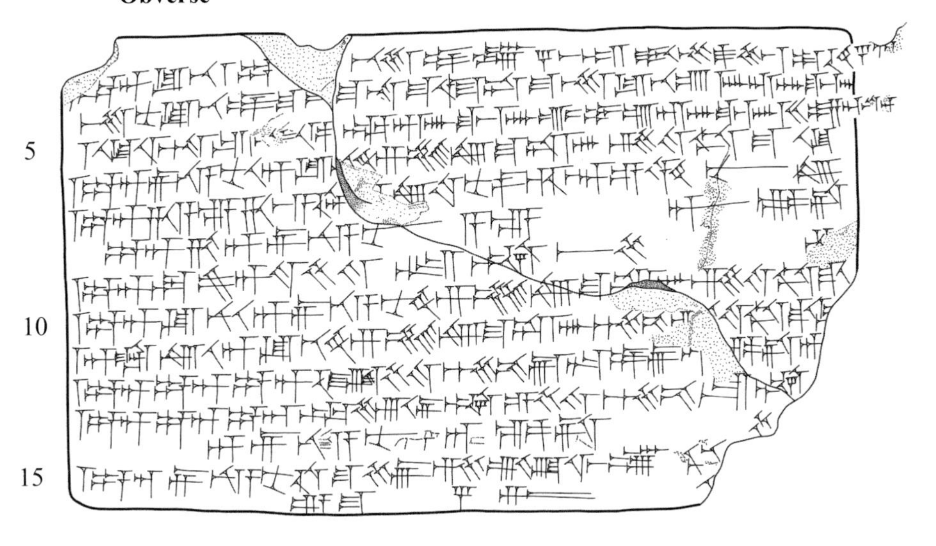

Plate 38

Rm 192 (1: 1)

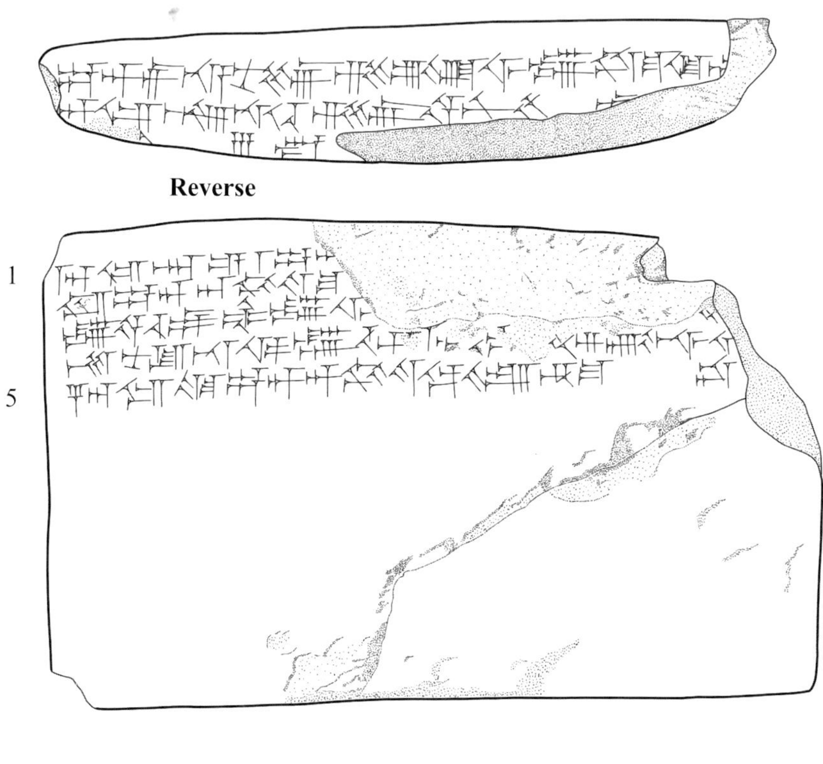

Plate 39

K 1494a (+) K 1494b (+) K 1522 + K 3594 (1: 1)

Obverse

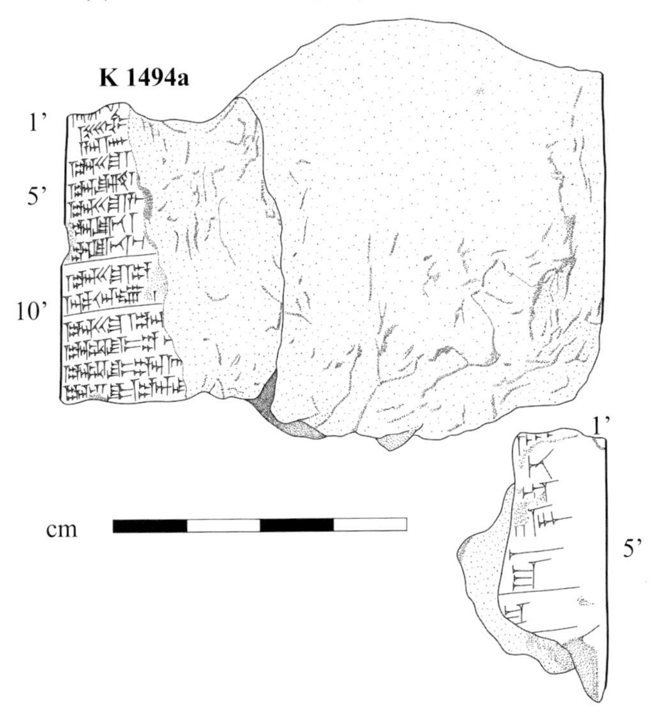

K 1494b

Plate 40

K 1494a (+) K 1494b (+) K 1522 + K 3594 (1: 1)

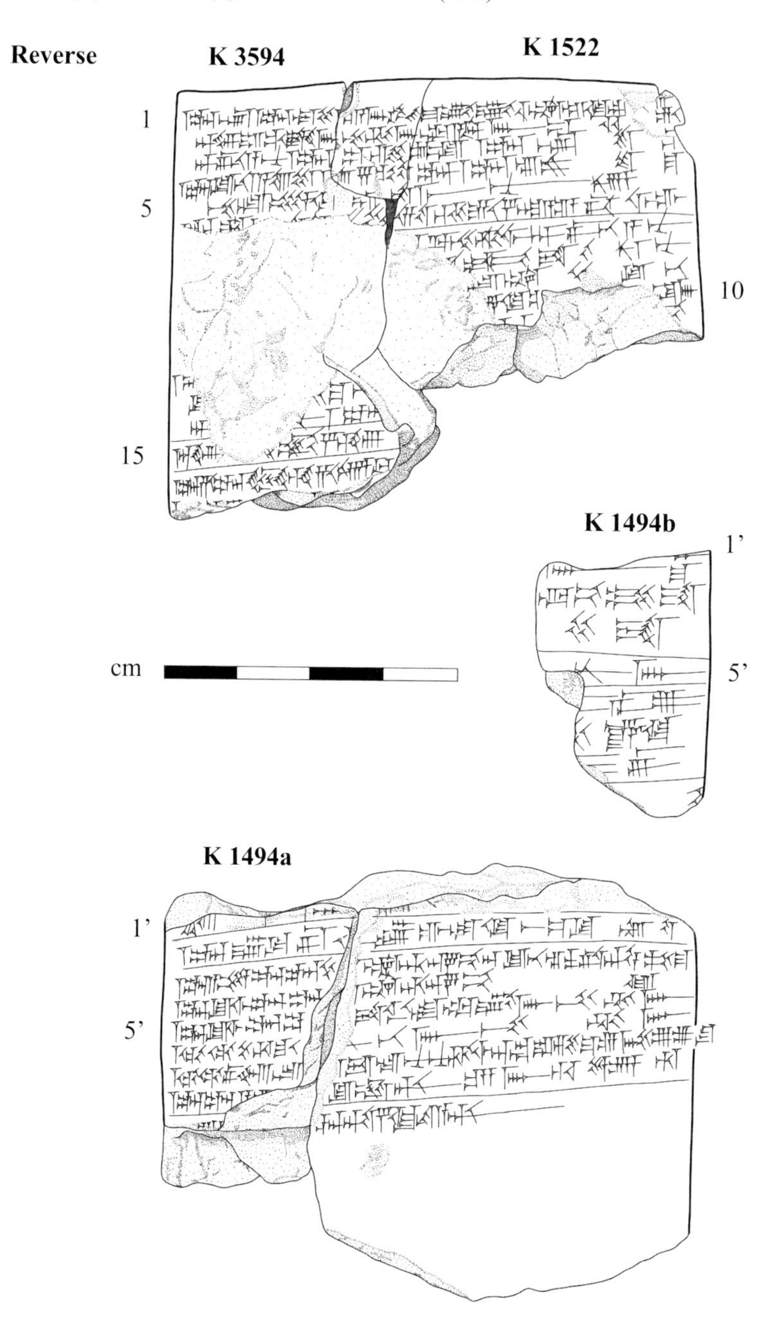

Plate 41

K 3780 (1: 1)

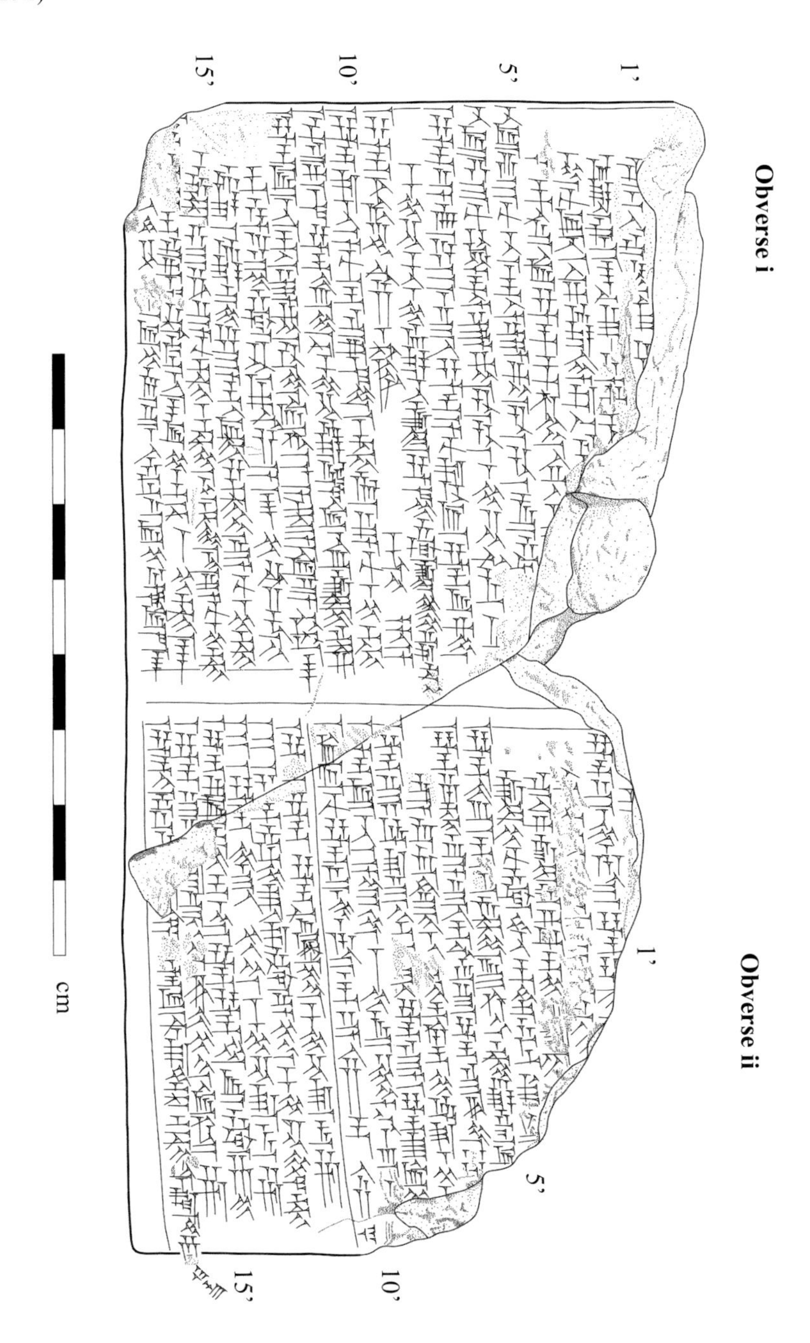

Plate 42

K 3780 (1: 1)

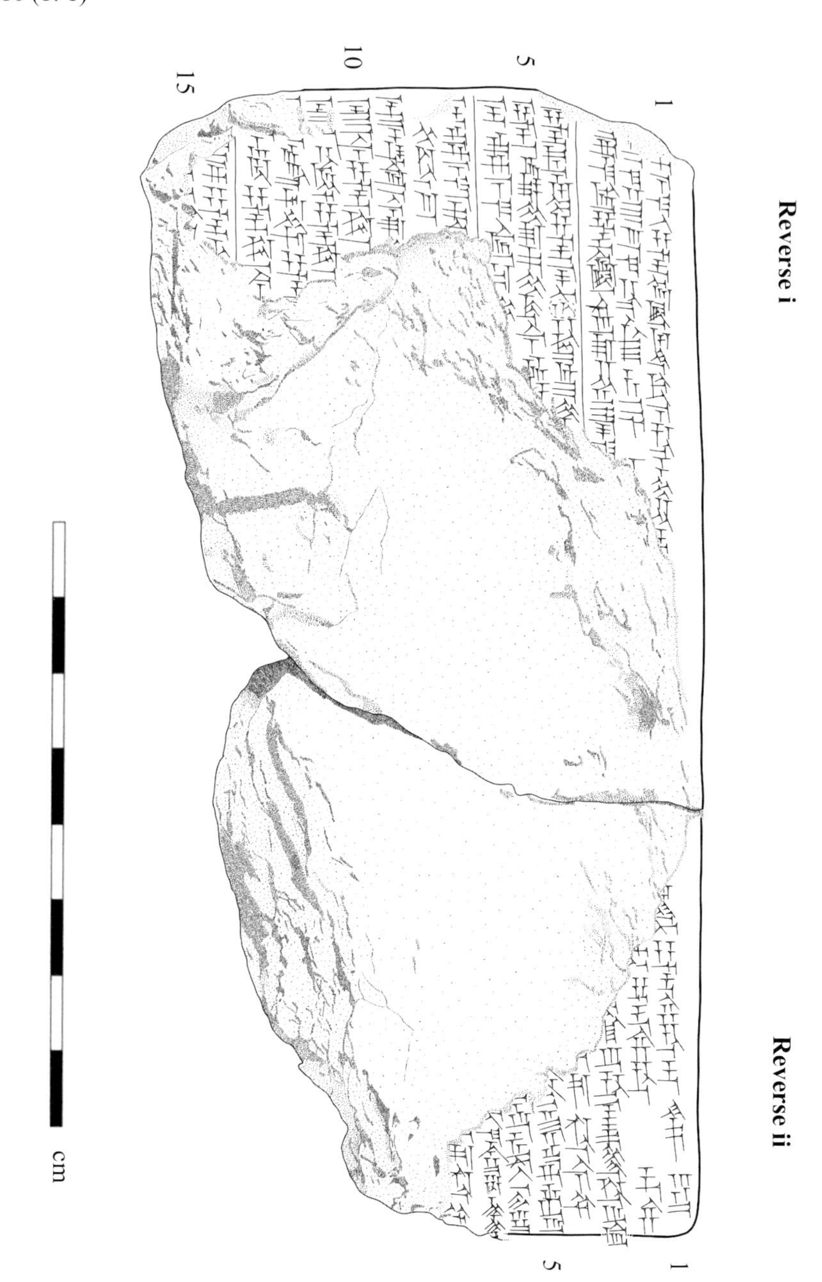

Plate 43

K 3123

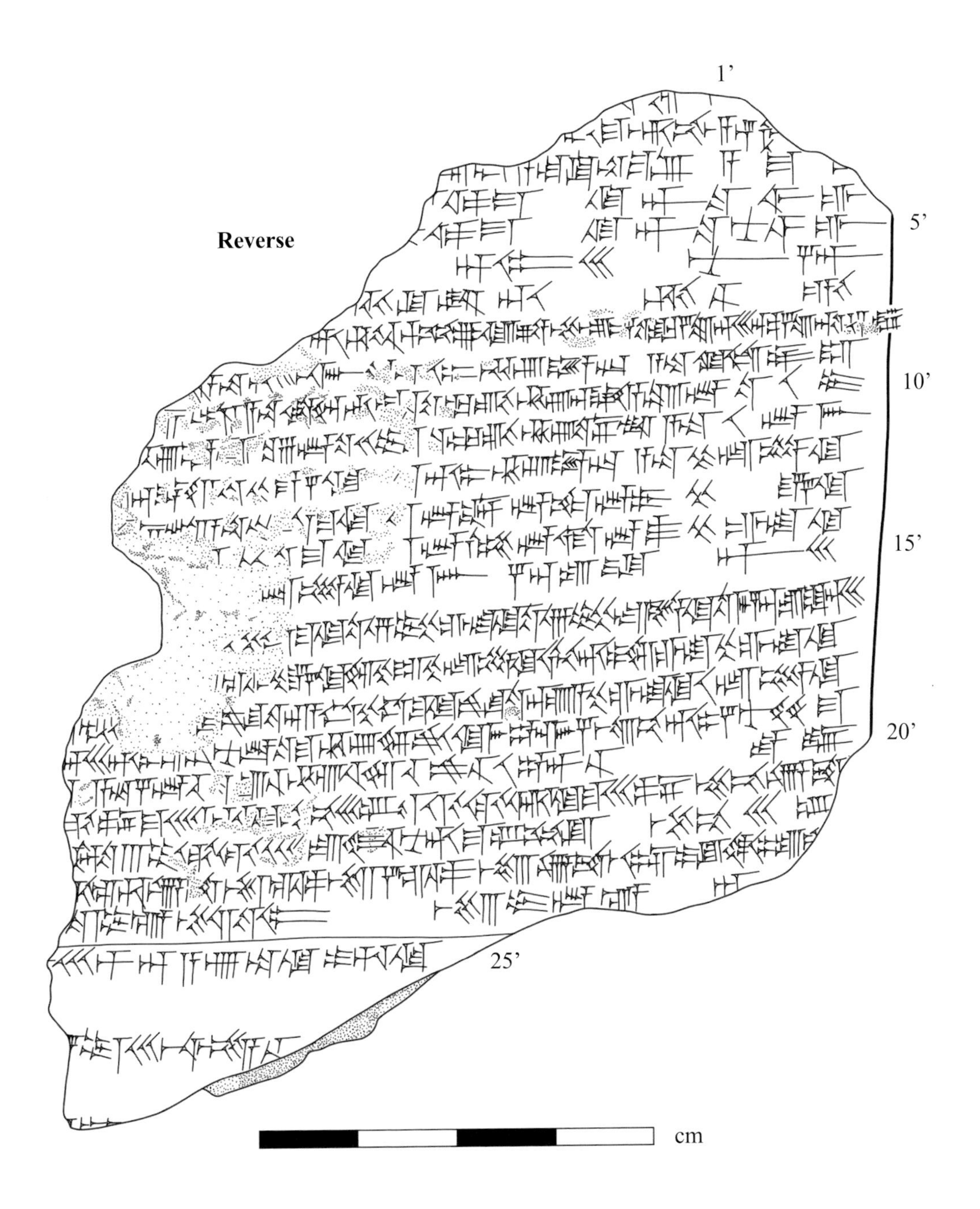

Plate 44

K 2254 (1: 1)

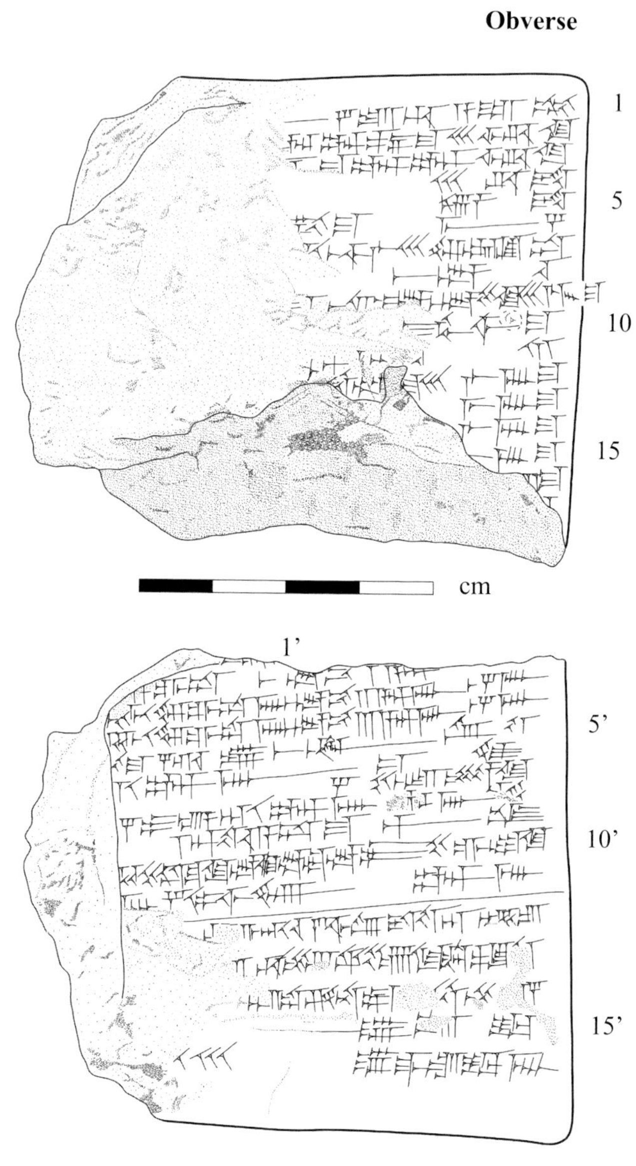

Plate 45

K 6686 + K 9234 (1: 1)

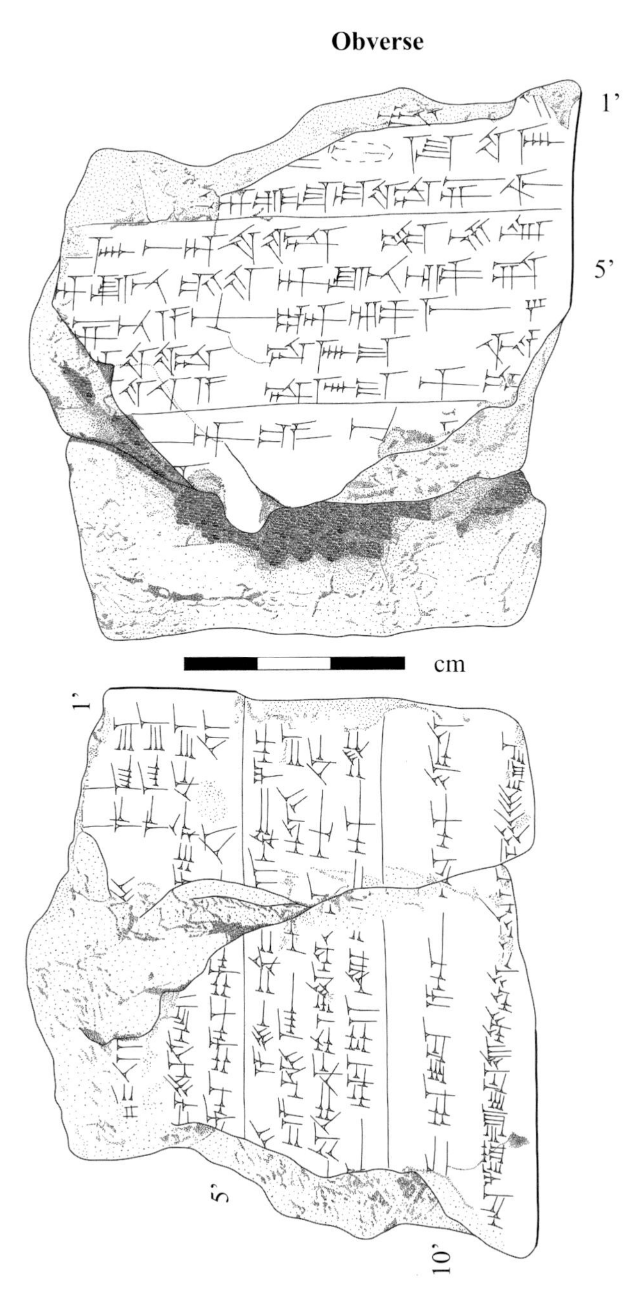

Plate 46

VAT 7850 (1: 1)

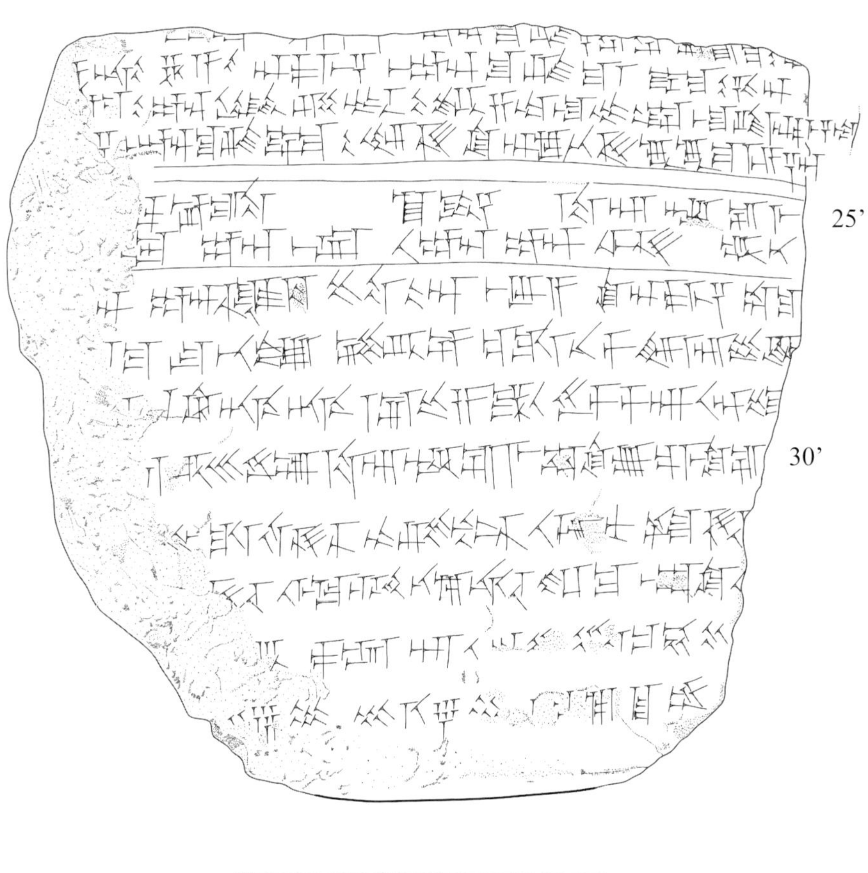

Logograms in the Reconstructed Tablets EAE 52, Commentary EAE 53, and SIT 6

Only the logograms which are found in a translatable context have been included. The English translations are given in the index of words and names (see pp. 458-473).

Logogram	**Akkadian**
A^{meš}	*mû*
Á	*idu*
A.AB.BA	*ajabba*
A.GÀR	*ugāru*
A.RÁ	*alaktu*
A.ŠÀ	*eqlu*
ÁB.GU$_{4}^{\text{ḫi.a}}$	*lâtu*
AB.SÍN	*šerʾu*
AN	*šamû*
AN.GE$_6$	*attalû*
BA.ÚŠ	*mâtu*
BABBAR	*peṣû*
BAD	*petû*
BAD$_4$	*dannatu*
BAD-*ma*	*šumma*, “if”
BAL	*nabalkutu*
BALA	*palû*
BÁRA	*parakku*
BI	*šū*, *šī*, *šuātu/i* etc. (third person anaphoric pronoun)
BIR	*sapāḫu*
BÚR	*napšartu*, *pišertu*
BURU$_{14}$	*ebūru*
DAGAL	*rupšu*
DIM$_4$	*sanāqu*
DINGIR	*ilu*

DIRI	*malû*
DIŠ	*šumma*, "if"; ¶
DU	*alāku*
DÙ.A.BI	*kalāma*
DU_{11}	*qibītu*
DÙG.GA	*ṭābu*
DUGUD	*kabtu*
DUMU	*māru*
DUMU.MUNUS	*mārtu*
È	*aṣû*
EDIN(.NA)	*ṣēru*
EGIR	*arki/u*, *arkutu*
(KUR) ELAM.MA$^{(ki)}$	*elam*
EN	*bēlu*
EN.TE.NA	*kuṣṣu*
ERÍN	*ummānu*
ERÍN-*man-da*	*ummān-manda*
GABA.RI	*meḫru*
GAL	*rabû*
GÁL	*bâšû*
GÁN.BA	*maḫīru*
GÁN.ZI	*mērešu*
GAR	*šakānu*
GE_6	*ṣalāmu, ṣalmu*
GI.NA	*kânu*
GIB	*parāku*
GIM	*kīma*
GIŠ.ḪUR	*uṣurtu*
^{še}GIŠ.Ì (NIM)	*šamaššammu* (*ḫarpu*)
^{giš}GIŠIMMAR	*gišimmaru*
GÙ (in GÙ-*šú* ŠUB-*di*)	*rigmu* (in *rigimšu iddi*)

GU_7	*akālu*
GU_7(*-ti*)	*ukultu*
GUB	*i/uzuzzu*
GÙB, 2,30	*šumēlu*
GUN	*biltu*
GUR	*târu*
GUR_4	*baʾālu*
GURUŠ	*eṭlu*
ḪÁD	*abālu*
Ì.GÁL	*ibašši* (*bašû*)
ÍD	*nāru*
IDIM	*kabtu*
IGI	*amāru*, *nanmuru*, *pānu*, *panû*
IGI.DU_8	*amāru*, *nanmuru*
ÍL	*našû*
ILLU	*mīlu*
IM	*šāru*
IM.KUR.RA, IM.3	*šadû*
IM.MAR.TU, IM.4	*amurru*
IM.SI.SÁ, IM.2	*iltānu*
IM.U_{18}.LU, IM.1	*šūtu*
ITI	*arḫu*
KA	*pû* (see *ina pî*)
KA.SES	*pû marru*
KÁ	*bābu*
KAM, KÁM, KAM*	(it is used to denote the ordinal numbers)
KAN_5 (KAxMI)	*adāru*
KAR	*mašāʾu*
KASKAL 20, $^{(d)}$UTU	*ḫarrān šamši*
KI	*itti*
KI.LAM	*maḫīru*

KI.SIKIL	*ardatu*
KI.TA	*šaplānu*, *šapliš*
KI.TUŠ	*šubtu*
KIMIN	ditto; variant
KU_4	*erēbu*
KU_6	*nūnu*
KUR	*kašādu*, *mātu*
KÚR	*nakāru*, *nakru*
$^{(munus)}$KÚR	*nukurtu*
LAL	*qalālu*, *šaqālu*
LI.DUR	*abunnatu*
LÚ	*amēlu*
LUGAL	*šarru*
MA.DAM	*ḫiṣbu*
(KUR) MAR.TU$^{(ki)}$	*amurru*
MAŠ.SÌLA	*naglabu*
MÁŠ.ANŠE	*būlu*
MIN	ditto; variant
MU	*aššu*, *šattu*
MÚ	*napāḫu*
MUL; MUL_4	*kakkabu*
MÚRU	*qablu*
MUŠEN	*iṣṣūru*
NAM	*pīḫātu*
NAM.KÚR	*nukurtu*
NAM.LÚ.U_{18}.LU	*amēlūtu*
NAM.ÚŠ	*mūtu*
NÍ	*ramānu*
NÍG.ZI.GÁL	*nammaššû*
NIGIN, NÍGIN	*lamû*; *saḫāru*
NINDA	*akalu*

NU	*lā*, *ul* (negation)
NUNUZ	*pelû*
^{munus}PEŠ$_4$	*a/erītu*
RA	*raḫāṣu*
RI.RI.GA	*miqittu*
SA$_5$	*sāmu*
SAG	*rēšu*
SI	*qarnu*
SI.SÁ	*ešēru*
SIG$_5$	*damāqu*, *dumqu*
SIG$_7$	*arāqu*, *arqu*
SILIM	*šalāmu*
(KUR) SU.BIR$_4$$^{(ki)}$	*subartu*
SU.GU$_7$	*ḫušaḫḫu*
SUD	*rūqu*
SUKKAL	*šukkallu*
ŠÀ	*libbu*
ŠEŠ	*aḫu*
ŠU.BI.AŠ.A.AN	ditto
ŠU.GI	*šību*
^{munus}ŠU.GI	*šībtu*
ŠUB	*maqātu*, *nadû*
ŠUB(-*ti*)	*miqittu*
ŠUR	*ṣarāru*
TA	*ištu* (see *ištu libbi*)
TE	*ṭeḫû*
TIL	*qītu*
TIN	*balṭūtu*
^{giš}TUKUL	*kakku*
TUKU	*išû*
TUR	*ṣeḫēru*, *ṣeḫru*

$^{(lú)}$TUR	*ṣuḫāru*
(É.)TÙR	*tarbaṣu*
Ú.GUG	*sunqu*
Ù.TU	*ilittu*
UB	*karmu*
UBUR	*tulû*
UD	*ūmu, šumma*, "if"
UD.DUG$_4$.GA	*adannu*
UGU	*elēnu, eli*
UN^{meš}	*nišū*
UR.BI	*ištēniš*
URU	*ālu*
UŠ	*redû*
ÚŠ	*mâtu*
USDUḪA	*ṣēnu*
ÚŠ^{meš}	*mūtānu*
ZABAR	*siparru*
ZAG, 15	*imittu*
ZÁḪ	*ḫalāqu, ḫalqu*
ZAL	*lazāzu*
ZI	*tebû*
ZI(*-ut*)	*tibûtu*

Names of Celestial Bodies and Gods

Logogram	**Akkadian**
20	*šamaš*
30	*sîn*
AB.SÍN	*absinnu*
AL.LUL	*alluttu*
AN	logogram for *ṣalbatānu* (only in LB)
AŠ.GÁN	*ikû*
ÁZAG	*asakku*
ÉLLAG	*kalītu*
EN.TE.NA.BAR.ḪUM	*ḫabaṣīrānu*
GAL	*rabû*
(giš)GÁN.ÙR	*maškakātu*
GE_6	*ṣalmu*
GÍR.TAB	*zuqaqīpu*
GÚ.ḪAL	*ur'udu*
GU_4.AN.NA	*alû, is lê*
IMIN.BI	*sebettu*
KA.MUŠ.Ì.KÚ.E	*pāšittu*
KAK.SI.SÁ	*šukūdu*
KU_6	*nūnu*
LU.LIM	*lulīmu*
MAN-*ma*	*šanûmma*
MAR.GÍD.DA	*ereqqu*
MUŠ	*nirāḫu*
NUN^{ki}	*eridu*
PAN	*qaštu*
SA_5	*sāmu*
SIM.MAḪ	*šinūnūtu*
SIPA.ZI.AN.NA	*šitaddaru*

ŠU.GI	*šību*
ŠUDUN	*nīru*
ŠUL.PA.È(.A)	*Šulpaea*
$TI_8^{(mušen)}$	*a/erû*
TIR.AN.NA	*manzât*
U.GUR	*nergal*
UD.AL.TAR	*dāpinu*
UD.KA.DUḪ.A	*ūmu nāʾiru*
UDU.IDIM	*bibbu*
UG_5.GA	*āribu*
$UGA^{(mušen)}$	*āribu*
UR.BAR.RA	*barbaru*
UR.MAḪ	*nēšu*
UTU	*šamaš*
ÙZ	*enzu*

Index of Words and Names in the Reconstructed Tablets EAE 52, Commentary EAE 53, and SIT 6

The determinatives are added in brackets, as well as parts of composite logograms which are not written in all the different sources used for the reconstructed texts. Only the words which are found in a translatable context have been included.

Words

Akkadian	Logogram	Translation	Reference
adannu	UD.DUG$_4$.GA	specified time	Commentary EAE 53: 9
abālu	ḪÁD	to dry	SIT 6: 14
abunnatu	LI.DUR	navel	Commentary EAE 53: 43–44
adāru,	KAN$_5$ (KAxMI)	to be(come) dark,	Commentary EAE 53: 10, 45, 48, 64
nanduru (N-stem)	–	to become worried	EAE 52: 7
a/erītu	^{munus}PEŠ$_4$	pregnant woman	EAE 52: 40–41, 44
aḫītu	–	misfortune	SIT 6: 40
aḫu	ŠEŠ	brother	EAE 52: 15
ajabba	A.AB.BA	sea	SIT 6: 14
akalu	NINDA	bread	SIT 6: 16
akālu	GU$_7$	to eat	EAE 52: 13; SIT 6: 18
ākilu	–	pest (lit. devourer)	SIT 6: 6
akkad	(KUR) URI$^{(ki)}$	Akkad	EAE 52: 5, 17–18, 26, 32, 47, 52, 57
alaktu	A.RÁ	course, (oracular) decision	EAE 52: 37
alāku	DU	to go	EAE 52: 12; Commentary EAE 53: 8, 14, 15; SIT 6: 35
ālu	URU	city	EAE 52: 21
amāru,	IGI	to see,	
nanmuru (N-stem)		to be visible	EAE 52: 1–11, 19–29; Commentary EAE 53: 42; SIT 6: 31
	IGI.DU$_8$	to be seen	EAE 52: 39–42
amēlu	LÚ	man	EAE 52: 15
amēlūtu	NAM.LÚ.U$_{18}$.LU	people, mankind	EAE 52: 46–47
amurru	(KUR) MAR.TU$^{(ki)}$	Amurru or	EAE 52: 18, 50, 55, 60

ana libbi	*ana* ŠÀ	in, into	EAE 52: 12–29; Commentary EAE 53: 25
ina libbi	*ina* ŠÁ	inside (lit. to the inside of)	EAE 52: 12, 15, 45–46; Commentary EAE 53: 26, 30; SIT 6: 8, 15–17, 28
ištu libbi	TA ŠÀ	out of (something)	Commentary EAE 53: 11
ša libbišina	*ša* ŠÀ-*ši-na*	in their wombs	EAE 52: 40–41
i/uzuzzu	GUB	to stand (still)	EAE 52: 12–18, 30–34, 43, 46, 51–55; Commentary EAE 53: 7–9, 14, 16, 24a–30, 50, 52, 58–61, 63–64; SIT 6: 15–18, 22–24, 39
magal	–	very	EAE 52: 44; SIT 6: 36
maḫīru	GÁN.BA, KI.LAM	market rate or business	EAE 52: 26; Commentary EAE 53: 1, 5; SIT 6: 1–2, 17, 22–23
malû	DIRI	to be(come) full	Commentary EAE 53: 49, 51
manzāzu	KI.GUB, GUB.BA	position	EAE 52: 34; Commentary EAE 53: 9, 24a–d, 45, 50–52
	GIŠGAL		Commentary EAE 53: 41, 57
maqātu,	ŠUB	to fall,	EAE 52: 41
šumqutu (Š-stem)	–	to overpower	EAE 52: 15
mārtu	DUMU.MUNUS	daughter	EAE 52: 51
māru	DUMU	son	EAE 52: 51
mašāḫu	–	to flare up, to produce a glow (with *mešḫu*)	Commentary EAE 53: 11, 32, 49
mašā'u	KAR	to rob	EAE 52: 15
mašrû	–	prosperity	SIT 6: 29
mâtu	BA.ÚŠ, ÚŠ	to die	EAE 52: 51; Commentary EAE 53: 49
mātu	KUR	country	EAE 52: 6–7, 10, 12–15, 18–20, 25, 30, 32, 39–42, 44, 46, 51, 56; Commentary EAE 53: 20, 41, 45–46, 48, 50, 55, 63; SIT 6: 11, 13, 19, 27, 30, 37
meḫru	GABA.RI	equivalent, or, correspondingly	Commentary EAE 53: 2; SIT 6: 4, 20
mērešu	GÁN.ZI	cultivation	EAE 52: 26, 32; SIT 6: 27

mešḫu	–	glow or meteor	Commentary EAE 53: 11, 32, 49
mīlu	ILLU	flood	EAE 52: 26; Commentary EAE 53: 8; SIT 6: 40
miqittu	RI.RI.GA, ŠUB(*-ti*)	downfall, epidemic	EAE 52: 46–47; Commentary EAE 53: 8; SIT 6: 19–20
mû	$\text{A}^{\text{meš}}$	water	SIT 6: 28
mūtānu	$\text{ÚŠ}^{\text{meš}}$	pestilence	EAE 52: 27, 39; Commentary EAE 53: 2, 53; SIT 6: 4
mūtu	NAM.ÚŠ	death	Commentary EAE 53: 51
nabalkutu *šubalkutu* (Š-stem)	BAL	to cross over, to overthrow a rule(r)	Commentary EAE 53: 50
nabāṭu, *itanbuṭu* (Ntn-stem)	–	to glow, to (constantly) shine brightly	Commentary EAE 53: 43
nadû	ŠUB	to sprinkle	Commentary EAE 53: 55
nagālu (stative *naglu*)	–	to *glisten*	Commentary EAE 53A: 54
naglabu	MAŠ.SÌLA	scapula	EAE 52: 30; Commentary EAE 53: 27
nakāru, *nakru*	KÚR	to be in enmity, enemy	EAE 52: 7; EAE 52: 14; Commentary EAE 53: 24a
namarkû	–	to lag behind	Commentary EAE 53: 19
namāru	–	to dawn	SIT 6: 23
nammaššû	NÍG.ZI.GÁL	herd (of wild animals)	EAE 52: 13, 18
napāḫu	MÚ	to glow	Commentary EAE 53: 8
napalsuḫu	–	to fall down or to be apart	Commentary EAE 53: 25
napālu, **nappulu* (N-stem)	–	to tear down, to be torn down	Commentary EAE 53:15
napšartu	BÚR	barley for shipment	EAE 52: 44
nāru	ÍD	river	EAE 52: 41
našû	ÍL	to bear	Commentary EAE 53: 20
nēḫu	–	peaceful	SIT 6: 30
nentû	–	to go parallel to each other	SIT 6: 11
nišū	$\text{UN}^{\text{meš}}$	people	EAE 52: 11; SIT 6: 13, 23, 36
nuḫšu	–	plenty	SIT 6: 36
nukurtu	$^{\text{(munus)}}$KÚR,	hostility	Commentary EAE 53: 47

	NAM.KÚR		Commentary EAE 53: 52
nūnu	KU$_{6}$	fish	EAE 52: 41; SIT 6: 28
palû	BALA	reign, dinasty	Commentary EAE 53: 49, 52
panû	IGI	to move ahead	Commentary EAE 53: 20
pānu	IGI	front	Commentary EAE 53: 12, 19, 24b; SIT 6: 23, 33
ina pāni	*ina* IGI	in front	Commentary EAE 53: 7, 14, 53–55, 60–61
parakku	BÁRA	sanctuary	Commentary EAE 53: 45
parāku	GIB	to lie across	EAE 52: 43
parāru, *purruru* (D-stem)	–	uncertain, to break up or to disperse	Commentary EAE 53: 21
pelû	NUNUZ	egg	EAE 52: 41
peṣû	BABBAR	white	Commentary EAE 53: 34, 54
petû	BAD	to open	EAE 52: 44, 47–50, 52–55, 57–60
pīḫātu	NAM	district	Commentary EAE 53: 6; SIT 6: 7
pišertu	BÚR	release	EAE 52: 44–45; Commentary EAE 53: 64
pû marru	KA.SES	bitter mouth	SIT 6: 11
qablu	MÚRU	middle	SIT 6: 34
qalālu, *qullulu* (D-stem)	LAL	to become weak, to diminish	Commentary EAE 53: 6; SIT 6: 7
qarnu	SI	horn	EAE 52: 31–32; Commentary EAE 53: 28–29, 58–59
qerēbu	–	to come near	Commentary EAE 53: 46–47
qibītu	DU$_{11}$	speech	EAE 52: 42
qītu	TIL	end	EAE 52: 26
rabû	–	to set	SIT 6: 39
rabû	GAL	great	EAE 52: 12
raḫāṣu	RA	to devastate	Commentary EAE 53: 2–3; SIT 6: 4–6, 8, 23
ramānu	NÍ	self	EAE 52: 7
redû	UŠ	to follow	Commentary EAE 53: 62
rēšu	SAG	head, top	Commentary EAE 53: 7, 11

šamaššammu (*ḫarpu*)	^{še}GIŠ.Ì (NIM)	(early) sesame	SIT 6: 9, 15, 25–26
šamû	AN	sky	EAE 52: 41; SIT 6: 39
šaplānu, *šapliš*	KI.TA	below	EAE 52: 33
šaqālu, *šitqulu* (Gt-stem)	LAL	to weigh, to (be in) balance	Commentary EAE 53: 16
šaqû	–	to become high, high (adj.)	Commentary EAE 53: 20 Commentary EAE 53: 42
šarru	LUGAL	king	EAE 52: 16, 18–19, 39, 50
šāru	IM	wind	Commentary EAE 53: 22, 41, 49
šatāḫu	–	to be elongated	Commentary EAE 53: 20
šattu	MU	year	EAE 52: 13, 26, 30, 32, 42, 45–60; Commentary EAE 53: 1–4, 15, 19–24d, 42, 53–55, 57; SIT 6: 1–5, 8, 18–20, 31, 39
šegunû	ŠE.GÙN.NU	crop	EAE 52: 26; Commentary EAE 53: 8
šērtu	–	misdeed	Commentary EAE 53: 49
šerʾu	AB.SÍN	furrow	SIT 6: 31
šeʾu	ŠE, ŠE.PAD	barley	Commentary EAE 53: 4; SIT 6: 3, 17
šībtu	^{munus}ŠU.GI	old woman	EAE 52: 13
šību	ŠU.GI	old man	EAE 52: 13
šitqultu	–	balancing	Commentary EAE 53: 19–20
šubtu	KI.TUŠ	abode	SIT 6: 30
šukkallu	SUKKAL	vizier	SIT 6: 27
šulpu	–	lit. stalk, but also *destruction*	EAE 52: 15
šumēlu	2,30 / GÙB	left (side)	EAE 52: 32; Commentary EAE 53: 29, 47, 59; SIT 6: 23
šūtu	IM.U_{18}.LU, IM.1	south (lit. southern wind)	EAE 52: 47, 52, 57; SIT 6: 23
tabāku	–	to pour out	SIT 6: 34
tabālu	–	to disappear	Commentary EAE 53: 9
tālittu	–	offspring	Commentary EAE 53: 26
tarbaṣu	(É.)TÙR	halo, cattle-pen	EAE 52: 46–55
târu	GUR	to recede	SIT 6: 40
tebû	ZI	to rise	EAE 52: 14, 16; Commentary EAE 53: 22
tibûtu	ZI(-*ut*)	revolt	EAE 52: 29, 43, 57–60

tulû	UBUR	breast	EAE 52: 51
ṭābu	DÙG.GA	good	Commentary EAE 53: 22; SIT 6: 16
ṭeḫû	TE	to come close	EAE 52: 40, 46–47; Commentary EAE 53: 6; SIT 6: 1, 7, 13–14, 19, 27
ṭeḫûtu	–	close approach	EAE 52: 45
ugāru	A.GÀR	meadow	Commentary EAE 53: 20
ukultu	GU$_7$(*-ti*)	consumption	Commentary EAE 53: 11; SIT 6: 10, 16
ummān-manda	ERÍN-*man-da*	enemy horde	EAE 52: 43, 56–60
ummānu	ERÍN	army	EAE 52: 16
ummu, *ummātu* in pl.	–	heat, summer	Commentary EAE 53: 2–3; SIT 6: 4–5
umšu	–	heat, blaze	Commentary EAE 53: 2; SIT 6: 4
ūmu	UD	day	SIT 6: 30, 36, 40
uṣurtu	GIŠ.ḪUR	drawing or (divine) design	Commentary EAE 53: 10
unnutu	–	to be faint	SIT 6: 40
wamālu	–	to scintillate	Commentary EAE 53: 23
zunnu	ŠÈG	rain	EAE 52: 26; Commentary EAE 53: 8

Names of Celestial Bodies and Gods

The determinatives for star (MUL, *kakkabu*) and god (DINGIR, *ilu*) are omitted.

		Sun”	EAE 53: 2, 5–6, 24a, 24c; SIT 6: 2, 4, 6–7, 11, 17, 23–25, 27
ḫabaṣīrānu	EN.TE.NA.BAR.ḪUM	Mouse	EAE 52: 18; Commentary EAE 53: 24c; SIT 6: 32
ikû	AŠ.GÁN	Field	EAE 52: 1–36; Commentary EAE 53: 6, 24a; SIT 6: 6, 7, 14, 18
is lê	GU$_4$.AN.NA	Jaw of the Bull	Commentary EAE 53: 26–32, 48
ištar	–	Ištar	Commentary EAE 53: 50
kajamānu	(UDU.IDIM.)SAG.UŠ	Saturn, lit. “Steady”	Commentary EAE 53: 1–2, 4–5, 7; SIT 6: 1–4, 7, 9–10, 23–25
kalītu	ÉLLAG	Kidney	Commentary EAE 53: 3; SIT 6: 5, 25
lulīmu	LU.LIM	Stag	SIT 6: 10, 11
lumnu	–	Evil One	SIT 6: 38
manzât	TIR.AN.NA	Rainbow	EAE 52: 56–60
marduk	AMAR.UTU	Marduk-star	EAE 52: 22; Commentary EAE 53: 2; SIT 6: 4, 8, 9, 25, 31
maškakātu	$^{(giš)}$GÁN.ÙR	Harrow	SIT 6: 20
(see *zappu*)	MUL.MUL	Pleiades, lit. “Stars”	EAE 52: 1, 12; Commentary EAE 53: 1–25, 46–47; SIT 6: 1–2, 4–8, 10–13, 33
nakaru	–	Hostile One	Commentary EAE 53: 52
nameru	ŠU.PA	Resplendent One	EAE 52: 16; SIT 6: 37
nēšu	UR.MAḪ	Lion	SIT 6: 30
nirāḫu	MUŠ	Snake	Commentary EAE 53: 5
nīru	ŠUDUN	Yoke	Commentary EAE 53: 1; SIT 6: 1–2
nūnu	KU$_6$	Fish	EAE 52: 25; SIT 6: 28
pāšittu	KA.MUŠ.Ì.KÚ.E	Pāšittu, lit. “Obliterator”	Commentary EAE 53: 4; SIT 6: 3, 32
qaštu	PAN	Bow	SIT 6: 15–17, 28
rabû	GAL	Great One	EAE 52: 39; Commentary EAE 53: 60
–	SAG.ME.GAR	Jupiter	EAE 52: 11, 5, 20, 22; Commentary EAE 53: 13; SIT 6: 4, 16
sāmu	SA$_5$	Red One	SIT 6: 27, 38
sebettu	IMIN.BI	Sebettu	SIT 6: 10, 34

simut	–	Simut	Commentary EAE 53: 24c, 53–56
sîn	30	Moon, Sin	EAE 52: 13, 30–34, 46–47; Commentary EAE 53: 16, 19, 26–30; SIT 6: 33
ṣalbatānu	AN (only in LB)	Mars	EAE 52: 12, 16, 25–26, 40; Commentary EAE 53: 1–6, 12–15, 24d; SIT 6: 1–5, 7, 11–15, 19–20, 38–40
ṣalmu	GE_6	Black One	SIT 6: 23
šamaš	20 / UTU	Sun, Šamaš	EAE 52: 13; Commentary EAE 53: 2, 13, 48; SIT 6: 4, 36
šanûmma	MAN-*ma*	Other One	Commentary EAE 53: 24d; SIT 6: 13–14
šību	ŠU.GI	Old Man	EAE 52: 42, 44–50
šinūnūtu	SIM.MAḪ	Swallow	SIT 6: 14, 28
šitaddaru	SIPA.ZI.AN.NA	True Shepherd of Anu	EAE 52: 13, 7; Commentary EAE 53: 33–49
šukūdu	KAK.SI.SÁ	Arrow	EAE 52: 14; Commentary EAE 53: 2; SIT 6: 17
–	ŠUL.PA.È(.A)	Šulpaea, lit. “Lord of the Bright Rising”	EAE 52: 20; SIT 6: 16, 21, 24
ūmu nā᾿iru	UD.KA.DUḪ.A	Demon with the Gaping Mouth	EAE 52: 28
ur᾿udu	GÚ.ḪAL	Throat	EAE 52: 24
zappu	(see MUL.MUL)	Bristle, i.e. Pleiades	Commentary EAE 53: 2–3, 19; SIT 6: 4–6, 8–10, 15
zuqaqīpu	GÍR.TAB	Scorpion	EAE 52: 19

Index of Relevant Subjects

Index of Quoted Texts

Postscript

In 2023, prior to the publication of this book, the Electronic Babylonian Literature (eBL) Project's database was made accessible online (https://www.ebl.lmu.de). Directed by Prof. Dr. Enrique Himenez at Ludwig Maximilian University of Munich, the project aims to compile preliminary transliterations of all fragments stored in museum cabinets. Additionally, its Fragmentarium (https://www.ebl.lmu.de/fragmentarium) serves as an online research tool, assisting scholars worldwide in the reconstruction of Babylonian literature.

The present author conducted a search through the Fragmentarium database, accessing it from 25 to 28 July 2023, and identified relevant parallel fragments and joins provided by the eBL team concerning the editions of cuneiform fragments presented in this book. It is hoped that the following lists of fragments (cited by their respective museum or accession numbers) will serve as a foundational resource for further expansion and improvement of the corpus of celestial divinatory omens of Mesopotamia.

Parallel fragments identified and transliterated by eBL: 79-7-8, 271 l. 5' // K 1872 + K 12062 ii' 23'–25'; BM 33772 l. 10'–16' // EAE 52: 44–50; K 1522 + K 3594 obv. 20'–22' // 81-7-27, 137 obv. 23'–24'; K 1522 + K 3594 obv. 7'–8' // K 7180 + 81-2-4, 347 + 81-2-4, 416 ll. 5'–7'; K 3099 + K 18689: rev. 11' // K 6589 l. 8'; K 3123 rev. 23'–24' // K 10654 ll. 3–4; K 3123 rev. 9'–13' // K 2721 ll. 5'–14'; K 3923 + K 6140 + 81-7-27, 149 + 83-1-18, 479 rev. 3–5 // K 7037 ll. 6'–9'; Rm 192 obv. 19, rev. 3–5 // K 9647 l. 11', 81-7-27, 137 obv. 8', and K 6687 obv. 3''–6''; K 11324 + K 12705 rev. 1 // K 12646 ll. 9'–12'; Sm 247 obv. 4'–5' // Koch-Westenholz 1999: 157, 54–55.

Joins identified and transliterated by eBL (marked in italics): K 1522 + K 3594 +? *Sm 1510* +? *K 12349*; K 2138 +? *K 17597*; K 2170 + K 3629 +? *K 6484*; K 2177 + K 7869 + Rm 473 +? *K 7275*; K 2254 +? *K 8280* + *K 11129* +? *K 11927*; K 3099 + K 18689 + *K 19461*; K 3558 +? *K 19218*; K 3780 *(+) K 6227*; K 3780 *(+) K 6227*; K 3918 + K 6239 +? *K 7986*; K 10845 +? *K 17893*; K 11324 + K 12705 *(+)? K 11088.*